Philosophy and Medicine

Founding Editors

H. Tristram Engelhardt Jr.
Stuart F. Spicker

Volume 132

The Philosophy and Medicine series is dedicated to publishing monographs and collections of essays that contribute importantly to scholarship in bioethics and the philosophy of medicine. The series addresses the full scope of issues in bioethics and philosophy of medicine, from euthanasia to justice and solidarity in health care, and from the concept of disease to the phenomenology of illness. The Philosophy and Medicine series places the scholarship of bioethics within studies of basic problems in the epistemology, ethics, and metaphysics of medicine. The series seeks to publish the best of philosophical work from around the world and from all philosophical traditions directed to health care and the biomedical sciences. Since its appearance in 1975, the series has created an intellectual and scholarly focal point that frames the field of the philosophy of medicine and bioethics. From its inception, the series has recognized the breadth of philosophical concerns made salient by the biomedical sciences and the health care professions. With over one hundred and twenty five volumes in print, no other series offers as substantial and significant a resource for philosophical scholarship regarding issues raised by medicine and the biomedical sciences.

Tomas Zima · David N. Weisstub
Editors

Medical Research Ethics: Challenges in the 21st Century

Springer

Editors
Tomas Zima
Charles University
Prague, Czech Republic

David N. Weisstub
International Academy of Law and
Mental Health
Montreal, QC, Canada

ISSN 0376-7418   ISSN 2215-0080   (electronic)
Philosophy and Medicine
ISBN 978-3-031-12691-8   ISBN 978-3-031-12692-5   (eBook)
https://doi.org/10.1007/978-3-031-12692-5

This Springer imprint is published by the registered company Springer Nature Switzerland AG
The registered company address is: Gewerbestrasse 11, 6330 Cham, Switzerland

Preface

Governments for three quarters of a century have elucidated legislation with the aim of reflecting public attitudes while simultaneously accommodating to scientific needs. In the earliest period following the Second World War, there were signals towards the international community that respect for persons was the uncompromising value for every form of experimentation. The grand statement of this value occurred in the trials of the Nazi doctors and became a reference point for subsequent codes and judicial decisions. Regrettably though it behoves us to confront the actual impact of this much-acclaimed protection because a close look reveals some disturbing features.

The ethical guidelines proclaimed in the Weimar Republic were well articulated in the period preceding the Nazi regime. In fact, the 'Richtlinien' was binding law through the end of 1945. The law was stricter and in greater detail than the Nuremberg Code of 1947 and the Helsinki Code of 1964. What then is the significance of such a powerful detail? It is that legal regulations without application are irrelevant if the political arm is disrespectful and even worse.

The celebration of the Nuremberg Code is not borne out by the subsequent conduct of American authorities. Approximately, 760 Austrian and German scientists were employed by the American 'Project Paperclip' that ended in 1955. Four of the accused, although exonerated at the Nuremberg trials, were employed by the US military. There were a number of defendants taken on by the US Air Force until their arrest. Among them was Hermann Becker-Frieyzing, who was sentenced for his experiments on freezing, high altitude, and sulfanilamide. Blome, who infamously did plague experiments in the concentration camps, was given a job in 1951 with the US Chemical Corps on Project 63. Having been denied US immigration clearance, he then was taken on as a camp doctor at Oberusal at the European Command Intelligence Center.

In addition, a substantial number of the accused were integrated into the pharmaceutical companies of post-war Germany. Finally, in the Tokyo war crime trials, Japanese doctors who experimented on prisoners of war in China during World War II to test the effects of biological weapons were not prosecuted because an agreement was made for them to disclose the results of their experiments to the US military officials.

Are the aforementioned a testimonial to the propensity even of liberal democracies to identify worthiness wherever needed in the so-called public interest, with technical science achieved at the expense of vulnerable citizens?

The Nuremberg trial was mainly avoided in medical journals because the horrors of Nazi Germany were perceived as being far away from the practices of post-war scientists in the Western democracies. Researchers distinguished themselves away from the unspeakable acts and thought that there was little to learn in the consolidation of rules that could be relevant to their ongoing work. The national interests prevailed in the US and elsewhere.

On the judicial side of the equation, the record of the lack of identification with the harmed parties substantiates this. The US Supreme Court decision of 1987, under-lined by Justice Scalia for the majority, pushed away 'Stanley'[1] who was a victim of LSD experimentation. This was the infamous CIA project named MKUltra, where brainwashing techniques involving some two hundred researchers were conducted in eighty institutions. Incidentally, the MKUltra records were evacuated by the head of the CIA in 1973. Justice Scalia concluded that 'Stanley' suing the armed forces would undermine military discipline and decided that the CIA activities were protected in the name of national defence and security.

Admittedly, there were a number of stinging dissents in this case. Justice Brennan noted that 'the government of the United States treated thousands of its citizens as though they were laboratory animals'. In his view, this exposed the extent to which American legal bodies at the highest level sacrificed the abuse of citizens under the penumbra of national security. Sadly, the perpetration of indefensible acts against 'Stanley' and a host of others were most likely conducted by civilians whose activities were covered up by the US military.

There has been much publicity given to the radiation experiments which involved the Navajo Indian Miners, who were not warned of the dangers of uranium. The US public health service failed to inform the miners even after the dangers were known; this was also done with the cooperation of the mine owners. The Court in Begay versus USA[2] superseded the claim to human rights violations by rendering its decision in favour of the Federal Tort Claims Act which protected defendants as acting based on a discretionary function. In addition to the Navajo community, radiation exposure cases involved prisoners, patients with a limited lifespan, and citizens at large.

Canada no less participated in horrific experimentation at the Royal Victoria Hospital connected to McGill University. Deep patterning experiments were conducted by Prof. Donald Ewen Cameron of the Allan Memorial Institute. This was denounced in recent decades as highly destructive, unprofessional, and tanta-mount to torture. Although there were reparation payments made by the Government

---

[1] US vs. Stanley, 107 US 3054 (1987).

[2] Begay vs. US, 768F. 2D. 1049, 4th Cir. 1985.

of Canada in the late eighties, there appears to have been no settlements from the American government.

Dr. Cameron held such positions as president of both the Canadian and American Psychiatric Associations as well as the World Psychiatric Association. His case has stood out as what can occur when a powerful figure is given official grant money, in this case by the CIA, which led to extraordinary violations. It is interesting to observe that in his defense leading psychiatric experts came forward to excuse his excesses as errors or misadventures but not rising to the level of negligence or professional misconduct.

Following these post-war histories that reached into the 1960s, there were certain milestone contributions such as Jay Katz's pioneer 'Experimentation with Human Beings' published in 1972. The late sixties and seventies heralded the onset of extensive works by bioethicists dedicated to elevating the principle of autonomy as the core value for the protection of vulnerable populations.

In this regard, we should take note of the National Commission for the Protection of Human Subjects of Biomedical and Behavioural Research, identified as the Belmont report of 1979. No less the Council for International Organizations of Medical Sciences and the World Health Organization issued guidelines in 1982. Moving further in time the European Council in what was called the Oviedo Convention in 1997 enunciated that the human being must supersede the interests of society and/or science.

The modern world has taken us into the sphere of breakthrough technology, rapidly evolving, for example, in the instance of gene editing. Leading institutions have jumped into the fray, such as the UK Nuffield Council in 2018 and the US National Academies in 2017, demanding that regulatory frameworks be put into place to guard against unsatisfactory violations with regard to risks.

Medical interventions which have had a massive play in the public media have announced a new era begging for international accords and for liberal democracies to take the lead in controlling what may be termed questionable 'morally defensible advances'. Shadowing this has been imposing political movements involving gender and race. This has frequently been accompanied by multicultural claims connected to religious identity and so-called ethnic minorities. Such pressures have contemporaneously amounted to a mushrooming almost to the level of the industrialization of bioethics.

With the rapid advancement of biotechnology, areas such as genetics, neuroscience, and artificial intelligence have put into place a dramatic upheaval of the well-worn bioethical liberal mantras, now reduced to very practical questions of the applicability of abstractions to practicality.

This has been exacerbated by a turn towards autocratic political regimes and the leap frogging, for example, of the Chinese economy, which many observers now believe can potentially lay waste to the consensus morality of Western liberal democracies on questions of human rights in the medical sphere.

What was accomplished in a period of anti-paternalism and the respect for autonomy may conceivably, and even predictably, go into retreat in a world of shrinking economic efficacy. One at this point can be justified to be fearful of dominance by newly emerging economic elites who might wish to produce a new morality or at least a profoundly different one from the prevailing liberal reference points.

Are we on the verge of transforming our very notion of personhood? For colleagues with a philosophical bent, the looming question may become how the core values of the enlightenment are in the course of being displaced by the reckoning of collective survival. In the world of bioethics, we can see this highlighted in the emotionally charged debates on the right to secure medically assisted dying.

Finally, in the face of managed care and corporate provider organizations, much of what we have known as the core values of the liberal traditions in bioethics are being put to demanding tests. If it is the case that the health systems will be compromised in being answerable to cost-saving governments and/or shareholders in insurance endeavours, the threats and burdens upon philosophers and practitioners alike can already be acknowledged to be formidable.

This volume is aimed to engage a diverse multidisciplinary readership in reviewing multinational perspectives on pressing ethical issues in medical research. The foundational essays are meant to engage readers at every level, from students to professional researchers, in revisiting philosophical premises that still remain both controversial and significant to working out the fundamentals of bioethics. Vulnerability, which is a concept that lies at the bedrock of applied bioethics, is treated historically and philosophically.

This is a mirror of contest and debate within research ethics committees, seminars from the beginning to the advanced, and thereby offering a research tool to lead the reader to associated writings. The ultimate purpose is to lead off discussions where answers are subject to reflection rather than uncontested conclusions.

A number of authors have ventured novel inroads to understanding how to approach the categorization of subjects, such as in neuroscience, showing the relevance of epidemiology for the future of research undertakings. In the field of reproductive research, which is an area of profound differentiation, authors have presented a point of view that is deserving of respectful responses from thinkers who will oppose the premises. What motivates medical research is dealt with once again as a position which should engage the perennial debates about the very foundations of the ethics of medical research.

Chapters are directed to tackle how the benefits and risks are weighed in the course of medical research. A powerful question is how this applies to the world of the non-consenting and what the significance of state-of-the-art areas such as stem cell research hold out for bioethics now and in future projections.

The pathway towards a sensitive reaction to dementia research is treated creatively in order to introduce justifiable inclusions of dementia suffers in research. Exaggerated claims of stem cell and regenerative medicine research are deconstructed and presented for cautious consideration. The problem of shared information which is an escalating intervention into privacy rights is so threatening in the short term that all citizens may become research subjects without acceptance or even awareness. Neuroscience research where enormous interest and investments are now normative is given an in depth look at the ethical mishaps that are now intrinsic to the evolution in process. A repetitious and gnawing ethical dilemma is addressed about the role of prisoners in the world of medical research. An innovative area to date under-researched is dealt with in investigating the special problems that occur in the field of surgery. The essential banning of research into the world of psychedelic experiences is revisited through an elaborate philosophical raising of compelling questions that invite major philosophical and empirical analyses. Risk assessment is a loaded inquiry, and the role and function of healthcare professionals is put under a magnifying glass in terms of the rights and obligations that occur in a world disturbed by rising acts of violence and terrorism. The limitations of the bioethical model are addressed by the manner in which PTSD is currently dealt with in research. The coronavirus pandemic is introduced through the intriguing vantage point of constitutional law and ethics. A number of chapters continue the timely response to COVID-19 and its association to the mounting of research and the ambiguities of physicians' moral duties about access to care.

This work has attempted to avoid parochialism and in this vein has reached out to introduce the readership to European, African, and Asian experiences. It is foreseen that these chapters in particular will introduce researchers and students alike to bodies of knowledge of law, philosophy, and social policy which are otherwise not readily available.

Prague, Czech Republic   Tomas Zima
Montreal, Canada   David N. Weisstub

Contents

Contributors

Adams Rachel Research ICT Africa, Cape Town, South Africa

Bunnik Eline Erasmus MC, Rotterdam, The Netherlands

Couderc Bettina Université de Toulouse, Toulouse, France

Cox Susan University of British Columbia, Vancouver, Canada

Duguet Anne-Marie University of Toulouse-Paul Sabatier, Toulouse, France

Eijkholt Marleen Leiden University Medical Centre, Leiden, The Netherlands

Freedman David International Academy of Law and Mental Health, Montreal, QC, Canada

George Robert Princeton University, Princeton, USA

Glannon Walter University of Calgary, Calgary, Canada

Hafetz Jonathan Seton Hall Law School, Newark, NJ, USA

Holm Søren University of Manchester, Manchester, UK

Horn Lyn Stellenbosch University, Stellenbosch, South Africa

Hutchison Katrina Macquarie University, North Ryde, NSW, Australia

Krimsky Sheldon Tufts University, Boston, MA, USA

Labuschaigne Melodie University of South Africa, Gauteng, South Africa

Lanzerath Dirk University of Bonn, Bonn, Germany

Łuków Paweł University of Warsaw, Warsaw, Poland

Macer Darryl R. J. American University of Sovereign Nations, Sacaton, USA

Matthews Kirstin R. W. Rice University, Houston, USA

Moreno Jonathan University of Pennsylvania, Philadelphia, USA

Napier Stephen Villanova University, Villanova, PA, USA

Novak David University of Toronto, Toronto, Canada

Puurveen Gloria University of British Columbia, Vancouver, Canada

Rogers Wendy A. Macquarie University, North Ryde, NSW, Australia

Rousset Guillaume Jean Moulin Lyon 3 University, Lyon, France

Różyńska Joanna University of Warsaw, Faculty of Philosophy, Warsaw, Poland

Shapiro David Nova Southeastern University, Fort Lauderdale, USA

Shaw Elizabeth University of Aberdeen, Aberdeen, Scotland

Staunton Ciara Middlesex University, London, England

Stuart Heather Queen's University, Kingston, Canada

Tollefsen Christopher University of South Carolina, Columbia, USA

Walker Lenore Nova Southeastern University, Fort Lauderdale, FL, USA

Waring Duff R. York University, Toronto, Canada

Part I
Philosophical Foundations

Chapter 1
Embryo Research Ethics

Robert George and Christopher Tollefsen

Research on human embryos promises several important scientific and medical benefits. Most directly, such research would make it possible to fill in significant gaps in our knowledge of human embryological development. Much of our current knowledge comes from the study of non-human embryogenesis in model species; such study is valuable but limited by the inevitable dissimilarities that exist between humans and non-human animals. In vitro human embryos are also studied, but for reasons to be explained below, rarely if ever beyond the 14-day mark post-conception; accordingly, there is a large gap in our knowledge of embryological development after the 14-day mark.[1] And, in certain countries, even research before 14 days is limited by lack of funds, or restricted or forbidden by law. Accordingly, much remains to be learned of embryological development even in the first two weeks of a human being's life and certainly beyond that point.

Further developing our understanding of early human development could yield numerous practical benefits. General interest in the study of human embryos was stimulated after the first test-tube baby was born; the potential benefits for infertility treatments have ranked among the primary justifications for human embryonic research ever since. New knowledge of embryological development could also lead to advances in medical treatment, in the form, for example, of new stem cell therapies. And knowledge of human embryogenesis could lead to greater insight into the

---

[1] See Wagner and Matthews (2019), for a helpful summary of the sources of our knowledge of embryological development, and the limitations of those sources for understanding the period between 14 days post-fertilization and 8 weeks of development.

R. George
Princeton University, Princeton, USA
e-mail: rgeorge@princeton.edu

C. Tollefsen (✉)
University of South Carolina, Columbia, USA
e-mail: christopher.tollefsen@gmail.com

© The Author(s), under exclusive license to Springer Nature Switzerland AG 2023
T. Zima and D. N. Weisstub (eds.), *Medical Research Ethics: Challenges in the 21st Century*, Philosophy and Medicine 132,
https://doi.org/10.1007/978-3-031-12692-5_1

variety of diseases that disable or kill human beings while they develop in the womb, and likely into conditions that affect those already born as well.

So research on and into human embryos is considered desirable among many scientists, and has strong political constituencies as well. But the history of such research is marked by moral and public policy controversy and disagreement, which this essay will map.

Most centrally, those controversies concern the so-called 14-day rule. By explicit legislation in some countries, such as the UK, or common agreement in others, such as the US, human embryos are not kept alive in vitro past the 14-day post-conception mark. For the past 40 or more years, that rule has been an occasional matter of contestation, with some arguing that 14 days is too permissive, others that it is too restrictive, and of course, some arguing, on various grounds, that 14 days marks the most reasonable point up to which research should be permitted but after which it should be restricted. These controversies have intensified recently, as scientists have pushed past practical boundaries beyond which it had hitherto been impossible to sustain an in vitro embryo's life; there are now calls to extend the 14-day limit to 21 or 28 days, or even further (Hyun et al. 2016).

A second controversy—of somewhat more recent origin but related in important ways to the first—concerns embryo-like or "embryoid" beings, synthetic entities which are designed to mimic some, many, or potentially all features of a developing human embryo. Should such entities be viewed with moral concern? May they be used to side-step controversy over the 14-day rule?

This essay considers both controversies. Part One addresses the 14-day rule. We first identify the origin of the 14-day rule, focusing on those attempts to articulate principled grounds for thinking that 14 days is both the morally correct dividing line between permissible and impermissible research and the best marker for public policy. We then look at the two competing positions, first, the view that 14 days is too restrictive, and then the view that it is too permissive. We offer our own arguments that human embryos deserve the same moral protections that are owed to more mature human beings; thus, we argue, the 14-day rule is too permissive.

Part Two addresses, albeit more briefly, the question of synthetic embryoid beings. We raise several moral questions that must be answered before such research can be approached with confidence. Nevertheless, we suggest that research on such embryoid beings might be a solution to the moral and policy disputes over human embryo research.

1.1 Research on Human Embryos

1.1.1 The 14-Day Rule

The claim that experimentation is permissible morally, and should be permitted legally, on embryos up to 14 days post-fertilization dates to deliberations in the

United States in the late 1970s in a Report of the Department of Health, Education, and Welfare (DHEW 1979); but the 14-day rule is primarily associated with the 1984 Report in the United Kingdom of the of the Committee of Inquiry into Human Fertilisation and Embryology, chaired by Dame Mary Warnock (later Warnock 1984). After a review of what Committee members considered the strongest arguments for and against the permissibility of research on human embryos, the report stated that a majority of the Committee "agreed that research on human embryos should continue," while recommending that such research be done only with permission from a licensing body (Warnock 1984, 64). The report then went on to consider two crucial questions: up to what point in an embryo's existence should research be permitted, and from what sources could embryos to be used in research be drawn?

The committee considered various options regarding the first question, including some that would have been quite permissive, such as until the onset of a capacity for the embryo to experience pain. The report then stated:

> …the objection to using human embryos in research is that each one is a potential human being. One reference point in the development of the human individual is the formation of the primitive streak. Most authorities put this at about fifteen days after fertilisation. This marks the beginning of individual development of the embryo…We accordingly recommend that no live human embryo derived from in vitro fertilisation, whether frozen or unfrozen, may be kept alive, if not transferred to a woman, beyond fourteen days after fertilisation, nor may it be used as a research subject beyond fourteen days after fertilisation (Warnock 1984, 66).

The Committee noted that there was internal disagreement about whether only embryos leftover from fertility treatments should be used for research, or whether embryos could also be generated specifically for research and recommended that "legislation should provide that research may be carried out on any embryo resulting from in vitro fertilisation, whatever its provenance, up to the end of the fourteenth day after fertilisation" (Warnock 1984, 69).

The Committee's report led, in 1990, to the United Kingdom's Human Fertilisation and Embryology Act, which established the 14-day limit in law. As noted above, other countries have either passed similar legislation, or maintain the 14-day limit as a consensual norm, as in the United States.[2]

Of greatest interest to us here is the rationale for the 14-day rule. That rationale is hinted at in the Warnock Report and made more explicit in a number of books and papers published defending the 14-day rule as a matter of principled public policy. Recall that the Warnock Report says of the development of the "primitive streak" that it "marks the beginning of individual development of the embryo." Why should that moment be taken as definitive of "individual development"?

There are two important answers to this question that have been given in the literature. The first concerns the phenomenon of human monozygotic twinning. Day fourteen or fifteen, and the beginning of the primitive streak, mark the outside point at which an embryo can, by symmetric or asymmetric division, give rise to two

---

[2] See Matthews and Marquez (2019), for a review of international policies regarding human embryo research.

embryos. Most embryos do not in fact divide; but even if twinning does not occur, it seems likely that every human embryo has the *capacity* to twin, even if that capacity can only be actualized by some external event or force acting on the embryo.

In the 1970s and 1980s, monozygotic twinning was made a premise in an influential argument by certain Catholic thinkers, prominently Joseph Donceel, and Norman Ford (Donceel 1970; Ford 1988). Their argument, also voiced by secular thinkers, was that because the early human embryo had a capacity to become two individual human beings, the early embryo with that capacity could not be considered properly to be one single individual organism; its individuality had not yet been established or fixed. But if not an individual, then not an individual human being, and if not an individual human being, then certainly not an individual human *person*, i.e., a being with fundamental rights not to be killed in the process of deliberate research aimed at providing no benefit to the embryonic organism itself. A second argument reached a similar conclusion. Some philosophers argued that until some point later in development, although not necessarily 14 days, the embryo is an undifferentiated or insufficiently differentiated collection of cells that do not compose a single organism (Smith and Brogaard 2003). Twinning is not essential to this argument; rather a perceived uniformity among the cells of the early embryo indicates that an early embryo is no more than a collection of cells held together by an external barrier, the *zona pellucida*. Only when the cells of the embryo begin to work together as a functional whole can the embryo be understood as a single organism. We will address the merits of both this argument and the argument from twinning in Sect. 1.1.3.

Here, however, we note the importance that the denial of individuality, especially on the basis of twinning, had for the public policy debate about human embryo research. In the context of the controversy over abortion, the distinction between "human being" and "human person" had been crucial: many philosophers engaged in the abortion debate had denied that being a human being was a sufficient condition for being a person, a subject of fundamental rights (Warren 1973). Such a position is obviously contentious, running contrary to important considerations of human equality; we will return to these considerations later. But a denial, on allegedly scientific grounds, that the early embryo was even a human being, because of its lack of individuality, offered a route around that largely philosophical disagreement about human persons. As philosopher George Khushf noted in a discussion about the United States Report of the Human Embryo Research Panel, which in 1994 also supported research up to 14 days,

> Drawing on scientific facts associated with the totipotency of the embryo cells before the appearance of the primitive streak, Panel members argue[d] that one could not have a unique individual, thus one could not have a person....[This argument] shifts the debate from the realm of philosophy and ethics (what are the criteria of personhood) to the realm of biological science; i.e., the "facts" of embryology are used to resolve whether or not developmental individuality is present (Khushf 1997, 504–505).

As mentioned above, this argument was endorsed by certain Catholic thinkers, members of a faith traditionally associated with strong moral protections on pre-natal human life. Accordingly, some participants in the debate have understood the 14-day rule to be not only a principled dividing line, given the strength of the argument from twinning, but also a good faith effort to find common ground between largely secular scientists, interested in research on human embryos beyond 14 days, and religious voices more cautious about such research. Laurie Zoloth has articulated this account:

> The establishment of the 14-day rule was the result of a careful, thoughtful and respectful conversation that took account of many competing moral and religious claims, and was worked out over a several-year period in the bioethics, theological and scientific communities…[A] compromise was reached with some Catholic theologians and nearly everyone else that would allow research for nontrivial purposes on embryos before 14 days… (Hurlbut et al. 2017)

Thus, Zoloth continues to support the 14-day rule not only on principled grounds such as the argument from twinning but also out of respect for a compromise that had been forged between people of a variety of faiths and shades of belief on an issue of critical importance for society.

Baroness Warnock also gave voice to the importance of treating the 14-day rule as a fixed matter (though late in her life she made statements suggesting she was not entirely opposed to eventual reconsideration): "Perhaps with the 14-day rule we erred on the side of caution. But you cannot successfully block a slippery slope except by a fixed and invariable obstacle, which is what the 14-day rule provided" (Warnock 2017). Warnock indicated her fears that "pro-life" opponents would seize on an extension to say "We always knew that the slippery slope would prove itself, and here it is, just as we said" (Ibid.).

The positions embraced by Zoloth and Warnock illustrate an important aspect of "public" bioethics concerned with biomedical research. There is a perception that science always runs ahead of the ethics and always drives public policy, a perception that is broadly damaging to the democratic legitimacy of the scientific enterprise, carried out as it is in public, with public money, for public ends. Even for those who disagree in principle with the 14-day rule, in either direction, there is a valuable good in seeing a publicly taken decision maintained despite pressure from one side to change without considerably more public deliberation. As Benjamin Hurlbut argues,

> Science is a social institution supported by—and celebrated by—societies committed to the project of enlightenment. Yet it is not self-justifying. Insofar as its projects risk running afoul of notions of human integrity and dignity that are socially important, even if not universally held, responsibility requires that scientists defer to society's judgments, and not the reverse. Therefore, although the 14-day rule is informal and voluntarily self-imposed by researchers in most of the US, asking these questions as though they are purely matters of internal scientific and ethical judgment…may allow certain experiments to be done and papers to be published, but it risks doing grave damage to the position of science in the project of democracy (Hurlbut et al. 2017).

1.1.2 Expanding Research Beyond 14 Days

Despite the concerns raised at the end of the previous section, there are nevertheless calls to expand the period in which human embryo research is permitted. In part, this is driven by practical advances that make it possible to keep an embryo alive in vitro for up to and potentially beyond the 14-day mark. Additionally, as noted earlier, research beyond 14 days offers the hope of filling in considerable lacunae in our knowledge of embryological development. The question for this section is whether there is some principled reason for thinking that 14 days is the wrong dividing point, such that some other marker could indicate a fixed point beyond which science could not pass, regardless of any increase in its practical capacities. To investigate this question, we turn to a recent statement of updated recommendations concerning human embryo research issued by the International Society for Stem Cell Research (ISSCR).

The ISSCR is a professional organization of researchers that periodically issues guidelines for research on stem cells and human embryos. The most recent iteration, the ISSCR Guidelines for Stem Cell Research and Clinical Translation, was issued in May of 2021, and drew attention, and both praise and criticism, for its unexpected suggestion that the 14-day rule could, under certain conditions, be breached. The ISCCR Guidelines often serve as de-facto rules in jurisdictions, such as the United States, that have not established formal regulation, and the new recommendation is anticipated to play an important role in shaping future human embryo research.

Regarding the question of human embryo research, the Guidelines say:

> Research goals must be assessed within an ethical framework to ensure that research proceeds in a transparent and responsible manner. The project proposal should include a discussion of alternative methods and provide a rationale for performing the experiments in a human rather than animal model system, for the proposed methodology, and if the studies involve preimplantation human embryos, a justification for the anticipated numbers to be used (International Society for Stem Cell Research. 2021, 2.1.2.c).

But in issuing their recommendation concerning the 14-day rule, the Guidelines say little if anything about an "ethical framework." Rather, according to Recommendation 2.2.2.1:

> Given advancements in human embryo culture, and the potential for such research to yield beneficial knowledge that promotes human health and well-being, the ISSCR calls for national academies of science, academic societies, funders, and regulators to lead public conversations touching on the scientific significance as well as the societal and ethical issues raised by allowing such research. Should broad public support be achieved within a jurisdiction, and if local policies and regulations permit, a specialized scientific and ethical oversight process could weigh whether the scientific objectives necessitate and justify the time in culture beyond 14 days, ensuring that only a minimal number of embryos are used to achieve the research objectives (International Society for Stem Cell Research 2021, 2..2.2.1).

Here, the 14-day limit is treated as a temporary impediment, while scientists and their supporters "lead public conversations" *aimed* at extending the deadline. The only justification offered is the "potential for such research to yield beneficial knowledge that promotes human health and well-being," and no ethical concerns regarding the moral status of human embryos are mentioned, much less discussed. The Guidelines radically fail to identify any other principled marker, and indeed, some scientists associated with the ISSCR have denied that there are any such markers. Robin Lovell-Badge, who chaired the ISSCR committee that drafted the guidelines, has said "I felt that it would be both difficult and a little pointless to propose any new limit, which would be arbitrary, much like 14 days" (quoted in Stein 2021b).

This is a remarkable position. At *some* point, clearly, the embryonic human is human being, and at *some* point, human beings are possessors of human dignity and subjects of fundamental rights that render them immune from unconsented to and lethal experimentation. To deny that there is *any* limit, to refuse to reckon with the need for a limit, is not just to initiate the slippery slope feared by Baroness Warnock; it is to abandon any genuinely ethical assessment of scientific research involving human subjects. One might respond that the ISSCR is asking precisely for more public deliberation given the impossibility of a principled account of human individuality or moral status; but then, what end or purpose could there be for such dialogue and deliberation in the absence of any possible correct or even non-arbitrary answer? Public deliberation for the sake of sound public policy cannot be adequately motivated if it is thought in advance that any limitation set on scientific research will be "arbitrary."

All this raises the question: Were the principles that generated the 14-day limit sound? Does a principled *and morally reasonable* approach to the question of human embryo research suggest something more, rather than less, restrictive? We turn to these questions in our next discussion.

1.1.3   Doubts About the 14-Day Rule

As we have seen, the 14-day rule, to the extent that it is not, as Lovell-Badge suggests, "arbitrary" but rather an attempt to establish a principled limit, is founded upon a claim about embryonic individuation. Whether because of the possibility of twinning, or because of facts about the functional organization of the early human embryo, or both, the early human embryo is not an individual organism of the human species, i.e., an individual human being. Thus, as we saw Khushf argue above, questions about personhood could be avoided entirely.

Yet there are significant scientific and philosophic reasons for rejecting both the denial of individuation, and the more "philosophical" denial of personhood. Longer treatments of both issues have been made elsewhere; we here only summarize the reasons for holding that from the point of embryogenesis itself, that is successful fertilization or some other process that generates a new organism of the species that is both genetically and functionally distinct from the gametes (or, in the case of somatic cell nuclear transfer, the enucleated ovum into which material from the nucleus of a somatic cell has been transferred), human embryos are both individual human beings, and human persons.[3]

The most significant objections to both the argument from twinning and the claim that the early embryo is merely an aggregate of cells come from contemporary embryology and developmental biology, which show unambiguously that from the zygote

---

[3] See George and Tollefsen (2011), for a fuller discussion of all the arguments presented here.

stage onward, a human embryo is an individual that undergoes self-directed development, repairs injury to itself, adapts to its changing environment, and manifests integrated functioning of parts that are differentiated by at least the four-cell stage, if not earlier. Maureen Condic, who summarizes the data in her book *Untangling Twinning*, describes the beginning of the embryo's life:

> Within minutes of sperm-egg fusion, the zygote initiates a specific molecular cascade, using elements derived from both sperm and egg, to direct its subsequent maturation. Substantial evidence indicates that by the four-cell stage (or earlier), individual blastomeres have distinct patterns of gene expression, different cellular functions, and unique developmental capabilities. This indicates that the integrated function of the embryo as a whole reflects collaboration of its parts to generate a normal developmental sequence. While all cells are capable of short-range communication, only embryos establish a global pattern of interaction that benefits the entity as a whole. Indeed, within the first minutes and days of life, the embryo initiates dozens of distinct, globally integrated events that are critical for its survival and healthy maturation (Condic 2020, 10–11).

This description suffices to rebut the claim that the embryo is insufficiently differentiated to be a single organism. But some defenders of the twinning argument hold that there is a principled concern regarding an allegedly individual entity that can twin, namely, that something that is genuinely *one* ought not to be able to divide to become two individuals. Thus, regardless of the appearances of unity, twinning poses a principled difficulty for defenders of embryonic individuation.

The argument is deeply flawed. Among its implications would be that an amoeba, or other single-celled organism, that reproduces by fission, is not a single living organism. But the implication of that is that there are no individual single-celled organisms at all. Further, there are multi-celled organisms that also are capable of fission, including various species of worm (such as the flatworm); it is implausible to deny that any of them are individual organisms. And so the principled argument from twinning seems weak; both for reasons of biological science and sound philosophy, the claim that a human embryo prior to 14 days is a single living organism is very strong.

It is, of course, still available to defenders of human embryo research to deny that human embryos are human *persons*, that is, beings owed the same forms of respect that are owed, for example, to the authors and readers of this chapter. This is a different kind of question than those of the individuation or humanity of the early embryo. The individuation question, and the question of whether the embryo is a human being, are questions of fact: is this putative entity a single being, and if so, what kind of being is it? We have answered these questions with a combination of philosophical analysis and scientific data. The question of personhood, by contrast, is not simply a philosophical question, and it goes beyond any question that can be answered by science alone: it is an ethical question. How *ought* we to treat human beings in general, and, more specifically, when they are in utero or in vitro? Are all such human beings to be treated with equal moral respect? Or is it the case that only some human beings are proper subjects of such respect, while others may be used to serve the research interests of scientists?

We encompass all these moral questions under the question of *personhood*: granting that only persons are subjects of full moral rights, and owed equal moral respect, are all human beings persons? Or only some? Numerous markers have been identified by those who defend experimentation on embryos or, in another context, the abortion of unborn human beings, to serve as a dividing point separating human beings who can from human beings who cannot be permissibly killed in service of research or other goals: the ability to feel pain, some level of self-awareness, the possession of conscious interests, the possession of self-conscious interests, all the way up to the ability to exercise the distinctively personal characteristics of reason and freedom.[4] Such thinkers hold that *not all human beings are persons*.

The readers and authors of this chapter are capable of these final, most exalted, capacities, the capacities for freedom and reason, and, like many who distinguish between human beings and human persons (we do not), we believe that these capacities do indeed mark off the domain of persons with rights: entities with these capacities possess intrinsic worth and dignity that entitles them to fundamental forms of moral respect; entities without those capacities are not owed the forms of respect that persons are.

The central question, then, is: which human beings do in fact possess these dignity-conferring capacities? For many who support research on human embryos (and other forms of destruction of unborn human beings), the answer is: "those beings that, like the readers and authors of this chapter right now, are more or less immediately capable of actualizing or exercising their powers of freedom, reason and whatever other capacities are essential to the dignity of persons." For those who answer in this way, many adult non-human mammals such as dogs or pigs are more like us than are human fetuses. But there is a way in which human fetuses are *vastly* more like us than *any* other creature of whose existence we are directly aware. Human fetuses—unless impeded by disease, injury, or violent attack—will grow and develop naturally (and indeed by an internally directed program and process) to the point of being able to exercise exactly those characteristics that we regard as grounding the fundamental dignity of persons. No other being of whose existence we are directly aware will *ever* do that, not even the highest functioning dolphin or ape.

That indicates that human embryos, fetuses, infants, and adults alike are equally possessed of something that is not at all possessed by any other earthly being: a nature that is such that beings with that nature develop to the point of being able to actively display their rationality. Call that a "rational nature." That is a radical commonality, the sort that should impress us when we are asking, "To which other beings like or unlike us in this or that way—bigger, smaller, darker, paler, younger, older—should we think that these fundamental rights belong? Which other beings are like us in the relevant way?" Possessing a rational nature is the *most* important similarity.

That similarity should additionally impress us for the following reason: possessing a rational nature is perhaps the only thing that all human beings at all developmental stages–embryos, fetuses, infants, toddlers, children adolescents, and adults—and in all conditions possess *equally*. The actualized abilities to, at any given time, sense,

---

[4] We address such challenges at greater length in George and Tollefsen (2011), Chaps. 4 and 5.

think, choose, speak, and the like are all possessed and realized by human beings to varying degrees and in varying ways. Yet it is central to our moral understanding of the most fundamental human rights, such as the right not to be killed at will, that they are always possessed equally by all who ever possess them. What property could possibly account for the possession of equal rights were it not a property that was itself equally possessed? A common rational nature thus seems foundational for the possession of equal dignity and equal rights, and that common rational nature is found in all human beings from conception onwards. Indeed, an affirmation along precisely these lines is presupposed and relied on by anyone who claims to believe in *human rights*—viz., rights people have simply in virtue of their humanity, rights that are possessed alike by young and old, by the weak no less than the strong, by the infirm as well as the healthy, by the lowborn as much as by the highborn, by people irrespective of race, ethnicity, sex, or any other factor or feature that distinguishes some human beings from others.

The conclusion to be drawn if such inferences are sound is that the 14-day rule is too permissive rather than the opposite. There is a principled norm that should govern human embryo research and it is the same boundary as for other members of the human community: research that provides no benefit to the human research subject is not to be performed on that subject without that subject's consent. But no human embryo gives its–truly, his or her (since, in the human, sex is established from the very point of embryogenesis)–consent to the research that ends its life; such research, we hold, is thus morally impermissible.

1.2   Research on Synthetic Embryoid Beings

The restrictions imposed by the 14-day rule, such as they are, have led to interest in research on embryo-*like* entities, typically, entities created from pluripotent stem cells that are used to generate multi-cellular structures that replicate embryos at some particular stage of development. Such entities might be used to model embryolog- ical development at the blastocyst, gastrulation, or neurulation stages, for example, without the entity undergoing the previous developmental stages of a natural, non- synthetic embryo. It is, at the very least, unclear whether and how the 14-day rule would even apply to such entities, since they are not embryos, and since their devel- opment does not commence (roughly) 14 days prior to the emergence of the primitive streak and indeed might commence at a stage subsequent to that emergence.

We believe that research on embryo-like entities offers a promising route through the moral problems raised by research on human embryos. Nevertheless, the use of such synthetic entities raises at least the following moral questions that also bear upon public policy.

First: insofar as synthetic embryos are created using human pluripotent stem cells derived from human embryos, then the process of their creation does not in any way avoid the moral issues raised by embryo research itself. However, embryoid bodies can also be created using induced pluripotent stem cells, i.e., stem cells created from somatic cells, and which do not involve the destruction of a human embryo (Malcolm 2019). So this difficulty appears to be surmountable.

Second, to have full confidence in the accuracy of embryoids as models past the 14-day stage, embryoids must, arguably, be compared to genuine human embryos at similar developmental stages. Thus, some scientists argue that synthetic embryo research must proceed hand in hand with expanded research on human embryos past the 14-day mark (Aach et al. 2017).

Third, concerns are raised that putatively embryo-*like* entities might be created that in fact have the very same developmental potential as a natural human embryo (Stein 2021a). Such an entity would be, by our account in the previous section, a human being, and hence a human person. It is thus essential that research on synthetic embryoids be conducted with an awareness of what does and does not constitute a human embryo and a commitment to avoiding the creation of what would in fact be human embryos.

Here, the discussion is continuous with the history of the discussion of the President's Council on Bioethics over human embryo research alternatives. Drawing on the then relatively new technique of somatic cell nuclear transfer (SCNT), more commonly known as cloning, William Hurlbut argued for the creation of pluripotent cells by means of altered nuclear transfer (ANT) as part of SCNT (Hurlbut 2006). Since it is possible to identify the epigenetic features and developmental trajectory of a genuine human zygote, cells created through ANT without those features and trajectory could be confidently known to be not human beings. Similar considerations should govern the creation of embryo-like entities intended to model later stages of development than the zygote.

Fourth, concerns have been raised about the development of synthetic embryos that, while not developmentally the same as early human beings, nevertheless possess the neurological underpinnings for features that are commonly thought to underlie some form of moral status (Aach et al. 2017; Matthews et al. 2019). For example, an embryoid created to model a late stage of development might have neurological properties characteristic of a being that can experience pain. Again, entities with such properties could be created without obviously violating the 14-day rule, since they would come into existence with these properties.

Addressing the concerns just raised goes beyond the scope of this chapter; but any full treatment of embryo research will need ultimately to grapple also with the moral questions raised by entities that are embryo-like, even if they are not, as human embryos are, human individuals—embryonic human beings.

References

Aach, John, Jeantine Lunshof, Eswar Iyer, and George M. Church. 2017. Addressing the ethical issues raised by synthetic human entities with embryo-like features. *eLife*, March 21. https://eli fesciences.org/articles/20674.
Condic, Maureen L. 2020. *Untangling twinning*. Notre Dame, IN: University of Notre Dame Press.

Donceel, Joseph. 1970. Immediate animation and delayed hominization. *Theological Studies* 31: 76–105.

Ford, Norman. 1988. *When did I begin? Conception of the human individual in history, philosophy, and science.* Cambridge: Cambridge University Press.

George, Robert P., and Christopher Tollefsen. 2011. *Embryo: A defense of human life*, 2nd ed. Princeton, New Jersey: Witherspoon Institute.

Hurlbut, J. Benjamin., et al. 2017. Revisiting the Warnock rule. *Nature Biotechnology* 35 (11): 1029–1042. https://doi.org/10.1038/nbt.4015.

Hurlbut, William. 2006. Framing the future: Embryonic stem cells, ethics and the emerging era of developmental biology. *Pediatric Research* 59: 4–11.

Hyun, Insoo, Amy Wilkerson, and Josephine Johnston. 2016. Revisit the 14-day rule. *Nature* 533 (7602): 169–171. https://doi.org/10.1038/533169a.

International Society for Stem Cell Research. 2021. *ISSCR guidelines for stem cell research and clinical translation.* Version 1.0, May 2021.

Khushf, George. 1997. Embryo research: The ethical geography of the debate. *Journal of Medical Philosophy* 22 (1997): 495–519.

Malcom, Kelly. 2019. Induced pluripotent stem cells harnessed to reliably create embryo-like structures. *M Health Lab.* September 11, 2019. https://labblog.uofmhealth.org/lab-report/induced-plu ripotent-stem-cells-harnessed-to-reliably-create-embryo-like-structures.

Matthews, Kirstin R.W., and Nuria G. Marquez. 2019. *The Warnock report and international human embryo research policies.* Center for Health and Biosciences, Rice University's Baker Institute for Public Policy. https://www.bakerinstitute.org/media/files/files/8a0b4eac/chb-pub-greenwall-intl-012219.pdf.

Matthews, Kirstin R.W., Jason Scott Robert, Ana S. Iltis, Immaculada de Melo-Martin, and Daniel S. Wagner. 2019. *Cell-culture models of early human development: Science, ethics, and policy.* Center for Health and Biosciences, Rice University's Baker Institute for Public Policy. https://www.bakerinstitute.org/media/files/files/4044f718/chb-pub-greenwall-sheef-021419.pdf.

Smith, Barry, and Berit Brogaard. 2003. Sixteen days. *The Journal of Medicine and Philosophy* 28: 45–78.

Stein, Rob. 2021a. Scientists create living entities in the lab that closely resemble human embryos. *All Things Considered*, March 17. https://www.npr.org/sections/health-shots/2021a/03/17/977 573846/scientists-create-living-entities-that-closely-resemble-human-embryos.

Stein, Rob. 2021b. Controversial new guidelines would allow experiments on more mature human embryos. *All Things Considered*, May 26. https://www.npr.org/sections/health-shots/2021b/05/ 26/1000126212/new-guidelines-would-allow-experiments-on-more-mature-human-embryos.

U.S. Department of Health, Education, and Welfare Ethics Advisory Board (DHEW). 1979. *Report and conclusions: HEW support of research involving human in vitro fertilization and embryo transfer.* Washington, D.C.: U.S. Government Office of Printing.

Wagner, Daniel S., and Kirstin R.W. Mathews. 2019. *Human embryo research: What do we know and how do we know it?* Center for Health and Biosciences, Rice University's Baker Institute for Public Policy. https://www.bakerinstitute.org/media/files/files/fcd4841d/chb-pub-greenwall-her-022519.pdf.

Warnock, Mary. 1984. *Report of the committee of inquiry into human fertilisation and embryology.* London: Her Majesty's Stationary Office.

Warnock, Mary. 2017. Should the 14-day limit on human embryo research be extended? *BioNews*, January 9. https://www.bionews.org.uk/page_95833.

Warren, Mary Anne. 1973. On the moral and legal status of abortion. *The Monist* 57: 43–61.

Robert George is McCormick Professorship of Jurisprudence and Director of the James Madison Program in American Ideals and Institutions at Princeton University. He has served as Chairman of the U.S. Commission on International Religious Freedom and on the U.S. Commission on Civil Rights and the President's Council on Bioethics. He has also served as the U.S. member of

UNESCO's World Commission on the Ethics of Scientific Knowledge and Technology. He was a Judicial Fellow at the Supreme Court of the United States, where he received the Justice Tom C. Clark Award. A Phi Beta Kappa graduate of Swarthmore, he holds the degrees of J.D. and M.T.S. from Harvard University and the degrees of D.Phil., B.C.L., D.C.L., and D.Litt. from Oxford University, in addition to twenty-two honorary doctorates. He is a recipient of the U.S. Presidential Citizens Medal, the Honorific Medal for the Defense of Human Rights of the Republic of Poland, the Canterbury Medal of the Becket Fund for Religious Liberty, the Bradley Prize, the Irving Kristol Award of the American Enterprise Institute, and Princeton University's President's Award for Distinguished Teaching. His books include *Making Men Moral: Civil Liberties and Public Morality* and *In Defense of Natural Law* (both published by Oxford University Press).

Christopher Tollefsen is Professor of Philosophy at the University of South Carolina; he has twice been a Visiting Fellow in the James Madison Program at Princeton University. He is the author of *Lying and Christian Ethics* and co-author of *Embryo: A Defense of Human Life*(with Robert P. George), and *The Way of Medicine: Ethics and the Healing Profession* (with Farr Curlin). He recently served as Commissioner on the US Department of State's Commission on Unalienable Human Rights.

Chapter 2
The Ethics of Medical Research

David Novak

2.1 Covid-19: Three Urgencies

The devastation brought on our society by the Covid-19 pandemic has forced us all to make very heavy demands on medical researchers to come up with scientifically attested means to combat the disease. Actually, combatting the disease itself is to be done by medical practitioners, but they depend on the results of accurate medical research nonetheless.

Their response will only be as effective as the skill of medical practitioners to apply it to cases of the disease at hand. Yet the short term demands we make on medical practitioners are only reasonable when we make longer term demands on medical researchers to provide the practitioners on the front lines of the battle against Covid-19 with the means to do their job well.

The specific demands we make on medical researchers relating to the Covid-19 pandemic are threefold. Let us look at them in the descending order of their urgency.

One, most urgent is the demand for an effective treatment of the symptoms of the disease so as to relieve as many current Covid-19 victims as possible of their suffering, and to save as many of the afflicted as possible from imminent death. Such demands for the effective treatment of symptoms of the disease rather than the eradication of the disease itself in the not so distant past did lead to the successful treatment of the symptoms of malaria, diabetes, and AIDS, even though actual cures for these diseases are yet to be found, if ever. We do hope that a similarly successful treatment of the symptoms of Covid-19 will come very soon, so that many more of the afflicted do not die, and that most of them will be able to resume living a normal life.

D. Novak (✉)
University of Toronto, Toronto, Canada
e-mail: david.novak@utoronto.ca

© The Author(s), under exclusive license to Springer Nature Switzerland AG 2023   17
T. Zima and D. N. Weisstub (eds.), *Medical Research Ethics: Challenges in the 21st Century*, Philosophy and Medicine 132,
https://doi.org/10.1007/978-3-031-12692-5_2

Two, of course the greater goal of this kind of medical research is the eradication of the disease itself. That goal will only be attained, however, when the real cause of the disease is precisely determined, and then an actual cure of the disease be devised so as to either eliminate the cause altogether or to greatly mitigate its effects. There is precedent for this anticipation. Long ago, bubonic plague was virtually eliminated when it was discovered it is caused by bacteria carried by rodents infected by fleas. Appropriate prophylactic measures were then taken. With the advent of antibiotics more recently, those afflicted with such highly infectious, lethal diseases could indeed be cured. Nevertheless, since the discovery of the cause of Covid-19 and devising a cure for it will take more time than coming up with a mere treatment of its symptoms will take, it is wise to make the search for an effective treatment of its symptoms the more urgent priority.

Three, of course the ultimate goal of medical research now is to come up with a vaccine to prevent all those vaccinated from ever contracting Covid-19 in the first place. In fact, demands for the prevention of the disease by vaccination did result in the prevention of smallpox in the nineteenth century and the prevention of polio in the twentieth century. Nevertheless, since the fulfillment of this goal is less likely to be achieved before the discovery of a more effective treatment of the disease's symptoms, and before an actual cure of the disease itself can be achieved, its urgency is less than that of the other two. That is so even though many of us hope for the vaccine to be devised sooner, as it promises the most complete return to "the way things were before" (*status quo ante*), although from past experience we have learned that will never be the case exactly. That it might never come, or might take years to come, means that it is prudent not to put all of our resources and all of our hopes into that urgent quest.

The seriousness and energy now devoted to all three specific goals by many medical researchers, plus the public encouragement of their efforts, indicates that medical research is an exalted pursuit because its overall purpose is the saving of many human lives and the enhancement of our quality of life. That is why most of us would say that all persons able and willing to engage in medical research are to be encouraged in every way to do so. Past experience has shown the beneficial effects of medical research for all society's members. That is why it deserves public support and the thanks of each and every one of us. Indeed, benefitting its members is any society's *raison d'être*, which needs no further rationale. No one could deny that devising and implementing measures that protect society's members from harm, here protection from harmful disease, are beneficial to society, both as a whole and to all its individual members. Yet that only answers the question of why society should support medical research. But we still need to ask: Why *would* any individual member of society actually *do* or engage in medical research? Why *would* anybody want to be a medical researcher? In other words, what *could* motivate a qualified person to engage in medical research? The fact that some activity like doing medical research is good for society (*bonum commune*) does not automatically tell anybody why doing this activity is good for any individual person (*bonum sibi*). What *could* motivate an individual person to do so? That is a psychological question. What *should* obligate one to become a medical researcher, though, is an ethical question.

2.2 Reasons and Motives

Now everything that *could* motivate an individual to do anything *should not* necessarily impel them to do so. Only ethically valid reasons or purposes can *obligate* any person to do medical research. Instrumental goals can only motivate a person to do so. However, we expect medical practitioners, and even more so the medical researchers who enable the practitioners to work effectively, to be ethically obligated, morally charged, to do their work, and to do it enthusiastically, with dedication.

This enthusiasm requires reasons and not just motives. The difference between a reason and a motive is that *the* reason for engaging in an activity is intrinsic to the activity itself. The activity itself and its intended consequences are its own intrinsic end. On the other hand, *a* motive makes the activity itself and its intended consequences the means to an extraneous end.[1] As such, *any* means that can accomplish that end is acceptable. In this perspective, there is only the criterion of usefulness, i.e., what is useful for some other purpose. Here the end or result justifies *whatever* means thereto work, i.e., whatever produces results that are useful for that extraneous end.[2]

This instrumentalist perspective is morally dangerous, especially when it comes to doing medical practice and medical research. In fact, it could even justify torturing prisoners in medical experiments so that information gained by the torturers be used in designing procedures to heal other humans.[3] Even if not justifiable prospectively (i.e., by actually citing it as a precedent for present and future action), it could still be justified retrospectively, i.e., when it can be shown that these harmful experiments did lead to beneficial results for others. (That is argued despite the fact that history is replete with examples of harm having been done and either no real benefit coming thereafter, or the harm done far outweighed any benefit that might have come as a consequence thereof.) Such retrospective justification comes quite close to actually exonerating the torturers after the fact. Conversely, ethical rejection of that kind of utilitarian justification of torture is that torture of others is inherently evil (*malum* per

---

[1] In *Nicomachean Ethics*, 1.1/1094a1-10, trans. J. A. K. Thomson (London: Penguin Classics, 1955), p. 3, Aristotle writes: "The good has been defined as that at which all things aim. … there is some difference between the ends [*telōn*] at which they aim: some are activities [*energeia*] and others results distinct from the actions [*ta erga*]. When there are ends distinct from the actions, the results are by nature superior to the activities. … the end of medical science is health." Now health is the result of medicine being done successfully. That is its intrinsic end, the reason *for* which medicine is practiced. It is also a *transaction* that is intended to affect somebody else by healing them, i.e., by *making* them healthy. However, if medicine is practiced (either directly in a clinic or indirectly in a research laboratory) primarily for the sake of its extraneous results (e.g., recognition *by others* of one's high social status as a medical researcher), then it is invalid ethically (although not necessarily immoral).

[2] Like Aristotle, Kant too is unconcerned with the objective consequences of an ethically valid act, which is obedience to what he calls a "categorical imperative." The act's ethical validity, its goodness, is subjective, i.e., "[it] consists in the mental disposition [*Gesinnung*], let the consequences [*der Erfolg*] be what they may." *Groundwork of the Metaphysic of Morals*, AK4:416, trans. H. J. Paton (New York: Harper and Row, 1964), p. 84.

[3] See Novak (2020).

se), just as it is inherently good (*bonum* per se) to heal others. Actions inherently evil are actions *for* which no convincing argument could be made. Actions inherently good are actions *against* which no convincing argument could be made. How could one argue for torture per se? How could one argue against healing per se?

Similarly, healing a prisoner so he or she can be sufficiently healthy for torturous medical experiments is as immoral as torturing a prisoner in a medical experiment so that the information gained therefrom might be useful in healing others. Good means to bad or evil ends are just as unacceptable as bad or evil means to good ends. Ends and means must always be correlated to each other. Also, it has been shown that despite the rationalization of torture as having been done for a good end, most torturers are sadists whose motive in torturing others is the perverse pleasure sadists take in torturing their victims and gloating over their agony. In other words, this is often evil for evil's sake, admitting of no ethical justification whatsoever. The intention of the torturers belies even any rationalization (i.e., lying about one's motives) they might offer when held responsible for what they did. Yet those for whom nothing is either good or evil per se, have no convincing arguments to condemn torturous medical experiments when it can be shown that they did lead to some good results for others nonetheless.

Now the usual motives for engaging in an activity like doing medical research are themselves benign, being neither good nor bad per se. These motives are one: intellectual curiosity; two: public recognition; three: remuneration. All three motives, though, are extrinsic to the activity of medical research. There are other ways to satisfy one's curiosity, or to gain public recognition, or to be remunerated. Moreover, one acts *from* motives that are always a means to something extrinsic, and the extrinsic goals towards which one is motivated are never final. About curiosity we need to ask further: "What *itself* is worthy of one being curious about?" About public recognition we need to ask further: "What *itself* is worthy of one being recognized for?" About remuneration we need to ask further: "What is the money to be *used for*?"[4] Lastly, motives vary from individual to individual. (For example, there are many individual factors motivating a person to pursue their intellectual curiosity, or public recognition, or wealth.) These three motives, then, are all essentially the pursuit of an individual's self-interest, when a person is more concerned with what benefits oneself than with what benefits others. Ethically valid reasons, conversely, are universal, enabling rational, morally earnest persons to rise above being primarily fixated on their individual, psychologically determined motives.[5]

Although we need ethically valid reasons for those qualified to engage in medical research to do so, reasons that are nobler than the three motives just mentioned, that does not mean the motives just noted for engaging in medical research should be dismissed, however.[6] Like anybody else engaged in work that benefits society, and whose work is intense, taking years of intense preparation, medical researchers deserve to be able to indulge their curiosity, they deserve public recognition, and

---

[4] See *Nicomachean Ethics*, 1.5/1095b15-1096a10; also, 5.5/1133b10-20.

[5] See *Groundwork of the Metaphysic of Morals*, AK4:421.

[6] See *Nicomachean Ethics*, 1.8/1099b1-5.

they deserve to be well remunerated. In fact, medical researchers actually need all three motivating factors in order to do their work well. Boredom, obscurity, and poverty detract rather than enhance the work of medical researchers. Curiosity, public recognition, and remuneration are not immoral goals that should be eschewed. Rather, these goals should always be subordinate but never superior, subsequent but never prior, to ethically valid reasons for doing medical research. Moreover, these goals are also more consistent with, and less contradictory of, ethically valid reasons for doing medical research especially when they entail negative imperatives rather than positive ones.

First, whereas the pursuit of one's curiosity can work against the dedication of medical researchers do so their job well, the avoidance of boredom is far less contradictory of the purposes of medical research, being more consistent with the positive purposes of medical research. That is because all one needs here are *some* opportunities for one's own intellectual stimulation.

Second, whereas the pursuit of public recognition can compete with the dedication of medical researchers to do their job well, the avoidance of obscurity is far less contradictory of and more consistent with the positive purposes of medical research: beneficence, truth, and communication. That is because all one needs here is *some* public recognition of the good work they are doing, lest they feel that they are working in total isolation from the society in which they live outside of their laboratories. (At this time, I think of how we are rightfully urged to thank those on the front lines of public service during the current pandemic, especially medical practitioners; although medical researchers seem to be largely forgotten in receiving public gratitude.)

The goal of receiving public recognition in the form of gratitude for one's socially beneficial activity poses a problem, however. That is, medical researchers do need to be gently reminded it was society that largely paid for their education, and that it is society (both the government and private corporations) that pays them to do their work full time. Also, most of them enjoy their work. Indeed, if they didn't enjoy it, there is little likelihood they would be able to do it, and do it well for any sustained period of time. Therefore, as much as medical researchers deserve public support (both financial and emotional), so does society deserve their gratitude for enabling them to do what most of them derive personal satisfaction from doing full-time. On the other hand, when researchers feel that the public has favoured others more than themselves, or that society has not adequately remunerated them or recognized their achievements sufficiently, often the gratitude expected by society soon turns into resentment of that society for one's perceived mistreatment by that society. Now resentful, embittered scientists can hardly be expected to engage in their research with any enthusiasm, any dedication, to the common good in a society they resent. As such, the quality or quantity of their research usually diminishes greatly. Therefore, gratitude here between medical researchers who serve society and society that serves them is a two-way street.

Third, whereas making money is a positive goal that can compete with beneficence as the positive goal of doing medical research, such is not the case with the negative goal of not becoming poor. That is because not becoming poor is not an end in itself, but rather the necessary negative means (*conditio sine qua non*), the clearing away of obstacles, to doing something else. Hence it can become a contributing means rather than a contradictory end to the positive end of doing beneficial, honest, and communicative medical research. So, one wants not to become poor, because without a sufficient livelihood (i.e., poverty by definition) one cannot do sustained medical research. Poverty becomes a distraction that must be removed for the sake of sustained medical research being done well or being done at all. Medical research cannot be done or done well when one's mind is constantly distracted by financial woes.

The dedication required for ethically valid medical research is not only needed in order to better respond to a particular emergency in the present like the Covid-19 pandemic. This dedication is needed even more for medical research to better enable medical practitioners to deal with health care challenges by anticipating them, and with better preparation than practitioners have had during this present health care emergency especially.

2.3   Beneficence, Truth, Communication

We need now ask: Since every significant human endeavor needs to be validated ethically, and certainly scientific research that has a direct effect on the well-being of many other humans is such an endeavor, how is the ethical validation of medical research the same as the ethical validation of other kinds of scientific research and how is it different?

Any scientific enterprise is ethically valid when its primary purpose is to explicate the intelligibility of what the scientists have discovered in the world they have not created.[7] That objective intelligibility is what is called *truth*. The theories scientists devise or invent whereby they explicate truth are only the subjective means that are subordinate to the end of allowing the truth of what they have not invented to become known. Therefore, we expect scientists to be dedicated or devoted to truth, i.e., to be truthful.[8] And that is why we are disappointed when they are not so dedicated, when the pursuit of truth is not their ultimate reason for doing what they do. Even more so, that is why we become incensed when some scientists actually lie about

---

[7] The French Catholic philosopher Jacques Maritain (d. 1973) expressed this very well in his *Existence and the Existent*, trans. L. Gallantiere and G. B. Phelan (Garden City, NY: Doubleday, 1956), p. 21: "*Veritas sequitur esse rerum...* Truth follows upon the existence of things ... Truth is the adequation of the immanence in the act of our thought with that which exists outside our thought ... which corresponds to the existence exercised or possessed by that other in the particular field of intelligibility which is its peculiar possession."

[8] The German Jewish philosopher Hermann Cohen (d. 1918) expressed this very well in his *Religion of Reason Out of the Sources of Judaism*, trans. S. Kaplan (New York: Frederick Ungar, 1972), pp. 421–22: "Truthfulness [*die Wahrhaftigkeit*] presupposes a foundation of truth [*die Wahrheit*] upon which it rests... the duty [*die Pflicht*] of truthfulness is enjoined in the Pentateuch by the prohibition of lying (Exod. 23:7)... it is also said positively: 'Speak the truth [*dabru emet*] to one another' (Zech. 8:16).".

what they claim to have discovered, deceitfully presenting or inventing a world they would like to be there as if re-presenting the world that is already there. Thus the primary purpose of explicating the truth entails the imperative "speak truthfully" as the proper means to its instantiation. Scientists who uphold this imperative are the most honest people in the world.

However, the primary purpose of scientific research is not beneficence, i.e., doing good. To be ethically valid, this activity need only avoid malfeasance, i.e., not doing any harm to anyone else.[9] That especially but not exclusively prohibits harm being done to the scientists' fellow humans. Other than avoidance of malfeasance, though, scientists ought not to be held to the political demand that their research directly lead to positive social benefits. To make beneficence the primary purpose of any scientific research would turn it into technology, whose ethical validity depends on its *good* public results or consequences, i.e., on its positive social utility. But, were that the primary purpose of scientific research, it would hamper the freedom of enquiry those engaged in "pure" science need to curiously explore nature so that they might discover truths that are unanticipated and hence unexpected. Most great scientific discoveries (I think of Galileo's, Newton's, Darwin's, and Einstein's) were not due to any publicly prescribed agenda. That is why, by the way, authentic scientific research is very much hampered in totalitarian societies. There scientists are technicians who are expected to deliver good or useful results to those in power, irrespective of whatever truth has been suppressed, and irrespective of whatever bad effects result for those oppressed by the powerful. The political culture there is not conducive to science. That is why scientists devoted to truth who live there often flee (or try to flee) to freer, more open societies.

Even though a society has the right to expect the medical researchers it is supporting to work towards doing tangible good (i.e., beneficence), that expectation should not entail a demand to produce specific results with a definite deadline for delivery. Such specific, timed demands often tempt or even frighten research scientists to compromise scientific truth standards in order to conform to the agendas of those who have political power over them. Free societies, though, should not be tempted to act like totalitarian societies where those having political power make truth whatever they want it to be. In totalitarian societies, a scientist is no different from any other technician. When they don't deliver what they have been ordered to deliver, they are not only not rewarded, they are most often penalized in one way or another.

Medical research's primary purpose is certainly to benefit (i.e., to do good for) others. That purpose makes it "applied" science, more akin to technology than it is to" pure" science. Yet its superiority to ordinary technology is that ordinary technology is concerned with making things better, whereas medical practice is concerned with making human lives better. Most of us think that things are valuable when they are useful for the enhancement of human life; and most of us think that things are

---

[9] See en.wikipedia.org/primum_non_nocere—Also, the great compendium of Roman law, the *Code of Justinian* begins with a definition of justice, one of whose three pillars is the norm "not to harm others" (*alterum non laedere*).

dangerous or bad when human lives are made to serve them. That is why society requires more of medical research than that it simply not be harmful. We require it to deliver positive public good to tangibly benefit *human* society as a whole among all its members.[10] That is why we require a greater dedication to the common good from medical researchers than we do from physicists, chemists, biologists, or mathematicians. Thus the primary purpose of beneficence entails the imperative "do good" as the proper means to its instantiation or realization.

The pursuit of truth is medical research's secondary purpose. Like beneficence for nonmedical scientific research, this criterion is negative, i.e., it is not so much that scientific medical research is required to certify its conclusive explanations of natural data, but rather that its conclusive explanations not be readily falsified.[11] In other words, as long as the scientists are not lying to the public, or passing off what they have imagined to be true without any independent evidence for it, their research is not violating society's trust in them to be truthful.

Finally, there is a third purpose, a tertiary goal of medical research: communication. That is because medical research is a social enterprise, involving a community of fellow researchers communicating with each other over the same noetic problems their research is supposed to solve, or at least deal with more perspicaciously. Medical research is not a private enterprise. It is expressed in a common language, which any medical researcher or group of medical researchers have neither been its first speakers nor its last. As Aristotle and Wittgenstein taught so persuasively, there is no private language. Language as the medium of human communication is essentially a social practice; it is inherently public.[12] That is why dedicated medical researchers, for whom language is the medium of rational discourse, should be prepared to defer to colleagues who make better scientific arguments just as their colleagues should defer to them when they make better arguments. (That is the synchronic character of this common discourse.) Moreover, dedicated medical researchers should be prepared to accept the likelihood that their research will be surpassed by their successors just as they have surpassed the research of their predecessors. (That is the diachronic character of this common discourse.) Moreover, researchers should communicate with the general public about their research for whose benefit this research is being conducted. Research scientists who think otherwise, who do not engage in communication that is dialogical, are often those whose primary concern for their own professional reputations frequently leads them to violate the ethical criteria of beneficence and truthfulness.[13]

This tertiary purpose of common research leads to the second purpose: the pursuit of truth, which ultimately leads to the primary purpose: doing good for others. Thus communication as an end entails the imperative "do communicate" as the proper means to its instantiation or realization.

---

[10] See Novak (2007).

[11] See Popper (1992).

[12] Aristotle, *Politics*, 1.1/1253a1-20. Ludwig Wittgenstein, *Philosophical Investigations*, 2nd ed., trans. G. E. M. Anscombe (New York: Macmillan, 1958), 1.256–69.

[13] See Habermas (1984).

We have now seen that thinking about the ethics of medical research means discerning medical research's teleology. What are its valid purposes? What would be ethically invalid purposes for it? What ethically invalid purposes are immoral and thus to be rejected in principle? And what purposes are not immoral but only pragmatic? As such, they are not to be rejected in principle, but must be shown to be appropriate complementary motivations serving the essential purposes of medical research: beneficence and truthfulness. These essential purposes alone are what makes medical research the dedicated, ethically charged enterprise we rightfully expect it to be. Conversely, there are inappropriate means to these ends.

An inappropriate means to the end of beneficence would be doing harm so that good will emerge thereafter as its intended consequence. An inappropriate means to the end of truth being spoken is lying to people whom one thinks will reject, misunderstand, or misuse truth if it is spoken to them explicitly. This has been called the "noble lie."[14] Lies are intentional errors; mistakes are unintentional errors. Science as a self-correcting enterprise has the capacity to recognize unintentional errors and correct them, without morally indicting those who have honestly made them. But because lying is an ethical issue, science itself cannot correct it. It is a political matter, for it is in society that significant scientific deception takes place in public discourse. Therefore, such deception requires a political corrective. That corrective is the encouragement of democratic debate through what has been called "public reason," so that errors can be corrected through free and open discussion.[15]

Unfortunately, though, those so deceived are unlikely to ever appreciate truth after they realize they have been lied to. Once those who have been deceived learn they have been deceived, "truth" becomes for them the rationalization of those who have the power to deceive them as a means of keeping them subordinate to their expertise, i.e., to keeping them in line. This knowledge most often does not lead to enlightenment, but rather to cynicism, or to the revolutionary desire of those awakened from deception to themselves replace their deceivers with their equally deceptive power. In other words, this confirms the "might makes right" ideology of their former oppressors.[16] In fact, their deceitfulness is a key factor in their oppression of others, fooling them into thinking they are actually being benefitted by their oppressors. Moreover, we have seen how totalitarian regimes have forced scientists to lie about their discoveries for the sake of bolstering the regime's power over those who are subordinate to it. (Need I mention how the government of China suppressed the truth about the outbreak of the Covid-19 pandemic in Wuhan?)

Finally, an inappropriate means to the end of scientific communication have been the attempts of lone scientists to look upon the scientific community as the means to their own professional advancement. Here there is no commonality, but there are only individual entrepreneurs. Now to be sure, individual ambition to be publicly recognized can be a useful motivation for medical research. It is only when ambition

---

[14] Plato, *Republic*, 414C–415C. For a powerful philosophical critique of Plato, especially on this point, see Popper (1966).

[15] See Habermas (1993).

[16] See Plato, *Republic*, 338C–339A.

or "careerism" becomes the end of communication rather than a contributor thereto that the means-end continuum here has been perverted thereby.

2.4   Conditional and Unconditional Imperatives

We have seen how the three ethically valid reasons (beneficence, truth, communication) for engaging in medical research are what we expect all those involved in medicine (whether practitioners or researchers) to be primarily committed to, i.e., dedicated to. On the other hand, if one's primary commitment is to any one of the three motives delineated above (money, recognition, curiosity), and even if this commitment doesn't contradict any of the three ethically valid reasons, we certainly cannot expect those so motivated to be truly dedicated to medical research. To require dedication of them is demanding too much of them. Dedication aims above their personal horizon. The motives that impel them cannot soar that high.

We must now ask whether the three imperatives (namely, do good; speak truthfully; do communicate) are conditional or unconditional. The answer to this question is crucial for determining why we seem to require (morally, though not legally) medical researchers to be dedicated to their vocation.

Conditional imperatives are obligations that can be avoided with impunity, because they are obligatory *only if* a person chooses to engage in the kind of activity these imperatives necessarily pertain to. However, a person is not obligated to engage in these activities to begin with (ab initio); and there are no punitive consequences if one chooses not to engage in the activity at all. So, we seem to be saying to candidates deliberating *whether or not* to do medical research: "*If* you want to do medical research in an ethically valid way, *then* these three imperatives must be kept therefore." However, there is no imperative at this level that any individual person herself or himself *ought* to do medical research (or medical practice). Doing medical research could just as easily be avoided as it could be engaged in. It is only immoral evasion of one's responsibility *when* one evades the norms inherent in the activity they have initially chosen to be committed to, thereby committing oneself to play by its rules. However, although the three imperatives specify the way medical research is to be done in an ethically valid way, they still do not impel anybody to engage in the enterprise itself. The fulfillment of these three specific imperatives seems to be contingent on one's general choice to commit oneself to the enterprise of medical research where they do pertain. The reason for making that general choice is thus prior to the reasons for following the three specific imperatives it impels the one making this choice to follow.

Now a society committed to ethically valid purposes is dependent on individual members to do good, to speak the truth, and to sincerely communicate. (In fact, some would call this "social justice.") Yet society still doesn't answer the question: Why should I volunteer to be society's agent to engage in this demanding work with dedication? Don't these volunteers have to believe themselves to be personally dedicated to doing so? Don't they have to say as Martin Luther is famously reported to have said, "I can do no other" (*Ich kann nicht anders*)? Surely, that is no ordinary choice, where one simply weighs possible options. Dedication to as exalted an enterprise as medical research has to be impelled by the sense that more than one chooses to engage in it, one is chosen or called to engage in it? And that calling is not one that can be accepted or rejected willy-nilly. But where does that calling come from, and where is it going towards?

2.5   Idealistic Dedication

Now many would say that those who are truly dedicated to something as noble as doing medical research are "idealists." Idealists imagine or envision an *idea* of what the world *could become* in the future, but which is not at all the way the world *really is* at present or the way it *really was* in the past. An *ideal* is the projection of that idea onto an unknown future horizon, but which idealists believe has to be a better world than the world at present nonetheless. Their medical ideal seems to be a world in which disease will be finally overcome, which is a world nobody has seen but could only imagine. Only then will their ideal be realized.

There is no doubt that such idealistic medical researchers are personally committed to their work, which for them is a vocation, calling them to obey an unavoidable, unconditional imperative *to be* totally engaged in what is infinitely more than making a good career choice. Such idealists are much more admirable than those whose personal goals are much lower.

Nevertheless, cynics (who often call themselves "realists") charge that these idealists are engaging in wishful thinking, that they are pursuing a deceptive fantasy that diverts them from seeking more realistic, more reasonable goals. For "wishful thinkers" want something infinitely better than the present situation in the world to simply *happen*, but through no effort of their own.[17]

However, these idealists can well counter the charge they are engaged in wishful thinking by arguing how they actively move *forward* what they ideally envision, rather than passively waiting for it to occur by itself. They declare their efforts to be their active contribution to a progressive trajectory, one that is future oriented. Since that trajectory is a cooperative venture, it has a political dimension. That is why most idealists today consider themselves to be politically "progressive," seeing their idealistic project as entailing such public imperatives as "follow the science" (especially following those scientists who are on their political wave length). Idealists who heed such an imperative often consider themselves to be in the vanguard of an unstoppable movement or evolution into an inevitable future. Indeed, many ideological progressives often boast of their being "on the right side of history."

The problem with idealism as the source of a cogent unconditional imperative to be *personally* committed to medical research (or any other "progressive" enterprise) as a vocation is its "futurism," i.e., its belief that the future *will* be much better than the

---

[17] *Nicomachean Ethics*, 3.2/1111b20-30.

present because it *must* be so. Why must it be so? That is inferred from the fact that the present has been better than the past, which certainly in true about both medical practice and medical research. Yet that belief as a source of an ethical imperative is very much debatable on two counts.

One, as Hume convincingly argued, the fact that something did occur in the past does not require that it has to occur in the present, and even more so in the future.[18] Only the past has such necessity insofar as only past events couldn't be other than what they were. The present and even more the future have no such necessity. Considering themselves beholden to evolution, many progressive idealists believe they can predict what will, indeed what must, occur in the future, based on their assumption of what necessarily occurred in the past. But that is a claim no sober scientific evolutionist would make, nor would any sober student of historical development. Both of them engage in retrospective description, not prospective prognostication, and not moral prescription. The future yet to evolve is not at all inevitable, hence it is highly unpredictable. It certainly cannot be commanded into being. So, idealists committed to working towards a better future, might well be chasing a phantom. They might actually be working towards a future totally contrary to anybody's hopeful expectations, predictions, or prescriptions. Isn't just as likely that the future might be much worse than the past or the present rather than being very much better? This cannot but haunt such idealistic medical researchers. Couldn't it be that what they have so very much committed themselves to might in fact be an exercise in futility, that it might be a contribution to medical regression rather than to medical progress?

Two, even if such idealists as a group are convinced that they are in the vanguard of medical progress and not in the rearguard of medical regression, aren't there always individual researchers who feel they haven't made any noteworthy contribution to this progress, who are resentful of the great likelihood they will certainly be forgotten in the future as they are mostly ignored in the present? In fact, many researchers are already ignored in the present, not only by the general public, but by their better known scientific colleagues (some of whom are more skilled self-promoters)?

Now this is more than the problem of "burnout" (or "midlife crisis") that afflicts many people in other highly demanding and highly competitive fields. Frequently, that problem is due to a particular emotional crisis that varies from individual to individual, but which they project as a more general psychological problem, especially a problem for most professionals. This projection is their way of denying their painful disappointment with significant other persons in their nonprofessional life. Nevertheless, even if a medical researcher is not suffering from burnout, there are the metaphysical or metaethical, existential questions a person must ask oneself: What is the *ultimate* significance of my work? What is the *prime* source of the dedication that not only is expected of me by others, but which I must expect of myself, lest I sink into indifference or despair? Any attempt to make these two great metaphysical questions psychological questions is blind to the inextricable metaphysical component of human nature.

---

[18] *A Treatise of Human Nature*, 2.3.1.

Just as we humans are essentially biological beings, who are necessarily concerned with surviving in the biosphere, and just as we humans are essentially social beings, who are necessarily concerned with our interpersonal relationships, so are we humans essentially metaphysically-oriented beings, who are necessarily concerned with what transcends our being in the ordinary world, but which gives this world its source (*archē*) and its end (*telos*) nonetheless. That is what makes the rejection of metaphysics, in modernity especially, an example of the irrational suppression of what is true about human nature, however much it doesn't fit into myopic views of human nature.

The sense of dedication we expect medical practitioners to have, and the medical researchers who back them up to have, that is what causes all of us to ask them the two metaphysical questions stated above. But only those *dedicated to* their work could seriously ask these existential questions of themselves. Only the answers to these questions can tell one *why* their dedication to medical research isn't arbitrary or accidental, even though this dedication is usually sensed by enthusiastic medical researchers before they question what is its source and what is its end. To be sure, idealism does provide answers to these questions. I only question the existential adequacy of its answers.

2.6 Religious Dedication

No matter how secularized the doctrine of dedication has become (as has been done most cogently by idealism), its historical origins are certainly religious. Moreover, the doctrine is clearly still operative in Judaism, Christianity, and Islam; and it is unlikely to disappear in the future. (Its strongest and most resilient adherents are found among less secularized Jews, Christians, and Muslims.) Since I can see with no other eyes than my own, let me explore how this concept operates in the Judaism that I and other faithful Jews still live by (and I am well aware of parallels in Christianity and Islam). Let me also clearly emphasize at the outset that I am not saying that the religious way, let alone the Jewish way, is the only way a person can be dedicated to such an exalted vocation as engaging in medical research. I am not trying to argue anybody into faith, nor am I invoking religion or Judaism as a kind of *deus ex machina* that is called upon when all else fails. I am not saying *that* dedication must be religious, nor *why* dedication is essentially religious, I am only saying *how* religious dedication operates in a particular tradition.

Now such dedication seems to require that one be dedicated *to* what is absolute, and that such dedication must be unconditional in order to be worthy of its object. Dedication to an absolute seems to be a religious pursuit or some facsimile of a religious pursuit. As such, that dedication could only be *to* God alone, because only God is the One "*whom* nothing greater can be conceived" (*id quo maius cogitari nequit*), thus alone deserving such unconditional dedication.[19] Dedication to anything

---

[19] See Barth (1960).

less than God is idolatry. And the unconditional imperative to be so dedicated could only come *from* God, who is both the source and the end of this unconditional imperative. "I am the first (*ri'shon*) and I am the last (*aḥaron*) and other than me there is no God." (Isaiah 44:6) The imperative is begun by God alone, and its full consequences will be finalized by God at a time unknown to us. Human persons act at some point in-between the beginning and the end, namely, by actively expressing this dedication by their imperfect keeping of the categorical imperatives of this absolute God. Unlike secular idealization whose proponents are convinced of their ability to realize their ideal in this world, the advocates of religious dedication are less certain of the realization of their hopes, only believing they will be fulfilled by Someone else at another time and in another place.[20]

The acceptance by the Jewish people of the covenantal authority of God revealed in the Torah is unconditional.[21] That is true even if many Jews do not keep all the Torah's commandments, and even if many Jews cannot make such a total commitment.[22] Without such an unconditional commitment to the Torah as a whole, one could not be cogently dedicated to keeping any of its commandments. Nevertheless, one who has not denied the Torah's authority altogether should still be engaged in keeping its commandments however haphazardly, hoping to be dedicated to it, for dedication often comes in the wake of one's observance of the commandments rather than before it.[23] One should only not misrepresent the Jewish tradition, claiming to be committed to it while denying its theological foundations.

In the Jewish tradition, an unconditional imperative is called a "duty" (*ḥovah*), which may not be avoided, but which everybody is urgently required to pursue it to fulfill it. One is obliged to actively seek out opportunities to fulfill this duty rather than avoiding or fleeing opportunities to do so. One is to be dedicated to it. On the other hand, there are conditional imperatives that one can avoid by avoiding situations that call for their observance.[24] However, only when one takes such a conditional imperative and makes it for oneself unconditional by seeking opportunities for its observance, only then can one be truly dedicated to keeping this commandment.[25]

However, what kind of unconditional duty is the imperative to become a medical researcher? Is it a duty that comes as an ad hoc command directly from God to a particular individual? Or, is it a specification of a more general commandment (*mitsvah*) that is incumbent on everybody in the community so commanded? If it is this kind of general commandment, its immediate source is the tradition of the commanded covenantal community that mediates the commandment, even though its original source could only be God, who is the Giver of the body of all commandments which is the *Torah*. In fact, any commandment that cannot be seen as having an

---

[20] Esther 4:14; Ps. 27:13–14 and 84:6–8; Job 19:25.

[21] *Babylonian Talmud*: Shabbat 88a re Exod. 24:7.

[22] See Novak (1995).

[23] *Babylonian Talmud*: Nazir 23b.

[24] Maimonides, *Mishneh Torah*: Blessings, 11.2.

[25] *Babylonian Talmud*: Nedarim 7b-8a re Ps. 119:106.

original divine source is considered invalid by the normative Jewish tradition.[26] It is "a commandment of men [*mitsvat anashim*] learned by rote" (Isaiah 29:13). Thus what the source of the commandment is, that is obvious. What needs to be ascertained is what the medium through which the specific commandment was given.[27]

The theological problem with taking this imperative to be an ad hoc command is that anyone so commanded has to have experienced a direct divine revelation of the command from God. That means that this person is a prophet (*navi*); yet the Talmud indicates that after the destruction of the Jerusalem Temple in 70 C.E., there are no more prophets among the Jewish people.[28] Therefore, even if one believes he or she has experienced such a direct divine revelation (and who can belie anybody's experience?), they should try to justify their religious pursuit of medical research as their vocation, both to their community and to themselves, as being their own pursuit of a general commandment given to all the members of the covenantal community. I mention all this as many critics of religion wrongly presume that religious people who obey commandments as God-given, believe themselves to have personally experienced God commanding them to do so de novo, rather than their being members of a tradition that claims to have experienced divine revelation at its inception.

What is the general commandment here? According to Maimonides (d. 1204), it is: "You shall love your neighbour as yourself" (Leviticus 19:18).[29] The commandment to love one's neighbour is not dictating to a person how to feel about her or his neighbour. (Emotions cannot be commanded, only elicited.) Rather, the commandment is to benefit one's neighbour by doing something good for your neighbour, specifically attending to their needs. One of the specifications of this commandment is the duty of "visiting the sick" (*biqqur holim*). That means attending to their needs. The general commandment and its subordinate specifications are duties, unconditional imperatives, that cannot be avoided or ignored with impunity. Moreover, these duties can only be fulfilled personally by those so duty-bound. As such, they may not be delegated to one's agents. In the Jewish tradition, they are what I would call "personal benefactions" (*gemilut hasadim*). Unlike the giving of charity, these personal benefactions may not be delegated to a third party acting on behalf of a donor.

Now ordinary people, who are non-specialists, can fulfill their duty to visit the sick by literally visiting them or by calling upon them (whether in hospital or at home), or by caring for them, or by praying for them, or by all of the preceding deeds.[30] And specialists like medical practitioners can only fulfill their duty to visit the sick by treating them in their illness with the healing tools medical research has provided the practitioners with. (As we have seen earlier, the practitioners and the researchers work in tandem.) The fulfillment of the duty to visit the sick or attending to their needs (both physical and spiritual) requires unusual preparation and unusual dedication. That is why, it seems to me, these specialists need the sort of inspiration

---

[26] *Mishneh Torah*: Kings, 8.11.

[27] See *Palestinian Talmud*: Sukkah 3.4/53d.

[28] *Babylonian Talmud*: Baba Batra 12b.

[29] *Mishneh Torah*: Mourning, 14.1.

[30] *Mishneh Torah*: Mourning, 14.4–6.

that impels them to engage in their exalted, demanding work and keep its three operative moral norms as divine commandments (i.e., do good; speak truthfully; communicate) persistently and happily, with the full intention (*kavvanah*) of their source and of their end. These specialists need to be inspired so that their work be a true participation in a cosmic drama that God began and God will end. Of course, all who keep God's commandments need this kind of inspiration, for without it their keeping of the commandments becomes mere behaviour rather than sublime human action.[31] Nevertheless, even though the difference between the specialists and the non-specialists is a difference of degree and not a difference of kind, the specialists need more such inspiration. That is because the temptation of despairing that their labours are ultimately futile needs the strong antidote of inspiration.

Such inspiration might well be needed to sustain one in doing medical research as a divinely authorized vocation. That is why Nahmanides (d. 1270) emphasizes that the practice of medicine requires divine authorization in the Torah, precisely because it is never an ordinary profession.[32] And earlier, the Talmud emphasizes that God is the original physician (*rofē*), who is to be imitated but not usurped by fallible, mortal humans.[33] Also, all this sustains one's hope that their work contributes significantly to the divine cosmic drama, even if that plan is one that "no eye but God's has seen" (Isaiah 64:3).[34] Unlike idealism, however, this participation in the divine cosmic drama does not pretend to know its trajectory; and its participants cannot be sure how their efforts are contributing to its final consummation.[35]

Needless to say, the Jewish tradition does not require anybody, let alone any Jew, to be so inspired as a prerequisite for engaging in medical practice or medical research. There are lesser reasons for doing so, and for doing so quite well. Indeed, were this sort of metaphysical perspective to be such a requirement, too few people would enter a field where more not less people are needed. As the Talmud puts it, "the Torah was not given to ministering angels."[36] And as the Talmud also puts it, "one who grasps too much grasps nothing; one who grasps much less grasps something."[37]

All that notwithstanding, in extraordinary times such as these, when such great demands are being made of medical practitioners and medical researchers, it is good to know that there are resources for doing work that can have extraordinary, metaphysical significance by inspiring those who take these religious resources seriously to do the work that calls for truly extraordinary effort and dedication.

---

[31] *Mishnah*: Rosh Hashanah 3.7–8.

[32] *Commentary on the Torah*: Lev. 26:11 re *Babylonian Talmud*: Baba Kama 85a re Exod. 15:26 and 21:19, quoted in Novak (1992).

[33] Sotah 14a re Deut. 13:5 and Gen. 18:1.

[34] *Babylonian Talmud*: Berakhot 34b.

[35] *Mishnah*: Avot 2.1 and 2.16.

[36] Babylonian Talmud: Kiddushin 54a.

[37] Babylonian Talmud: Rosh Hashanah 4b.

References

Barth, Karl. 1960. *Fides Quaerens Intellectum*, trans. I. W. Robertson. London: SCM Press.

Habermas, Jürgen. 1984. *The theory of communicative action*, trans. T. McCarthy, 243–337. Boston: Beacon Press.

Habermas, Jürgen. 1993. *Moral consciousness and communicative action*, trans. C. Lenhardt and S.W. Nicholsen, 43–115. Cambridge, MA: MIT Press

Novak, D. 1992. *The theology of Nahmanides systematically presented*, 86. Atlanta, GA: Scholars Press.

Novak, D. 1995. *The election of Israel: The idea of the chosen people*, 189–199. Cambridge: Cambridge University Press.

Novak, D. 2007. *The sanctity of human life*, 91–110. Washington, DC: Georgetown University Press.

Novak, D. 2020. Is the use of Nazi medical experiments justifiable or not? *Ethics, Medicine, and Public Health* 12: 1–7.

Popper, Karl R. 1966. *The open society and its enemies*, 5th ed., 140–144. Princeton, NJ: Princeton University Press.

Popper, Karl R. 1992. *The logic of scientific discovery*, 40–48. New York and London: Routledge.

David Novak A renowned philosopher and theologian, David Novak is the J. Richard and Dorothy Shiff Professor of Jewish Studies and Philosophy at the University of Toronto, and Fellow of the Royal Society of Canada. He is the Vice-President of the Institute on Religion and Public Life. The author of nineteen books, he is also the subject of five books devoted to his theories. He recently delivered the Gifford lectures now published by the University of Toronto Press. Two separate festschrift volumes in his honour are scheduled for publication shortly.

Chapter 3
Genopolitics: Biotechnology Norms and the Liberal International Order

Jonathan Moreno

According to the LIO, international relations are to be organized according to principles of open markets, liberal democracy, and multilateral organizations. The Atlantic Charter, signed by President Franklin Roosevelt and Prime Minister Winston Churchill in 1941, is often taken be the first statement of the order, followed by the founding of the UN in 1945. The LIO can be said to have three aspects: one is financial (e.g., the World Bank, the International Monetary Fund, and the World Trade Organization), a second is security (e.g., NATO), and a third is human rights. The deep relationship between modern bioethics and the liberal order, the global system of the rule of law and the organizations created to implement the rule of law since World War II, is not widely appreciated. That liberal order is now under threat in various ways, including the ascent of "strongmen" as heads of government, U.S. withdrawals from several international agreements, China's efforts to extend its influence, Russia's actions in the Crimea and its online influence operations, and the British decision to withdraw from the European Union.[1] Today, the LIO has its anxious defenders as well as its harsh critics. Its critics see it as a cynical excuse for U.S. hegemony and sense of moral superiority (Stokes 2018). Its defenders argue that, "although America's hegemonic position may be declining, the liberal international characteristics of order—openness, rules, multilateral cooperation—are deeply rooted and likely to persist" (Ikenberry 2018, p. 18).

But America's "hegemonic position" has long been viewed by many as less than benign, and by others as downright threatening. During the cold war the Soviet Union created various entities intended to resist U.S. dominance, from a military alliance of the USSR and its eastern European client states called the Warsaw Pact to

---

[1] This manuscript was prepared before the Russian invasion of Ukraine. I consider the implications of this episode for my argument in subsequent publications.

J. Moreno (✉)
University of Pennsylvania, Philadelphia, USA
e-mail: morenojd@pennmedicine.upenn.edu

© The Author(s), under exclusive license to Springer Nature Switzerland AG 2023   35
T. Zima and D. N. Weisstub (eds.), *Medical Research Ethics: Challenges in the 21st Century*, Philosophy and Medicine 132,
https://doi.org/10.1007/978-3-031-12692-5_3

confront NATO forces; to Comecon, a trade group consisting of Soviet bloc nations. In fact these organizations were mainly Soviet paper tigers, never truly competitive with their Western counterparts beyond the strength of the USSR itself. There is no question that the U.S. attempted to use its dominant position in fields like trade to effect its geopolitical goals, though sometimes these efforts were embarrassing failures. For example, President Carter's grain embargo did nothing to discourage the Soviet invasion of Afghanistan in 1980 and instead fueled a propaganda coup and might have aided his loss to Ronald Reagan (Parrlberg 1980). In the post-cold war era, U.S. positioning in various sectors of international competition is far more vulnerable, as the People's Republic of China brings much more strength to the table in fields like science and technology than the Soviet Union ever did.

3.1 The Age of Biotechnology

The various outputs of biotechnology and their promise for breakthroughs in agriculture, energy, and health care have stimulated a highly competitive global industry, one that did not exist at the creation of the LIO but is now a key factor in economic growth and national prestige. Measured by the number of biotech firms, patent approvals, and biomedical treatment approvals, the most important national players are the U.S., South Korea, Germany, the United Kingdom, Japan, Mexico, New Zealand, and Belgium. In terms of research and development, the U.S. is still the leader by far, followed by China, France, Switzerland, South Korea, Japan, Germany, and Denmark. But China began doubling its R&D spending in 2008, and the OECD expects China to be the leading spender this year. In biotechnology, one should add relatively substantial investments on the part of Singapore as well. Investment in biotechnology R&D is generally funneled through an academic-industrial research complex, although the details vary greatly from one country to another, such as how grants are disseminated and how incentives for intellectual property are managed. Military establishments in some of these countries also play a role, for example in funding defensive research on select agents and "dual-use" projects that could have both military and civilian applications.

Human beings have engaged in a traditional biotechnology called agriculture for at least 10,000 years. The era of modern biotechnology began in 1971, when Paul Berg and colleagues at Stanford succeeded in using recombinant DNA technology to create novel strands of genes. Just as farming produced a qualitative change in the nature of human life in nearly every respect, many have expressed concerns that modern biotechnology stands to create another radically new phase in the human experience. Novelists like Aldous Huxley in *Brave New World*, writing at the height of the eugenic era, speculated about the possibility that, whatever "genes" really are, they could be manipulated in a laboratory to produce different kinds or castes of individuals, Alphas, Betas, Deltas and Epsilons developed in a Social Predestination room. These concerns were vastly sharpened in the nearly two decades between the publication of James Watson and Francis Crick's 1953 paper in which they reported the decoding of the DNA molecule and the advent of modern biotechnology two decades later. Questions about the implications of genetic control were among those that led to the field called bioethics in the late 1960s, just around the time that Berg and colleagues were approaching their milestone. That subject has been a touchstone of bioethics since the founding of the first bioethics research institutes, professorships, professional organizations, and journals in the late 1960s and early 1970s.

3.2 Bioethics and the Liberal Consensus

Modern bioethics can be viewed as part of the human rights aspect of the liberal order, although bioethical values may also inform and guide economic development (e.g., global pharmaceutical research endeavors and public health programs) and security (through the laws of armed conflict and the treatment of prisoners of war). The deep relationship between modern bioethics and the liberal order is not widely appreciated. In the years immediately following the war, the most important milestone was an ethics document produced by the judges at the medical war crimes tribunal at Nuremberg, one that posterity has come to call the Nuremberg Code. The Code was a product of the judges' dissatisfaction with international recognition of ethico-legal standards for the conduct of human experiments.[2]

Underlying the Code and other articulated standards to follow was a certain view of the human person rooted in Enlightenment philosophies. As the bioethicist Robert Baker puts it, the liberal order that flourished after WWII shaped bioethics "as an ethics of covenants and conventions grounded in human rights." United Nations documents are redolent of these themes. The 1945 U.N. Charter "reaffirmed faith in fundamental human rights and dignity and worth of the human person" and committed member states to promote "universal respect for, and observance of, human rights and fundamental freedoms for all without distinction as to race, sex, language, or religion." In 1948 Article I of The Universal Declaration of Human Rights asserted that "[a]ll human beings are born free and equal in dignity and rights. They are endowed with reason and conscience and should act towards one another in a spirit of brotherhood."[3]

---

[2] Although the questions of guilt or innocence of the 23 defendants did not turn on the ethics of human experiments, the defense lawyers did succeed in calling attention to questionable experiments that were a matter of public record. The first proposition of the Code, that "[t]he voluntary consent of the human subject is absolutely essential," as well as provisions concerning legal capacity and professional responsibility, have shaped virtually every national and international document on the subject. This is true despite continuing disagreement about the legal status or precise reference of the Code. The Code was not of course addressed to modifications of human cellular material, but it implicitly established the notion that the international community could set ethical rules for medical science.

[3] The Universal Declaration on Bioethics and Human Rights (UDBHR 2006) states, "human rights constitute the tangible elements for achieving … freedom, and therefore, peace as progress, as transformation and as deepening in dignity, equality, and freedom of all human beings". The UDBHR established a wide range of principles (bioethicists love principles), including personal autonomy,

Regardless of the source, these themes run like a red line through the decades since World War II. The World Medical Association Declaration of Helsinki (the first version of which was produced in 1964) was in part an effort by the international medical community to reclaim control over the ethics of human experiments in a manner more acceptable to actual practices. In the United States, the National Commission for the Protection of Human Subjects of Biomedical and Behavioral Research, in a 1979 summary document known as the Belmont Report, expresses the same underlying ethos as the Nuremberg Code and the Declaration of Helsinki. Besides certain foundational principles (respect for persons, beneficence, and justice) the Belmont Report insists on the moral relevance of the distinction between medical research and medical practice. In 1982, the Council for International Organizations of Medical Sciences (CIOMS) in association with the World Health Organization (WHO) issued "guidelines [...] to provide internationally vetted ethical principles and detailed commentary on how universal ethical principles should be applied, with particular attention to conducting research in low-resource settings." In 1997, the European Council's Oviedo Convention stipulated that "The interests and welfare of the human being shall prevail over the sole interest of society or science," proceeding then to a number of bioethics standards.

A U.S. presidential bioethics commission was among the first to address the moral implications of the new biotechnology in a 1982 report called *Splicing Life*. The report identified themes that would reappear over the next decades and are now familiar, including the scientific community's self-regulation, the threat of "biohazards," the "Frankenstein factor," genetics and drug development, and various forms of direct manipulation of genetic material. As would be common for analyses produced in cooperation with the scientific community, the report downplayed vague worries about "playing God" and creating new life forms, although it allowed for worries about over-determining a "designed" child's future, undermining traditional conventions about the biological family and equality of opportunity, and even changing the very meaning of being human. The report warned that genetic modifications are no easy fix for social problems. And it hewed to a distinction between modifying an individual's DNA and that of their descendants, discouraging the latter out of fears of a modernized, government managed eugenic program but otherwise not drawing a bright line. The potential benefits of gene splicing for the growth of human knowledge, for medical science, and for virtually every other domain in which genes are implicated were viewed as simply too great to give up with some permanent prohibition.[4] One might say that, even as the technology has progressed, the bioethical

---

consent, privacy, equality, and benefit sharing. Two UNESCO entities, the International Bioethics Committee and the World Commission on the Ethics of Science and Technology have published reports on such topics as non-discrimination and non-stigmatization, bioethics and refugees, and adaptation to climate change. The Council for International Organizations of Medical Societies, established by UNESCO and the World Health Organization (WHO), publishes specific ethics guidance documents on topics like drug promotion.

[4] In the years since *Splicing Life* international governmental organizations such as UNESCO, including guidance documents on the human genome and human rights (1997) and human genetic data (2003), reflect similar topics, principles, and concerns.

themes have appeared again and again, adjusted for specific cases and types of innovation. As a far more efficient system for modifying DNA than its predecessors, the gene editing technology known as CRISPR has become especially salient.

Since 2017, the U.S. National Academies and the U.K. Nuffield Council have published consensus reports on gene editing, both concluding that such modifications should only be done within strict regulatory frameworks and setting out several specific conditions including data on risks and benefits. Again, consistent with the history, somatic cell editing is generally viewed as manageable under current regulatory regimes for innovative medical interventions, with off-target effects a particular hazard. So-called enhancements are discouraged as unjustifiably risky and poor investments of scarce research resources as compared to efforts to mitigate diseases. Germline modifications, while not wholly ruled out if safety and efficacy can be established (a major obstacle considering the countless variables involved), are not seen as justifiable at this time.

The commercial and health possibilities implicit in modern biotechnology (in the largest sense the "bioeconomy"), particularly in precision and personalized medicine, have stimulated their own ethical issues. But these have mainly concerned privacy and data ownership, as vast quantities of patient information will increasingly be entered into databanks so that various datapoints (family history, genetics, response to medication, and other personal information) can be digested by machine-learning systems and translated into individualized treatment plans. Someday all patients might be "on protocol," a status historically reserved for participants in clinical trials, thus blurring the line between being a patient and being a research participant. In the case of cancer, a forerunner of commercial investment in personalized medicine is Roche Pharmaceutical's $2 billion purchase of the data-mining company Flatiron in 2018. With the important exception of well-defined, data-rich, and information-accessible diseases like cancer that have a certain range of treatment options and predictable progression, those systems are some way off. Thus, government investment in the basic science that can lead to such systems will continue to be critical in order to attract market players that see a plausible investment. And even in the case of cancer, the several hundred-thousand-dollar per-patient annual price tag for the promising new immunotherapies in which CRISPR can be utilized is giving pause to both the pharmaceutical industry and private and public insurers.

3.3 Along Came Dr. He

By contrast, spectacular incidents like the "CRISPR babies" are far more immediate. And unlike cancer treatment trials, which are fairly well regulated, this episode has run up against bioethical norms. Dr. He Jiankui justified his use of CRISPR to modify the CCR5 gene in two live-born female babies in November 2018, a gene associated with susceptibility to a dominant strain of HIV, as a beneficent prevention of disease. But its loss could expose persons to life-threatening reactions to influenza, and there are of course other ways to prevent HIV infection, which is now treatable as a chronic rather than fatal illness. Dr. He lost his academic position at a prestigious science university and has reportedly been placed under house arrest. He has been accused of violating reporting rules, but it appears that some of his work was supported by state grants, though seemingly without the knowledge of authorities. The international science community has been unsure how to respond to Dr. He's announcement. Paradoxically, although publication is the gold standard of objective scientific assessment, because of the prima facie violation of international norms, the data were not published in a peer-reviewed journal.[5]

The international community was caught off-guard by Dr. He's announcement, and despite the previously articulated norms I've cited, its response has not been uniform. Some prominent scientists have called for a moratorium, reminiscent of the famous Asilomar moratorium on recombinant DNA research in 1975, one led by Paul Berg himself. But other distinguished researchers have resisted, apparently worrying that a moratorium would be unenforceable (the global science world is quite different in 2019) or that it would add significantly to the international opprobrium, including that of China's scientific community itself. The World Health Organization formed an advisory group that has so far recommended a global registry of human genome editing with publishing and grants dependent on participation. China's health ministry has proposed that a new commission review any experiments on human DNA including for the first-time penalties for violators. Also, proposed amendments to the civil code, already undergoing revision, bring genes and human embryos under the heading of personality rights.

For most of 2019 it was unclear what official sanctions would be applied to He or any of his associates. Writing in *Science* in summer 2019, Jon Cohen wrote that "the institutional response in China has been feeble. The central government or other bodies have yet to conduct the transparent investigation many people had wanted. SUStech declared in its one statement on the matter that it wanted 'international experts to for an independent committee to investigate this incident, and to release the results to the public.' But no such inquiry has occurred" (Cohen 2019, p. 437). Finally, at the end of the year He and two colleagues were sentenced to prison, a far more aggressive punishment than would have been available in the United States. Yet, as two prominent Chinese bioethicists immediately noted, the legal basis cited by the court—practicing medicine without a license—was an attempt to fill a "legal gap" that should be addressed by the Chinese legislature (Lei and Qui 2020).

Although the production of living children surely sets Dr. He's experiment apart from other uses of gene editing, the international outrage it has attracted is something of a distraction from less boundary-breaking but nonetheless striking work in China. Taken together, these incidents suggest a pattern. A clinic in China has conducted gene editing in patients with esophageal cancer (Rana and Fan 2018). Another group created macaque monkeys with genes known to be active in human intelligence and speech (Regalado 2020). These practices would have required a far longer approval

---

[5] A scientist has published a book about CRISPR that included data from the experiment obtained through his communications with Dr. He: Kiran Musunuru, *The CRISPR Generation: The Story of the World's First Gene-Edited Babies* (BookBaby, 2019).

process if proposed elsewhere, if they would even be approved at all. For example, a University of Pennsylvania cancer immunotherapy trial using CRISPR as an enhancer for the drug effects enrolled its first two patients in 2019, having jumped through various regulatory hoops over a couple of years.

However, the efforts to build China's life sciences capacity have also become the tip of a legal spear in responses by U.S. law enforcement against suspected unlawful activity. The Thousand Talents program, established in 2008, aims to establish China as the world's technology leader by creating financial and other incentives for talented scientists to remain in the country, to return from a period abroad, or for non-Chinese scientists to establish labs there. (Dr. He himself was a beneficiary of a related effort called the Peacock Program, with $6 million to launch a startup to sequence single DNA molecules [Cohen 2019, p. 437].) Thousand Talents has been implicated in the arrest of several scientists in the U.S. on charges of espionage (Chung 2018). More than a dozen scientists working at U.S. universities have been referred to federal officials due to concerns about foreign influence and dozens of institutions have been contacted about these suspicions, stoking anxiety among Chinese scientists in the U.S. (Joseph 2019). A former director of the U.S. National Institutes of Health, Elias Zerhouni, has warned that years of productive collaboration between the two countries is under threat if these sanctions are indiscriminately applied. Writing in the journal *Science*, Zerhouni argued that "The United States should not risk losing critical intellectual assets such as productive foreign-born scientists and engineers to global competitors to serve short-term security concerns at the expense of long-term national interests" (Hvistendahl 2019). Yet no legal intervention could have under-lined the fact that not only "foreign-born scientists" are under suspicion than the January 2020 arrest of Harvard chemistry department chair Charles M. Lieber on charges of lying to authorities about payments from Wuhan University (Subbaraman 2020). The charges against him were ultimately dropped. Again, however, the inti-macy of the global network of relationships among life scientists and funders is hard to overstate.

It is far from absurd to imagine that the lag between the house arrest of Dr. He and the failure to file specific charges against him reflected some hesitation on the part of authorities about whether flexibility with respect to international norms is required in order to innovate, perhaps worth a certain amount of short-term national reputa-tional risk in the hopes that significant breakthroughs will happen later. Despite the sanctions levied against Dr. He, the loosening of assumptions about moral norms with regard to genome modification might already have occurred, as evidenced by a Russian scientist's statement that he will use a superior embryo-editing technique to address CCR5 and AIDS susceptibility (Cyranoski 2019), though Denis Rebrikov's target has since shifted to heritable deafness. Absent Russian laws governing the proposal, reactions to Rebrikov's aggressive timeline among Russian scientists has been predictably critical, including a confidential conference of Russian geneti-cists during the summer of 2019. Rebrikov himself dismisses international scientific community reviews of the matter as "just yammering." (Since then, Rebrikov has pledged to abide by the decision of the health ministry before proceeding [Dobrovi-dova 2019].) No better illustration of the geopolitical status of the issue can be offered than the reported presence at that Russian meeting of Vladimir Putin's putative daughter Maria Vorontsova, an endocrinologist said to be sympathetic to Rebrikov's plans. According to one news outlet Vorontosova's "views on bioethics are becoming increasingly influential" (Kravchenko 2019).

There's a reasonable if somewhat cynical case to be made here considering the "softer" and culturally contextual nature of ethical standards in esoteric fields of science, as compared to more rigid global norms that have been more formally adopted in other areas like armed conflict and trade. In that sense, adherence to the norms of biotechnology depends on a nation's willingness to abide by the spirit of prevailing international norms. The co-chair of the WHO panel and the president of the U.S. National Academy of Medicine have noted that there is no international mechanism for halting these kinds of experiments (Berke 2019).

3.4   Hard Bioethics

Moreover, if some authorities in China are reluctant to cede moral authority to the global hegemon (i.e., the U.S. and its traditional close allies) in the norms of science, they wouldn't be the first. For years, a scattered alliance of South American bioethicists has called for a "hard bioethics," a kind of weaponizing of the field, such as protesting N.I.H. funded speakers' appearance in Latin America to lecture on the rules of human research ethics. These critics argue that the North American agency's actual purpose is to create a respectable environment for the scientific exploitation of bodies in the developing world (Moreno 2020).

As the smoke clears around the case of these edited embryos, it now seems clear that China will subscribe to the international governance efforts now underway, including forthcoming recommendations by a World Health Organization Expert Advisory Committee. Notably, in the wake of the embarrassment of the gene-edited embryos—and before the emergence of the novel coronavirus—China released the draft of a new biosecurity law that focuses on laboratory safety and protections against bioterrorism and biological weapons. This law would cover roughly the same ground as dual-use research of concern (DURC) in the United States (Xinhua 2019).

An underlying complication is the tension between a globalized scientific community of which China is very much a part, as exemplified in the new dual-use law, and a governmental system that may have somewhat different goals. Perhaps China will ultimately propose its own rules as the basis for international standardization. In this way, it could assert itself as a leader in thinking about the ethics of genome modification as part of its role in defining a new international order. There is a precedent for this approach in a 2018 white paper on artificial intelligence (AI) by China's Standards Administration that sets rules on AI safety, ethics, privacy. In the words of the white paper, "establishing policies, laws, and a standardized environment in which AI technologies benefit society and protect the public interest are important prerequisites for the continuous and healthy development of AI technology" (Ding and Trolo 2018). As though to embody the competition for leadership, in July 2019 the U.S. National Institute of Standards and Technology, in response to a presidential executive order, issued a draft plan to establish "AI standards that articulate requirements, specifications, guidelines, or characteristics can help to ensure that AI technologies and systems meet critical objectives for functionality, interoperability, and trustworthiness, and perform accurately, reliably and safely" (Huergo 2019).

Unlike the implications of advances of various forms of AI without a global standards regime, the risks of systematic, lab-based human germline modification are at best highly speculative. Skeptics about the weight of moral norms against deliberately modifying heritable traits might also note that such alterations might not only be imperceptible but would pale beside natural "de novo" variations that take place in each individual genome and are passed down through generations. But what counts as a natural variation? Unlike humans, rapidly propagating species like mosquitoes are not only vastly easier to modify in large numbers, they are also of far greater interest for population health.

Two areas of biotechnology that demand a vibrant international normative and institutional order are those of laboratory security and biological weapons. Managing the results of experiments in microbes and in animals through safe laboratory practices, regimens to monitor and control "dual use" research in select agents, and cooperation to prevent the creation of new and virulent threats as weapons of terror should continue to motivate nations to maintain a basic normative regime for biotechnology. Whether such a regime can co-exist with one in which values concerning human reproductive materials are viewed quite differently is at the moment both an open question and a stress test for the normative scope of the liberal international order.

Finally, any confidence about China's commitment to international scientific norms of any sort has been shaken by the central government's suppression of information about the spread of the coronavirus in late 2019. This and other longstanding aggressively nationalistic policies, from Tibet to Xinjiang to Hong Kong, raise grave doubts about its capacity for moral leadership of any stripe in any domain. An era of Chinese hegemony under the banner of a Middle Kingdom could someday create nostalgia for the LIO, for all its limitations.

3.5 Postscript: The Death of an Article of Faith

Beyond the uncertainties of contemporary genopolitics, the emergence of China as a formidable player in the life sciences serves to decisively bury one artifact of cold war science policy, namely, that only democratic conditions could give rise to important scientific breakthroughs, for science, like democracy, requires the oxygen of transparency, and only democracy can provide that oxygen. This was an article of faith to which virtually every Western science policymaker and philosopher, from left to right, subscribed. The occasional apparent counter-examples—the Soviet development of atomic weapons, the first artificial satellite and man in space—could be written off as products of traitorous behavior by Western scientists or the results of massive, targeted and somewhat raw engineering projects. More typical, it was thought, were foolish "innovative" theories of agronomy like Lysenkoism, driven by ideology rather than science, or in the same field Mao's agricultural revolution that led to famine in the Great Cultural Revolution. In light of China's ascendance in the life sciences, who today would defend the view that a state that controls access to media and closely monitors laboratory activities and the movements of its scientists cannot compete and lead at a global scale?

References

Berke, Rick. 2019. Officials say they lack authority to halt 'CRISPR Babies' plan in Russia. *STAT*, STAT, June 24 June. www.statnews.com/2019/06/24/outraged-by-new-crispr-babies-plan-top-sci ence-figures-say-theyre-powerless-to-stop-it/.

Chung, Li-Hau. 2018. Action urged after FBI arrests scientists. *Taipei Times, The Taipai Times*, October 13. www.taipeitimes.com/News/front/archives/2018/10/14/2003702336.

Cohen, Jon. Inside the circle of trust. *Science* 365 (645): 437.

Cyranoski, David. 2019. Russian biologist plans more CRISPR-edited babies. *Nature News*, Nature Publishing Group, June 10. www.nature.com/articles/d41586-019-01770-x.

Ding, Jeffery, and Paul Triolo. 2018. Translation: Excerpts from China's 'White paper on artificial intelligence standardization'. *New America*, June 20. www.newamerica.org/cybersecurity-initia tive/digichina/blog/translation-excerpts-chinas-white-paper-artificial-intelligence-standardizat ion/.

Dobrovidova, Olga. 2019. Russian authorities seek to ease fears of a scientist going rogue. *STAT*, October 16. www.statnews.com/2019/10/16/russia-health-ministry-calls-human-embryo-editing-premature/.

Huergo, Jennifer. 2019. NIST releases draft plan for federal engagement in AI standards development. *NIST*, July 2. www.nist.gov/news-events/news/2019/07/nist-releases-draft-plan-federal-engagement-ai-standards-development. Released July 2, 2019, Updated January 15, 2020.

Hvistendahl, Mara. Exclusive: Major U.S. cancer center ousts 'Asian' researchers after NIH flags their foreign ties. *Science*, American Association for the Advancement of Science, June 28. www.sciencemag.org/news/2019/04/exclusive-major-us-cancer-center-ousts-asian-res earchers-after-nih-flags-their-foreign.

Ikenberry, G. 2018. Why the liberal world order will survive. *Ethics & International Affairs* 32 (1): 17–29. https://doi.org/10.1017/S0892679418000072.

Joseph, Andrew. 2019. NIH has referred 16 allegations of foreign influence to investigators. *STAT*, June 5. www.statnews.com/2019/06/05/nih-has-referred-16-allegations-of-foreign-influence-on-u-s-research-to-investigators/.

Kravchenko, Stepan. 2019. Future of genetically modified babies may lie in putin's hands. *Bloomberg.com, Bloomberg*, September 29. www.bloomberg.com/news/articles/2019-09-29/fut ure-of-genetically-modified-babies-may-lie-in-putin-s-hands.

Lei, Ruipeng, and Renzong Qiu. 2020. Chinese bioethicists: He Jiankui's crime is more than illegal medical practice. *The Hastings Center*, February 3. www.thehastingscenter.org/chinese-bioethici sts-he-jiankuis-crime-is-more-than-illegal-medical-practice/.

Moreno, J. 2020. The Helsinki declaration and the "American Stamp." In *Ethical research: The declaration of Helsinki, and the past, present, and future of human experimentation*, ed. U. Schmidt, A. Frewer, and D. Sprumont, 351–369. New York, NY: Oxford University Press.

National Academies of Sciences, Engineering, and Medicine. 2017. *Human genome editing: Science, ethics, and governance*. Washington, DC: The National Academies Press. https://doi.org/10.17226/24623.

Nuffield Council on Bioethics. (2018). *Genome editing and human reproduction: social and ethical issues.* Retrieved from https://www.nuffieldbioethics.org/assets/pdfs/Genome-editing-and-human-reproduction-short-guide.pdf.

Parrlberg, Robert L. 1980. Lessons of the Grain Embargo. *Foreign affairs*, 144–162.

Rana, Preetika, and Wenxin Fan. Chinese gene-editing experiment loses track of patients, alarming technology's inventors. *The Wall Street Journal*, December 28. www.wsj.com/articles/chinese-gene-editing-experiment-loses-track-of-patients-alarming-technologys-inventors-11545994801.

Regalado, Antonio. Chinese scientists have put human brain genes in monkeys-and yes, they may be smarter. *MIT Technology Review*, April 2. www.technologyreview.com/2019/04/10/136131/chinese-scientists-have-put-human-brain-genes-in-monkeysand-yes-they-may-be-smarter/.

Stokes, Doug. 2018. Trump, American hegemony and the future of the liberal international order. *International Affairs* 94 (1): 133–150. https://doi.org/10.1093/ia/iix238.

Subbaraman, Nidhi. 2020. Harvard chemistry chief's arrest over China links shocks researchers. *Nature News*, February 3. www.nature.com/articles/d41586-020-00291-2.

Xinhua. China starts making biosecurity law. Edited by Hauxia. *Xinhua*, October 21. www.xinhua net.com/english/2019-10/21/c_138491234.htm.

Jonathan D. Moreno is the David and Lyn Silfen University Professor at the University of Pennsylvania where he is a Penn Integrates Knowledge (PIK) Professor. At Penn, he is also Professor of Medical Ethics and Health Policy, of History and Sociology of Science, and of Philosophy. His most recent books are *Everybody Wants to Go to Heaven but Nobody Wants to Die: Bioethics and the Transformation of Healthcare in America*, co-authored with Penn president Amy Gutmann; and *The Brain in Context: A Pragmatic Guide to Neuroscience*, written with neuroscientist Jay Schulkin. Among Moreno's previous books are *Impromptu Man: J.L. Moreno and the Origins of Psychodrama, Encounter Culture, and the Social Network*; *The Body Politic,* which was named a Best Book of 2011 by Kirkus Reviews. He has published more than a thousand papers, articles, reviews and op-eds. He been translated into German, Japanese, Korean, Portuguese and Romanian. Moreno is senior consultant to a six-year, 10-million-euro project on cold war medical science on both sides of the iron curtain, funded by the European Research Council. Moreno frequently contributes to such publications as *The New York Times, The Wall Street Journal, ScienceNature, Slate, Foreign Affairs, Axios.com, The Huffington Post,* and *Psychology Today.* The *American Journal of Bioethics* has called him "the quietly most interesting bioethicist of our time." In 2018 the American Society for Bioethics and Humanities presented him with its Lifetime Achievement Award.

Part II
Vulnerability

Chapter 4
Persons and Groups: Protection of Research Participants with Vulnerabilities as a Process

Paweł Łuków

4.1 Introduction

Part of the difficulty of conceptualization of the ethics of research involving humans is that it must be responsive to the needs and circumstances of research participants and confront the practicalities of research. Ethical concepts that do not respond to or downplay the needs or significance of the circumstances of research participants are blind to moral problems; ethical standards that employ such concepts are ineffective as instruments of protection of research participants. The evolution of the concept of vulnerability in ethics of research involving humans is a story of a gradual adjustment of a conceptual framework to such fundamental ethical values as respect for persons and protection against harm. It shows how the initially less than adequate conceptualisation has been modified without rejecting the initial intuitive idea. It also helps to reconceptualise the idea of protection of vulnerable research participants.

The process started with recognition of susceptibility to harm in response to unethical treatment of members of groups, subsequently shaping the conceptualisation of vulnerability in terms of group membership (the categorical approach). Later developments led to recognition of the weaknesses and the potential to cause harm of the categorical approach, which was largely replaced by the analytic approach, which focuses on individual vulnerability. However, despite its limitations and potential for harm, the concept of vulnerability understood in terms of group membership has not been completely removed from regulations. It is thus important to explain and understand the role of the concept of group-membership vulnerability in the ethics of research involving humans. This chapter argues that a restricted employment of the categorical approach to vulnerability can be justified if it is seen within the dynamics of study design, review, and implementation, and if the recognition and protection

P. Łuków (✉)
University of Warsaw, Warsaw, Poland
e-mail: p.w.lukow@uw.edu.pl

© The Author(s), under exclusive license to Springer Nature Switzerland AG 2023
T. Zima and D. N. Weisstub (eds.), *Medical Research Ethics: Challenges in the 21st Century*, Philosophy and Medicine 132,
https://doi.org/10.1007/978-3-031-12692-5_4

of human participants with vulnerabilities are construed as a process, rather than a labelling device, that extends over these stages and engages researchers and members of ethics review bodies.

To argue for this perspective on the recognition and protection of vulnerable research participants, the first section of this chapter outlines the transformations of the conceptualisation of participant vulnerability in selected ethics guidelines for research involving humans. [A more comprehensive overview of the treatment of vulnerability in research ethics regulations, excluding CIOMS 216, can be found in (Bracken-Roche et al. 2017).] These transformations, it will be argued, underwent three stages. First, vulnerabilities have been recognised from the beginning of the regulations of medical research involving humans, although the term "vulnerability" was not used. Despite this, the need for protection of persons belonging to some groups perceived as vulnerable was recognised due to the realities of research at the time. In the second stage the word vulnerability was introduced but was still understood mostly as group membership of participants, again due to the ethical concerns arising in response to the harm sustained by participants identified as members of a group. It was in the third stage that vulnerability was decisively linked to individuals in relative independence from their group membership. However, the idea of group-membership has not been left behind completely.

The second section includes conceptual considerations which are intended to pinpoint the differences between the two approaches to research participant vulnerability: the individual vulnerability (or analytic) approach and the group-membership vulnerability (or categorical) approach. The key ethically significant difference between the two conceptual approaches turns out to be that the analytic approach is more specific than the categorical one, and so makes identification of vulnerabilities more likely and accurate. In effect it is less prone to such moral errors as stigmatisation, stereotypisation, or discrimination. The suitability of either approach will however depend on the available information on participants.

The third section contextualises these observations in the dynamics of the process of study design, review, and implementation. The analysis recognises different users of the ethics standards (in this case: the obligation to protect participants with vulnerabilities) and different stages in the project lifetime. This 'processual dynamic' suggests that in different contexts and when employed by different users, the analytic approach and the categorical approach can play their protective roles in a complementary way. For this purpose they need to be seen as elements of a process that comprises the stages of study design, review, and implementation and engages researchers and members of ethics review bodies.

4.2 The Transformations of the Concept of Vulnerability

Vulnerability of research participants was not conceptualised at the beginning of the ethical regulation of human experimentation, although to an extent it was recognised in some of the very first ethics regulations of human experimentation. The 1931

Rundschreiben of the minister of the interior of the *Reich*, which regulated human medical experimentation stipulated that "Medical ethics rejects any exploitation of social or economic need in conducting New Therapy" (Sass 1983, 105) or therapeutic experimentation. Although the circular recognised that the participant's social or economic situation must not be taken advantage of in the pursuit of medical progress, the problem addressed was not conceptualised as vulnerability.

A similar approach can be found in the first international regulation of medical research known as *The Nuremberg Code* (The Nuremberg Code 1949). Here, vulnerability of research participants is related to their social or political circumstances (imprisonment) and membership of an ethnic group (Jews, Poles etc.). In both cases the research participants' compromised ability to protect their own interests, while not conceptualised as vulnerability, was understood via group membership rather than as a characteristic of a person or their circumstances or relationships. Such a collective view of participant vulnerability was later adopted in *The Declaration of Helsinki* (WMA 1964), which recognizes dependent relationships as requiring special protection of the participant's personal integrity: "The investigator must respect the right of each individual to safeguard his personal integrity, especially if the subject is in a dependent relationship to the investigator."

These early recognitions of research participant vulnerability were significantly limited due to the lack of appropriate conceptualisation. First, by relying on group membership of research participants they contained the potential for indiscriminate treatment of individuals. In this way, important differences between persons participating in research could be left out of the perspective. More importantly, this perspective can remove from the researcher's view the actual reasons for the ethical concern and the grounds for the special protections of participants with vulnerabilities. The protections may be insufficiently precise or comprehensive to respond adequately to the needs of individual participants. Secondly, the understanding of participant vulnerability which was limited to the "dependent relationship to the investigator" can suggest that vulnerability can be overcome by exclusion of persons with vulnerabilities from the study. Such a move would not only rule out studies that target an issue which is specific to a population of persons with particular vulnerabilities; it could also lead to unfair exclusion of some such persons from the future benefits from a study.

The Tuskegee study stimulated conceptualisation of vulnerability. Due to the design of that research, the Presidential Commission's construal of vulnerability relied on group membership. *The Belmont Report* (National Commission for the Protection of Human Subjects of Biomedical and Behavioral Research 1978) focused on participants' membership in a racial group, which had been linked to the participants' social status related further to their race. The Commission introduced "vulnerability" as the conceptual instrument of tackling the relative limitation of the participant's ability to protect their own interests.

The *Report* continues the group membership perspective in its treatment of the issues of justice and exposure to risk. It stipulates that: "When vulnerable *populations* are involved in research, the appropriateness of involving them should itself be demonstrated." (p. 17) and: "Certain groups, such as racial minorities, the economically disadvantaged, the very sick, and the institutionalized may continually be sought as research subjects, owing to their ready availability in settings where research is conducted. Given their dependent status and their frequently compromised capacity for free consent, they should be protected against the danger of being involved in research solely for administrative convenience, or because they are easy to manipulate as a result of their illness or socioeconomic condition" (p. 19–20). The second passage above links the already familiar group-membership perspective to the capacity for free consent and fairness in allocation of the burdens of participation in research. This approach to vulnerability is further elaborated in the *Report's* focus on voluntariness of consent in relation to "inducements that would ordinarily be acceptable [and] may become undue influences if the subject is especially vulnerable" (p. 14).

Apart from the introduction of the concept of vulnerability, the *Report* adds three significant innovations to the group-membership perspective on vulnerability. First, it relates vulnerability to the assessment and justification of the participants' exposure to the risks of research. Thus, the *Report* indirectly recognises that vulnerability relates to the participant's (limited or compromised) ability to protect their own interests. However participant's interests are defined, limited or compromised capacity to consent makes a potential participant especially vulnerable to harm. Secondly, and relatedly, the potential participant's capacity to protect their own interests is linked closely to their capacity for informed consent. In this way the *Report* highlights two central dimensions of vulnerability: capacity to protect oneself against harm and the capacity for informed consent. Thirdly, and in close relation to the capacity for informed consent, the two dimensions of vulnerability are put in the context of justice or distribution of the burdens and benefits of participation in research. In this way, the basic building blocks of the concept of vulnerability were in place by 1979 when the *Report* was first published.

This focus on group membership in the understanding of vulnerability was accompanied by the CIOMS *Guidelines'* concept of "vulnerable persons", who are defined as "those who are relatively (or absolutely) incapable of protecting their own interests" (CIOMS 2002, Guideline 13). The concern that arises from vulnerability of the research participant relates to "inequitable distribution of the burdens and benefits of research participation", and so it is seen also as a matter of justice (CIOMS 2002, Guideline 13). The *Guidelines* reiterate the *Report's* concern with hierarchical relations. The incapacity to protect one's own interests stems, according to the *Guidelines*, from the "limited capacity or freedom to consent or to decline to consent." Patients who suffer from severe or life-threatening diseases and who participate in research in the context of "compassionate use" are deemed highly vulnerable (CIOMS 2002, Guideline 13), which suggests a relation to the freedom to consent or refusal to participate in research.

An important aspect of the CIOMS 2002's definition of vulnerability is that it relies on the distinction between a participant's relative and absolute incapacity to protect their own interests. Although the *Guidelines* do not clarify the distinction, one can hypothesise that absolute incapacity to protect one's own interests is inherent to the participant, whereas relative incapacity stems from the participant's situation (e.g. dependent relationship) or circumstances. Support for this hypothesis can be found in the references to Guidelines 14 and 15, which concern children and persons incapable of adequately informed consent due to mental or behavioural disorder respectively. Thus, CIOMS (2002) maintains the group membership approach but its definition of vulnerability refers to individuals.

In the period since the *Belmont Report* to CIOMS (2002) vulnerability of participants has been given its name and definition. It has been seen as relating to the issues of justice, protection of one's own interests and capacity for consent. Such a broad view of vulnerability suggests that every research participant could be considered vulnerable (Forster et al. 2001; Levine et al. 2004; Leavitt 2006). In combination with the group-membership conceptualisation, this broad view of vulnerability may encourage uniform and indiscriminate treatment of persons who need special protection, or even protection of those who do not require it at all (Resnik 2004). Without appropriate amendments, this conceptualisation of vulnerability may turn out not only to be useless—because it may be unable to recognise the individuals who need special protection—but also ineffective or even harmful, because by requiring protections for those who do not need special protection it may impose unjustified restrictions on, or misallocate protection of, such persons (Campbell 2004).

This problem of taking too broad a view has been indirectly recognised in the *Declaration of Helsinki* of 2013, which asserts that some "groups and individuals are particularly vulnerable and may have an increased likelihood of being wronged or of incurring additional harm" (WMA 2013, art. 19). This statement can be interpreted as implying that all humans are vulnerable and some of them are *particularly* vulnerable. The *Declaration* predicates vulnerability on persons and groups and suggests that lack of protection for such persons or groups implies infringement of their rights ("wronged"), or harm. This view of vulnerability has been adopted in CIOMS (2016) as an alternative to the view of CIOMS (2002). According to the guidelines, the definition from the *Declaration* "implies that vulnerability involves judgments about both the probability and degree of physical, psychological, or social harm, as well as a greater susceptibility to deception or having confidentiality breached" (CIOMS 2016, Guideline 15). Thus, the interests that need to be protected in the context of research have been specified in some more detail. The interplay between vulnerability and "the ability to provide initial consent to participate in research" is highlighted by stressing the continuing nature of consent in the lifetime of a research project.

Guideline 15 is critical of the understanding of vulnerability as membership in a group. Vulnerability is to be predicated on individuals, not groups, in relation to a situation in which a person finds themselves. Accordingly, Guideline 15 lists various characteristics that can make persons vulnerable: limited capacity to consent or decline to consent to participation in research, being in hierarchical relations, being institutionalized. Also listed are women in certain situations of research contexts, pregnant women in certain circumstances, and other potentially vulnerable individuals (including members of stigmatised groups). However, the Guideline recognizes that groups may be disadvantaged in some social or political circumstances, and so membership in such a group can imply the vulnerability of a person. The Guideline stresses in addition that vulnerability judgments, whether with respect to an individual or a group, must be made on the basis of contextualized empirical evidence. Unlike individual vulnerability, which (one can assume) is to be recognised by both researchers and RECs, group vulnerability is presented as being of concern for committees. This is not to say that only committees are instructed to pay attention to group vulnerability. It seems however, that a focus on groups is deemed appropriate in the context of ethics review of research proposals.

This outline of the transformations of the concept of vulnerability in selected research ethics guidelines suggests a dual view of vulnerability according to which it is a component of the human condition, which calls for the protection of research participants by ethics standards, and can also be a special characteristic of some human beings who thus require special protections in the context of research. Every human being is susceptible to violations of their rights or interests, to various kinds of physical, psychological, or social harm, and to limitations of their capacity for consent or refusal to participation in various activities (Goodin 1985; MacIntyre 1999; Rendtorff 2002; Kottow 2003, 2004; Mackenzie et al. 2014) or in research. Some human beings may be particularly vulnerable in the context of research due to illness, disability, low education, immaturity, difficult social or economic situation etc. These forms of vulnerability can arise, become salient, exacerbated or effective during, or due to participation in, research, and so they raise concerns regarding the voluntariness of the decision to participate in a study, confidentiality, misconception or deception. While vulnerability judgments must be context-dependent, vulnerability can have different degrees and can take a number of forms.

The degrees of a research participant's vulnerability can be gauged with the distinction between vulnerability of humans as such and special vulnerability in the context of research, human vulnerability being a reference point for research-related vulnerability judgments. This distinction provides background for the distinction between absolute (participant-dependent, inherent, intrinsic) and relative (context-dependent, situational, extrinsic) vulnerability (Silvers 2004; Rogers and Ballantyne 2008). The ethics of research as a whole involving humans is a response to the recognition of the various kinds and degrees of vulnerability (Hurst 2008). Since some forms or degrees of vulnerability of human beings can arise, become salient, exacerbated or effective in, or due to involvement in, research, the ethical standard of special protections for participants with such vulnerabilities needs to be included in the ethics of research involving humans.

One can also look at vulnerability of research participants from the point of view of the *Belmont Report* principles of respect for autonomy, beneficence and justice. First, a potential research participant's vulnerability can stem from their limited capacity for informed consent, which may compromise or restrict their ability to exercise their right to self-determination, or the researcher's discharge of their duty of respect for participant autonomy. The second form of vulnerability relates to the principle of justice or fairness. Due to their special characteristics or circumstances, participants may bear unfair burdens of research (Nickel 2006). In extreme cases this form of vulnerability may engender exploitation of research participants (Macklin 2003). The third form of vulnerability relates to the principle of beneficence and involves the research participant's limited ability or inability to avoid harm.

As indicated above, the shift from group vulnerability to individual vulnerability has not been accompanied by rejection altogether of the concept of vulnerable groups. While group membership remains in the vocabulary of protection of vulnerable persons, vulnerability which pertains to individuals is moved to the forefront. Group vulnerability is treated with suspicion as potentially distorting the judgement of both researchers and REC members by perpetuating or encouraging, among other things, stereotyping, stigmatization, and discrimination (Danis and Patrick 2002; Levine et al. 2004; Hurst 2008; Luna 2009).

In view of the criticisms and risks related to the construal of vulnerability in terms of group membership, the concept of vulnerability understood as a characteristic of persons rather than groups seems unquestionably superior. A person's vulnerability is not a function of their membership in a group. Rather, it is the characteristics of individuals, the circumstances in which they find themselves, or their relations to others that make them more or less vulnerable in research and in other contexts. A focus on characteristics, circumstances, and relations makes it possible to avoid uniform treatment of different participants and neglect of the vulnerability of those who do not fit into a group description. It can also appreciate the kinds of vulnerability that stem from relations in which participants find themselves. One can be vulnerable in some circumstances and not in others (Levine et al. 2004). The individualised approach seems clearly more reliable and ethically acceptable than the group-membership approach as an instrument of protection of research participants. It would seem therefore that research ethicists "should stop focusing on vulnerable groups and begin thinking about vulnerabilities." (Kipnis 2004). It is vulnerabilities that should be identified, and the vulnerable persons who should be identified and protected.

4.3 Some Conceptual Considerations

Despite the disadvantages of the group-membership approach, CIOMS (2016) provides that "circumstances exist that require research ethics committees to pay special attention to research involving certain groups. In some resource-limited countries or communities, lack of access to medical care and membership in ethnic and racial minorities or other disadvantaged or marginalized groups can be factors that constitute vulnerability" (CIOMS 2016). In view of its justification in terms of group membership, this provision could be criticized on the grounds mentioned above, despite the warning that "the judgment that groups are vulnerable is context dependent and requires empirical evidence to document the need for special protections." Even if limited in scope and supplementary, the group-membership approach to vulnerability can be seen as questionable because it still carries with it the dangers of this approach. However, before proceeding further it will be helpful to clarify the difference between the individual and the group-membership approaches.

Obviously, the concept of vulnerable groups should not be interpreted as relying on the idea that groups can be vulnerable, apart perhaps from investigations of groups in terms of their dynamics, cohesion etc. It is individuals who can be characterised as vulnerable and they can be characterised collectively as a group. The focus of research ethics standards is on persons who are identified as vulnerable in view of certain characteristics that they are believed to share. The essence of the difference between the group-membership and individual approaches is thus different selections of descriptors of vulnerable persons. While the group membership approach tends to rely on descriptors that cover wide spectra of participants (presumably, with many subtypes), the individual approach would usually look for vulnerabilities to identify subtypes of participants with fewer numbers in each subtype. Accordingly, the individual approach will tend to offer more finely grained descriptions of participants, whereas the group membership approach will offer more unspecific or less discriminate descriptions. For this reason the group-membership approach is often labelled 'categorical'—because it identifies large categories of participants—and the individual approach 'analytic'—because it provides tools for the analysis of vulnerability in terms of participants' characteristics, circumstances or relationships, offering in effect more precise, and so more reliable identifiers of vulnerabilities which can be found in relation to persons (DeBruin 2004; Levine et al. 2004; Luna 2009; Silvers 2004). This terminology will be used in the reminder of this chapter.

The categorical approach sees individuals as belonging to categories which may rely on such descriptors as ethnicity, nationality, gender, disease, physiological state, social situation etc. The original reason for the introduction of vulnerability in the *Belmont Report* aptly illustrates this categorical view. The *Report* responded to the treatment of underprivileged members of a racial group enrolled in a study. The participants were identified by their race but the reason why they were taken advantage of were *inter alia* racial prejudice and the social and economic situation of the participants. It is not their race that made them vulnerable but the researchers' prejudice and abuse. The categorical approach ("vulnerable group") was a convenient means of identification of the persons who were maltreated. Thus, at least initially, it seemed that the categorical approach could be useful in designations of persons who are in a situation that makes them more vulnerable than others. It seemed useful because the link between (in this case) racial identification and vulnerability (due to the racist prejudice) was clear and systematic.

The situation can be different in the case of other persons with vulnerabilities where the link between a descriptor and vulnerability is not as reliable as in the original case. In such cases the categorical approach may lead to uniform and indiscriminate treatment of persons who may differ in many respects, such as social or financial situation, which may be responsible for their vulnerabilities. A helpful example is pregnant women, who were considered vulnerable in the pre-2018 Common Rule (45CFR46 Subpart B). While it is likely that, due to their concern for the welfare of their foetuses or babies, pregnant women may be more susceptible to harm or exploitation, it is not necessarily so. Pregnant women can become especially vulnerable in a society that is oppressive to women, when they live in poverty, or when there is insufficient medical knowledge regarding pregnant women's responses to medications or treatment options (van der Zande et al. 2017). Women who live in a society that protects and respects their rights, who are free from poverty or dependence on others, who have access to adequate healthcare and do not experience serious health problems etc. usually will not be vulnerable whether they are pregnant or not. What can make pregnant women vulnerable are their circumstances, social situation, or the state of medical knowledge. The categorical description "pregnant women" is too general to identify reasons that would justify a judgment about their vulnerability. In other cases such a general categorisation may be evidence of, or an incentive to, stereotypisation, stigmatisation, or discrimination.

The two examples illustrate that a key problem with the categorical approach is that if categories are used carelessly the result may be labelling, which can draw the researchers' or REC members' attention away from actual vulnerabilities. There are two main weaknesses of the categorical approach. First, the categories may be ill-conceived in that they may assume that participants have characteristics that they do not have or describe participants as being in circumstances or relations in which they are not. In this way participants may be placed in a category that does not characterise them correctly. For example, the category "citizens of a low income country" may incorrectly suggest that they share a set of characteristics, which can require, for example, that "one should provide the same protections for a white, male, urban businessman from South Africa and a black, female villager from Ethiopia" (Resnik 2004). The problem with such a category is that being insufficiently specific it erroneously ascribes certain characteristics etc. to some members of the categorised group.

Secondly, a category may describe participants correctly but misconceptions, ignorance, or unwarranted generalisations, may suggest that they also share some other characteristics in virtue of which they need special protection. An example of such an employment of a category is "persons suffering from depression", which may correctly identify persons suffering from a disorder, but is used to identify persons whose capacity for informed consent is compromised. The problem here is that, since depression does not necessarily imply decisional incapacity, it is located in a context of beliefs that are mistaken or prejudiced.

By contrast, the analytic approach, which relies on more specific concepts, does not leave much room for the first type of problem because it begins with identification of the vulnerabilities of persons. However, even well-designed concepts can be used sloppily or misleadingly, and so the analytic approach cannot prevent the second type of error. A participant may be believed to have a characteristic they do not have or to be in circumstances or relations in which they are not. For example, a person who is in a dependent relationship to the researcher may be seen as vulnerable despite the fact that the research is risk-free and the participant is a liberated and self-governing person.

From the point of view of the possibility of researchers' or REC members' errors, the key difference between the two approaches, which translates into a practical and moral difference, is that the categorical approach's focus on groups tends to shift its users' attention away from persons and their actual characteristics, circumstances, or relations, which may constitute or reveal vulnerabilities. The categories of the categorical view resemble shortcuts in the thinking about the vulnerabilities of research participants. Thus, whereas the categorical approach holds a higher risk of harm resulting from uniform and indiscriminate treatment of participants, neither of the two approaches is immune to erroneous use of concepts which stem from the beliefs or attitudes of their users.

A conclusion from this conceptual discussion is that, if appropriately used, the analytic approach can be more sensitive, and encourages attention to the information on participants. It is clear that the more information on participants is available to researchers or REC members, the less defensible the categorical approach could be, because it is less accurate as a device for identification of vulnerabilities and allows for less fine-tuned protection of participants with them. And vice versa: the less information on participants is available, the more limited is the opportunity to apply the analytic approach. Consequently, to secure a fine-tuned and adequate protection of participants with vulnerabilities, the researchers and REC members should seek knowledge on the participants of the projects they design, conduct or review.

4.4   Different Users, Different Stages, Different Information

Different users of the ethics standards of research involving humans may have access to different types and scope of information necessary for the application of these standards. Even in the hands of the same type of user, a standard may work differently due to the complexity and dynamics of the design process, ethics review, and implementation of a study.

There are two main types of users of research ethics standards. The first are researchers who are expected to follow the ethics standards in the process of study design and implementation. Each of these stages may afford them different kinds or scope of information regarding the study participants. In consequence, at one stage the duty to protect participants with vulnerabilities may call for activities which are not appropriate for some other stage. In particular, the details on prospective participants, which are available before the study, may differ from those on actual participants after its commencement. For example, even though during the designing of a study it may be clear that persons with particular vulnerabilities will be recruited, their other vulnerabilities may be revealed only after the study has started.

The second type of user of the ethics standards are REC members who are responsible for the ethics review of projects, and do not have contact with individual participants (DeMarco 2004). Depending on the regulatory framework, the standards can also be used by members of supervisory agencies or bodies which conduct ethics review during the implementation of the study or after its conclusion e.g. by conducting a review of deliverables. (Since in some regulatory frameworks this role can be played by RECs, for the sake of simplicity, all bodies responsible for ethics supervision will be referred to as RECs.) In their supervisory role, REC members will usually have access to information which is different from that available to them at the ethics review phase. Again, the information on prospective participants may be different from what is known about actual participants.

If the observations above are correct it can be expected that the information on study participants, which is relevant for the evaluation of their vulnerability and which is available for some users of ethics standards at some stages of the research design, review, or implementation, can be significantly different from the vulnerability-relevant data available to other users at other stages of the process. In effect, the information that is necessary for the analytic approach may be unavailable to some persons involved at some stages of the process, and for this reason the analytic approach to vulnerability judgments may not always be feasible. The processual reality and dynamics of the research design process, review, and implementation, suggests that both analytic and categorical approaches to vulnerability can be useful. More precisely, whenever the available information concerning participants does not allow for the analytic approach, the categorical one may be appropriate.

To see some of the consequences of the differences in the kinds of information relevant for vulnerability judgments, one can follow the typical sequence of a project's lifetime from its design, through ethics review, implementation, to supervision. Clearly, at the stage of project design the type and scope of information on study participants will depend on the question studied and the problem addressed. Certainly, the clinical inclusion and exclusion criteria will afford fairly specific information on the participants. This information alone will not, however, reveal much about the specific characteristics, circumstances or relations of the prospective participants, any or all of which may be responsible for their vulnerabilities, unless the study is planned to be conducted in a setting in which its participants are found in circumstances or relationships that make them vulnerable in certain respects. For example, a study of a novel therapy for a medical condition whose prevalence is not linked to environmental factors, which is planned to be conducted in a community of economically underprivileged persons, allows for identification of this particular kind of vulnerability. In view of the availability of the information on the community, in which the study is to be conducted, the categorical approach is inappropriate to the extent to which it prevents the making of fine-tuned vulnerability judgments. The analytic approach would be possible and for this reason, required.

The analytic approach will be similarly appropriate at the stage of design of studies that target problems endemic to particular populations (e.g. members of an exploited group) or problems which are invariably linked to characteristics that are responsible for participants' vulnerability (e.g. studies of genetic conditions which cause cognitive deficits). If the problem under investigation is responsible for, or invariably linked to, a vulnerability, this fact alone is sufficient for the application of the analytic approach to be possible. To the extent to which the information on the prospective participants is available to the researchers at the design stage of a study, the analytic approach can also be used by RECs. The necessary data can be submitted to the REC, which can assess the adequacy of the protections included in the project.

However, some information may not be available before the beginning of project implementation. The analytic approach may be feasible only when recruitment starts. One can think of a study from our first example, which is not planned to be conducted in a community that is economically underprivileged. Some participants of the study may, however, lack financial resources, which will make them significantly more vulnerable than other participants. In such cases it would be unreasonable and excessively burdensome to design protections of vulnerable persons in advance relying on the analytic approach. The categorical approach could in such cases rely on a procedure similar to that described by Hurst (Hurst 2008) or on a list offered by Kipnis (Kipnis 2006; cf. Council of Europe 2005, § 69). In such cases the categories could be used as reminders of possibilities rather than parts of a plan for protection of persons with vulnerabilities. However, the research team might be advised to be prepared for the emergence of factors that will require protection of persons. For example, researchers may need to address the issue of the ability to continue in the study of economically underprivileged participants who need to commute to the study site.

Information on the grounds for vulnerability judgments that was not available during the designing stage and ethics review process can be revealed at the stage of implementation of the study. In some cases the researchers can be criticised for not arranging for certain protections, that is, for being unprepared to respond to incoming data on prospective participants or for not making sufficient effort to collect data which can be relevant for vulnerability judgments based on the analytic approach. It is possible, however, that some vulnerability-relevant information will be available only after commencement of the study. This new information can prompt the analytic approach to vulnerability judgments.

Although the analytic approach is superior to the categorical one, the latter can be used as a supplementary method for identification of broad types of vulnerabilities and development of participant protections. In some cases and depending on the context, group membership may be a guide for provisional identification of vulnerabilities. For example, the history of potential participants may be the first step to identification of a particular form of socially determined vulnerability (cf. Silvers 2004). Well-designed categories of potentially vulnerable participants may signal an increased probability of the need for protection (DeMarco 2004), and so they may be useful in the process of research design and implementation. Attention paid to categorical designators during the study design stage can therefore improve the researchers' awareness of the potential for harm associated with their study. Such designators can serve as a checklist of issues that need to be paid heed to during implementation of the study. Perhaps such a checklist can become a part of a "vulnerability policy" that would list vulnerability-engendering factors and provide context for the analytic approach. Such a policy might also undergo an ethics review to further improve protection of participants with vulnerabilities.

A vulnerability policy would be an appropriate response to the processual nature and dynamics of research, being an instrument of awareness of, and responsiveness to, vulnerabilities that emerge in research. Such a policy could combine the two approaches discussed here, and in this way it could help researchers to direct their focus in their identifications of vulnerability-engendering factors during the implementation of studies. It could also provide assistance to REC members in the ethics review process. To play its role and to identify the vulnerability-relevant items, the policy must not function in isolation. It would have to be accompanied by an awareness of, and sensitivity to, various forms of vulnerability or to context-dependent vulnerabilities (DeBruin 2001, 2004). This combination of the two-layered conceptual apparatus with the skills and knowledge of researchers and REC members can make them better at looking for vulnerabilities and more knowledgeable in implementing protections (cf. Kipnis 2004).

4.5 Conclusion

An appreciation of the processual nature and dynamics of study design, review, and implementation suggests that recognition and protection of participants with vulnerabilities should also be construed as a process, rather than a labelling device. The ethics requirement to protect research participants with vulnerabilities needs to combine the analytic and categorical approaches to assure a better scrutiny of the participants' vulnerabilities. The special protections for participants with vulnerabilities is to be seen as an ongoing process with various actors who learn about participants as they go and respond to emerging issues related to vulnerability.

The processual perspective on vulnerabilities of research participants requires that the recognition of vulnerabilities and the protection of participants with vulnerabilities rely on both the categorical and the analytic approaches, depending on the availability and extent of participant information, the stage in the sequence of study design, review, and implementation, and the role a person plays in the process. Such a process involves learning by researchers during the design and implementation of studies as well as openness to new factors that reveal participants' vulnerabilities.

Acknowledgements Research for this chapter was funded by the National Science Centre, Poland, under Grant no. 2014/15/B/HS1/03829

References

Bracken-Roche, D., Bell, E., Macdonald, M.E., and Racine, E. 2017. The concept of 'vulnerability' in research ethics: an in-depth analysis of policies and guidelines. *Health Research Policy and Systems.* 15 (1):8

45CFR46. U.S. Department of Health and Human Services. Code of Federal Regulations—Title 45 Public Welfare CFR 46.

Campbell, A.T. 2004. "Vulnerability" in context: Recognizing the sociopolitical influences. *The American Journal of Bioethics* 4 (3): 58–59; discussion W32. https://doi.org/10.1080/152651604 90497100

CIOMS. 2002. *Council for International Organizations of Medical Sciences (CIOMS). World Health Organization: International ethical guidelines for biomedical research involving human subjects.* Geneva: CIOMS.

CIOMS. 2016. *Council for International Organizations of Medical Sciences (CIOMS). World Health Organization: International Ethical Guidelines for Health-related Research involving Humans.* Geneva: CIOMS.

Council of Europe. 2005. *Explanatory report to the additional protocol to the convention on human rights and biomedicine concerning biomedical research*. Strasbourg.

Danis, M., and D.L. Patrick. 2002. Health policy, vulnerability, and vulnerable populations. In *Ethical dimensions of health policy*, ed. M. Danis, C. Clancy, and L. Churchill, 310–334. New York, NY: Oxford University Press.

DeBruin, Debra. 2001. Reflections on "vulnerability". *Bioethics Examiner* 5 (2): 1–4.

DeBruin, D.A. 2004. Looking beyond the limitations of "vulnerability": Reforming safeguards in research. *The American Journal of Bioethics* 4 (3): 76–78; discussion W32. https://doi.org/10.1080/15265160490497579.

DeMarco, J.P. 2004. Vulnerability: A needed moral safeguard. *The American Journal of Bioethics* 4 (3): 82–84; discussion W32. https://doi.org/10.1080/15265160490907366.

Forster, H.P., E. Emanuel, and C. Grady. 2001. The 2000 revision of the declaration of Helsinki: A step forward or more confusion? *Lancet* 358 (9291): 1449–1453. https://doi.org/10.1016/S0140-6736(01)06534-5.

Goodin, Robert E. 1985. *Protecting the vulnerable: A reanalysis of our social responsibilities*. Chicago: University of Chicago Press.

Hurst, S.A. 2008. Vulnerability in research and health care; describing the elephant in the room? *Bioethics* 22 (4): 191–202. https://doi.org/10.1111/j.1467-8519.2008.00631.x.

Kipnis, K. 2004. The limitations of "limitations". *The American Journal of Bioethics* 4 (3): 70–72; discussion W32. https://doi.org/10.1162/152651604323097916.

Kipnis, Kenneth. 2006. *Vulnerability in research subjects: A bioethical taxonomy (research involving human participants V2)*. Online Ethics Center for Engineering 6/15/2006 OEC.

Kottow, M.H. 2003. The vulnerable and the susceptible. *Bioethics* 17 (5–6): 460–471.

Kottow, M.H. 2004. Vulnerability: What kind of principle is it? *Medicine, Health Care and Philosophy* 7 (3): 281–287.

Leavitt, F.J. 2006. Is any medical research population not vulnerable? *Cambridge Quarterly of Healthcare Ethics* 15 (1): 81–88.

Levine, C., R. Faden, C. Grady, D. Hammerschmidt, L. Eckenwiler, J. Sugarman, and Consortium to Examine Clinical Research, Ethics. 2004. The limitations of "vulnerability" as a protection for human research participants. *The American Journal of Bioethics* 4 (3): 44–49. https://doi.org/10.1080/15265160490497083.

Luna, Florencia. 2009. Elucidating the concept of vulnerability: Layers not labels. *International Journal of Feminist Approaches to Bioethics* 2 (1): 121–139.

MacIntyre, Alasdair C. 1999. *Dependent rational animals: Why human beings need the virtues*. The Paul Carus Lecture Series, vol. 20. Chicago, IL: Open Court.

Mackenzie, Catriona, Wendy Rogers, and Susan Dodds. 2014. *Vulnerability: New essays in ethics and feminist philosophy*. Studies in Feminist Philosophy. New York: Oxford University Press.

Macklin, R. 2003. Bioethics, vulnerability, and protection. *Bioethics* 17 (5–6): 472–486. https://doi.org/10.1111/1467-8519.00362.

National Commission for the Protection of Human Subjects of Biomedical and Behavioral Research. 1978. *The Belmont report: Ethical principles and guidelines for the protection of human subjects of research*. In DHEW Publication No. (OS) 78-0012. Bethesda, MD, Washington.

Nickel, P.J. 2006. Vulnerable populations in research: The case of the seriously ill. *Theoretical Medicine and Bioethics* 27 (3): 245–264. https://doi.org/10.1007/s11017-006-9000-2.

Rendtorff, J.D. 2002. Basic ethical principles in European bioethics and biolaw: Autonomy, dignity, integrity and vulnerability–towards a foundation of bioethics and biolaw. *Medicine, Health Care and Philosophy* 5 (3): 235–244.

Resnik, D.B. 2004. Research subjects in developing nations and vulnerability. *The American Journal of Bioethics* 4 (3): 63–64; discussion W32. https://doi.org/10.1080/15265160490497155.

Rogers, W., and A. Ballantyne. 2008. Special populations: Vulnerability and protection. *RECIIS: Electronic Journal of Communication Information and Innovation in Health* 2: S30–S40.

Sass, H.M. 1983. Reichsrundschreiben 1931: Pre-Nuremberg German regulations concerning new therapy and human experimentation. *Journal of Medicine and Philosophy* 8 (2): 99–111. https://doi.org/10.1093/jmp/8.2.99.

Silvers, A. 2004. Historical vulnerability and special scrutiny: Precautions against discrimination in medical research. *The American Journal of Bioethics* 4 (3): 56–57; discussion W32. https://doi.org/10.1080/15265160490497353.

The Nuremberg Code. 1949. In *Trials of war criminals before the Nuremberg military tribunals*, t. II, s. 181–182. Washington, DC: US Government Printing Office.

van der Zande, I.S.E., R. van der Graaf, M.A. Oudijk, and J.J.M. van Delden. 2017. Vulnerability of pregnant women in clinical research. *Journal of Medical Ethics* 43 (10): 657–663. https://doi.org/10.1136/medethics-2016-103955.

WMA. 1964. *World Medical Association (WMA): Declaration of Helsinki, ethical principles for medical research involving human subjects (1964)*. Helsinki, Finland.

WMA. 2013. *World Medical Association (WMA): Declaration of Helsinki, ethical principles for medical research involving human subjects (2013)*. Fortaleza, Brazil, October 2013.

Paweł Łuków is the Dean of the Faculty of Philosophy at the University of Warsaw where he also serves as the chairperson of the Department of Ethics, and the Center for Bioethics and Biolaw. He is the author of books on Kantian ethics, bioethics, and the philosophy of medicine. His publications have concentrated on bioethical regulation, the legal status of the body, transplant medicine, human dignity, biomedical research and the ethical education of medical professionals.

.

Chapter 5
Centring the Human Subject: Catalyzing Change in Ethics and Dementia Research

Gloria Puurveen, Jim Mann, and Susan Cox

5.1 Introduction

Dementia is a progressive and irreversible syndrome, a collection of symptoms, that can take on many different forms depending on the cause. Common traits include impairment and deterioration in memory and other mental abilities, communication, behaviour, and performance of everyday activities. It is a major and increasing cause of disability among older adults worldwide. The number of people worldwide diagnosed with dementia in 2015 is estimated at 46.8 million and this number is expected to increase to 131.5 million by 2050 (Prince et al. 2015). The magnitude of related costs (psychological, social, and financial) coupled with the growing challenge of providing optimal person-centred care for people living with dementia drives home the necessity of understanding and respecting individual values and preferences from diagnosis to the end of life. Yet, people with dementia are often denied their right to participate in decisions that impact their lives, and their values and preferences for care may be unknown or disrespected.

This exclusion from civic participation goes hand-in-hand with the historic exclusion of people with dementia from participation in health and social research. While important developments have been made in the last two decades in centring their perspectives and subjective experiences, significant barriers to meaningful involvement in research persist. As the authors of a review of articles published in 2008 and

G. Puurveen
School of Nursing, University of British Columbia Okanagan, Vancouver, Canada

J. Mann
Centre for Research on Personhood and Dementia, University of British Columbia, Vancouver, Canada

S. Cox (✉)
School of Population and Public Health, University of British Columbia, Vancouver, Canada
e-mail: susan.cox@ubc.ca

© The Author(s), under exclusive license to Springer Nature Switzerland AG 2023
T. Zima and D. N. Weisstub (eds.), *Medical Research Ethics: Challenges in the 21st Century*, Philosophy and Medicine 132,
https://doi.org/10.1007/978-3-031-12692-5_5

2009 in the *Journal of American Geriatrics Society* attest, frequent and unexplained patterns of exclusion of people with cognitive impairment from a broad range of geriatric research (including research on co-occurring conditions) reflect an ethos of research practice that "deserves critical appraisal and continuous updating and improvement" (Taylor et al. 2012, p. 4).

In Canada, the Tri-Council Policy (TCPS2, Canadian Institutes for Health Research 2018) is the guiding ethical framework for all research involving human subjects. In parallel with other jurisdictions (e.g., the International Ethical Guidelines for Health-Related Research Involving Humans), this policy recognizes the unique needs of people living with conditions that may impact their decision-making capacity and advocates for their inclusion. It is understood that progressive impairment in cognitive functioning may impact an individual's ability to participate in research, including their ability to protect their own self-interests and provide free and informed consent. However, the assumption that a diagnosis of dementia is valid grounds for exclusion from research is becoming less acceptable as a normative practice. Across disciplines and research communities, inclusive practices have been successfully adopted yielding relevant and impactful research that foregrounds the lived experience of dementia (Phillipson and Hammond 2018). Indeed, there is now wide recognition of the proactive ways and practical toolkits available to promote inclusive research practices (see Scottish Dementia Working Group Research Sub-Group 2014) for engaging people with dementia in research (Dewing 2007; Keady et al. 2018).

Despite this shift, however, researchers' efforts to cultivate more inclusive research practices often encounter seemingly insurmountable barriers. As Brooke (2019) questions, "why on the one hand are people with dementia involved in research exploring dementia as participants, collaborators, co-researchers and through co-production, and on the other hand are being excluded from the opportunity to participate in wider health research?" (p. 3724). In a recent National forum on "Dementia-Friendly Research" it was observed that enacting inclusive research practices is a multi-faceted and complex process, highlighting systemic, institutional, and individual factors (Alzheimer Society of British Columbia 2018). These factors include prevailing biases and assumptions about the legal capacity and general competence of persons with dementia, a lack of coordination between different institutional research ethics boards, and a paucity of research activities that are well-suited to persons with dementia and thus preclude meaningful research participation. Moreover, initiatives related to patient and public involvement (PPI), which are gaining firmer purchase in dementia research (see Mann and Hung 2019; Poland et al. 2019), are met with similar barriers (Poland et al. 2019).

Yet, these developments in dementia research reflect a growing ethical imperative—that relevant and responsive health and social research must place patients (people) at the centre of research endeavours (Canadian Institutes for Health Research 2018; Cox and McDonald 2013). And, in the context of end-of-life dementia research, this imperative becomes even more important. If people with dementia are excluded from research, it follows that subsequent health and social care will not be aligned with their preferences.

In this chapter, we interrogate why the problem of research exclusion persists and offer specific recommendations for what can be done about it. We first present relevant background information about the project informing this chapter and our corresponding commitments to social citizenship and relational autonomy as overarching concepts guiding our approach. We then discuss the notion of tokenism in both human subject research and PPI. This leads to a discussion about unconscious bias as it relates to persons with dementia and how this may be at the heart of research exclusion. Finally, we draw upon examples of practices we have employed in our current research on living well with dementia to illustrate how adopting a strength-based approach to person-centred research may help to overcome tokenism and unconscious bias and support meaningful and respectful inclusion of people living with dementia in research.

5.2 Background to Project

The impetus for this chapter arose within the context of a current research project on living well with dementia to the end of life.[1] The project employs individual in-depth interviews and arts-based group workshops to elicit the perspectives of people living with dementia and their care partners, on a range of relevant issues including advance care planning. Fortunately, we had concluded interviews with 35 participants[2] prior to the restrictions imposed on research related activities in the wake of COVID-19 in March 2020. In addition, we held four in person arts-based workshops and were able to engage participants in four additional on-line workshops in a modified format after the research restrictions.

Our project also provided fertile ground to identify and explore a range of ethical issues related to research participation and partnership with persons with dementia. As part of this, Dr. Jim Mann (co-author), a person living with dementia, was invited to be the inaugural Community-Based Scholar at the W. Maurice Young Centre for Applied Ethics (CAE), University of British Columbia. During Dr. Jim's one-week residency in July 2019, we held individual conversations and round-table discussions with ethicists, research ethics board (REB) members, dementia researchers, advocacy groups, faculty and graduate students to unpack some of the persistent barriers that thwart meaningful inclusion of people living with dementia in health and social science research. These conversations served as a catalyst for us to probe more deeply into the complexities of ethics in dementia research and prompted us to re-examine

---

[1] This research is funded by the Canadian Institutes for Health Research (CIHR). Dr. Puurveen is supported by the Alzheimer Society of Canada postdoctoral award and the Michael Smith Foundation for Health Research trainee award.

[2] We employed purposive and snowball sampling to recruit participants. We gave presentations and distributed study information through community-based partners such as: the Alzheimer's Society of B.C., Burnaby Senior Services, Neighbourhood Houses, Alzheimer's/Dementia Cafes, and other community senior centres.

why exclusionary practices remain so deeply entrenched. Figure 5.1 illustrates the key issues arising from our discussions.

The underlying rationale for our focus stems from the fundamental realization that everyday societal discourse and attitudes towards persons with dementia diminishes and constrains the range of opportunities available to them. As such, our work is organized around two central concepts. First, *social citizenship* is a rights-based perspective in which "the person with dementia is entitled to experience freedom from discrimination and to have opportunities to grow and participate in life to the fullest extent possible" (Bartlett and O'Connor 2010, p. 37). In terms of research ethics, this concept prompts us to place persons with dementia at the centre of decision-making about research participation, including supporting the right to make choices that might be deemed risky. As such, it promotes a shift in thinking from an ethos of protectionism or safeguarding and its allied threat of paternalism, to an ethos of 'dignity of risk', and its potential to support the agency of persons with dementia.

Second, relational ethics and specifically the notion of *relational autonomy* suggests that "autonomy emerges *within* and *because* of relationship" (Ells et al. 2011, p. 89); that is, people are bonded in relationship and are shaped by their connections with the social environment and web of relationships, as well as inter-secting socio-cultural factors and political structures and norms (Ells et al. 2011; Sherwin 2008). This frame prompts us to think about how people with dementia's agency to make decisions related to research participation is nurtured through and within intersubjective conditions. The focus then is less on cognitive or intellectual "competencies" and more on the relational and dialogical context that impacts the individual's ability to express understanding and reasoning.

In short, both concepts (social citizenship and relational autonomy) allow us to carefully consider tacit as well as more overt ways in which the research process, and those involved (e.g., researchers and REBs) may enable and promote, or constrain and even undermine the autonomy of persons with dementia through the distribution of power, the sharing of information, and the circumstances that may or may not respect the moral agency and accrued wisdom of people living with dementia.

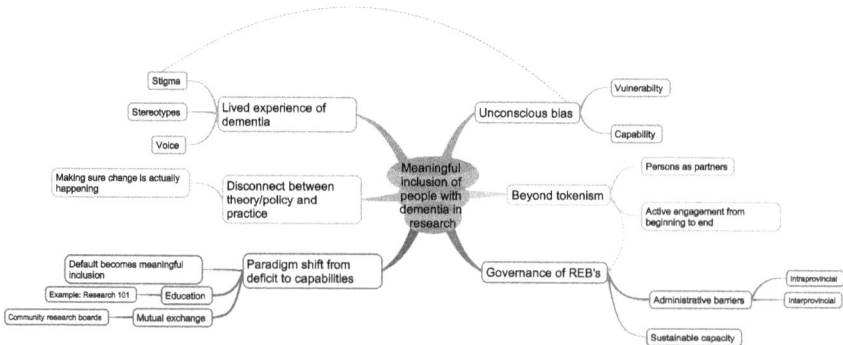

Fig. 5.1 Meaningful inclusion of people with dementia in research

5.3 Problems of Exclusion and Tokenistic Gestures

5.3.1 Exclusionism in Human Subject Research

Up until the 1990s, the perspectives of people with dementia were ignored in research, with an over-reliance on proxy accounts. Over the years a growing body of research has shown that proxy responses cannot credibly 'stand-in' for the perspective of the person with dementia. For example, advance care planning research demonstrates that proxy or family carers do not necessarily know their relative's preference for end of life care, and may base their decision on their own caregiving experiences and the quality of the relationship with their relative. A sense of obligation may lead to carers to wish for more treatment for their relative than they would choose for themselves. Moreover, they may make decisions based on their assumptions of preference rather than confirming with the person with dementia directly (Ayalon et al. 2012; Dening et al. 2012; Smebye et al. 2012).

There is now much scholarship that problematizes the practice of using proxy responses in research and offers recommendations and strategies to maximize the inclusion of people with dementia in research (see Murphy et al. 2014), including how proxy reports might be used in cases where self-report is not possible (Smith et al. 2020). This extends to the practice of soliciting consent for research participation from proxies, or in some cases, even from the person with dementia's physician (O'Connor and Mann 2019). Again, a wealth of scholarship offers guidance on how to solicit consent differently (see Beattie et al. 2019; Dewing 2007; Rubright et al. 2010). This includes thinking about consent as an ongoing, relational and embodied process that highlights a persons' agency in the research relationship rather than relying solely on proxy consent, when they might be deemed legally incapable to provide informed consent; for example, someone in advanced stages of dementia (Dewing 2007; Puurveen et al. 2015). Yet, as we heard in our round-table discussions at UBC, some researchers are still challenged to 'make the case' for the inclusion of people with dementia in research; and, some REBs are challenging researchers to 'make the case' for the reliance on proxy reports. While this might represent a practice lag, whereby researchers or REBs alike are unaware of best practices, a more insidious issue (taken up below in a subsequent section) relates to the notion of unconscious bias that may be present in the development of research protocols.

While relying on proxies overtly excludes the perspectives of people with dementia, inflexible research practices can also inadvertently lead to exclusion, or pose a barrier to people's successful involvement. Rigid adherence to research protocols, failing to talk about the research process in lay language, and not tailoring methods to the person *in the moment* can hinder meaningful or effective involvement. "Misfitting" the research methods (Webb et al. 2020) is as problematic as outright exclusion; as one of our research participants critiqued "researchers are more interested in getting the numbers up, so it looks good on paper" than ensuring that the person can effectively participate. That is, without adapting the research practice in preparation for data collection as well as in situ at each stage of the research, researchers create a potentially incongruent research relationship. However, adapting protocols and practice in the moment can also lead to a misfit between what the REB has approved and how the research approach has evolved. The process of submitting ethics amendments and receiving approval can be time-consuming, which might result in missing a window of opportunity for effective engagement. While some research ethics policies (see TCPS2 2018) recognize the fluid nature of research, particularly qualitative and arts-based research, others constrain the involvement of people with dementia and this has "moral, cultural, and political consequences" (Webb et al. 2020).

A comment from a research participant raises the issue of integrating knowledge translation into the research process and leaving the lines of communication open. The participant stated that when researchers fail to share the results of a study in which he contributed; it signals to him that his contribution was not of value. To him, his participation felt tokenistic because the feedback loop had not been closed. People participate in research for multiple reasons, one of which is making a contribution to knowledge (Cox and McDonald 2013). As such, how research results are communicated back to participants is as crucial as gathering their perspectives in the first place. An aspect of integrated knowledge translation (iKT) is that the cycle of communication between researchers and knowledge users remains open and alive throughout the research process (CIHR 2015). The central premise of iKT is that knowledge users, who, in many cases, are the research participants, are "equal partners alongside researchers" (CIHR 2015); not closing the feedback loop suggests that equal partnership may be in name only.

5.3.2   Tokenism in Patient and Public Involvement

Over the past 25 years there has been a growing shift in the conduct of health research that incorporates those with lived experience of different health conditions to advise on or carry out different aspects of the research process. Often framed as research "being carried out with or by members of the public rather than to, about, or for them" (Staniszewska et al. 2017), patient- and public-oriented (PPI) research seemingly accommodates greater power sharing between researchers and end-users of research resulting in research that has (or could have) greater resonance with patient populations and potentially, greater validity in scientific terms (Cox et al. 2011).

In dementia research, public partners are involved in a variety of ways—from advisory roles that guide the research process to active involvement in recruitment, data collection, analysis and dissemination of research findings (Pickett and Murray 2018). Yet, with increased pressure for researchers to demonstrate direct engagement with public partners in the research process (Staniszewska et al. 2017), there comes an increased risk for their participation in research to be tokenistic and driven by the needs of researchers to fulfill their personal, political or institutional agendas. This is amply illustrated by Dr. Jim's experience with a research advisory group that invited him to give guidance on various aspects of the research process that might impact people living with dementia. At the first meeting, which he flew all the way across the country for, he was not asked for his perspective or even invited to speak. He subsequently withdrew his participation from this group.

In reflecting on two decades of conducting research with people living with dementia, Charlesworth (2018) identifies sources of both overt and covert tokenism. She writes, "Over the years I have been troubled to witness PPI representatives being silenced, sidelined, or ignored where their views do not concur with investigator beliefs, and PPI representatives' views being appropriated, or being misused as ammunition to 'shoot down' alternative perspectives" (p. 1065). Other tokenistic gestures include identifying projects as patient-oriented in name only for the purpose of checking off mandatory fields on funding applications as well as the inclusion of care partners as 'representatives' of the voice of the person with dementia to the exclusion of the person themselves.

One of the challenges in adopting effective PPI roles centres on whether individuals are given adequate resources and support to be *enabled* to successfully enact a particular role. That is, researchers have a duty to help public partners develop literacy related to the tasks and responsibilities associated with different aspects of the research process, including how these roles might differ from being a research subject/participant for data collection purposes. Capacity building and training for those who may have insufficient research experience to adopt instrumental roles takes time and resources—something that may be out of reach for students or early career scholars.[3] If this is not done, claiming that advisors or co-researchers contributed in significant ways without clarity in roles or scaling up their capacity is exploitive, reifies power imbalances and advances the researchers' own agenda rather than truly reflecting an egalitarian process (Cowden and Singh 2007; Stevenson and Taylor 2019).

Additionally, an important tension to acknowledge is that researchers and members of ethics committees have been socialized in their role, including training to conduct rigorous research according to quality standards identified by their scientific community. It seems inevitable that projects that include people living with dementia as co-researchers would receive push-back from researchers' peers and/or ethics review boards, who may not understand the value of enabling and encouraging public partners to adopt meaningful roles in activities such as data analysis or when raising concerns regarding the validity in the interpretation of findings (Waite et al. 2019). As such, it is imperative to undertake ongoing dialogue within research disciplines and ethics committees to learn from other successful partnerships to develop guidelines and best practices related to inclusion across the range of research roles. Indeed, it has taken great effort on the part of advocates, both persons with dementia and dementia researchers, to convince the scientific community of the validity of including people with dementia as research subjects rather than defaulting to the use

---

[3] The issue of adequate funding from competitive sources is important to consider. We recognize the different cultures that operate in various peer review committees and how this impacts whether research projects that aim to be more participatory will be funded. While this issue is beyond the scope of the chapter, we refer the reader to the mindmap (Fig. 5.1) that illustrates this as an intersecting macro-issue.

of proxy responses. It is our hope that the uptake of people with dementia as co-researchers will not face similar resistance; as researchers observe, involvement of people with dementia across the research process enhances the quality of the research and adds much needed context to the findings (Miah et al. 2019).

5.4 Stigma and Tackling Unconscious Bias

5.4.1 Stigma

The concept of stigma was originally identified in the late 1960's by sociologist Erving Goffman as possession of an attribute that discredits its possessor (Goffman 1963). Goffman's work has spawned much subsequent scholarship including, for example, the work of Scambler and Hopkins (1986) who challenged the emphasis on individual attributes in order to distinguish experiences of felt versus enacted stigma and Link and Phelan (2001: 363) who draw upon critical disability studies to define stigma as "the co-occurance of its components—labeling, stereotyping, separation, status loss, and discrimination" and underscore that power must be exercised in order for stigmatization to occur. Accordingly, stigma is not just about the way that one person treats another who possesses an attribute, it is about the structural and institutional factors that shape and reproduce power.

Stigma directed towards people living with dementia is prevalent and has profound social implications. The wrongful perception that a diagnosis of dementia renders the person incapable and incompetent in all aspects of their lives is dehumanizing. It erases them as persons and makes them invisible, accorded with the status of a 'nonperson'. While positive personal experiences with people living with dementia (or other neurological conditions such as Huntington Disease or epilepsy) may temper individual assumptions about capability, when left unexamined patholog-ical aspects (or the deficit model) of dementia become the primary lens through which an individual is perceived (Dupuis et al. 2011).

Benbow and Jolley (2012) argue that stigma distorts all levels of social participa-tion. As persons with dementia may feel embarrassed or even untrustworthy because of their diagnosis, they may feel a corresponding need to hide their diagnosis from others (Milne 2010) thus precluding themselves from seeking assistance (Benbow and Jolley 2012). Stigma impacts how others care for individuals with evidence showing that the stigma of dementia is sometimes associated with inaccurate diag-nosis and/or delays in diagnosis (Koch et al. 2010). These sweeping categorizations about dementia co-mingle with negative associations of ageing, homogenizing a large group of people as a vulnerable group "leaving them without strong social rights or the ability to participate in shaping the measures undertaken for their support" (Mattsson and Giertz 2020, p. 144).

Ageist practices and stigmatizing behaviours directed towards individuals living with dementia should be considered in relation to how we are socialized to think about dementia in the first place. Stereotypes of bizarre behaviours, of mindlessness, and 'zombie metaphors' (Bchuniak 2011) are commonplace and sensationalized in the media (Brookes et al. 2018; Gerritsen et al. 2018). Also contributing is the lack of dementia training in health disciplines and the undesirability of geriatrics and gerontology as a career path for health professionals. It is not surprising then that the deficit model of dementia persists, contributing to broader social narratives of decline, incapacity and precarity in ageing (Grenier 2020) and fuelling assumptions and biases that can insidiously creep into how researchers approach research with people with dementia or REB's assess the ethics of research involving persons with dementia.

5.4.1.1   Unconscious Bias in Research Practice

Unconscious, or implicit, bias occurs where spontaneous judgements or assessments of people and situations arise without consideration of their impact and implications. For example, seeing a person's name alongside the word Alzheimer's or dementia communicates, for many, that the individual will be unable to perform a task or assume responsibility. In terms of research practice, unconscious bias may manifest in terms of how those involved in the research enterprise (REBs, researchers, and other gatekeepers) make decisions in relation to the involvement of people living with dementia in a variety of research roles.

Ageist and stigmatizing perceptions of dementia fuel unconscious biases, leading to discriminatory research practices such as exclusion and/or tokenism. As Dr. Jim glibly asks, "why take the time to recruit, educate, and meet separately with persons with dementia when you know that, without asking (or probing), they will not meaningfully contribute to research and your efforts will go unrewarded?"

5.4.2   Unconscious Bias Manifested in Informed Consent Processes

The process of free and fully informed consent, as a core pillar of research ethics, is intended to safeguard research participants from exploitation and undue harm. Yet, consent processes also have the potential to lead to the overprotection of participants who are considered particularly vulnerable, such as those with dementia, when it is presumed that they do not have the capacity to provide free and informed consent. In the effort to protect human subjects, REB's and researchers inadvertently pose barriers to the full inclusion of people with dementia in research. Unconscious bias becomes evident in the consent process when researchers or REBs make assumptions about individuals' capacity based largely upon a medical diagnosis, with the implication that consent is then solicited from a proxy without due process with persons themselves. This is a multi-layered issue that arises in several interrelated ways.

First, often times assumptions about capacity are based on specific thresholds assessed by cognitive screening tools such as the Mini Mental Status Exam (MMSE; Reisberg et al. 1984). This is problematic as these tools were not intended as screening for research consent (Ries et al. 2020), yet are seemingly given much authority to determine inclusions and exclusions without adequate explanation for such decisions (Taylor et al. 2012). As many have pointed out, thinking about consent in this manner suggests that decisional capacity is a binary concept (Kong et al. 2020), and while a score may tell us something about how a person is cognitively functioning *in that moment*, it does not tell us whether the person can meaningfully consent to participate in research.

Similar issues emerge with capacity to consent to research assessments. Although there are several validated assessments of capacity to consent to research specifically designed to assess a person's understanding, reasoning, and appreciation related to different aspects of the research, these are utilized less frequently than the afore-mentioned cognitive screening tools and they are not used consistently (Ries et al. 2020). Further, while these tools have been shown to promote inclusion in research, including participation of nursing home residents with dementia (Beattie et al. 2019), there is a lack of transparency in terms of how they are used in practice and a more general lack of awareness that such tools exist. As Kong et al. (2020) argue, individualistic assumptions grounding both types of assessment (cognitive screening tools and capacity to consent tools) suggest that, individuals can be presented with relevant information and it is up to them to process it, deliberate about its potential harms and benefits, and communicate a decision. Capacity is viewed as a placeholder for autonomy where autonomy concerns the *individual* and his or her choices, expressed through *cognitive abilities* (internal intellectual skills and abilities of reasoning). This paints a picture of the ideal autonomous agent as a bounded, self-sufficient, independent and rational mind (p. 139).

Notwithstanding the incredible stress 'testing' places on a person with dementia, thinking about consent in this manner suggests that failing to meet screening thresholds means that people with dementia cannot provide meaningful research consent. Yet importantly, the question is not whether to exclude if people are not able to proceed through traditional consent procedures, rather it is a question of what mechanisms have been provided to support understanding of the research. Consent involves evaluating where a person is within that moment, assessing their understanding of the benefits and harms related to their research participation, and evaluating possible barriers to informed consent. Given this dynamic situation in which consent occurs, rigid protocols do not work. Therefore, incapacity should not be assumed without making necessary adaptations and efforts to minimize how the information presented might impact various difficulties with memory, communication and abstract reasoning. What is needed therefore is a shift in thinking: consent can be promoted through relational support, dialogue, facilitated learning, and creative and appropriate communication tools.

Second, while there is a trend to solicit consent differently including developing processes for securing ongoing consent, (Dewing 2007; Rubright et al. 2010), these processes are not widely adopted and there is a general lack of awareness on the part of REBs and researchers that there is precedent to adopt more inclusive consent processes. This trend is grounded in the understanding that standardized or template-style research information and consent forms do little to promote decision-making capacity. Lengthy forms that are filled with discipline-specific jargon do little to promote research literacy and set individuals at a disadvantage when assessing their capacity to consent. As one of our co-authors on a poster presentation (about ethics and dementia research) said in reference to her role in assisting us to construct a meaningful consent process (Puurveen et al. 2018), "when you're faced with forms like this, you feel too stupid to even attempt to read them. The whole consent process exercise is stigmatizing." Yet, while plain language consent forms are preferred, the practical reality of health and social science research involves complex scientific concepts and research study elements that people may have insufficient health and research literacy to understand. This exacerbates existing social inequities and entrenches feelings of powerlessness thus undermining the agency of persons with dementia who may pre-emptively assume they will not understand so give up trying. Further, while researchers have posed different ways of sharing research materials, very little guidance exists around how these materials can be most effectively shared, to what degree they can be tailored to the suit the capacities of the individual, and/or at what point the assessment of capacity is to be undertaken.

A third issue relates to the lack of clarity in legal requirements related to research consent itself and the role of proxies as enduring or temporary decision-makers. As Ries and colleagues (2020) identify, many researchers are not clear about legal requirements related to research consent; for example, whether family members have the legal authority to provide consent on the person's behalf. While researchers need clear guidance in this regard, some researchers have also advocated for advanced directives related to research consent (Kim 2011; Ries et al. 2019).

These interrelated issues underscore the pressing need to educate researchers and REB members alike on assessing cognitive capacity to consent, the legal aspects related to soliciting proxy consent, and appropriate means of avoiding unconscious bias as it relates to practices related to informed consent. Yet, without unpacking assumptions about a related issue—vulnerability—efforts to reconsider informed consent will remain tokenistic at best.

5.4.3 Unconscious Bias Manifested in Assumptions About Vulnerability

The notion of vulnerability arises from bioethics to identify research participants who may be in need of particular protection during the course of the research (Katz et al. 2019; Luna 2009). While some research ethics policies and guidelines[4] explicitly

---

[4] In an analysis of eleven different research ethics policies, Bracken-Roche et al. (2017) found only three out of the eleven policies defined vulnerability; the TCPS2 (2014 edition) being one of the policies. Since Bracken-Roche and colleagues' (2017) analysis, the TCPS2 has been updated.

define vulnerability or vulnerable subjects in relation to research, (Bracken-Roche et al. 2017), generally, the concept itself is somewhat ambivalent—vaguely defined and lacking scholarly consensus regarding its central features (Bracken-Roche et al. 2017; Katz et al. 2019; Luna 2009). In practical terms, and to ensure "careful inclusion rather than outright exclusion…from research" (Bracken-Roche et al. 2017, p. 11), a definition of vulnerability should provide sufficient guidance as to the appropriate protection for those who may need their interests safeguarded without being overly protectionist with regard to those for whom this is unnecessary. Yet, there is little evidence about how researchers and REBs understand or apply this guidance in the context of dementia research. As Mattsson and Giertz (2020) contend, people with dementia are "in many ways trapped in their categorization as a 'vulnerable group,' as traditionally understood" (p. 142). In addition to belonging to the blanket group "people with dementia", they could also belong to a myriad of other groups made explicit in research ethics policies; including "institutionalized people", "persons in nursing homes", "persons with mental illness or mental health problems, "the elderly", and "those who lack decision-making capacity" (Bracken-Roche et al. 2017). As it pertains to individuals with dementia research ethics guidelines presuppose, a subject that is independent, self-sufficient and able to pursue his or her own interests. When this ideal is transferred to persons with different degrees of dementia who may have declining abilities in these regards, they are often stigmatized as a group seen as vulnerable, irrespective of their individual abilities (Mattsson and Giertz 2020, p. 142).

Vulnerability then is a "fixed label" (Luna 2009, p. 124) attached to people with dementia. They are seen as a homogenous, vulnerable group because they are assumed to be likely lack the capacity to provide free and informed consent (see previous section). In this context, decision-making *in*capacity is both inevitable and immutable. In some ways using the incapacity argument to operationalize vulnerability in dementia research is a somewhat simplistic answer to a complicated issue, tacitly discouraging researchers to dig deeper (or confront their own assumptions). As Taylor and colleagues (2010) demonstrate, this broad labeling is rarely defined further, often providing the basis for exclusion from research or the default to proxy consent. Yet it is clear that leaving the term 'vulnerability' vague or unidimensional (i.e., vulnerable because of incapacity) fails to assess an individual's *particular* situation beyond biological mechanisms and other contextual factors that contribute to vulnerability. That is, a label of dementia does not automatically come with the label of vulnerable. What it should do however is to prompt one to consider situations that contribute to a person's vulnerability.

Mattsson and Giertz (2020), drawing upon Fineman's Vulnerability Theory, posit that vulnerability is an "inevitable aspect of the human condition" (p. 143) and not just an attribute of specific groups or individuals. This suggests that everyone is

---

The TCPS2 (2018) defines vulnerability as "a diminished ability to fully safeguard one's own interests in the context of a specific research project. This may be caused by limited decision-making capacity or limited access to social goods, such as rights, opportunities and power. Individuals or groups may experience vulnerability to different degrees and at different times, depending on their circumstances" (p 202).

vulnerable at some point in their life, and therefore it is our collective responsibility to witness and respond accordingly. Western conceptions of autonomy fail to recognize this interdependency, in addition to the ebbs and flows of dependencies that can be impacted by multiple intersecting factors such as age, illness, family situation, and economic resources. It stands to reason then that a more layered approach to understanding vulnerability in dementia research is warranted. Luna (2009) argues for a richer, more nuanced conceptualization of vulnerability; one that allows us to examine "new vulnerabilities that arise from conditions of economic, social, and political exclusion" (p. 123). From this perspective, vulnerability is seen as changing and dynamic, which allows us to move beyond categorical notions of vulnerable groups. It emphasizes interdependency in social institutions, underscores equal opportunity by offering different ways in which situations of dependency can be overcome by legislation or practice, and emphasizes the need for respecting diversity among people along the life course (Mattsson and Giertz 2020).

If left undefined, vulnerability in relation to dementia research is an individual characteristic rooted in disease/disability (i.e., greater susceptibility to being vulnerable due to a diagnosis of dementia). This ignores the other possibilities, including the risk of vulnerability made manifest by attitude/assumptions of the researcher, an REB member, or a reader. When researchers fail to explicate the different types of vulnerability and how they intersect, for example in their research ethics applications, "the reader is left to imagine what the authors are implying, and to apply their own notions of vulnerability and risk" (Katz et al. 2019, p. 7) potentially turning readers to draw upon dangerous narratives to fill in the blanks. Yet, even if researchers clearly explain their reasons for inclusion/exclusion, readers may still apply their own unconscious biases related to dementia.

To respond appropriately then, requires case-by-case solutions for each person as an individual with unique needs, not solutions based on their membership as "people with dementia". After identifying layers of vulnerability, including that which arises in relation to research protocols and the research environment, researchers can then "analyze the situation in a more refined way by requiring that we unfold the different layers involved" (Luna 2009, p. 141). Accessing and implementing guidance from people living with dementia themselves (e.g., Scottish Dementia Working Group's "Core principles for involving people with dementia in research") and taking cues from researchers and disciplinary communities (e.g., nursing, social work) who champion participatory approaches (e.g., Dewing 2007; Murphy et al. 2014; O'Connor and Mann 2019) coheres well with approaches to meaningful research participation that engage with the *particular* person and the nuances of how they will participate, including how they might consent to the research.

5.4.4 Unconscious Bias Manifested as Epistemic Hierarchy

Researchers need to acknowledge and be proactive about the limits of their own knowledge and expertise. Yet, academic systems, steeped in tradition, can fuel intellectual overconfidence and lack of epistemic humility. While there have been growing calls for research to be more inclusive, this has been met with resistance from the scientific community and patient contributions have been cited as 'anecdotal' and secondary to the researchers' perspective (Quennell 2003; see also Richards 2020).[5] If this is the value placed on patient perspectives, it is easy to fall into tokenistic gestures—a researcher may have given a person with dementia a seat at the table, but if their perspective is not solicited or taken seriously, it reflects the assumption that their contributions and knowledge are not valued or seen as credible. While the person does not, and should not be assumed to, represent the broader patient population, they hold "insider" knowledge and experience not readily available to the research team (Poland et al. 2019). As Dr. Jim articulates, "if you have met one person with dementia, you have met one person with dementia."

At the heart of this provocation is the issue of power. As Flyvberg (2001) asserts, power defines what gets counted as knowledge and what gets attended to; knowledge therefore can be repressed depending on who holds the power. In the context of dementia, preconceptions about the disease progression shape how we think about it and this holds much sway when recruiting people for research (as in conversations about consent discussed above). It also holds much affective power in terms of whether people with dementia will even be given a seat at the research table, let alone being empowered to contribute in a variety of ways that go well beyond being a research participant.

People with dementia have much to offer in terms of their experiences and understandings of their illness, their relationships, and ways of being in this world. Valuing the perspectives of those with lived experience not only opens up the opportunity for dementia research to close this gap by engaging holistically with what it means to live with dementia, it also enables researchers to confront their own tacit assumptions and grapple with how these assumptions narrow the possible horizons of dementia research.

5.5   Adopting a Strength-Based Approach

Thus far we have emphasized two guiding principles grounded in relational ethics and social citizenship that promote the centring of persons living with dementia as patient-partners in research: (1) tackling unconscious bias; (2) moving beyond tokenistic participation. Here we turn to some examples drawn from various stages

---

[5] We also refer readers to Jennifer Johannesen's blogpost about her experience of writing an invited essay in the British Medical Journal, from her perspective as a patient (https://johannesen.ca/2018/01/the-time-i-didnt-get-published-in-the-bmj/).

in the lifecycle of our current project 'Living well with dementia to the end of life (EOL)' to illustrate what it means to embrace a strengths-based research paradigm as a third guiding principle.

The project includes three phases. In the first, we employed open-ended interviews to learn what persons with dementia and their care partners (i.e., dyads) have to say about living well with dementia generally and in relation to present and future care needs. We also explored how shared decision-making occurs (or not) in real life settings. The second phase included arts-elicitation workshops, which aimed to explore the meanings of living well with dementia in creative, non-verbal as well as verbal ways (such as the use of images in making collage or words and phrases in found poetry). In the third phase, we are working with participants to share their perspectives on living well with dementia to EOL with a wider audience through a curated online art exhibition in which we will display the stories, perspectives and co-created artworks of persons with dementia and their care partners. The exhibition will also open up space to have a broader dialogue about EOL care amongst a broad spectrum of stakeholders.[6]

Throughout the design and implementation of each of the three phases, we have priorized the importance of *fitting the research timeframe and activities to the capabilities of the person* (Webb et al. 2020) as opposed to trying to shoehorn participants into a predetermined schedule and approach. This was facilitated through our choice of qualitative and arts-based methods of data collection which are inherently less rigid than surveys and scales in terms of structure and better suited to a flexible and emergent design that meets the needs of participants (Novek and Wilkinson 2017; Phillipson and Hammond 2018). Key aspects of our approach also ensure that we integrate the research process within a timeframe that optimizes the abilities of participants to engage meaningfully (Murphy et al. 2014; Scottish Dementia Working Group 2014) and that we avoid unduly fatiguing participants by, for example, expecting to complete all components of the interview in one visit. Offering choices in terms of how participants may wish to express themselves, through verbal or nonverbal forms, is another important component exemplifying a commitment to fit the method with the person, rather than stringently sticking to one method that may not be suited to the individual. In our arts-based workshops, we provided many options for how participants might engage creatively and often found that it was possible to identify activities that mirrored important dimensions of participants' lives. In one workshop, for example, we collectively created a mandala made from natural materials such as pinecones, cedar bows, and twigs (see Fig. 5.2). During this activity, one participant who was deeply passionate about the natural world, spoke about his knowledge of and love for trees. This gave him confidence to express his ideas about living well with dementia through other means throughout the rest of the workshop.

In adopting a strengths-based approach, we recognized early on that it would be essential to view *consent as a person-centred, and ongoing process*. This raised a number of relevant considerations such as how to best communicate essential

---

[6] The virtual art exhibition is titled "Out Here, In There: Art making Space for Living Well with Dementia" and can be found online at: https://www.artmakingspace.com/.

Fig. 5.2 Mandala, natural and found materials, collective creation

information about the study and when and how it would be most appropriate to ask for consent. Persons with dementia played a pivotal role in assisting us with this process by providing extensive input as we developed visual and narrative-based study information sheets and consent forms. Figure 5.3 provides an example of our study brochure and illustrates how we integrated a visually appealing format with a narrative approach that relayed how the study would unfold and what would be asked of participants.

Meaningful consent requires having a good rapport with participants and attending closely to the environment in which consent is being sought as well as the specifics of what is being asked. As Kong et al (2020, p. 139) suggest, "a singular focus on the individual's performance obfuscates the other half of the consenting process— namely the supportive efforts of others and the environment *around the individual* (such as study recruiters) to explain and support. One need not be wholeheartedly committed to the social model of disability to recognize the importance of environmental, societal, and relational factors in contributing to exclusion and inequality." In our study, we often found it was helpful to schedule a separate meeting to get to know participants, build rapport and discuss research participation and consent.

Given the nature of our qualitative and arts-based approaches to data collection, we anticipated that it would not be possible for participants to provide meaningful consent about issues such as whether a photograph or artwork could be publicly displayed, until after the creative activity was complete and participants could offer consent in light of the specific work they had created (Cox et al. 2014). Thus, in preparation for our virtual art exhibition we have revisited the consent process several times to ensure that each artwork along with the narrative about it reflects the participant's intended presentation. And, before launching our exhibition publicly, we will have

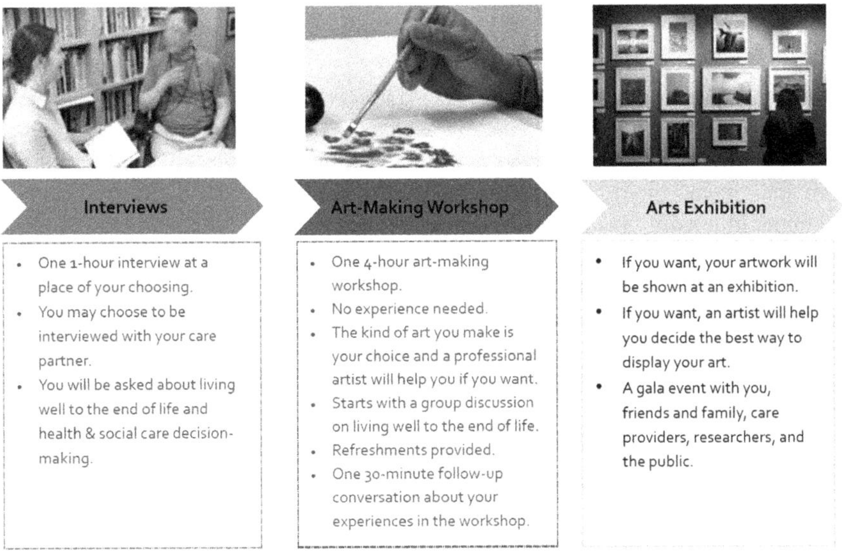

Interviews

- One 1-hour interview at a place of your choosing.
- You may choose to be interviewed with your care partner.
- You will be asked about living well to the end of life and health & social care decision-making.

Art-Making Workshop

- One 4-hour art-making workshop.
- No experience needed.
- The kind of art you make is your choice and a professional artist will help you if you want.
- Starts with a group discussion on living well to the end of life.
- Refreshments provided.
- One 30-minute follow-up conversation about your experiences in the workshop.

Arts Exhibition

- If you want, your artwork will be shown at an exhibition.
- If you want, an artist will help you decide the best way to display your art.
- A gala event with you, friends and family, care providers, researchers, and the public.

Fig. 5.3 Brochure illustrating 3 stages of research on living well with dementia

a special opening that is for our participants only so that they can view their work within the context of the overall exhibition and assess their level of comfort with how it is displayed before finalizing consent.

This latter example of how we are ensuring meaningful consent for the virtual exhibition also highlights the importance of *providing participants with the opportunity to learn about and respond to study findings*. Our online exhibition creation process includes a preview session in which our participants are invited to see all of the works and discuss how they are presented, what key messages are being portrayed and how these might best be conveyed to viewers. We are also actively consulting participants on what they would most like to hear back about from viewers. This feedback loop has evolved from our experience of presenting poetry about living well with dementia at a recent academic conference (Cox et al. 2019). One of our co-authors, a woman with dementia, could not travel to the conference with us but allowed us to show a video of her reading her poetry as part of our presentation. At the conclusion of our chapter, we then invited the audience to write down on a file card several words or phrases that captured their responses to our co-author's poetry. We collected these and shared them with the co-author so she could experience something of the impact her work had for this audience.

5.6   Conclusion

Despite the potential benefits of engaging persons with lived experience of dementia, there are challenges unique to dementia research that cross-cut ethical and methodological domains. These challenges include soliciting informed and ongoing consent from individuals themselves rather than defaulting to proxy consent along the dementia trajectory; balancing opportunities for engagement with paternalistic requirements from research ethics boards; including care partners' perspectives whilst promoting individual agency of the person with dementia; and developing person-centred methods that accentuate the individual's capabilities when communication may be impeded.

In this chapter we have presented three guiding principles grounded in relational ethics and social citizenship that promote the centring of individuals with dementia as partners in research: (1) tackling unconscious bias; (2) moving beyond tokenistic participation; and (3) embracing a strengths-based paradigm. We have illustrated these principles with examples from our project 'Living well with dementia to the end of life' and hope these examples will provide concrete and practical strategies for enacting meaningful and respectful inclusion of people with dementia in research adopting a range of methodological approaches. Although it is apparent that qualitative methods more readily provide researchers with opportunities to creatively adapt planned approaches to data collection to suit the immediate context and the particular strengths of the research participant, there is no reason to think that the model of social citizenship we advocate here is any less relevant for quantitative or clinical research that is about or conducted *with* persons living with dementia.

References

Alzheimer Society of British Columbia. 2018. Strategies for engaging people with dementia in research. In *Pre-conference Workshop, Canadian Association on Gerontology Annual Scientific Meeting*, Vancouver, BC.

Ayalon, L., Y.G. Bachner, T. Dwolatzky, J. Heinik. 2012. Preferences for end-of-life treatment. Concordance between older adults with dementia or mild cognitive impairment and their spouses. *International Psychogeriatrics* 24 (11): 1798–1804. https://doi.org/10.1017/S10416102 12000877

Bartlett, R., and D. O'Connor. 2010. *Broadening the dementia debate: Towards social citizenship*, 151. Bristol, England: The Policy Press.

Beattie, E., M. O'Reilly, D. Fetherstonhaugh, M. McMaster, W. Moyle, and E. Fielding. 2019. Supporting autonomy of nursing home residents with dementia in the informed consent process. *Dementia* 13 (7–8): 2821–2835. https://doi.org/10.1177/1471301218761240.

Behuniak, S. 2011. The living dead? The construction of people with Alzheimer's disease as zombies. *Ageing and Society* 31: 70–92. https://doi.org/10.1017/S0144686X10000693.

Benbow, S.M., and D. Jolley. 2012. Dementia: Stigma and its effects. Neurodegenerative Disease Management 2 (2). [Published online] April 16, 2012. https://doi.org/10.2217/nmt.12.7

Bracken-Roche, D., E. Bell, M.E. Macdonald, and E. Racine. 2017. The concept of 'vulnerability' in research ethics: An in-depth analysis of policies and guidelines. *Health Research Policy and System* 15 (8). [Published online] February 07, 2017. https://doi.org/10.1185/s12961-016-0164-6

Brookes, G., K. Harvey, N. Chadborn, and T. Dening. 2018. "Our biggest killer": Multimodal discourse representations of dementia in the British press. *Social Semiotics* 28 (3): 371–395. https://doi.org/10.1080/10350330.2017.1345111.

Brooke, J. 2019. Equity of people with dementia in research. Why does this issue remain? *Journal of Clinical Nursing*, 28: 3723–3724. https://doi.org/10.1111/jocn.14957

Canadian Institutes of Health Research (CIHR). 2015. *Guide to knowledge translation planning at CIHR: Integrated and end-of-grant approaches.* Retrieved from: https://cihr-irsc.gc.ca/e/45321. html

Canadian Institutes of Health Research, Natural Sciences and Engineering Research Council of Canada, and Social Sciences and Humanities Research Council of Canada. 2018. *Tri-council policy statement: Ethical conduct for research involving humans.* Retrieved on 12/08/2015 from: http://www.pre.ethics.gc.ca/pdf/eng/tcps2-2014/TCPS_2_FINAL_Web.pdf

Charlesworth, G. 2018. Public and patient involvement in dementia research: Time to reflect? *Dementia* 17 (8): 1064–1067. https://doi.org/10.1177/2397172X18802501.

Cowden, S., and G. Singh. 2007. The 'User': Friend, foe or fetish? A critical exploration of user involvement in health and social care. *Critical Social Policy* 27: 5–23. https://doi.org/10.1177/0261018307072205.

Cox, S.M., and M. McDonald. 2013. Ethics is for human subjects too: Participant perspectives on responsibility in health research. *Social Science & Medicine* 98: 224–231. https://doi.org/10.1016/j.socscimed.2013.09.015.

Cox, S.M., K. Ross, A. Townsend, and R. Woodgate. 2011. From stakeholders to shareholders: Consumer collaborators in health research. *Health Law Review* 19 (3): 63–71.

Cox, S.M., S., Drew, M., Guillemin, C., Howell, D. Warr, and J. Waycott. 2014. *Guidelines for ethical visual research methods.* Melbourne: The University of Melbourne. ISBN 987-0-7340-4907-0

Cox, S.M., G. Puurveen, and M. Norman. 2019. Imbrication: (Re)imagining conversations on memory and loss. Paper Presentation, *6th Poetic Inquiry.* Halifax, Nova Scotia.

Dening, K.H., L. Jones, and E.L. Sampson. 2012. Preferences for end-of-life care: A nominal group study of people with dementia and their family carers. *Palliative Medicine* 27 (5): 409–417. https://doi.org/10.1177/0269216312464094.

Dewing, J. (2007). Participatory research: A method for process consent with persons who have dementia. *Dementia: The International Journal of Social Research and Practice* 6, 11–25. https://doi.org/10.1177/1471301207075625

Dupuis, S.L., E. Wiersma, and L. Loiselle. 2011. Pathologizing behavior: Meanings of behaviors in dementia care. *Journal of Aging Studies* 26 (2): 162–173. https://doi.org/10.1016/j.jaging.2011.12.001

Ells, C., M.R. Hunt, and J. Chambers-Evans. 2011. Relational autonomy as an essential component of patient centered care. *International Journal of Feminist Approaches to Bioethics* 4 (2): 79–101. https://doi.org/10.3138/ijfab.4.2.79.

Flyvberg, B. 2001. *Making social science matter: Why social inquiry fails and how it can succeed again.* Cambridge: Cambridge University Press.

Gerritsen, D.L., J. Oyebode, and D. Gove. 2018. Ethical implications of the perception and portrayal of dementia. *Dementia* 17 (5): 596–608. https://doi.org/10.1177/1471301216654136.

Goffman, E. 1963. *Stigma: Notes on the management of spoiled identity.* New Jersey: Prentice Hall.

Grenier, A. 2020. Rereading frailty through a lens of precarity: An explication of politics and the human condition of vulnerability. In *Precarity and ageing: Understanding insecurity and risk in later life,* ed. A. Grenier, C. Philipson, and R.A. Settersten, 69–90. Bristol, UK: Bristol University Press.

Katz, A.S., B.J. Hardy, M. Firestone, A. Lofters, and M.E. Morton-Ninomiya. 2019. Vagueness, power and public health: Use of 'vulnerable' in public health literature. *Critical Public Health.* https://doi.org/10.1080/09581596.2019.1656800

Keady, J., L.C. Hydén, A. Johnson, and C. Swarbrick, eds. 2018. *Social research methods in dementia studies: Inclusion and innovation.* Oxon, UK: Routledge.

Kim, S.Y.H. 2011. The ethics of informed consent in Alzheimer disease research. *Nature Reviews Neurology* 7: 410–414. https://doi.org/10.1038/nrneurol.2011.76.

Koch, T., S. Iliffe, and E.E. Project. 2010. Rapid appraisal of barriers to the diagnosis and management of patients with dementia in primary care: A systematic review. *BMC Family Practice* 11, 52. https://doi.org/10.1186/1471-2296-11-52

Kong, C. M. Efrem, and M. Campbell. 2020. Education versus screening: The use of capacity to consent tools in psychiatric genomics. *Journal of Medical Ethics* 46, 137–143. https://doi.org/10.1136/medethics-2019-105396

Link, B., and J. Phelan. 2001. Conceptualizing stigma. *Annual Review of Sociology* 27: 363–385.

Luna, F. 2009. Elucidating the concept of vulnerability: Layers not labels. *International Journal of Feminist Approaches to Bioethics* 2: 121–139. https://doi.org/10.3138/ijfab.2.1.121.

Mann, J., and L. Hung. 2019. Co-research with people living with dementia for change. *Action Research* 17 (4): 573–590. https://doi.org/10.1177/1467503318787005.

Mattsson, T., and L. Giertz. 2020. Vulnerability, law, and dementia: An interdisciplinary discussion of legislation and practices. *Theoretical Inquiries in Law* 21 (1): 139–159. [Published online] February 26, 2020. https://doi.org/10.1515/til-2020-0007

Miah, J., P. Dawes, S. Edwards, I. Leroi, B. Starling, and S. Parsons. 2019. Patient and public involvement in dementia research in the European Union: A scoping review. *BMC Geriatrics* 19: 220. https://doi.org/10.1186/s12877-019-1217-9.

Milne, A. 2010. The 'D' word: Reflections on the relationship between stigma, discrimination and dementia. *Journal of Mental Health* 19 (3): 227–233. https://doi.org/10.3109/0963823100037 28166.

Murphy, K., F. Jordan, A. Hunter, A. Conney, and D. Casey. 2014. Articulating strategies for maximising the inclusion of people with dementia in qualitative research studies. *Dementia* 14 (6): 800–824. https://doi.org/10.1177/1471301213512489.

Novek, S., and H. Wilkinson. 2017. Safe and inclusive research practice for qualitative research involving people with dementia: A review of key issues and strategies. *Dementia* 18 (3): 1042–1059. https://doi.org/10.1177/1471301217701274.

O'Connor, D., and J. Mann. 2019. The meaning of 'collaboration': A candid conversation between a researcher and a dementia advocate. In *Everyday citizenship and people with Dementia*, ed. A.C. Nedlund, R. Bartlett, and C.L. Clarke, 75–90. London, UK: Dunedin.

Phillipson, L., and A. Hammond. 2018. More than talking: A scoping review of innovative approaches to qualitative research involving people with dementia. *International Journal of Qualitative Methods* 12: 1–13. https://doi.org/10.1177/1609406918782784.

Pickett, J., and M. Murray. 2018. Editorial: Patient and public involvement in dementia research: Setting new standards. *Dementia* 17 (8): 939–943. https://doi.org/10.1177/1471301218178929O.

Poland, F., G. Charlesworth, P. Leung, and L. Birt. 2019. Embedding patient and public involvement: Managing tacit and explicit expectations. *Health Expectations* 22: 1231–1239. https://doi.org/10.1111/hex.12952.

Prince, M., A. Wimo, M. Guerchet, G.C. Ali, Y. Wu, and M. Prina. 2015. *World Alzheimer report 2015. The global impact of dementia: An analysis of prevalence, incidence, cost and trends.* London: Alzheimer's Disease International. Accessed January 20, 2018 from https://www.alz.co.uk/research/world-report-2015

Puurveen, G., A. Phinney, S.M. Cox, and B. Purves. 2015. Ethical issues in the use of video observations with people with advanced dementia and their caregivers in nursing home environments. *Visual Methodologies* 3 (2): 16–26. https://doi.org/10.7331/vm.v3i2.49.

Puurveen, G., S.M. Cox, C.A. Courneya, M. Gregorio, and M. Norman. 2018. (Re)storying informed consent for persons with dementia: The role of narrative and visual thinking. Poster Presentation, *9th Narrative Matters Conference.* Enschede, The Netherlands.

Quennell, P. 2003. Getting a word in edgeways? Patient group participation in the appraisal process of the National Institute for Clinical Excellence. *Clinical Governance: An International Journal* 8 (1): 39–45. https://doi.org/10.1109/14777270310459968

Reisberg, B., S.H. Ferris, R. Anand, M.J. De Leon, M.K. Schneck, C. Buttinger, and J. Borenstein. 1984. Functional staging of dementia of the Alzheimer Type. *Annals of the New York Academy of Sciences* 435 (1): 481–483. https://doi.org/10.1111/j.1749-6632.1984.tb13859.

Richards, D.P. 2020. Don't call my experience a patient "story". *BMJ Opinion.* Published January 9, 2020. Retrieved from: https://blogs.bmj.com/bmj/2020/01/09/dawn-p-richards-dont-call-my-experience-a-patient-story/

Ries, N.M., E. Mansfield, and R. Sanson-Fisher. 2019. Planning ahead for dementia research participation: Insights from a survey of older Australians and implications for ethics, law and practice. *Bioethical Inquiry* 16, 415–429. https://doi.org/10.1007/211673-019-09929-x

Ries, N.M., E. Mansfield, and R. Sanson-Fisher. 2020. Ethical and legal aspects of research involving older people with cognitive impairment: A survey of dementia researchers in Australia. *International Journal of Law and Psychiatry* 68. [Published online] December 3, 2019. https://doi.org/10.1016/j.ijlp.2019.101534

Rubright, J., P. Sankar, D.J. Casarett, R. Gur, S.X. Xie, and J. Karlawish, 2010. A memory and organizational aid improves Alzheimer disease research consent capacity. Results of randomized controlled trial. *American Journal of Geriatric Psychiatry* 18 (12): 1124–1132. https://doi.org/10.1097/JGP.0b013e3181dd1c3b

Scambler, G., and A. Hopkins. 1986. Being epileptic: Coming to terms with Stigma. *Sociology of Health and Illness* 8 (1): 26–43.

Scottish Dementia Working Group Research Sub-Group. 2014. Core principles for involving people with dementia in research: Innovative practice. *Dementia* 13: 680–685. https://doi.org/10.1177/1471301214533225 5.

Sherwin, S. 2008. Wither bioethics? How feminism can help reorient bioethics. *International Journal of Feminist Approaches to Bioethics*, *1*(1), 7–27. https://doi-org.ezproxy.library.ubc.ca/; https://doi.org/10.3138/ijfab.1.1.7

Smebye, K.L., M. Kirkevold, and K. Engedal. 2012. How do persons with dementia participate in decision making related to health and daily care? A multi-case study. *BMC Health Service Research* 12, 241. https://doi-org.ezproxy.library.ubc.ca/; https://doi.org/10.1186/1472-6963-12-241

Smith, S.C., A.A.J. Hendriks, S.J. Cano, and N. Black. 2020. Proxy reporting of health-related quality of life for people with dementia: A psychometric solution. *Health and Quality of Life Outcomes.* 18: 148. https://doi.org/10.1186/212955-020-01396-y.

Staniszewska, S., J. Brett, I. Simera, K. Seers, C. Mockford, S. Goodlad, D.G. Altman, D. Moher, R. Barber, S. Denegri, A. Entwistle, P. Littlejohns, C. Morris, R. Suleman, V. Thomas, C. Tysall. 2017. GRIPP2 reporting checklists: Tools to improve reporting of patient and public involvement in research. *Research Involvement and Engagement* 3(13). https://doi.org/10.1186/s40900-017-0062-2

Stevenson, M., and B.J. Taylor. 2019. Involving individuals with dementia as co-researchers in analysis of findings from a qualitative study. *Dementia* 18 (2): 701–712. https://doi.org/10.1177/1471301217690904.

Taylor, J.S., S.M. DeMers, E.K. Vig, and S. Borson. 2012. The disappearing subject: Exclusion of people with cognitive impairment and dementia from geriatrics research. *Journal of the American Geriatric Society* 60 (3): 413–319. https://doi.org/10.1111/j.1532-5415.2011.03847.

Waite, J., F. Poland, and G. Charlesworth. 2019. Facilitators and barriers to co-research by people with dementia and academic researchers: Findings from a qualitative study. *Health Expectations* 22: 761–771. https://doi.org/10.1111/hex.12891.

Webb, J., V., Williams, M., Gall, and S. Dowling. 2020. Misfitting the research process: Shaping qualitative research "in the field" to fit people living with dementia. *International Journal of Qualitative Methods*, 19: 1–11. https://doi.org/1-.1177/1609406919895926

Gloria Puurveen is a research associate with the Centre for Research on Personhood in Dementia at the University of British Columbia (UBC) and a project manager at the School of Nursing (UBC). Her current work supports a hospice volunteer intervention that promotes a compassionate community approach to end of life care for older adults, persons living with dementia and family caregivers.

Jim Mann, LL.D. Diagnosed with dementia in 2007, Dr. Mann has been acknowledged as perhaps the single-most influential person in Canada for reducing the stigma of Alzheimer's and other forms of dementia, and for promoting that persons with dementia can make an active & meaningful contribution. For this work, Jim was awarded an honorary Doctor of Laws degree in 2020 by the University of British Columbia.

He is a researcher and a published author. Current research projects focus on the reduction of stigma and the promotion of social inclusion of people with dementia, consent and engagement in research with people with dementia, ethically and legally, and technology and robots in long-term care.

At the invitation of UBC's W. Maurice Young Centre for Applied Ethics, Dr. Mann was the inaugural Visiting Community Research Partner in 2019 where his focus was on consent & ethics in dementia research.

Dr. Mann was one of two members with dementia on the federal Ministerial Advisory Board on Dementia, responsible for the release of Canada's first National Dementia Strategy in 2019, *A Dementia Strategy for Canada – Together We Aspire*. He was co-chair of a Canada-wide advisory group that led to the development of both an ethical framework, and the development of the *Canadian Charter of Rights for People with Dementia*.

Susan Cox is Associate Professor at the School of Population and Public Health and the W. Maurice Young Centre for Applied Ethics, University of British Columbia. Her research and teaching focus on ethics and methodological innovation, especially the use of arts-based methods as both a form of inquiry and a form of knowledge translation in health research. Recent projects include: a multi-phase study of the meaning and experience of being a human subject in health research; a research-based theatre performance addressing issues of graduate supervisory relationships, mental health and wellbeing, and a qualitative, arts-based project on what it means to live well with dementia to the end of life.

Part III
Genetics

Chapter 6
Unproven Stem Cell-Based Interventions: Addressing Patients' Unmet Needs or Causing Patient Harms?

Kirstin R. W. Matthews

Scholars began reporting about this marketplace in the mid-2000s; however, there is no sign of it dissipating despite efforts from regulators and scientific societies. While there have been periodic and repeated calls for a global approach to combat the unproven stem cell intervention (SCI) market, little action has been taken.

In this chapter, I will define unproven SCI, describe the marketplace, and discuss the risks to patients, clinical research, and regulatory authorities as well as the discontent between patients and regulators regarding what is safe to use in the clinic. I will also review current regulatory efforts to curb the market's growth. Finally, I will argue for a series of strategies will be presented that can better inform patients and harmonize regulatory efforts between countries.

6.1 Introduction

In the early 2000s, many countries were assessing whether isolating human embryonic stem cells (hESCs) were morally and ethically appropriate, if research utilizing them should be permitted, and if funding should be appropriated towards hESC research. Scientists, doctors, patients, and their advocates promoted hESCs and more broadly stem cell research as being a panacea that could potentially to treat or cure most chronic and debilitating diseases, in the effort to spur government funding, especially in the United States. As the media perpetuated hyperbolic assertions that stem cell-based interventions (SCIs) would be available within the decade, new clinics began to emerge already offering them (Baker 2005; Enserink 2006; Kiatpongsan and Sipp 2008; Lau et al. 2008; Sipp 2017a).

K. R. W. Matthews (✉)
Rice University, Houston, USA
e-mail: krwm@rice.edu

© The Author(s), under exclusive license to Springer Nature Switzerland AG 2023   89
T. Zima and D. N. Weisstub (eds.), *Medical Research Ethics: Challenges in the 21st Century*, Philosophy and Medicine 132,
https://doi.org/10.1007/978-3-031-12692-5_6

An emerging and growing global marketplace of unproven SCIs now exists. Clinics now offer autologous and/or allogenic stem cells from various tissues for a range of diseases such as autism, heart disease, diabetes, multiple sclerosis and spinal cord injuries (Sipp 2017a; Master and Ogbogu 2012). Yet, limited to no data exists to show that these procedures would be effective or safe, since the majority of SCI clinical trials are still in initial stages (phases I and II) (Bubela et al. 2012; Daley 2012; Li et al. 2014). Most clinics offer SCIs provide little information on the actual procedures themselves. They compensate for this lack of data by presenting patient testimonials and anecdotal evidence, appealing to patients' hope and desire for cures (McMahon 2014). SCIs can look like a viable option for patients who desire to be active agents in their healthcare, especially those with conditions that lack a standard therapy or who dislike or distrust the therapies currently available (Hyun 2013; Petersen et al. 2013; Waldby et al. 2020). Unfortunately, patients and caregivers often are not fully aware of the specifics of these interventions or the risks they involve, including physical, financial, and emotional harms as well as how they can undermine of clinical research (Petersen et al. 2013; The PEW Charitable Trusts 2021; Zarzeczny et al. 2020).

Unproven SCIs offered are typically physically benign, having no effect on the patient or the disease progression. However, as more patients have received SCIs, more reports of serious side effects and complications have emerged, including cases of sepsis, cancer, blindness, and death (The PEW Charitable Trusts 2021). Besides physical harms, the procedures cost thousands of dollars, which most patients must cover themselves as public healthcare or private insurance do not cover them.

Scholars began reporting about this marketplace in 2008 after noticing patients traveling from wealthier nations (such as Australia, Canada, the United States, the United Kingdom) to countries with less developed healthcare and regulatory infrastructures (such as Mexico, India, and Thailand), coining the term "stem cell tourism" to describe the phenomenon (Baker 2005; Enserink 2006; Kiatpongsan and Sipp 2008; Lau et al. 2008; Sipp 2017a; Master and Ogbogu 2012). As stem cell tourism became more popular, similar businesses were identified in more developed economies including the United States (Berger et al. 2016; Turner and Knoepfler 2016). Scholars now estimate this market is worth \$2.4 billion and impacts 60,000 patients annually (Master et al. 2021).

There is no sign of the SCI market dissipating, despite efforts from science and medical societies to raise awareness and from regulators to close unscrupulous clinics. Little progress has been made to curtail this industry despite repeated calls for a global approach to combat the it. Activities to regulate this marketplace may be undermined by the divide in what many patients consider sufficient evidence to warrant the marketing of SCIs compared to what scientists and regulators deem is safe and effective. Further issues emerge when regulations are unclear, inconsistently enforced, and require different levels of oversight depending on national jurisdiction. Despite these challenges, stakeholders should work to find ways to eliminate dangerous practices, minimize questionable procedures, and promote clinical research on more promising SCIs. This chapter will define unproven SCI, describe their marketplace, and discuss their risks to patients, clinical research, and regulatory authorities as well as the discontent between patients and regulators regarding what is safe to use in the clinic. Current regulatory efforts to curb the market's growth will be reviewed and a series of strategies that can better inform patients and harmonize regulatory efforts between countries will be argued.

6.2 The Unproven SCI Marketplace Landscape

Unproven SCI has been defined as a stem cell-based medical procedure that has an unclear scientific rationale and mechanism of action, insufficient or nonexistent preliminary data, unstandardized cell handling and administration, inadequate risk information, and no procedural controls (Dominici et al. 2015; Srivastava et al. 2016). In contrast, "proven" treatments are backed by clinical trial data providing scientific evidence of safety and efficacy, with the traditional randomized control trials (RCTs) being the gold standard. RCTs prevent unintended biases that can be found in observational and real-world studies, even those using large datasets, and allow us to understand the effectiveness of interventions and their side-effects (Collins et al. 2020; Gunter et al. 2010).

Only a few stem cell-based therapies have been approved by regulators to date, most related to blood diseases and cancers (Li et al. 2014). The majority of clinical trials using stem cells are still in the initial stages (phases I and I) to assess dosage (how many cells to use), evidence of effectiveness, and side effects on small populations of patients (Bubela et al. 2012; Daley 2012; Li et al. 2014; Kabat et al. 2020). However, the safety and efficacy are not truly tested until interventions progress to a phase III trial, where the number of patients receiving the interventions is significantly high enough. Those under clinical investigation often use autologous or allogenic hematopoietic or mesenchymal stem cells to target cardiovascular, neurological, cancer, and bone conditions (Li et al. 2014). These trials do not require the patient to pay for the experimental intervention, a major different between clinical research and the unproven SCI clinics. They also include an informed consent process that advises the patient of all the potential risks and side-effects of the interventions. The number of clinical trials using stem cells have been increasing annual, still only a small number of phase III trials have been approved (Kabat et al. 2020).

Unlike traditional medicine where one receives a pill or dose of medication that impacts the health of the individual in a transient nature, SCIs are often intended to be integrated in the body and focus on restoration of tissues long-term (Li et al. 2014). As a result, regulators can be more conservative when reviewing this data and require more preliminary evidence before approving a new phase trial. Even though the clinical application of stem cell and regenerative medicine is promising, much has yet to be understood about how stem cells can be used to treat different disease states before they can be approved for marketing and used outside of a clinical trial or setting (Daley 2012).

Unfortunately, determining what is and is not an unproven SCI is not always transparent, especially since countries can have varying national priorities and implement and enforce slightly or significantly different regulations (Lindvall and Hyun 2009; Sleeboom-Faulkner 2016; Sleeboom-Faulkner et al. 2016). For example, in 2013, while many countries were closing down SCI clinics, the Italian Parliament earmarked €3 million for an unproven SCI clinic despite the lack of peer-reviewed data detailing its protocol, safety, and efficacy (Cattaneo and Corbellini 2014; Margottini 2014). After protests from scientists, the Italian Health Ministry withdrew funding and later the founder was convicted on charges of conspiracy and fraud. While this is a particularly egregious case, countries can differ in their views regarding whether unproven SCI are medical innovations or what minimal cell manipulation is permitted, resulting in variations in what SCIs can and cannot be provided without regulatory approval (Zarzeczny et al. 2018). In addition, some doctors also challenge whether RCTs are required for procedures using autologous stem cells to treat for pain or musculoskeletal disorders that can be seen as low or lower risk (Centeno et al. 2021).

Much of what we know about the unproven SCI market comes from empirical studies that analyze provider websites, patient blogs, and media reports or interviews with providers and patients (Master and Ogbogu 2012). However, these studies are limited by the methodologies used and because the field is continually evolving as clinics open, close, relocate, and change names (Sipp 2017a; Knoepfler and Turner 2018). The first reports of the unproven SCI clinics were written by science journalists detailing stem cell tourism (Baker 2005; Enserink 2006). Starting in 2008, more systematic studies characterized the hallmarks of the market including using online direct-to-consumer (DTC) advertising; promoting cures for numerous cosmetic, neurological, and cardiac conditions; and listing of benefits but not risks (Lau et al. 2008; Regenberg et al. 2009). Over time, clinic websites became more "aesthetically appealing" and provided lists of questionable publications, accreditation, advisory and board members, and patents to signal their legitimacy (Connolly et al. 2014; Knoepfler 2019; Ogbogu et al. 2013; Sipp et al. 2017). Despite lacking FDA approval, several clinics also register their interventions as a trial on the US clinical trial database ClinicalTrials.gov to gain additional legitimacy (Turner 2018a, 2017; Wagner et al. 2018). ClinicalTrials.gov works on the honor system. Registered studies listed on the database are not rigorously reviewed or screened, which unproven SCI clinics exploit.

By 2016 scholars identified more than 400 unique English websites for SCI services and clinics around the world (Berger et al. 2016; Erikainen et al. 2020; Munsie et al. 2017; Turner 2018b). Additional clinic websites were identified after searches were conducted in Japanese (74 clinics) and Chinese (more than 2500) (Fujita et al. 2016; Lv et al. 2020). This suggest that current figures might be underestimating the market. Providing and exact numbers is complicated by the fluidity of the market, limited non-English websites assessments, and reliance predominate on website analysis.

The majority of SCI clinics identified provide autologous adult stem cells that are easily obtained from adipocytes or peripheral blood, although some also provide allogeneic stem cells (Berger et al. 2016). Websites advertise treating a range of diseases and conditions with the most common being diabetes, orthopedic injuries, multiple sclerosis and Parkinson's disease (Regenberg et al. 2009; Connolly et al. 2014). In addition, at the beginning of the COVID-19 pandemic several clinics promoted questionable uses of SCIs to treat or prevent SARS-CoV2, despite no data proving SCIs would be effective (Turner 2020; Turner et al. 2021).

In the United States, estimates range from 215 to 2754 clinics offering unproven SCIs in hotspots such as Los Angeles, New York City, and San Antonio as well as the US-Mexico border town of Tijuana, Mexico (Berger et al. 2016; Turner and Knoepfler 2016; Zarzeczny et al. 2018; Connolly et al. 2014; Chavez et al. 2021; Frow et al. 2019; Turner 2021). US clinics predominantly focus on orthopedic, pain, and sports conditions using autologous adipose or bone marrow-derived stem cells, since allogenic stem cells require FDA approval (Turner 2018a; Turner 2021). While most are run by physicians, only half are formally trained in the areas they are treating (Fu et al. 2019).

Unproven SCI clinics have also been identified through lawsuits and court cases (Horner et al. 2018; Martinho and Turner 2017). Nine US cases filed, with 19 named injured patients, found claims of false advertising, fraud, product liability, deception, and unfair trade practices (Horner et al. 2018). Ironically, in some cases the Court dismissed claims of fraud, due to the lack of evidence that SCIs did not work despite a lack of evidence that they did work, which was the reason for the claims. In contrast, a Japanese court ruled in favor of the plaintiff in 2015 because the clinic failed to disclose the risks associated with the procedure (Ikka et al. 2015).

6.3   The True Costs and Harms of Unproven SCIs

Unproven SCIs are sometimes promoted as a harmless way for patients to try a new intervention when other treatments are not appealing or available. However, since 2004 there have been approximately 360 reports of individuals harmed after receiving unproven SCIs (The PEW Charitable Trusts 2021). These physical harms include tumors, strokes, sepsis, blindness, and deaths (The PEW Charitable Trusts 2021; Amariglio et al. 2009; Bauer et al. 2018; Dlouhy et al. 2014; Kuriyan et al. 2017; Lysaght et al. 2017; Olmedo-Reneaum et al. 2019). One systematic study of 2965 orthopedic patients published in 2021 identified 241 complications and 82 hospitalizations in the group receiving SCIs, compared with only 61 complications and eight hospitalizations in the similarly size control group (Pritchett 2021). These complications were not reported by the administering physician, but identified by subsequent treating physicians, suggesting incidents might be underreported. Other studies have noted that of the reports of SCI harms, many do not contain enough information to determine whether or not the SCI was FDA-approved (The PEW Charitable Trusts 2021; Toyserkani et al. 2017).

Harms can also occur in the administration of SCIs. As the clinics are unregulated, the procedures can be conducted improperly, in unsafe, non-sterile conditions, or by staff practicing without a license or outside of the scope of their training (The PEW Charitable Trusts 2021; Fu et al. 2019). For example, in 2018, 12 patients across three states contracted *Enterobacter cloacae* and/or *Citrobacter freundii* infections after receiving non-FDA–approved stem cell products derived from umbilical cord blood for pain or orthopedic conditions (Beil 2020; Perkins et al. 2018).

Physical harms are not the only loss patients incur. SCIs can be expensive, with several studies estimating the cost to be between $3500 and $400,000, not including travel [Ballantyne 2019; Chennells and Steenkamp 2019; McLean et al. 2015; Zarzeczny et al. 2010). The 2021 study of orthopedic patients found costs ranging from $1200 to $13,000 (Pritchett 2021). While some patients only undergo the treatment once, other get treated multiple times, increasing the total costs. In addition, since the procedures are unproven, SCIs are not covered by public health-care or private insurance plans in most countries (McLean et al. 2015). As a result, patients are personally responsible for paying the cost of the intervention and any cost related to side-effects.

Scientists are also concerned that unproven SCIs could erode recruitment of patients for clinical trials and tarnish the reputation of the field (Gunter et al. 2010; Cossu et al. 2018; Matthews and Iltis 2015). Patients might opt to obtain an unproven SCI instead of participating in a RCT where they might only receive a placebo. If a patient does receive an unproven SCI, this could undermine their eligibility to participate in a future trial, especially if it's unclear what the SCI protocol was.

Furthermore, by professing legitimacy, unproven SCI marketing can sell false hope to patients about the effectiveness of interventions, potentially leading to emotional harm (Waldby et al. 2020). Some patients falsely assume that the SCIs are approved by government regulatory bodies or, in some cases, performed by a licensed provider, blurring the line between realistic hope and unrealistic optimism (Michie et al. 2018).

6.4 Perpetuation of Unproven SCI Claims

To encourage hESC funding, scientists and the media promoted hyperbolic claims about SCIs including "a world without sickness and disease," with cures and treatments only a decade away (Caulfield et al. 2016; Kamenova and Caulfield 2015; Master and Resnik 2013). When cures took longer than promised, patients became frustrated at the lack of progress and looked for cures from unproven SCI clinics that employ polished websites, patient testimonials espousing hope, and other DTC marketing tools to sell their claims that created "hype" around the field. Noted health communication scholar Timothy Caulfield often decries the practices of science hype, which he defines as when "the state of the scientific progress, the degree of certainty in the models or bench results, or the potential applications of research are exaggerated" (Caulfield et al. 2016). Ultimately, all major stakeholders—the media, academics, patients, businesses, and governments—bear some responsibility for creating unrealistic expectations of SCIs that helped created the unproven SCI market.

Many argue that the overpromotion of SCI originated at the academic level (Caulfield and Condit 2012). Exaggerated claims surround new scientific innovations or findings often are a result of "the pressure to publish, patent, promote, and commercialize research results, as well as to secure funding for future research" (Sipp et al. 2017; Caulfield et al. 2016). Scientists and universities produce scientific abstracts and press releases with inflated results and overly optimistic conclusions using sensational language to attract attention and funding. The media, in turn, uses these inflated assertions, leading the public to believe therapies and cures are closer to being completed, even when the studies are only proof of concepts, preliminary results, or animal data (Caulfield et al. 2016; Master and Resnik 2013).

The media is often cited as the primary source of scientific information for the general public as well as the exaggerated stem cell claims (Bubela et al. 2009). Traditional print and online media, predominately write articles on stem cells when featuring major advancements in the field; clinics offering unproven SCI; patient narratives, including those of celebrities; and crowdfunding efforts. Articles on unproven SCIs tend to take a positive, optimistic, non-critical tone, with only a few mentioning the risks and efficacy, at times misleading the public into believing these treatments have regulatory approval (Master and Ogbogu 2012; Zarzeczny et al. 2010; Caulfield et al. 2016; Kamenova and Caulfield 2015; Murdoch et al. 2019).

Celebrities promoting clinics and publicly discussing receiving SCI exacerbate the problems. Several Hollywood personalities used their fame to promote hESC research funding in the early 2000s, including Christopher Reeves and Michael J Fox, both of whom created foundations to fund the work. Later, celebrities, athletes, and prominent doctors were associated with questionable SCIs including Kim Kardashian, Peyton Manning, Gordon Howe, and Dr. Oz (Michie et al. 2018; Caulfield 2012; Caulfield and McGuire 2012; Matthews and Cuchiara 2014). Celebrities have a strong and outsized influence on the public's health-related knowledge, attitudes, and behavioral intentions with the media coverage of their SCIs imply that the procedures are legitimate, safe and effective, only rarely noting the lack of evidence that the procedure is effective (Kamenova and Caulfield 2015; Bubela et al. 2009; Caulfield 2012; Caulfield and McGuire 2012; Matthews and Cuchiara 2014; Brown et al. 2003). Adding another layer of complication, fake news outlets disseminate grossly misleading or false information surrounding SCIs, overstating their efficacy and casting doubt about conventional medicine and the healthcare industry (Marcon et al. 2017).

Social media increases the durability and distribution of the unfounded claims of SCI clinics (Petersen et al. 2015). Blogs, social media posts, and YouTube videos detail patients' experiences and provide testimonials that are personal, emotional, and overwhelmingly positive—sometimes even when the intervention resulted in no noticeable difference (Hawke et al. 2019; Murdoch and Scott 2010; Ryan et al. 2009; Levine 2010; Rachul 2011). Posts are often connected and sometimes produced by SCI clinics, which employ them in their DTC marketing efforts. Patient testimonials videos are especially effective as they can present information to viewers with a variety of health literacy levels (Hawke et al. 2019). Of note, more than 40% of these SCI patient blogs describe minors undergoing SCIs (Ryan et al. 2009; Levine 2010; Rachul 2011).

Patients that undergo SCIs also use social media to create a narrative around their condition, post crowdfunding campaigns, and find online communities to share with and support them (Petersen et al. 2015). They sometimes see themselves as "medical pioneers," believing they are adding to scientific knowledge (Rachul 2011; Sipp 2018). These communities, often led by former patients, emphasize hope while downplaying the risks.

One interesting example is Gordon Howe, a famous hockey player how announced in 2014 that he received SCIs in a clinic in Mexico. Previously, it was reported that Howe suffered from dementia, therefore he might not have been able to understand the risks associated with the procedure. Regardless, media reports focused on the hope, families' comments on his improvement, and the procedures, with little to no detail related to the risks (Snyder et al. 2018). In addition, of the more than 2700 tweet responses to media on Howe's procedure, only one noted the intervention was unproven and three warned it lacked scientific evidence.

To cover the high costs of SCIs, some clinics suggest that patients try crowdfunding through sites like GoFundMe.com, although funding unproven interventions is against their policies (Marcon et al. 2020). Similar to social media posts, these crowdfunding campaigns, set up by patients and their families, regularly overstate the efficacy and underemphasize the risks by copying language from the providers' websites, contributing to misinformation around SCIs (Snyder and Turner 2021). Patients often suggest that their treatment has been proven safe and effective, with some suggesting the treatment is approved by a regulatory agency and the funding is to participate in a clinical trial (Snyder and Turner 2018; Du et al. 2016).

Fortunately, more recent media articles are mentioning risks and side-effects of unproven SCIs (Beil 2020; Marcon et al. 2020). Journalists have started exposing predatory SCI clinics, leading to their closure (Sipp et al. 2017; Wan and McGinley 2019). However, exaggerated and unrealistic claims still continue to sneak into stem cell reporting.

6.5 The Role of Scientists, Physicians, and Patient Advocates

Scientists, physicians, and patient advocates have unique and important roles to play in addressing unproven SCI problems, but each group perceives the risks, benefits, priorities, and goals differently (Matthews and Iltis 2017). Scientists and scientific societies are concerned with how the unregulated SCI clinics can impact public perception of research and researchers in the field and compete with clinical research. Medical doctors how get questions from patients must balance respecting patient autonomy with providing the best evidence-base advice in addition to having the resources and knowledge to do so. Patient advocates are often reconciling desires from patients for treatments and hope, when provided with limited options by the medical community. Collaboration and partnerships between these stakeholders as well as regulators will be required to minimize risk to patients and the stem cell research community (Dominici et al. 2015; Matthews and Iltis 2015).

6.5.1   Professional Societies

Professional scientific and medical societies are well-positioned to promote the appropriate uses and discourage inappropriate use of SCIs—and arguably have an obligation to hold their members accountable (Bowman et al. 2015; Fears et al. 2021; Sugarman et al. 2018). Using members expertise, many have developed educational materials to provide information on how to determine if an SCI is appropriate or risky (Taylor et al. 2010; Weiss et al. 2018).

The most vocal in this space is the International Society for Stem Cell Research (ISSCR), a global science society that represents more than 4000 stem cell researchers. ISSCR developed it's first series of recommendations in 2008 on how stem cell research should be translated from the research bench to the clinic, which were updated in 2016 and 2021 (Guidelines for Stem Cell Research and Clinical Translation 2021; ISSCR 2016, 2008). The 2021 ISSCR guidelines have more than 60 recommendations related to the clinical translation of SCIs including guidance on informed consent, preclinical studies, cell manufacturing and processing, clinical research and application, and unproven SCIs (Guidelines for Stem Cell Research and Clinical Translation 2021). Similar to regulation in the United States and European Union (described below), ISSCR Recommendation 3.1.1 states that "Stem cells, cells, and tissues that are substantially manipulated or used in a non-homologous manner must be proven safe and effective for the intended use before being marketed to patients or incorporated into standard clinical care" (Master et al. 2021; Guidelines for Stem Cell Research and Clinical Translation 2021; Turner 2021). In addition, the guidelines mention that early data should precede clinical studies (Recommendation 3.3.1.2) and calls out the practice of using unproven SCIs (Recommendation 3.5.1), suggesting they only be used in clinical trials or within regulatory guidelines. ISSCR has information sheets on SCIs available at the website www.closerlookat stemcells.org. While these guidelines and documents do not have regulatory oversight or enforcement capabilities, they do provide the field with ethical and scientific standards (Turner 2021).

The International Society for Cellular Therapy (ISCT), which represents approximately 2400 cell therapy expert in more than 60 countries, also developed guidance on SCIs (Dominici et al. 2015; Gunter et al. 2010). ISCT focused on eight recommendations, which include implementing a long-term program to harmonize regulation from different countries and regions, establishing a stem cell registry, and promoting scientific research. The society also proposed working with stakeholders to ensure patient protections, provide tools for patients to evaluate treatments, and enable ethical early access to patients.

The InterAcademy Partnership (IAP), which represents more than 140 science academies including the US National Academy of Science and the UK Royal Society, published a statement in 2021 promoting regulation for SCIs (Fears et al. 2021; Inter-Academy Partnership (IAP) 2021). IAP's priorities were: "to use advances in research and development as rapidly as possible, safely and equitably, to provide new routes to patient benefit" and "to support medical claims by robust and replicable evidence so that patients and the public are not misled" (Fears et al. 2021; InterAcademy Partnership (IAP) 2021).

Several other societies have released statements condemning unproven SCIs including the American Thoracic Society (ATS), the Chinese Diabetes Society, the International Spinal Cord Society, the Australia and New Zealand Spinal Cord Injury Network, and the Japanese Society for Regenerative Medicine (McMahon 2014; Ikonomou et al. 2016). Still other societies that do not have statements, have codes of ethics related to social responsibility, which could be used to constrain the marketing of unproven SCIs (Master and Resnik 2011). For example, the American Medical Association (AMA) has a series of 'opinions,' one related to medical tourism (which can be applied to stem cell tourism) promoting physicians to talk with their patients to help them understand the risks associated with SCIs (American Medical Association (AMA); Levine and Wolf 2012). However, they "are not intended to establish standards of clinical practices or rules of law" and do not address unproven SCIs provided within the United States.

6.5.2   Individual Scientists and Physicians

Individual scientists and physicians also have a role in challenging misinformation about SCIs. Scientists should manage materials and information to make sure it is used appropriately. For example, researchers should carefully examine requests for cells and materials, only providing them to responsible investigators and clinicians (Master and Resnik 2011). Researchers should also disclose if clinical trials are pay-to-participate, which are essentially schemes to have patients, instead of the company, pay for the clinical trial (Sipp 2012a; Turner and Snyder 2021).

Physicians have competing obligations regarding SCIs. While they should respect patient autonomy, they also should provide them with evidence-based medicine, stress the risk of unproven SCIs, and suggest alternative treatments if available (Bowman et al. 2015; Sugarman et al. 2018; Levine and Wolf 2012). Many patients seek information about SCIs to their personal physicians prior to obtaining them. A 2018 survey of academic neurologists across the United States found that 89% received questions about SCIs and 65% had patients that received SCIs (Julian et al. 2020). Another student noted that academic physicians who received questions about SCIs, advocated to their patients that SCIs were unproven and unsafe, with only ortho-pedic specialists suggesting SCIs may be appropriate for patients with limited options

(Smith et al. 2021). In both surveys, a number of physicians felt under-prepared and did not have enough information to address their patients' questions (Julian et al. 2020; Smith et al. 2021).

Physician awareness and education is vitally important to ensure they provide up-to-date information regarding SCIs (Master and Resnik 2011). Physician education programs can include continued education modules or an academic medical fellowship training program to prepare physicians for treating patients with SCIs (Knoepfler 2013). Physicians should be aware of location of education materials or other experts, such as stem cell counselors, to answer patients' questions.

In addition to outreach tools, physician regulatory bodies should sanction or remove from its ranks members that provide unproven and unsafe SCIs in the United States and abroad (Liang and Mackey 2012). Already a several US physicians have been sanctioned by professional regulatory bodies, including fines and suspensions (Zarzeczny et al. 2014). But the process often requires the injured patient to submit a complaint, which can be complicated if the incident was in a foreign country where there might be language and cultural differences. In addition, jurisdictional variations allow clinics to move to less regulated locations following actions that restrict the market (Berfield 2013).

6.5.3 Patient Outreach and Engagement

Patients are confronted with conflicting information about unproven SCIs and may not know where to obtain reliable information or determine which claims are credible or trustworthy (Smith et al. 2020). Therefore, education materials directed towards patients as well as working with patient advocate groups are an important part of hindering misinformation about SCIs while still promoting legitimate research and addressing patients' needs (Matthews and Iltis 2015, 2017; Bauer et al. 2017). Patient advocate groups are stakeholders seeking to inform patients and provide them with choices as well as helping translate medical information for a broad public (Bauer et al. 2017). They can help curb dangerous SCIs by increasing patient awareness of their risks, stressing the need for regulatory oversight and enforcement, promoting clinical research, managing patient hope and expectations, and helping push regulatory agencies and scientists to prioritize patient needs.

Providing educational information to help patients is one of the most common recommendations for limiting patient demand for unproven SCIs, and recent studies have found it effective (Taylor et al. 2010; Unsworth et al. 2020). However, not all educational materials adequately address unproven SCIs. A survey of online stem cell educational materials determined that most patient advocate materials included no information on the clinical translation process or unproven SCIs (Master et al. 2014). To be effective, scholars suggest that educational materials include: differences between established legitimate clinical research and pseudo-medicine; a review of the clinical trial process; descriptions of different types of evidence and why testimonials are not given much weight; identification of many of the hallmarks of an unproven SCIs clinics (such as using one type of cell to treat many diseases); and explanation of risks associated with unproven SCIs (Master et al. 2013).

However, patients' perceptions of unproven SCIs can be different from physicians' and scientists' views, especially if the patients feel they have limited or no alternatives (Matthews and Iltis 2017). In addition, educational materials are a one-way form of engagement, experts providing information to patients. As an alternative, stem cell counselors, modeled after genetic counseling, could address patient questions and explain the risks and benefits of SCIs to help patients make a more informed decision (Scott 2015; Tanner et al. 2017). These services allow for two-way communication, provide the patient with a knowledgeable expert, and remove the burden from the physician to be the only source of expertise (Smith et al. 2020; Scott 2015). While counseling may not eliminate all SCI clinics, those patients who decide to pursue an unproven SCI will understand the risks, costs and potential harms (Smith et al. 2020).

6.6   National Regulations

While self-regulation and engagement efforts might help, ultimately intra- and international regulation and oversight is needed to curb unproven and unsafe SCIs. Several countries have regulations which require approval of all SCIs by an authority, such as Canada (Caulfield and Murdoch 2019). Others allow for exemptions, such as allowing unproven SCI procedures in hospitals (Australia and Singapore) or permit autologous, minimally-manipulated procedures (Australia, United States and European Union) (Master et al. 2021; Lysaght et al. 2017, 2013; Ghinea et al. 2020; Lysaght and Campbell 2011). Yet although these policies protect against the majority of unproven SCI clinics, the numbers of clinics continue to rise in both developing and developed countries (Berger et al. 2016).

SCI policies, unfortunately, often compete with other political agendas. Unproven SCIs are now at the center of the propagation of medical commercialism, which is dominated by neoliberal and libertarian ideologies that promote individual choice and patient autonomy over producing generalizable data and therapeutic evidence for the public good (Sugarman et al. 2018; Sipp et al. 2019). Patients perceive autologous stem cells as their own property to be used as they see fit, arguing against what they describe as "your body is a drug" regulatory policies (Bianco and Sipp 2014; Sipp 2017b, 2012b). These arguments are linked to deregulation and downsizing of government campaigns, including the Right to Try movement in the United States (Sipp et al. 2019; Sipp 2012b; Matthews et al. 2018). Furthermore, the potential economic benefits of stem cells, including returns on research investment, job creation, and gaining an early-advantage in an emerging market, incentivizes governments to support unproven SCIs. This leads nations to compete with each other to become leaders in regenerative medicine, sometimes creating more permissive SCI policies instead of enforcing existing regulations (Sipp and Sleeboom-Faulkner 2019).

6.6.1 Developing Countries and Stem Cell Tourism

Rapidly modernizing countries with little medical regulatory oversight constitute an ideal environment for the unproven SCI market. Emerging economies are often branded as hubs for stem cell tourism due to a lack of international and national regulation as well as poor support for national and local authorities to enforce existing regulations (Sipp 2017a; Holzer and Mastroleo 2019). Two Latin American countries, Brazil and Argentina, for example, have limited or no regulatory authority over SCI procedures (Rosemann et al. 2019). However, it has been argued that these countries should not be responsible for enforcing regulations when the providers travel from developed countries to perform the procedures there (Holzer and Mastroleo 2019; Chan et al. 2017). For example, a clinic shutdown in the United States moved their procedures to the Cayman Islands, which had less regulatory oversight (Sipp 2017a). Instead, the home countries, the United States in this example, should be obligated to oversee and sanction their citizens when procedures result in harms. Furthermore, both stem cell tourism and potential regulations created to restrict SCIs could have unintended impacts on scientists in the country, making it more challenging to conduct research and promoting brain drain.

6.6.2 Tightening of Rules and Regulatory Loopholes

The growing number of unproven SCI clinics described previously has caused several nations to review and tighten their policies. These changes include developing more explicit policies (China and India), clarifying existing policies (United States and European Union), or removing exemptions (Australia).

The US FDA's oversight of autologous SCIs was challenged and upheld in 2008 by Regenerative Sciences (Regenexx) in the court case *US vs Regenerative Sciences, LLC* (Lysaght and Campbell 2011). The United States allows FDA-approval exemptions for SCIs: (1) conducted in the same surgical procedure (as the stem cell removal), (2) using autologous stem cells, (3) manipulated only minimally, and (4) used in a manner homologous to their original function (Master et al. 2021; Knoepfler 2015). While the courts ruled in favor of the government, it became apparent that the FDA guidelines were ambiguous and the industry needed more clarity on what types of procedures were permitted without approval and what required clinical data (Sipp and Turner 2012; Taylor-Weiner and Zivin 2015). In 2017, the FDA released industry guidance documents that provided explicit definitions for terms such as 'same surgical procedure,' 'minimal manipulation,' and 'homologous use' with examples of how common stem cell techniques or protocols would be regulated (Master et al. 2021). The FDA also published a series of commentaries reiterating the need for clinical trials and risks of unproven SCIs (Marks and Gottlieb 2018, 2020; Marks et al. 2017). In 2018, the FDA began renewed efforts to block unproven SCIs (Zarzeczny et al. 2018; Knoepfler 2018; Turner 2015). Beyond the FDA, the

US Federal Trade Commission (FTC) has charged several companies with deceptive practices related to unproven claims that SCIs can cure disease, and several state Attorneys General have filed suits against clinics for making misleading claims (Bauman 2021; Richardson et al. 2020).

The European Union's European Medical Agency (EMA) also updated its guidelines creating a new regulatory classification "advanced therapy medicinal products" (ATMPs) in 2007, and added more details in 2015 (Master et al. 2021; Lysaght et al. 2017; Takashima et al. 2021). Any cell or tissue product manipulated and/or intended for non-homologous use falls under the category of ATMP and is regulated (requiring clinical safety and efficacy data) (Master et al. 2021). Minimal manipulation and homologous use were defined, using definitions similar to the FDA guidelines. In December 2020, the UK announced that despite leaving the European Union, they would maintain the European Union's ATMP regulations, which would be overseen by the UK Medicine and Healthcare products Regulatory Agency (MHRA).

To combat the growing number of SCI clinics in China, the Chinese National Health and Family Planning Commission (NHFPC) created SCI guidelines in 2015, which allow only institutions that provide medical care, education, and research to conduct clinical trials that must be approved by the NHFPC and the China Food and Drug Administration (CFDA) as well as registered on the Chinese Medicine Registry and Management System (Rosemann and Sleeboom-Faulkner 2016). However, China's geographic size and population make enforcement challenging and, in 2020, more than 2500 unproven SCI clinics were identified on Baidu (China's largest search engine) (Lv et al. 2020; Sleeboom-Faulkner et al. 2018; Zhang 2017).

India and Australia have also tightened standards that previously allowed autologous SCIs to go unregulated. The Indian Council for Medical Research (ICMR) revised autologous SCI guidelines in 2017 to require approval by the Central Drugs Standard Control Organization (CDSCO) prior to use in a clinical setting (Rosemann et al. 2019; Jose et al. 2020). In Australia, the Therapeutics Goods Administration (TGA) also updated its regulations in 2017 creating three categories of autologous SCIs—excluded, exempt, and fully regulated (Ghinea et al. 2020). Only procedures in hospitals can qualify for the exclusion category and exempt procedures have to be minimally-manipulated and intended for homologous use. All other autologous SCIs fall under the "fully regulated" category. In addition, Australia banned DTC advertising of autologous SCIs, albeit with limited success as clinics still use online advertising to patients (Rudge et al. 2021).

6.6.3 Economic Competition and Conditional Approval

While some countries have tightened their regulations, others have arguably loosened restrictions as part of efforts to promote innovation and national prosperity. South Korea was one of the first countries that created an accelerated pathway for SCIs and approved several stem cell-based products for the market (Sipp and Sleeboom-Faulkner 2019). Following the Korean lead, Japan created a new medical category, "regenerative medicine products" with conditional approval, allowing products to be marketed for up to seven years while safety and efficacy data is collected (Sipp et al. 2017; Sipp and Sleeboom-Faulkner 2019; Takashima et al. 2021; Azuma 2015; Konomi et al. 2015; Lysaght 2017; Sipp 2015; Tobita et al. 2016). Some scholars regard conditional approval as a pay-to-participate scheme to get patients to fund the clinical trial research and assumes clinical data is similar or better than clinical trial data, which has not been shown to be the case (Sipp 2012a; Lee and Lysaght 2018). In a similar effort, the United States created the new Regenerative Medicine Advanced Therapy (RMAT) designation in 2018 to accelerate stem cell therapies through the clinical review process (Sipp and Sleeboom-Faulkner 2019; Knoepfler 2018).

6.6.4 Beyond National Regulations

National regulations are limited to enforcement only in the territory under their jurisdiction, which allows clinics to exploit regulatory differences to find ways to provide unproven SCIs (Sipp 2017a; Sipp et al. 2017; Kuriyan et al. 2017; Master and Resnik 2011). Solving this constraint requires coordination between national agencies through a global organization, such as the World Health Organization (WHO) (Sipp 2017a; Master et al. 2021). A WHO Expert Advisory Committee on Regenerative Medicine similar to that developed to address heritable genome editing could harmonize national regulations, promote regulation responsive to unmet patient needs, and develop a patient and physician education campaign (Master et al. 2021).

6.7 Conclusion

Finding ways to encourage innovation while hindering unproven and unsafe interventions has proved challenging around the globe (Lindvall and Hyun 2009). Policy scholar Doug Sipp views unproven stem cell clinics as a malignant tumor, while tumors are sustained in environments "that support and protect malignant cells, thus conferring a competitive advantage against both healthy cells and therapeutic interventions," so too has the SCI industry "developed a number of self-protective strategies that support its survival and growth" (Sipp 2017a). By promoting hope to patients, using exaggerated claims about the promise of stem cell research, promoting an unrealistic timeline for innovation, scientists found their research vulnerable both to actors working in the patients' interest, trying to provide an innovative but untested therapy, as well as less ethical characters interested only in profit (Sleeboom-Faulkner et al. 2016). As a result, the unproven SCI market has been expanding for more than two decades leading to physical, financial, and emotional harms to patients (The PEW Charitable Trusts 2021; Bauer et al. 2018; Pritchett 2021).

Efforts to restrict unproven SCI clinics need to proceed at all levels. Local officials should work to shut down clinics that are proven unsafe and limit DTC marketing claims to patients. At the national level, regulatory agencies need to set clear and comprehensive guidelines for what SCIs can and cannot be performed and work with local officials to properly enforce them (Zarzeczny et al. 2018; Sipp 2013). And at the international level, national regulatory agencies should collaborate to develop a comprehensive governance framework and work to enforce it consistently, perhaps relying on the WHO to convene interested stakeholders (Master et al. 2021; Zarzeczny et al. 2018). Coordination should focus on patient protection efforts and discourage national competition in the regenerative medicine space that often leads to loosening instead of enforcing regulation. It should also remove loopholes that allow unscrupulous actors to move between jurisdictions to avoid regulations (Sipp et al. 2017).

In addition, better tools are needed to identify and remove misleading and scientifically incorrect information from social media outlets, video repositories, clinical trial databases, and crowdfunding campaigns. While Google and a few internet-based search companies have started to promote reliable sources (such as regulatory agencies and science societies), social media platforms are notorious for perpetuating science myths and misinformation. Finding mechanisms to flag, tag, or report misinformation online could help remove claims by or for unproven SCI clinics. In the United States, the ClinicalTrials.gov reporting system should be updated to limit its manipulation by unproven SCI clinics and improve the ability to report questionable or unethical studies listed. An improved ClinicalTrials.gov system should provide: straightforward mechanisms for reporting questionable studies; information on the regulatory status of trials listed; processes for flagging informed consent and Institutional Review Board (IRB) issues within studies; resources for reporting including state boards; and the studies' investigational new drug (IND) application number (to verify it was approved by the FDA) (Zarzeczny et al. 2018; Wagner et al. 2018).

Furthermore, scientists and medical societies along with patient advocate groups should work with the media to spotlight the harms that unproven SCI clinics are causing and promote more legitimate forms of SCIs, such as clinical trials. In addition, increased physician access to education material such as websites, outreach materials, and counseling services is needed to help physicians explain risk to patients. Medical societies should also penalize members who provide unproven SCIs, especially those who try to avoid regulation by traveling to less regulated countries to conduct procedures not permitted in their home country (Chan et al. 2017).

Ultimately, regulating this industry will require engagement from stakeholders including scientists, physicians, patients and their advocates, regulatory entities, and funders (Sipp et al. 2017; Matthews and Iltis 2015). Engagement between these groups will be necessary to determine how to harmonize international policies in order to satisfy patient needs, encourage robust clinical data and oversight of the industry, and ensure all relevant advocates are working towards a common goal: to create proven, effective and safe SCIs.

References

Amariglio, N., A. Hirshberg, B.W. Scheithauer, et al. 2009. Donor-derived brain tumor following neural stem cell transplantation in an Ataxia Telangiectasia patient. *PLoS Medicine* 6 (2): e1000029. https://doi.org/10.1371/journal.pmed.1000029.

American Medical Association (AMA). Opinion 1.2.13 medical tourism. In *Code of medical ethics: Patient-physician relationships.* https://www.ama-assn.org/delivering-care/ethics/med ical-tourism

Azuma, K. 2015. Regulatory landscape of regenerative medicine in Japan. *Current Stem Cell Reports* 1: 118–128. https://doi.org/10.1007/s40778-015-0012-6.

Baker, M. 2005. Stem cell therapy or snake oil. *Nature* 23: 1467–1469. https://doi.org/10.1038/nbt 1205-1467.

Ballantyne, A. 2019. Adjusting the focus: A public health ethics approach to data research. *Bioethics* 33 (3), 357–366. https://doi.org/10.1111/bioe.12551

Bauer, G., M. Abou-el-Enein, A. Kent, et al. 2017. The path to successful commercialization of cell and gene therapies: empowering patient advocates. *Cytotherapy* 19 (2): 293–298. https://doi.org/ 10.1016/j.jcyt.2016.10.017.

Bauer, G., M. Elsallab, and M. Abou-el-Enein. 2018. Concise review: A comprehensive analysis of reported adverse events in patients receiving unproven stem cell-based interventions. *Stem Cells Translational Medicine* 7 (9): 676–685. https://doi.org/10.1002/sctm.17-0282.

Bauman J. 2021. Stem cell clinics face mounting attack from Federal Enforcers. *Bloomberg Law.* August 20, 2021. https://news.bloomberglaw.com/pharma-and-life-sciences/stem-cell-cli nics-face-mounting-attack-from-federal-enforcers

Beil, L. 2020. How unproven stem cell therapies are costing desperate patients. *Texas Monthly.* January 2020. https://www.texasmonthly.com/news-politics/how-unproven-stem-cell-therapies-costing-desperate-patients/

Berfield, S. 2013. CellTex says it's moving its stem cell business to Mexico. *Bloomberg Businessweek.* January 31, 2013. http://www.businessweek.com/articles/2013-01-31/celltex-wants-to-bring-its-stem-cell-business-to-mexico

Berger, I., A. Ahmad, A. Bansal, et al. 2016. Global distribution of businesses marketing stem cell-based interventions. *Cell Stem Cell* 19 (2): 158–162. https://doi.org/10.1016/j.stem.2016. 07.015.

Bianco, P., and D. Sipp. 2014. Regulation: Sell help not hope. *Nature* 510 (7505): 336–337. https:// doi.org/10.1038/510336a.

Bowman, M., M. Racke, J. Kissel, and J. Imitola. 2015. Responsibilities of health care professionals in counseling and educating patients with incurable neurological diseases regarding "stem cell tourism" *JAMA Neurology* 72 (11): 1342–1345. https://doi.org/10.1001/jamaneurol.2015.1891

Brown, W., M. Basil, and M. Bocarnea. 2003. The influence of famous athletes on health beliefs and practices: Mark McGwire, child abuse prevention, and androstenedione. *Journal of Health Communication* 8 (1): 41–57. https://doi.org/10.1080/10810730305733.

Bubela, T., M. Nisbet, R. Borchelt, et al. 2009. Science communication reconsidered. *Nature Biotechnology* 27 (6): 514–518. https://doi.org/10.1038/nbt0609-514.

Bubela, T., M. Li, M. Hafez, et al. 2012. Is belief larger than fact: Expectations, optimism and reality for translational stem cell research. *BMC Medicine* 10 (1): 133. https://doi.org/10.1186/ 1741-7015-10-133.

Cattaneo, E., and G. Corbellini. 2014. Stem cells: Taking a stand against pseudoscience. *Nature* 510 (7505): 333–335. https://doi.org/10.1038/510333a.

Caulfield, T. 2012. What does it mean when athletes get 'stem cell therapy'? *The Atlantic* October 22, 2012. https://www.theatlantic.com/health/archive/2012/10/what-does-it-mean-when-athletes-get-stem-cell-therapy/263875/

Caulfield, T., and C. Condit. 2012. Science and the sources of hype. *Public Health Genomics* 15 (3–4): 209–217. https://doi.org/10.1159/000336533.

Caulfield, T., and A. McGuire. 2012. Athletes' use of unproven stem cell therapies: Adding to inappropriate media hype? *Molecular Therapy* 20 (9): 1656–1658. https://doi.org/10.1038/mt.2012.172.

Caulfield, T., and B. Murdoch. 2019. Regulatory and policy tools to address unproven stem cell interventions in Canada: The need for action. *BMC Medical Ethics* 20: 51. https://doi.org/10.1186/s12910-019-0388-4.

Caulfield, T., D. Sipp, C. Murry, et al. 2016. Confronting stem cell hype. *Science* 352 (6287): 776–777. https://doi.org/10.1126/science.aaf4620.

Centeno, C.J., M.A. Jerome, S.M. Pastoriza, et al. 2021. Use of Bone Marrow Concentrate to Treat Pain and Musculoskeletal Disorders: An Academic Delphi Investigation. *Pain Physician* 24 (3): 263–273.

Chan, S., C. Palacios-González, and M. Arellano. 2017. Mitochondrial replacement techniques, scientific tourism, and the global politics of science. *Hastings Center Report* 47 (5): 7–9. https://doi.org/10.1002/hast.763.

Chavez, J., N.A. Shah, S. Ruoss, et al. 2021. Online marketing practices of regenerative medicine clinics in US-Mexico Border Region: A web surveillance study. *Stem Cell Research & Therapy* 12 (1): 189. https://doi.org/10.1186/s13287-021-02254-4.

Chennells, R., and A. Steenkamp. 2018. International genomics research involving the San People. In *Ethics dumping*, ed. D. Schroeder, J. Cook, F. Hirsch, S. Fenet, and V. Muthuswamy, pp. 15–22. Springer International Publishing. https://doi.org/10.1007/978-3-319-64731-9_3

Collins, R., L. Bowman, M. Landray, and R. Peto. 2020. The magic of randomization versus the myth of real-world evidence. *New England Journal of Medicine* 382 (7): 674–678. https://doi.org/10.1056/nejmsb1901642.

Connolly R., T. O'Brien, G. Flaherty. 2014. Stem cell tourism—A web-based analysis of clinical services available to international travelers. *Travel Medicine and Infectious Disease* 12 (6, Part B): 695–701. https://doi.org/10.1016/j.tmaid.2014.09.008.

Cossu, G., M. Birchall, T. Brown, et al. 2018. Lancet commission: Stem cells and regenerative medicine. *Lancet* 391 (10123): 883–910. https://doi.org/10.1016/S0140-6736(17)31366-1.

Daley, G. 2012. The promise and perils of stem cell therapeutics. *Cell Stem Cell* 10 (6): 740–749. https://doi.org/10.1016/j.stem.2012.05.010.

Dlouhy, B.J., O. Awe, R.C. Rao, et al. 2014. Autograft-derived spinal cord mass following olfactory mucosal cell transplantation in a spinal cord injury patient: Case report. *Journal of Neurosurgery* 21 (4): 618–622. https://doi.org/10.3171/2014.5.SPINE13992.

Dominici, M., K. Nichols, A. Srivastava, et al. 2015. Positioning a scientific community on unproven cellular therapies: The 2015 international society for cellular therapy perspective. *Cytotherapy* 17 (12): 1663–1666. https://doi.org/10.1016/j.jcyt.2015.10.007.

Du, L., C. Rachul, Z. Guo, et al. 2016. Gordie Howe's "miraculous treatment": Case study of Twitter users' reactions to a sport celebrity's stem cell treatment. *JMIR Public Health and Surveillance* 2 (1): e8. https://doi.org/10.2196/publichealth.5264.

Enserink, M. 2006. Selling the stem cell dream. *Science* 313 (5784): 160–163. https://doi.org/10.1126/science.313.5784.160.

Erikainen, S., A. Couturier, and S. Chan. 2020. Marketing experimental stem cell therapies in the UK: Biomedical lifestyle products and the promise of regenerative medicine in the digital era. *Science and Culture* 29 (2): 219–244. https://doi.org/10.1080/09505431.2019.1656183.

Fears, R., H. Akutsu, L. Alentajan-Aleta, et al. 2021. Inclusivity and diversity: Integrating international perspectives on stem cell challenges and potential. *Stem Cell Reports* 16 (8): 1847–1852. https://doi.org/10.1016/j.stemcr.2021.07.003.

Frow, E., D. Brafman, A. Muldoon, et al. 2019. Characterizing direct-to-consumer stem cell businesses in the Southwest United States. *Stem Cell Reports* 13 (2): 247–253. https://doi.org/10.1016/j.stemcr.2019.07.001.

Fu, W., C. Smith, L. Turner, et al. 2019. Characteristics and scope of training of clinicians participating in the US direct-to-consumer marketplace for unproven stem cell interventions. *JAMA* 321 (24): 2463–2464. https://doi.org/10.1001/jama.2019.5837.

Fujita, M., T. Hatta, R. Ozeki, et al. 2016. The current status of clinics providing private practice cell therapy in Japan. *Regenerative Medicine* 11 (1): 23–32. https://doi.org/10.2217/rme.15.64.

Ghinea, N., M. Munsie, C. Rudge, et al. 2020. Australian regulation of autologous human cell and tissue products: Implications for commercial stem cell clinics. *Regenerative Medicine* 15 (2): 1361–1369. https://doi.org/10.2217/rme-2019-0124.

International Society for Stem Cell Research (ISSCR). 2021. Guidelines for stem cell research and clinical translation. https://www.isscr.org/policy/guidelines-for-stem-cell-research-and-clinical-translation

Gunter, K., A. Caplan, C. Mason, et al. 2010. Cell therapy medical tourism: Time for action. *Cytotherapy* 12 (8): 965–968. https://doi.org/10.3109/14653249.2010.532663.

Hawke, B., A.R. Przybylo, D. Paciulli, et al. 2019. How to Peddle Hope: An analysis of YouTube patient testimonials of unproven stem cell treatments. *Stem Cell Reports* 12 (6): 1186–1189. https://doi.org/10.1016/j.stemcr.2019.05.009.

Holzer, F., and I. Mastroleo. 2019. Innovative practice in Latin America: Medical tourism and the crowding out of research. *The American Journal of Bioethics* 19 (6): 42–44. https://doi.org/10.1080/15265161.2019.1602189.

Horner, C., E. Tenenbaum, D. Sipp, et al. 2018. Can civil lawsuits stem the tide of direct-to-consumer marketing of unproven stem cell interventions. *NPJ Regenerative Medicine* 3 (1): 1–5. https://doi.org/10.1038/s41536-018-0043-6.

Hyun, I. 2013. Therapeutic hope, spiritual distress, and the problem of stem cell tourism. *Cell Stem Cell* 12 (5): 505–507. https://doi.org/10.1016/j.stem.2013.04.010.

Ikka, T., M. Fujita, Y. Yashiro, et al. 2015. Recent court ruling in Japan exemplifies another layer of regulation for regenerative therapy. *Cell Stem Cell* 17 (5): 507–508. https://doi.org/10.1016/j.stem.2015.10.008.

Ikonomou, L., R. Freishtat, D. Wagner, et al. 2016. The global emergence of unregulated stem cell treatments for respiratory diseases. Professional societies need to act. *Annals of the American Thoracic Society* 13 (8): 1205–1207. https://doi.org/10.1513/AnnalsATS.201604-277ED.

InterAcademy Partnership (IAP). 2021. IAP statement on regenerative medicine. https://www.interacademies.org/statement/iap-statement-regenerative-medicine

ISSCR. 2008. Guidelines for the clinical translation of stem cells. https://www.isscr.org/docs/default-source/all-isscr-guidelines/clin-trans-guidelines/isscrglclinicaltrans.pdf?sfvrsn=fd1fa5c8_6

ISSCR. 2016. Guidelines for stem cell research and clinical translation. https://www.isscr.org/policy/guidelines-for-stem-cell-research-and-clinical-translation/guidelines-archive

Jose, J., T. George, and A. Thomas. 2020. Regulation of stem cell-based research in India in comparison with the US, EU and other Asian countries: Current issues and future perspectives. *Current Stem Cell Research & Therapy* 15: 492–508. https://doi.org/10.2174/1574888X15666200402134750.

Julian, K., N. Yahasz, R. Widjan, et al. 2020. Complications from "stem cell tourism" in neurology. *Annals of Neurology* 88 (4): 661–668. https://doi.org/10.1002/ana.25842.

Kabat, M., I. Bobkov, S. Kumar, and M. Grumer. 2020. Trends in mesenchymal stem cell clinical trials 2004–2018: Is efficacy optimal in a narrow dose range? *Stem Cells Translational Medicine* 9 (1): 17–27. https://doi.org/10.1002/sctm.19-0202

Kamenova, K., and T. Caulfield. 2015. Stem cell hype: Media portrayal of therapy translation. *Science Translational Medicine* 7 (278): 278ps4. https://doi.org/10.1126/scitranslmed.3010496

Kiatpongsan, S., and D. Sipp. 2008. Offshore stem cell treatments. *Nature Reports Stem Cells*. https://doi.org/10.1038/stemcells.2008.151.

Knoepfler, P. 2013. Call for fellowship programs in stem cell-based regenerative and cellular medicine: New stem cell training is essential for physicians. *Regenerative Medicine* 8 (2): 223–225. https://doi.org/10.2217/rme.13.1.

Knoepfler, P. 2015. From bench to FDA to bedside: US regulatory trends for new stem cell therapies. *Advanced Drug Delivery Reviews* 82–83 (March): 192–196. https://doi.org/10.1016/j.addr.2014.12.001.

Knoepfler, P. 2018. Too much carrot and not enough stick in new stem cell oversight trends. *Cell Stem Cell* 23 (1): 18–20. https://doi.org/10.1016/j.stem.2018.06.004.

Knoepfler, P. 2019. Rapid change of a Cohort of 570 unproven stem cell clinics in the USA over 3 years. *Regenerative Medicine* 14 (8): 735–740. https://doi.org/10.2217/rme-2019-0064.

Knoepfler, P., and L. Turner. 2018. The FDA and the US direct-to-consumer marketplace for stem cell interventions: A temporal analysis. *Regenerative Medicine* 13 (1): 19–27. https://doi.org/10.2217/rme-2017-0115.

Konomi, K., M. Tobita, K. Kimura, and D. Sato. 2015. New Japanese initiatives on stem cell therapies. *Cell Stem Cell* 16 (4): 350–352. https://doi.org/10.1016/j.stem.2015.03.012.

Kuriyan, A.E., T.A. Albini, J.H. Townsend, et al. 2017. Vision loss after intravitreal injection of autologous "stem cells" for AMD. *New England Journal of Medicine* 376 (11): 1047–1053. https://doi.org/10.1056/NEJMoa1609583.

Lau, D., U. Ogbogu, B. Taylor, et al. 2008. Stem cell clinics online: The direct-to-consumer portrayal of stem cell medicine. *Cell Stem Cell* 3 (6): 591–594. https://doi.org/10.1016/j.stem.2008.11.001.

Lee, T., and T. Lysaght. 2018. Conditional approvals for autologous stem cell-based interventions: Conflicting norms and institutional legitimacy. *Perspectives in Biology and Medicine* 61 (1): 59–75. https://doi.org/10.1353/pbm.2018.0027.

Levine, A.D. 2010. Insights from patients' blogs and the need for systematic data on stem cell tourism. *American Journal of Bioethics* 10 (5): 28–29. https://doi.org/10.1080/15265161003686571.

Levine, A.D., and L.E. Wolf. 2012. The roles of responsibilities of physicians in patients' decisions about unproven stem cell therapies. *Journal of Law, Medicine & Ethics* 40 (1): 122–134. https://doi.org/10.1111/j.1748-720x.2012.00650.x.

Li, M.D., H. Atkins, and T. Bubela. 2014. The global landscape of stem cell clinical trials. *Regenerative Medicine* 9 (1): 27–39. https://doi.org/10.2217/RME.13.80.

Liang, B., T. Mackey. 2012. Stem cells, Dot-Com. *Science Translational Medicine* 4 (151): 151cm9. https://doi.org/10.1126/scitranslmed.3004030.

Lindvall, O., and I. Hyun. 2009. Medical innovation versus stem cell tourism. *Science* 324 (5935): 1664–1665. https://doi.org/10.1126/science.1171749.

Lv J, Y. Su L. Song, et al. 2020. Stem cell "therapy" advertisements in China: Infodemic, regulations and recommendations. *Cell Proliferation* 53 (12): e12937. https://doi.org/10.1111/cpr.12937.

Lysaght, T. 2017. Accelerating regenerative medicine: The Japanese experiment in ethics and medicine. *Regenerative Medicine* 12 (6): 657–668. https://doi.org/10.2217/rme-2017-0038.

Lysaght, T., and A.V. Campbell. 2011. Regulating autologous adult stem cells: The FDA steps up. *Cell Stem Cell* 9 (5): 393–396. https://doi.org/10.1016/j.stem.2011.09.013.

Lysaght, T., I. Kerridge, D. Sipp, et al. 2013. Oversight for clinical uses of autologous adult stem cells: Lessons from international regulations. *Cell Stem Cell* 13 (6): 647–651. https://doi.org/10.1016/j.stem.2013.11.013.

Lysaght, T., W. Lipworth, T. Hendl, et al. 2017. The deadly business of an unregulated global stem cell industry. *Journal of Medical Ethics* 43 (11): 744–746. https://doi.org/10.1136/medethics-2016-104046.

Marcon, A.R., B. Murdoch, and T. Caulfield. 2017. Fake news portrayals of stem cells and stem cell research. *Regenerative Medicine* 12 (7): 765–775. https://doi.org/10.2217/rme-2017-0060.

Marcon, A.R., D. Allan, M. Barber, et al. 2020. Portrayal of umbilical cord blood research in the North American Popular Press: Promise or hype? *Regenerative Medicine* 15 (1): 1228–1237. https://doi.org/10.2217/rme-2019-0149.

Margottini L. 2014. Final chapter in Italian stem cell controversy? *Science* October 7, 2014. https://www.science.org/content/article/final-chapter-italian-stem-cell-controversy

Marks, P.W., and S. Hahn. 2020. Identifying the risks of unproven regenerative medicine therapies. *JAMA* 324 (3): 241–242. https://doi.org/10.1001/jama.2020.9375.

Marks, P.W., C.M. Witten, and R.M. Califf. 2017. Clarifying stem-cell therapy's benefits and risks. *New England Journal of Medicine* 376 (11): 1007–1009. https://doi.org/10.1056/NEJMp1613723.

Marks, P.W., and S. Gottlieb. 2018. Balancing safety and innovation for cell-based regenerative medicine. *New England Journal of Medicine* 378 (10): 954–959. https://doi.org/10.1056/NEJ Msr1715626.

Martinho, A., and L. Turner. 2017. Stem cells in court: Historical trends in US legal cases related to stem cells. *Regenerative Medicine* 12 (4): 419–430. https://doi.org/10.2217/rme-2017-0002.

Master, Z., and U. Ogbogu. 2012. Stem cell tourism in the era of personalized medicine: What we know, and what we need to know. *Current Pharmacogenomics and Personalized Medicine* 10 (2): 106–110. https://doi.org/10.2174/187569212800626340.

Master, Z., and D. Resnik. 2011. Stem-cell tourism and scientific responsibility. *EMBO Reports* 12 (10): 992–995. https://doi.org/10.1038/embor.2011.156.

Master, Z., and D. Resnik. 2013. Hype and public trust in science. *Science and Engineering Ethics* 19: 321–335. https://doi.org/10.1007/s11948-011-9327-6.

Master, Z., A. Zarzeczny, C. Rachul, and T. Caulfield. 2013. What's missing? Discussing stem cell translational research in educational information on stem cell "tourism." *The Journal of Law, Medicine & Ethics* 41 (1): 254–268. https://doi.org/10.1111/jlme.12017.

Master, Z., K. Robertson, and D. Frederick. 2014. Stem cell tourism and public education: The missing elements. *Cell Stem Cell* 15 (3): 267–270. https://doi.org/10.1016/j.stem.2014.08.009.

Master, Z., K.R.W. Matthews, and M. Abou-el-Enein. 2021. Unproven stem cell interventions: A global public health problem requiring global deliberation. *Stem Cell Reports* 16 (6): 1435–1445. https://doi.org/10.1016/j.stemcr.2021.05.004.

Matthews, K.R.W., and M. Cuchiara. 2014. U.S. National Football League Athletes seeking unproven stem cell treatments. *Stem Cells and Development* 23 (Suppl 1): 60–64. https://doi.org/10.1089/scd.2014.0358.

Matthews, K.R.W., and A. Iltis. 2017. Unproven stem cell–based interventions: Advancing policy through stakeholder collaboration. *Texas Heart Institute Journal* 44 (3): 171–173. https://doi.org/10.14503/THIJ-17-6244.

Matthews, K.R.W., B. Kunisetty, and K. Sprung. 2018. Texas H.B. 810: Increased access to stem cell interventions or an increase in unproven treatments? *Stem Cells and Development* 27 (21): 1463–1465. https://doi.org/10.1089/scd.2018.0148.

Matthews, K.R.W., and A. Iltis. 2015. Unproven stem cell-based interventions and achieving a compromise policy among the multiple stakeholders. *BMC Medical Ethics* 16 (1): 75. https://doi.org/10.1186/s12910-015-0069-x.

McLean, A., C. Stewart, and I. Kerridge. 2015. Untested, unproven, and unethical: The promotion and provision of autologous stem cell therapies in Australia. *Stem Cell Research & Therapy* 6 (1): 33. https://doi.org/10.1186/s13287-015-0047-8.

McMahon, D. 2014. The global industry for unproven stem cell interventions and stem cell tourism. *Tissue Engineering and Regenerative Medicine* 11 (1): 1–9. https://doi.org/10.1007/s13770-013-1116-7.

Michie, M., M. Allyse, K.A. Stoll, et al. 2018. Weaponizing hope: Sources of hope, unrealistic optimism, and Denial. *American Journal of Bioethics* 18 (9): 25–27. https://doi.org/10.1080/152 65161.2018.1498946.

Munsie, M., T. Lysaght, T. Hendl, et al. 2017. Open for business: A comparative study of websites selling autologous stem cells in Australia and Japan. *Regenerative Medicine* 12 (7): 777–790. https://doi.org/10.2217/rme-2017-0070.

Murdoch, C.E., and C.T. Scott. 2010. Stem cell tourism and the power of hope. *American Journal of Bioethics* 10 (5): 16–23. https://doi.org/10.1080/15265161003728860.

Murdoch, B., A.R. Marcon, D. Downie, et al. 2019. Media portrayal of illness-related medical crowdfunding: A content analysis of newspaper articles in the United States and Canada. *PLoS ONE* 14 (4): e0215805. https://doi.org/10.1371/journal.pone.0215805.

Ogbogu, U., C. Rachul, and T. Caulfield. 2013. Reassessing direct-to-consumer portrayals of unproven stem cell therapies: Is it getting better? *Regenerative Medicine* 8 (3): 361–369. https://doi.org/10.2217/rme.13.15.

Olmedo-Reneaum, A., I. Garcia-Juarez, L. Toapanta-Yanchapaxi, et al. 2019. Rogue "stem cell clinic" leads to Mycobacterium abscessus infection. *The Lancet* 393 (10174): 918. https://doi.org/10.1016/S0140-6736(19)30299-5.

Perkins, K.M., S. Spoto, D.A. Rankin, et al. 2018. Notes from the field: Infections after receipt of bacterially contaminated umbilical cord blood–derived stem cell products for other than hematopoietic or immunologic reconstitution—United States, 2018. *Morbidity and Mortality Weekly Report* 67 (50): 1397–1399. https://doi.org/10.15585/mmwr.mm6750a5.

Petersen, A., K. Seear, and M. Munsie. 2013. Therapeutic journeys: The hopeful travails of stem cell tourists. *Sociology of Health and Illness* 36 (5): 670–685. https://doi.org/10.1111/1467-9566.12092.

Petersen, A., C. MacGregor, and M. Munsie. 2015. Stem cell miracles or Russian Roulette? Patients' use of digital media to campaign for access to clinically unproven treatments. *Health, Risk & Society* 17 (7–8): 592–604. https://doi.org/10.1080/13698575.2015.1118020.

Pritchett, J.W. 2021. The debit side of stem-cell joint injections: A prospective cohort study. *Current Orthopaedic Practice* 32 (2): 118–123. https://doi.org/10.1097/BCO.0000000000000961.

Rachul C. 2011. "What have I got to lose?" An analysis of stem cell therapy patients' blogs. *Health Law Review* 20 (1): 5–12. https://heinonline.org/HOL/P?h=hein.journals/hthlr20&i=5.

Regenberg, A., L. Hutchinson, and B. Schanker. 2009. Medicine on the fringe: Stem cell-based interventions in advance of evidence. *Stem Cells* 27 (9): 2312–2319. https://doi.org/10.1002/stem.132.

Richardson, E., F. Akkas, and Z. Master. 2020. Evaluating the FDA regenerative medicine framework: Opportunities for stakeholders. *Regenerative Medicine* 15 (7): 1825–1832. https://doi.org/10.2217/rme-2020-0073.

Rosemann, A., G. Bortz, and F. Vasen. 2019. Regulatory developments for nonhematopoietic stem cell therapies: Perspectives from the EU, the USA, Japan, China, India, Argentina and Brazil. In *A roadmap to nonhematopoeitic stem cell-based therapeutics*, ed. X.-D. Chen, 463–492. London: Academic Press. https://doi.org/10.1016/B978-0-12-811920-4.00019-7.

Rosemann, A., and M. Sleeboom-Faulkner. 2016. New regulation for clinical stem cell research in China: Expected impact and challenges for implementation. *Regenerative Medicine* 11 (1): 5–9. https://doi.org/10.2217/rme.15.80.

Rudge, C., N. Ghinea, M. Munsie, and C. Stewart. 2021. Regulating autologous stem cell interventions in Australia: Updated review of the direct-to-consumer advertising restrictions. *Australian Health Review* 45 (4): 507–515. https://doi.org/10.1071/ah20217.

Ryan, K.A., A.N. Sanders, D.D. Wang, et al. 2009. Tracking the rise of stem cell tourism. *Regenerative Medicine* 5 (1): 27–33. https://doi.org/10.2217/rme.09.70.

Scott, C.T. 2015. The case for stem cell counselors. *Stem Cell Reports* 4 (1): 1–6. https://doi.org/10.1016/j.stemcr.2014.10.016.

Sipp, D. 2012a. Pay-to-participate funding schemes in human cell and tissue clinical studies. *Regenerative Medicine* 7 (6S): 105–111. https://doi.org/10.2217/rme.12.75.

Sipp, D. 2012b. Converging ideological current in adult stem cell marketing phenomenon. *Ethics in Biology, Engineering and Medicine* 3 (4): 275–286. https://doi.org/10.1615/EthicsBiologyEngMed.2013007599.

Sipp, D. 2013. Direct-to-consumer stem cell marketing and regulatory responses. *Stem Cells Translational Medicine* 2 (9): 638–640. https://doi.org/10.5966/sctm.2013-0040.

Sipp, D. 2015. Conditional approval: Japan lower the bar for regenerative medicine products. *Cell Stem Cell* 16 (4): 353–356. https://doi.org/10.1016/j.stem.2015.03.013.

Sipp, D. 2017a. The Malignant Niche: Safe spaces for toxic stem cell marketing. *NPJ Regenerative Medicine* 2 (December): 33. https://doi.org/10.1038/s41536-017-0036-x.

Sipp, D. 2017b. Identity and ownership issues in the regulation of autologous cells. *Regenerative Medicine* 12 (7): 827–838. https://doi.org/10.2217/rme-2017-0063.

Sipp, D. 2018. Stem cell mismarketing: Implications for the transfusion community. *ISBT Science Series* 14 (1): 45–48. https://doi.org/10.1111/voxs.12457.

Sipp, D., and M.E. Sleeboom-Faulkner. 2019. Downgrading of regulation in regenerative medicine. *Science* 365 (6454): 644–646. https://doi.org/10.1126/science.aax6184.

Sipp, D., and L. Turner. 2012. Stem cells: U.S. regulation of stem cells as medical products. *Science* 338: 1296–1297. https://doi.org/10.1126/science.1229918.

Sipp, D., L. Turner, and J.E.J. Rasko. 2019. Stem cell businesses and right to try laws. *Cell Stem Cell* 25 (3): 304–305. https://doi.org/10.1016/j.stem.2019.08.012.

Sipp D, T. Caulfield, J. Kaye, et al. 2017. Marketing of unproven stem cell–based interventions: A call to action. *Science Translational Medicine* 9 (397): eaag0426. https://doi.org/10.1126/scitra nslmed.aag0426.

Sleeboom-Faulkner, M.E. 2016. The large grey area between 'Bona Fide' and 'Rogue' stem cell interventions—Ethical acceptability and the need to include local variability. *Technological Forecasting and Social Change* 109 (August): 76–86. https://doi.org/10.1016/j.techfore.2016. 04.023.

Sleeboom-Faulkner, M., C. Chekar, A. Faulkner, et al. 2016. Comparing national home-keeping and the regulation of translational stem cell applications: An international perspective. *Social Science and Medicine* 153 (March): 240–249. https://doi.org/10.1016/j.socscimed.2016.01.047.

Sleeboom-Faulkner, M., H. Chen, and A. Rosemann. 2018. Regulatory capacity building and the governance of clinical stem cell research in China. *Science and Public Policy* 45 (3): 416–427. https://doi.org/10.1093/scipol/scx077.

Smith, C., C. Martin-Lillie, J.D. Higano, et al. 2020. Challenging misinformation and engaging patients: Characterizing a regenerative medicine consult service. *Regenerative Medicine* 15 (3): 1427–1440. https://doi.org/10.2217/rme-2020-0018.

Smith, C., A. Crowley, M. Munsie, et al. 2021. Academic physician specialists' views toward the unproven stem cell intervention industry: Areas of common ground and divergence. *Cytotherapy* 23 (4): 348–356. https://doi.org/10.1016/j.jcyt.2020.12.011.

Snyder, J., and L. Turner. 2018. Selling stem cell 'treatments' as research: Prospective customer perspectives from crowdfunding campaigns. *Regenerative Medicine* 13 (4): 375–384. https://doi. org/10.2217/rme-2018-0007.

Snyder, J., and L. Turner. 2021. Crowdfunding, stem cell interventions and autism spectrum disorder: Comparing campaigns related to an international 'stem cell clinic' and US Academic Medical Center. *Cytotherapy* 23 (3): 198–202. https://doi.org/10.1016/j.jcyt.2020.09.002.

Snyder, J., L. Turner, and V.A. Crooks. 2018. Crowdfunding for unproven stem cell-based interventions. *JAMA* 319 (18): 1935–1936. https://doi.org/10.1001/jama.2018.3057.

Srivastava, A., C. Mason, E. Wagena, et al. 2016. Part 1: Defining unproven cellular therapies. *Cytotherapy* 18 (1): 117–119. https://doi.org/10.1016/j.jcyt.2015.11.004.

Sugarman, J., R.A. Barker, I. Kerridge, et al. 2018. Tackling ethical challenges of premature delivery of stem cell-based therapies: ISSCR 2018 annual meeting focus session report. *Stem Cell Reports* 11 (5): 1021–1025. https://doi.org/10.1016/j.stemcr.2018.10.020.

Takashima, K., M. Morrison, and J. Minari. 2021. Reflection on the enactment and impact of safety laws for regenerative medicine in Japan. *Stem Cell Reports* 16 (6): 1425–1434. https://doi.org/ 10.1016/j.stemcr.2021.04.017.

Tanner, C., A. Petersen, and M. Munsie. 2017. No one here's helping me, what do you do?' Addressing patient need for support and advice about stem cell treatments. *Regenerative Medicine* 12 (7): 791–801. https://doi.org/10.2217/rme-2017-0056.

Taylor, P.L., R. Barker, K.G. Blume, et al. 2010. Patients beware: Commercialized stem cell treatments on the web. *Cell Stem Cell* 7 (1): 43–49. https://doi.org/10.1016/j.stem.2010.06.001.

Taylor-Weiner, H., and J.G. Zivin. 2015. Medicine's Wild West—Unlicensed stem-cell clinics in the United States. *New England Journal of Medicine* 373: 985–987. https://doi.org/10.1056/NEJ Mp1504560.

The PEW Charitable Trusts. 2021. Harms linked to unapproved stem cell interventions highlight need for greater FDA enforcement. *The PEW Charitable Trusts* June 1, 2021. https://www.pewtrusts.org/en/research-and-analysis/issue-briefs/2021/06/harms-linked-to-unapproved-stem-cell-interventions-highlight-need-for-greater-fda-enforcement

Tobita, M., K. Konomi, Y. Torashima, et al. 2016. Japan's challenges of translational regenerative medicine: Act on the safety of regenerative medicine. *Regenerative Therapy* 4: 78–81. https://doi.org/10.1016/j.reth.2016.04.001.

Toyserkani, N.M., M.G. Jørgensen, S. Tabatabaeifar, et al. 2017. Concise review: A safety assessment of adipose-derived cell therapy in clinical trials: A systematic review of reported adverse events. *Stem Cells Translational Medicine* 6 (9): 1786–1794. https://doi.org/10.1002/sctm.17-0031.

Turner, L. 2015. US stem cell clinics, patient safety, and the FDA. *Trends in Molecular Medicine* 21 (5): 271–273. https://doi.org/10.1016/j.molmed.2015.02.008.

Turner, L. 2018a. The US direct-to-consumer marketplace for autologous stem cell interventions. *Perspectives in Biology and Medicine* 61 (1): 7–24. https://doi.org/10.1353/pbm.2018.0024.

Turner, L. 2018b. Direct-to-consumer marketing of stem cell interventions by Canadian businesses. *Regenerative Medicine* 13 (6): 643–658. https://doi.org/10.2217/rme-2018-0033.

Turner, L. 2020. Preying on public fears and anxieties in a pandemic: businesses selling unproven and unlicensed 'stem cell treatments' for COVID-19. *Cell Stem Cell* 26 (6): 806–810. https://doi.org/10.1016/j.stem.2020.05.003.

Turner, L. 2021. The American stem cell sell in 2021: U.S. businesses selling unlicensed and unproven stem cell interventions. *Cell Stem Cell* 28(11): 1891–1895. S1934590921004203. https://doi.org/10.1016/j.stem.2021.10.008

Turner, L. 2021. ISSCR's guidelines for stem cell research and clinical translation: Supporting development of safe and efficacious stem cell-based interventions. *Stem Cell Reports* 16 (6): 1394–1397. https://doi.org/10.1016/j.stemcr.2021.05.011.

Turner, L., and P. Knoepfler. 2016. Selling stem cells in the USA: Assessing the direct-to-consumer industry. *Cell Stem Cell* 19 (2): 154–157. https://doi.org/10.1016/j.stem.2016.06.007.

Turner, L., and J. Snyder. 2021. Ethical issues concerning a pay-to-participate stem cell study. *Stem Cells Translational Medicine* 10 (6): 815–819. https://doi.org/10.1002/sctm.20-0428.

Turner, L., M. Munsie, A.D. Levine, and L. Ikonomou. 2021. Ethical issues and public communication in the development of cell-based treatments for COVID-19: Lessons from the pandemic. *Stem Cell Reports*. https://doi.org/10.1016/j.stemcr.2021.09.005.

Turner, L. 2017. ClinicalTrials.gov, stem cells and 'pay-to-participate' clinical studies. *Regenerative Medicine* 12 (6): 705–719. https://doi.org/10.2217/rme-2017-0015.

Unsworth, D.J., J.L. Mathias, D.S. Dorstyn, et al. 2020. Are patient educational resources effective at deterring stroke survivors from considering experimental stem cell treatments? A randomized controlled trial. *Patient Education and Counseling* 103 (7): 1373–1381. https://doi.org/10.1016/j.pec.2020.02.012.

Wagner, D., L. Turner, A. Panoskaltsis-Mortari, et al. 2018. Co-opting of ClinicalTrials.gov by patient-funded studies. *The Lancet Respiratory Medicine* 6: 579–581. https://doi.org/10.1016/S2213-2600(18)30242-X.

Waldby, C., T. Hendl, I. Kerridge, et al. 2020. The direct-to-consumer market for stem cell-based interventions in Australia: Exploring the experiences of patients. *Regenerative Medicine* 15 (1): 1238–1249. https://doi.org/10.2217/rme-2019-0089.

Wan W, and L. McGinley. 2019. Clinic pitchers unproven treatments to desperate patients, with tips on raising cash. *The Washington Post*. December 1, 2019.

Weiss, D.J., L. Turner, A.D. Levine, et al. 2018. Medical societies, patient education initiatives, public debate and marketing of unproven stem cell interventions. *Cytotherapy* 20 (2): 165–168. https://doi.org/10.1016/j.jcyt.2017.10.002.

Zarzeczny, A., C. Rachul, M. Nisbet, et al. 2010. Stem cell clinics in the news. *Nature Biotechnology* 28 (12): 1243–1246. https://doi.org/10.1038/nbt1210-1243b.

Zarzeczny, A., H. Atkins, J. Illes, J. Kimmelman, et al. 2018. The stem cell market and policy options: A call for clarity. *Journal of Law and the Biosciences* 5 (3): 743–758. https://doi.org/10.1093/jlb/lsy025.

Zarzeczny, A., C. Tanner, J. Barfoot, et al. 2020. Contact us for more information: An analysis of public enquiries about stem cells. *Regenerative Medicine* 14 (12): 1137–1150. https://doi.org/10.2217/rme-2019-0092.

Zarzeczny, A., T. Caulfield, and Ogbogu, et al. 2014. Professional regulation: A potential valuable tool in responding to "stem cell tourism." *Stem Cell Reports* 3 (3): 379–384. https://doi.org/10.1016/j.stemcr.2014.06.016.

Zhang, J.Y. 2017. Lost in translation? Accountability and governance of clinical stem cell research in China. *Regenerative Medicine* 12 (6): 647–656. https://doi.org/10.2217/rme-2017-0035.

Kirstin R. W. Matthews is a fellow in science and technology policy at the Baker Institute and a lecturer in the Department of BioSciences at Rice University. Her research focuses on ethical and policy issues at the intersection between traditional biomedical research and public policy. Specifically, she focuses on regulation and ethical issues associated with emerging biotechnology, including synthetic biology, stem cells, and genomic medicine. Kirstin Matthews also leads a project to review scientific advice in and to the federal government, including the White House Office of Science and Technology Policy and the President's Council of Advisors for Science and Technology. Matthews has a doctorate in molecular biology from The University of Texas Health Science Center at Houston.

Chapter 7
Genetic Privacy in the Age of Consumer and Forensic DNA Applications

Sheldon Krimsky

7.1 Introduction

When the human genome was sequenced at the turn of the millennium, it brought high expectations that new scientific and medical advancements were forthcoming. Allan Maxam and Walter Gilbert developed the first sequencing method in 1977. Also, in that year Frederick Sanger introduced a DNA sequencing technique that he and his team used to sequence the first full genome of a virus. Segments of the human genome were sequenced in the 1980s and 1990s to detect mutations in certain genes. But sequencing the entire human genome was a milestone that some called "the genetic moon shot," comparing it to humans reaching the Earth's natural satellite.

As the DNA sequencing technology advanced, it became more rapid and less expensive. The Human Genome Project was launched in 1990 and took 13 years to complete. The estimated cost was $2.7 billion. Within a decade of the first sequenced human genome, the $1000 genome was considered a realistic goal.

In the mid-1980s a British scientist, Alec Jeffreys, a geneticist at the University of Leicester, discovered a method for comparing segments of the human genome for purposes of identification. The technique of electrophoresis was used to compare short tandem repeats (i.e., actg actg) on certain sites on the chromosome where the number of repeats varies among individuals. The length of the segments can be compared when exposed to an electric field. An electric current is used to move DNA molecules placed on a gel. Smaller molecules move faster in the electric field than larger molecules, allowing for their separation and a comparison of weights and thus the number of repeats. For the first time, one's DNA could be used to exonerate or implicate an individual who was a suspect of a crime.

S. Krimsky (✉)
Tufts University, Boston, MA, USA
e-mail: krwm@rice.edu

© The Author(s), under exclusive license to Springer Nature Switzerland AG 2023
T. Zima and D. N. Weisstub (eds.), *Medical Research Ethics: Challenges in the 21st Century*, Philosophy and Medicine 132,
https://doi.org/10.1007/978-3-031-12692-5_7

In the United Kingdom and the United States police began applying Jeffrey's method to search for serial killers and rapists, sometimes collecting DNA from thousands of men on a voluntary basis, the first of what would be known as DNA dragnets. Privacy issues immediately arose. These DNA dragnets were executed without a court warrant, probable cause, or individual suspicion. Civil liberties activists questioned whether police coercion to collect the DNA was an inescapable part of the dragnet.

7.2 Genetic Privacy in the United States

It is generally recognized that the collection of one's DNA without a warrant, which requires probable cause, is a violation of the Fourth Amendment of the U.S. Constitution, which protects individuals against searches and seizures of their property. If police took non-voluntary samples of a suspect's DNA, it would meet the conditions of an illegal search.

In a police investigation of a serial rapist in Ann Arbor, Michigan in 1994, 600 African American men were asked to submit DNA samples. The police informed the men that anyone who did not volunteer their DNA became a suspect in the crime being investigated.[1]

DNA sequencing took on a new role in criminal investigations. In 1993 the United States passed legislation establishing a national registry of DNA collected from convicted felons, whose DNA samples would remain on the database their entire lives.

Personal genetic information became part of an individual's medical records. The U.S, Health Insurance Portability and Accountability Act (HIPAA) was passed in 1996. The legislation establishes the privacy and confidential handling of protected health information. Initially, genetic information was not included in HIPAA. That changed in 2013 with the passage of the HIPAA Omnibus Rule, which included genetic information in the definition of protected health information.[2]

The amendment to HIPAA was prompted by the passage of another major piece of legislation that protects the privacy of genetic information. The Genetic Information Non-Discrimination Act (GINA) passed on May 21, 2008, protects individuals against discrimination in health coverage and employment based on one's genetic information. In 2007, Senator Edward Kennedy, speaking in favor of GINA stated: "It is difficult to image information more personal or more private than a person's genetic makeup."[3]

---

[1] Krimsky and Simoncelli (2011).

[2] Department of Health & Human Services. Modifications to the HIPAA Privacy, Security, Enforcement and Breach Notification Rules. *Federal Register* 78(17):5566 (January 25, 2013).

[3] Statement of Senator Edward Kennedy in support of the Genetic Information and Nondescrimination Act. *Congressional Record* Senate 153, No. 123 (January 22, 2007), S. 847.

In 1993 Congress passed the DNA Identification Act. The law established a national forensic laboratory under the authority of the Federal Bureau of Investigation (FBI) with a database called the Combined DNA Index System (CODIS), which was created to store the DNA samples of violent convicted offenders. Eventually, each state was linked into CODIS and allowed to establish its own laws governing DNA collection, namely, the crimes and the conditions that justify collecting DNA samples.

This became the backdrop for discussions and legal challenges against violations of genetic privacy, which were brought to new heights with the expansion of forensic genetics and the growth of DNA ancestry testing.

7.3   Genetic Privacy in the European Union

The European Union (EU) passed the General Data Protection Regulation (GDPR) in 2016. The regulation treats genetic information as personal and sensitive data, not only to an individual but also to their relatives. It distinguishes between pseudonymized and anonymized data. Pseudonymized data cannot reveal the identity of a specific subject without the use of additional information. To ensure privacy, such additional information is kept sequestered from those who have access to the genetic data so the data subject cannot be identified.

In contrast, anonymized data consists of information that does not relate to an identifiable person. The data are stripped of any information, which, however combined with other information, could reveal a person's identity. No matter what other means are used, it is not reasonably likely to identify the person. Since privacy is assured, anonymized data is not subject to GDPR regulations.[4] The questions regarding genomic data concerns whether it can be considered anonymized. Can some sequences or markers of a person's DNA, if combined with other information, reveal the person's identity?

7.4   Genetic Privacy in China

Direct-to-Consumer genetic testing (DTCT) has begun to grow rapidly in China as many new test providers have taken to advertising in social media for services including medical risk analysis and lifestyle guidance. The growth in this sector is estimated to increase in five years from 1.52 million in 2018 to 56.8 million in 2022.[5]

Since 2014, clinical genetic testing must be registered and approved by the China Food and Drug Administration (CFDA) before entering the marketplace. While privacy is one of the considerations for the oversight of DTCT in China, there is

---

[4] Shabani and Marelli (2019).

[5] Du and Wang (2020).

no comparable consumer protection for tests involving ancestry analysis and lifestyle genomics such as nutrition. Privacy protections of consumer genetic data are left to individual companies and vary significantly on informed consent, sale of data, and transparency.

China currently does not have any national legislation specifically for the protection of personal genetic data but is expected to enact a comprehensive data protection law in 2022.[6] The legal basis for the new law is found in Article 111 of the General Provisions of Civil Law (2017), which established the protection of citizens' personal information stating: "Any organization or individual who needs to obtain another's personal information should obtain it lawfully and ensure its security."[7]

7.5   Privacy and Ownership of One's Cells

No one doubts that we own what is in our body. But what if something leaves our body? Do we still own it? In one of the most widely publicized cases, someone's cells were taken from her body after surgery and then were appropriated under private ownership. This is the story of Henrietta Lacks.

In 1951 Ms. Lacks went for treatment of cervical cancer at Johns Hopkins Hospital. After surgery, when her tumors were biopsied, and the cells were cultured they were found ideal for research. The cells reproduce indefinitely under the right conditions and became one of the most important cell lines, known as Hela cells, in medical research. Neither Ms. Lacks, nor her family, after she succumbed from her cancer, were notified, or compensated for the use of her cells despite their commercial value. At the time there were no consent issues or laws pertaining to the appropriation of one's cell in surgery. In 2013 researchers published the DNA sequence of a strain of Hela cells without permission of Ms. Lack's family. Her DNA was now in the public domain. Without ownership, there was no privacy. With the growing negative publicity, NIH signed an agreement in 2013 that would give the family of Ms. Lacks some control over the cell's DNA sequence.

A similar story, 25 years after Henrietta Lacks' cells were appropriated, begins with John Moore, who was a leukemia patient in 1975. Surgeons at the UCLA Medical Center removed Mr. Moore's diseased spleen. To the surprise of the medical team, his blood profile returned to normal after the operation. When researchers at UCLA examined Mr. Moore's blood cells, they learned that they produce a protein that stimulates white blood cells, which can fight infections.

In 1984 the Regents at the University of California patented the cell line they name "Mo." Mr. Moore filed a lawsuit seeking a share of the profits from his cells. His legal challenge was denied by the Los Angeles Superior Court. In appeal, the California Court of Appeal ruled that a patient's blood and tissues are his personal property, and that Mr. Moore had a right to share the profits of his cells. The case

---

[6] Feng (2019).

[7] Ibid.

was brought to the U.S. Supreme Court, which ruled that a hospital patient gives up any rights to his tissues taken from his body, whether or not they have commercial value.[8]

What about the ownership of genes? No agent of government in the United States can acquire your DNA and sequence your genes without your consent, except for a court warrant under the relevant laws. If your DNA is sequenced, say for medical purposes, can those who have access to the DNA patent your genes? As strange as it sounds, it has happened. The genetic mutations for breast cancer, known as BRCA1 and BRCA2 were sequenced and patented by Myriad Genetics, a company that developed a screening tests for the mutations.

While it is true that that Myriad did not invent the genes, the U.S. patent law allows patents for discoveries of new compositions of matter. The approved patent on the BRACA genes excluded other researchers from using the genetic sequence for research and development without paying fees to Myriad. The response from the public health and scientific communities against the monopoly ownership of a gene sequence was very strong. The Myriad Genetics patent case was eventually heard in the U.S. Supreme Court. The gene sequences (mutations) were in the bodies of millions of women. With its patent protection, Myriad controlled the only test for BRCA-type breast cancer.

The Court ruled that the entire gene sequence (the introns and the exons) were found in nature and could not be patented. But the Court held that once Myriad removed the introns (the non-coding regions of the DNA sequence), which the cells do naturally to synthesize proteins, and create copy DNA (cDNA), then Myriad could patent the cDNA. Thus, our DNA sequences are private and protected under the Fourth Amendment, but they can also be patented once they are altered by scientists. Patenting of genes was not universally accepted by the high courts of other nations. Australia's highest court ruled that a gene mutation linked to cancer could not be patented. This decision impacted over one thousand genes under patent in Australia.[9]

7.6 Forensic Genetics and Privacy

We have seen how genetic information has transformed criminal investigations. But how is forensic DNA balanced against personal privacy? As technology has produced new applications for gene sequencing, just how private are your genes?

As we have noted, in most cases, without a warrant U.S. police cannot force you to give them a DNA sample. There are two ways, however, that police can get around this. First, your DNA is on all sorts of objects you abandon. You throw away a drinking cup with your cells on the lip of the cup. Your trash may contain your cells on many items. With modern techniques, such as Polymerase Chain Reaction

---

[8] Krimsky (2003).

[9] Patrova (2015).

or PCA, even a miniscule sample of your DNA can be replicated millions of times and used to identify the DNA profile of the individual who discarded the item.

In one notable case, police were trying to solve a cold case of a murder that took place decades ago. They had the crime scene DNA. They also had a suspect, but no probable cause to obtain a warrant for his DNA. They used an unusual method to obtain their suspect's DNA. The police created a ruse to get the suspect to release his DNA to them. They wrote the suspect a letter from a law firm they fabricated stating that they represented clients in a class action suit against local authorities who overcharged them in parking violations. The letter stated that if he signed the class-action affidavit he would receive a rebate on his parking tickets. The suspect signed the form, licked the envelope, and sent it back to the phony law firm made up of police investigators. With DNA taken from the licked envelope, police were able to match the suspect's DNA with the crime scene DNA and charge their suspect with the crime. When the case was challenged, the Supreme Court of the State of Washington ruled in favor of the police despite their unusual and possibly illegal tactics pretending to be lawyers. The court ruled that the suspect had no privacy interest in his DNA that he left on the envelope—it was abandoned.[10]

The second way the police can get around DNA privacy is through indirect methods. When police obtain a DNA sample at a crime scene, the first thing they seek to determine is the identity of the source. Uploading the DNA identifier (20 pairs of numbers) on CODIS will reveal the source of the DNA if they get an exact match on the database. If there is a match it means that the person was previously arrested, charged and convicted of a serious crime. Also, the procedures for suspects who are not convicted to have their DNA profile deleted vary from state to state. Thus, exonerees or people charged but not prosecuted or convicted may remain on the database.

When the uploaded crime scene DNA does not match one of the more than 14 million offender DNA records (as of 2020) on CODIS, investigators can undertake "familial searching." For this method, they reduce the number of matched alleles from 20 for an exact match to a lower number (lower stringency match) expecting that they can identify family members of the person whose DNA was left at the crime scene. This allows investigators to inquire into the composition of the family using traditional genealogical sources to find a suspect of the crime.

With "familial searching" police are conducting suspicionless searches leading to surveillance of family members of someone on the database who has similar but not identical DNA sequences to what was found at the crime scene. Applying familial searches, developing a family tree, and then by obtaining abandoned objects from a suspect, police have been successful in apprehending offenders, years after the crime was committed.[11]

---

[10] *State of Washington v. Athan*, 160 Wn.2d354 (2007), Para 35 (Case No. 75312–1. Filed May 10, 2007.

[11] Ram et al. (2018).

7.7 DNA Privacy

While the concept of privacy does not appear in the U.S. Constitution or the Bill of Rights as a fundamental right, in 1965 the U.S. Supreme Court declared in *Griswold v. Connecticut* a constitutional right of privacy. The case involved a married couple who sued the State of Connecticut for its law prohibiting any person from using any substance, including contraceptives or drug for the purpose of preventing conception. The Court ruled the law unconstitutional on the grounds that it violated the "right to marital privacy." Justices Byron White and John Marshall Harlan II wrote in their opinion that personal privacy is protected by the Due Process clause in the Fourteenth Amendment of the Constitution.

Four privacy concerns pertaining to DNA were cited in Krimsky and Simoncelli (2011).[12]

1. Physical and bodily privacy. This pertains to collecting one's DNA. While extracting DNA from blood has been considered an intrusive process, obtaining cells, and thus DNA, from a person's cheeks (buccal swab) is hardly intrusive, but still an intimate personal space and would ordinarily require informed consent.
2. Informational privacy. The information on our DNA is considered a private matter. That information can reveal health risks. By 2000, around 1000 of the estimated 7000 single gene inherited diseases that can be revealed in people's DNA have been identified and characterized, including Huntington's disease and cystic fibrosis.[13]
3. Familial or relational privacy. One's DNA is shared through ancestry with relatives. The closer the relationships, the greater the shared DNA up to 50 percent of child to parent. Given the development of DNA ancestry, our DNA can locate the population group of our ancestral heritage and thus our genealogical ethnicity.
4. Locational privacy. Generally, people consider their spatial movements (places where they go) a private and personal matter. Government surveillance of an individual through a technology like cell phones is highly controversial. Yet by leaving our DNA at a site, it is possible to determine that we were there. Crime scene DNA is the way police learn that an individual was at the location of a crime and thus becomes a suspect. Parolees or suspects on bail lose their locational privacy when they are fitted with an ankle electronic monitoring device as part of their bail or probation conditions.

A person's DNA may be recognized as a private matter, but for a convicted felon, the expectation of privacy is significantly reduced. Under what conditions can the police collect your DNA, with or without a court warrant? How is individual privacy balanced against the interests of the state? What has been the trajectory of court decisions on extracting DNA from a suspect?

---

[12] Krimsky and Simoncelli (2011).

[13] Claussnitzer et al. (2020).

7.8   DNA Collection on Arrest: Personal Identity, Privacy, and Criminal Investigation

The fact that courts have ruled that buccal swabs are mildly invasive allowed states to pass legislation that gave police authority to collect DNA from individuals before they were charged with a criminal offense, before they were arraigned, and even if they were not convicted. U.S. federal and state courts have reached different conclusions over whether the Fourth Amendment prohibits the collection and analysis of DNA from an arrestee. In addition, the severity of the crimes justifying a DNA collection have diminished. Another factor that made DNA collection more accessible to law enforcement was the U.S. Supreme Court decision in *Maryland v. King*.

In 2009, Alonzo Jay King Jr. was arrested in Maryland on first-and second-degree assault charges for menacing a group of people with a shotgun. After he was booked at the police station, the booking personnel took a cheek swab to get cells for a sample of Mr. King's DNA. The Maryland DNA Collection Act allows law enforcement officers to collect DNA samples from an individual charged with a crime of violence or burglary, or an attempt to commit a crime of violence or burglary. Under the law, if criminal charges are dropped, the collected DNA sample of the arrestee is destroyed. Tests for familial matches are also prohibited.

Mr. King did not give his consent for the cheek swab. His DNA was collected and analyzed putatively for an identification, where it was entered into the CODIS database. There are two distinct databases in CODIS. One is for the DNA identification of felons. The other is for the DNA left at crime scenes, in which many of the crimes remain unsolved. The police uploaded Mr. King's DNA on both databases and found that his DNA matched the DNA found on a 2003 rape victim. There was no probable cause that Mr. King committed rape when his DNA was collected. If the state had probable cause for the rape, they could get a court order for a DNA sample.

The Court of Appeals of Maryland, which is the supreme court of the state of Maryland, ruled that the DNA taken when Mr. King was booked for the 2009 rape charge was an unlawful seizure and an unreasonable search of the person. It overturned the rape conviction. The Appeals Court found that taking Mr. King's DNA was an unreasonable search because King's expectation of privacy is greater than the state's interest in using his DNA to identify him.

There is no dispute that using a buccal swab on the inner tissues of a person's cheek to obtain DNA is a search. Any intrusion on the human body is an invasion and protected under the Fourth Amendment. The U.S. courts, however, consider the reasonableness of a search. The level of intrusion into the body is central to determining reasonableness although it is still a search. The key questions are: Is the intrusion with the cheek swab justified? Is the search reasonable?

The state of Maryland appealed the decision of the Maryland Appeals Court to the U.S. Supreme Court. The Court ruled 5–4 against King and in favor of the state of Maryland. Justice Kennedy wrote the majority opinion.

> …the Court concluded that DNA identification of arrestees is a reasonable search that can be considered part of a routine booking procedure. When officers make an arrest supported

by probable cause to hold for a serious offense and they bring the suspect to the station to be detained in custody, taking, and analyzing a cheek swab of the arrestee's DNA is, like fingerprinting and photographing, a legitimate police-booking procedure that is reasonable under the Fourth Amendment.[14]

The fact that the police had no prior suspicion of the rape case and used Mr. King's DNA to trawl its database of unsolved crimes did not factor into the majority opinion. It stated that "individual suspicion is not necessary…the fact of a lawful arrest, standing alone, authorizes a search…When probable cause exists to remove an individual from the normal channels of society and hold him in legal custody, DNA identification plays a critical role in serving those interests." The Court stated that a suspect's criminal history is a critical part of his identity and thus the state was justified in trawling its database to see if Mr. King was associated with any other offenses.

> "In some circumstances, such as "[w]hen faced with special law enforcement needs, diminished expectations of privacy, minimal intrusions, or the like, the Court has found that certain general, or individual, circumstances may render a warrantless search or seizure reasonable."[15]

Justice Anthony Scalia wrote a dissenting opinion on the case. He began with a categorical statement: "The Fourth Amendment forbids searching a person for evidence of a crime when there is no basis for believing the person is guilty of the crime or is in possession of incriminating evidence. The prohibition is categorical and without exception." He objected to the majority's assertion that the DNA was taken to identify the suspect. "The Court's assertion that DNA is being taken, not to solve crimes, but to identify those in State's custody, taxes the credibility of the credulous." He went on to argue: "Solving unsolved crimes is a noble objective, but it occupies a lower place in the American pantheon of noble objectives than the protection of our people from suspicionless law-enforcement searches. The Fourth Amendment must prevail." He then goes on to predict that, based on the majority decision of the Court, DNA searches will someday be standard for traffic violations. "When there comes before us the taking of DNA from an arrestee for a traffic violation, the Court will predictably (and quite rightly) say 'We can find no significant difference between this case and *King*."

The King case brings the United States to a new plateau in the balance between law enforcement interests and personal privacy. Taking one's DNA during booking or even during traffic stops may well become routine police procedures and treated like a "reasonable search."

Most Western nations have not reached that point. Nevertheless, familial searching, DNA profiling, and DNA dragnets are part of the investigative tools used in the United Kingdom and Australia. And in South Australia, police can take DNA samples from individuals merely suspected of having committed a crime, if the crime, no matter how trivial, is punishable by a prison sentence.[16]

---

[14] Supreme Court of the United States. *Maryland v. King* No. 12–207, Decided June 3, 2013.

[15] *Maryland v. King* 2013, pp.8–9.

[16] Gregoire and Nedim (2017).

7.9 DNA Ancestry Genealogy: What Happens to Your DNA?

Since 2000, when the first DNA ancestry company was formed, the business of direct-to-consumer DNA ancestry testing has grown exponentially as a lucrative business sector. As of 2018, 74 companies offered ancestry services. By sending a saliva sample or a cheek swab, a company will analyze a consumers' DNA and produce a profile of the region of the world in which their ancestors had lived.[17]

In 2020, 23andMe, one of the leaders in the field of ancestry testing founded in 2007, stated it had 10 million people signed up for an ancestry DNA analysis. Another leader in the field, AncestryDNA reported that it had sold tests to 16 million people. By 2019 the ancestry testing sector was valued at $1 billion.[18]

Companies entering the DNA ancestry marketplace realized that the leisure DNA testing consumer was only part of their business model. They began operating under a two-sided business model in which different user groups draw benefits from the company's data collection. First, there are the people seeking a personal genealogical analysis. Second, there are the companies and research institutions that want access to a company's DNA collection for research and drug development.

For example, 23andMe has many partnerships with drug companies who want access to their consumer data.[19] "In 2018, GlaxoSmithKline invested $300 million in 23andMe, giving the pharma giant access to 'large-scale genetic resources' …and 23andMe sold the rights to a drug it had developed in-house using customers' data to pharma company Almirall."[20]

People's DNA profiles are spread over forensic and DNA ancestry databases. What does this portend about genetic privacy? Many DNA customers allow companies to share their anonymized samples to research groups. Most companies will not, without a warrant, release DNA information to the criminal justice system. It is also recognized that anonymized DNA data can be de-anonymized thus subverting privacy considerations.

Other companies have emerged that provide open access to DNA information that DNA ancestry customers upload on their database. For example, GED Match, founded in 2010, and describing itself as an open-source genealogy site, provides a service of matching DNA profiles to determine potential family relationships. Customers who obtained DNA profiles from different companies upload the raw data on GED Match. Criminal investigators have posted crime scene DNA under a pseudonym and discovered familial DNA matches to others on the database. From this information, they create a genealogical tree, which allows them to generate suspects for an unsolved crime. Thus, by uploading their DNA ancestry results on an open-access database, individuals cannot only be identified but they could also

---

[17] Sheldon Krimsky (2022) Understanding DNA Ancestry. Cambridge University Press, Cambridge, UK.

[18] de Groot (2020).

[19] Stoekle et al. (2016).

[20] Hamzelou (2020).

be subject to investigation or surveillance. Police used this method to identify and prosecute the "Golden State killer," a highly publicized case that took place between 2019 and 2020.

Ehrlich et al. (2018) selected an anonymous genome from a publicly accessible database and uploaded the DNA file to GED Match. By utilizing social media and public records, they were able to identify the anonymous genome after a day's work.[21] Wallace et al. (2015) noted:

> Given expertise, resources and will, it is possible to re-identify individuals from anonymized family history data suggesting that procedures such as name removal and encoding are not sufficient to protect against privacy breeches.[22]

As noted by Shabani and Marelli (2019) "fewer than 100 single nucleotide polymorphisms (SNP) are sufficient to distinguish an individual's DNA record."[23] These results muddy the distinction between pseudonymized and anonymized genetic data and shows that under the right circumstances, the latter can be de-anonymized. In the article by Schmidt and Shawnecqua titled "How anonymous is 'anonymous'," the authors state: "reidentifiability of [personal anonymous] datasets can increase significantly through cross-reference with publicly available datasets."[24]

7.10 Searching for One's Biological Parents

One of the unintended outcomes of DNA ancestry testing is their role in the discovery of unknown family members as well as the revelation that a legal parent may not be one's biological parent. Even if one knows he or she was adopted, the thought about who the biological parents are is never erased.

Children given up for adoption do not always gain the opportunity to learn who their biological parents were or why they were abandoned by them. Before open adoption laws were passed in many jurisdictions, adoptees would go through life always wondering who the people were who brought them into the world.

When people submit their DNA for ancestry testing, their genetic profile can be compared to other individuals who have also submitted DNA samples to the same company. Upon the request of the customer, the company can report any samples which are close enough to be a relative of that person. Usually, customers are asked whether they wish to learn about other people who have submitted their DNA to the company database who might be related to them.

With the introduction of open-access companies like GEDmatch, individuals who have undertaken their ancestry test from different companies can upload their DNA data onto one site. They can be matched by similarities of their genetic profile as

---

[21] Erlich et al. (2018) at p. 691.

[22] Wallace et al. (2015), at p. 94.

[23] Shabani and Marelli (2019), p. 2.

[24] Schmidt and Shawneequa (2012).

being related. Eleni Liff always knew she was adopted. As a newborn, she was left in a shopping bag in the foyer of a Brooklyn, New York apartment building. At age 27, Ms. Liff learned who her biological mother was through DNA ancestry testing and sharing her genetic profile in an open access database. Eventually, through company algorithms that match DNA, she was able to locate cousins who uploaded their genetic profile on the database. The mystery of her biological mother was eventually solved.

'GEDmatch has also provided police with a new way to solve crimes. Crime scene DNA is uploaded to the open-access ancestry site through a pseudonym. Police then request any matches to the crime scene DNA. They use those matches to create a family tree. Within that family tree they identify any potential suspects who might match the crime scene DNA.

Sperm donors work with IVF clinics to help women have a successful pregnancy. The donors usually prefer to remain anonymous and choose not to have a relationship with the mother or the child, who they help bring into the world. Many countries have passed laws opening adoption records allowing adoptees to learn the identity of the donor. Children born from sperm donation, who do not have access to open files, have been known to use whatever means is available to them to uncover the identity of the donor.

On the assumption that many sperm donors are serial donors, adoptees post their DNA profiles on open access databases searching a sibling DNA match who had the same sperm donor. One such case was highlighted in an ABC news documentary 20/20 titled "Seed of Doubt": Eve Wiley's story. Eve learned that her mother went to a sperm bank when she and her husband had difficulty conceiving. As an adult, Eve was able to learn the identity of the sperm donor from the sperm bank. Then she used DNA ancestry in the hope of finding siblings from the same donor who had donated his sperm on several occasions. When she found a genetic cousin, whose DNA was not related to the sperm donor, she became suspicious. Eventually she traced her paternal DNA not to the sperm donor but to the fertility doctor who used his own sperm to fertilize her mother's egg. The fertility doctor mistakenly believed his anonymity was secure.

As stories like these become more plentiful, the idea of genetic privacy slowly begins to erode. While there are certainly limits on what an ancestry test can reveal, they have shown that family secrets have been unearthed through DNA analysis that would otherwise have been forever unknown.

When a person signs up for a DNA ancestry test, companies typically provide a consent form that gives the company certain rights over your data and explains the rights of the consumer. Of course, your data also involves other members of your family since they share a sufficient amount of your DNA to establish an ancestral connection.[25] Casa et al. discuss the risk of revealing one's siblings identity by placing one's DNA SNPs on a public database.[26] As noted by Wallace et al. (2015): "The ease with which genealogical and other personal data from the client, and by

---

[25] Ibid.

[26] Casa et al. (2008).

extrapolation from their relatives, can be shared, linked, and used, raises issues of who gives consent to provide the data and how well all parties are aware of the implications of participation."[27] By cross referencing DNA information presumed anonymous with publicly available databases, re-identifiability becomes more likely.

There are no uniform rules of consent and privacy among DNA ancestry companies. One company wrote: "We will disclose collected information without your permission when we believe in good faith that such disclosure is required by law or is necessary or desirable to investigate or protect against harmful activities to customers, employees or others or to property including this site."[28] AncestryDNA, established in 2007, writes in its privacy statement:

> …we share your DNA test results and DNA sample without your name and other common identifying information. You own your own DNA data. At any time, you can choose to download raw DNA data, have us delete your DNA Privacy Statement, or have us destroy your physical DNA test results as described in the AncestryDNA Privacy Statement, or have us destroy your physical DNA saliva sample.[29]

Krimsky and Johnson published the privacy statement of thirty DNA ancestry companies illustrating the variations in consumer privacy protections.[30]

Even when a customer signs an agreement with an ancestry company, which holds their genetic data, the privacy policy may change in the future without customers being informed. Often the burden is placed on customers to keep checking the company's website to see if the privacy policy has changed. And if the company is bought out by another company, the privacy policy you signed may be no longer in effect. Bankruptcy laws allow creditors to get maximum value of a company's resources and that may mean selling your data and making your personal data public, even though you were promised confidentiality.

And finally, there is the problem of stolen or hacked information. Companies often have you sign a warranty against liability from theft of your information. Here is the statement from 23andMe, established in 2006.

> Your genetic data, survey responses and/or personally identifying information may be stolen in the event of a security breach. In the event of such a breach, if your data is associated with your identity, they may be made public or released to insurance companies, which could have a negative effect in your ability to obtain insurance coverage.[31]

7.11   Conclusion

The growth of DNA sequencing has led to applications in medicine, forensics, and amateur genealogy, also known as leisure DNA testing. This has led to the widespread dissemination of personal genomic data. De Groot (2020) notes:

---

[27] Shabani and Marelli (2019).

[28] Ethno Ancestry, established in 2004. http://www.ethnoancestry.com.

[29] Krimsky and Johnson (2017).

[30] Krimsky and Johnson (2017), pp. 21–29.

[31] 23andMe. https://www.23andMe.com.

The 'leisure' DNA testing market has expanded greatly in the past ten years, with around 250 firms now offering some kind of test to consumers. The expanding commercialization and marketization of DNA testing—companies offer tests to match users to wine, to make art from their DNA, even offering Valentine's day offers—signals a shift into considering genetics outside of the medical and health sphere and very particularly related to leisure and commodity.[32]

The criminal justice system and the courts have slowly lowered the bar for acquiring a suspect's DNA, as leading judges compare DNA to fingerprints, which has, up until now, been the standard for confirming identification. But fingerprints are hardly definitive as an identifier compared to DNA. Fingerprinting a person has never been viewed as a search since one's hands are out in the open and not a hidden part of the body. Also, they do not reveal any traits of a person.

Given the widespread dissemination of personal DNA information, leakage is inevitable. And there is no law in the United States preventing people from sequencing someone's DNA from an abandoned object and making that information public. Since 2006, under the Human Tissue Act, it is a criminal offense in the United Kingdom to analyze someone's DNA without their consent. The exception to the Act is the DNA that is collected and analyzed by police for criminal investigation.[33] This is particularly relevant to abandoned DNA. By circumventing the very weak genetic privacy laws, some compromising parts of a public figure's DNA spread through the social media can place the most intimate of information in the hands of people who should not have it.

Where all of this is leading is still not clear. What is clear is that genetic privacy is no longer an idea that can be protected, largely in part because the Supreme Court compares DNA to fingerprints, because taking one's DNA is only mildly invasive and hardly intrusive, because people are willing to share their DNA with others on open-access databases and because the costs of sequencing has dropped considerably, and the speed of sequencing has increased. It is all leading to the warning issued by the late Justice Scalia that traffic stops will someday become a pretense for acquiring one's DNA. An editorial in the *New Scientist* wrote that if genetic privacy is to be protected, there needs to be an international agreement on regulating the testing of "abandoned DNA."[34]

References

Casa, C.A., B. Schmidt, I.S. Kohane, and K.D. Mandl. 2008. My sister's keeper? Genomic research and the identifiability of siblings. *BMC Medical Genomics* 1: 32.
Claussnitzer, M., J.H. Cho, R. Collins, N.J. Cox, E.T. Dermitzakis, and M.E. Hurles, et al. 2020, January. A brief history of human disease genetics. *Nature* 577 (7789): 179–189.

---

[32] De Groot (2020), p. 12.

[33] Aldous (2009).

[34] Editorial (2009).

de Groot, J. 2020, February. Ancestry.com and the evolving nature of historical information companies. *The Public Historian* 42 (1): 8–28.

Du, L., and E. Wang. 2020, April. Genetic privacy and data protection: A review of Chinese direct-to-consumer genetic test services. *Frontiers in Genetics* 11 (416): 1–10, p. 2.

Editorial. 2009, January 31. Why your DNA needs more protection. *New Scientist* 3.

Erlich, Y., T. Shor, I. Pe'er, and S. Carmi, 2018, November. Identity inference of genomic data using long-range familial searches. *Science* 362 (6495): 690–694

Feng, Y. 2019. The future of China's personal data protection law: Challenges and prospects. *Asia Pacific Law Review* 27: 62–82.

Gregoire, P., and U. Nedim. 2017, February 24. The national DNA database is watching you. https://www.sydneycriminallawyers.com.au/blog/the-national-dna-database-is-watching-you/. Accessed 9 May 2021.

Hamzelou, J. 2020, February 15. The business of DNA analysis. *New Scientist* 242 (3269).

Krimsky, Sheldon. 2022. *Understanding DNA ancestry*. Cambridge, UK: Cambridge University Press.

Krimsky, S., and T. Simoncelli. 2011. *Genetic justice*. New York: Columbia University Press.

Krimsky, S., and D.K. Johnson. 2017, March. *Ancestry DNA testing and privacy: A consumer guide*, 21. Council for Responsible Genetics. A project funded by the Rose Foundation. https://sites.tufts.edu/sheldonkrimsky/files/2018/05/pub2017AncestryDNAPrivacy.pdf. Accessed 10 May 2020.

Krimsky, S. 2003. *Science in the private interest*, 131–132. Lantham, MD: Rowman & Littlefield.

Patrova, S. 2015, October 6. High Court rules breast cancer gene cannot be patented. *The Conversation.*

Peter, Aldous. 2009, January 31. Could your DNA betray you? *New Scientist.*

Ram, N., C.J. Guerrini, and A.I. McGuire. 2018, June 8. Genealogy databases and the future of criminal investigations. *Science* 360 (63930): 1078–1079

Schmidt, H., and C. Shawneequa. 2012, May. How anonymous is 'anonymous'? Some suggestions towards a coherent universal coding system for genetic samples. *Journal of Medical Ethics* 38 (5): 304–309

Shabani, M., and L. Marelli. 2019. Re-identifiability of genomic data and GDPR. *Embo Reports* 20e48316: 1–5.

Stoekle, H.-C., M.-F. Manzer-Bruneel, G. Vogt, and C. Hervé. 2016. 23andMe: A new two-sided data-banking model. *BMC Medical Ethics* 17: 19–27.

Wallace, S.E., E.G. Gourna, V. Nikolova, and N.A. Sheehan. 2015. Family tree and ancestry inference: Is there a need for 'generational' consent? *BMC Medical Ethics* 16: 87–96.

Sheldon Krimsky research has focused on the linkages between science/technology, ethics/values and public policy. His areas of specialization include biomedical sciences, bioethics, science and technology studies, risk assessment and communication, social history of science, and environmental health. He is the author of over 220 articles and reviews and author, co-author or editor of 17 books including: In Press*Understanding DNA Ancestry* (Cambridge Univ. Press). Other books include*Genetic Alchemy: The Social History of the Recombinant DNA Controversy* (MIT Press) 1982, *Biotechnics and Society: The Rise of Industrial Genetics* (Praeger) 1991, *Hormonal Chaos: The Scientific and Social Origins of the Environmental Endocrine Hypothesis* (Johns Hopkins University Press, 2000), *Science in the Private Interest: Has the lure of profits corrupted biomedical research* (Roman Littlefield); *Genetic Justice: DNA Databanks, Criminal Investigations, and Civil Liberties*, Independent Publishers Award (Columbia University Press) 2011; *Stem Cell Dialogues: A Philosophical and Scientific Inquiry into Medical Frontiers* (2015) Columbia University Press; *Conflicts of Interest in Science* (2019) Skyhorse Pub.; GMOs Decoded (2019) MIT Press.

Part IV
Neuroscience

Chapter 8
Ethical Issues in Neuroscience Research

Walter Glannon

8.1 Introduction

Despite advances in neuroscience, we still have a limited understanding of how the brain enables physical and mental capacities and how it disables them in traumatic injury and neurodevelopmental and neurodegenerative disorders. Continued research and translating experimental findings into therapies for these conditions will increase functional independence and improve quality of life for the millions of people affected by them. Mapping the brain and intervening in it with electrical stimulation or neural interfacing in proof-of-concept studies and clinical trials entail certain risks to research subjects. But these risks must be weighed against the potential benefit of knowledge gained from these techniques and their application in clinical settings.

The risk in neuroscience research is arguably greater than in any other area of medicine and biotechnology because the brain is the source of the mind and the psychological properties that define us as persons. Research subjects must have the cognitive and emotional capacity to understand the design and purpose of clinical trials and give informed consent to participate in them. When patients lack this capacity, family members may give proxy consent for them. Investigators have an obligation to protect research subjects from neurological and psychological sequelae while undergoing procedures to monitor or alter the brain. They also have an obligation to protect subjects from unauthorized third parties disrupting or inappropriately using information about their brains.

This chapter consists of four thematic sections. In Sect. 8.2, I examine problems in using neuroimaging data to explain or predict neuropsychiatric disease or criminal behavior. I also discuss how this data could be used to discriminate against people

W. Glannon (✉)
University of Calgary, Calgary, Canada
e-mail: wglannon@ucalgary.ca

who participate in imaging research, and how this could be prevented. In Sect. 8.3, I consider the therapeutic potential of neuromodulation for patients with prolonged disorders of consciousness. I discuss some of the challenges faced by patients, families and investigators in clinical trials testing the safety and efficacy of this technique. In Sect. 8.4, I discuss how brain-computer interfaces (BCIs) may enable paralyzed patients to move or communicate. I consider how unauthorized third parties could disrupt electrical signals in implantable neural interfaces and prevent subjects from completing intended motor tasks. Like the unauthorized use of neuroimaging data, brainjacking of implanted BCIs could violate research subjects' neural and mental privacy. In Sect. 8.5, I examine questions about fairness in access to neuroscience research and explain why excluding some patients with brain injuries from it may be ethically justified. I summarize the main points in the conclusion. I briefly mention possible future directions in neuroscience research and how the ethical issues we anticipate should influence how it is conducted.

In response to the US Brain Research through Advancing Innovative Neurotechnologies (BRAIN) initiative, a bioethics commission focused on the ethical and social implications of "cognitive enhancement, consent capacity and neuroscience and the legal system" (Phillips 2015, Preface). There is some overlap between these issues and the ones I address. But the three neural technologies on which I focus have a broader range of ethical, legal and social implications for their use in research and how they can affect people with nervous system disorders now and in the foreseeable future.

8.2 Neuroimaging: Explaining and Predicting Disease and Behavior

Computed tomography (CT) uses three-dimensional x-rays and magnetic resonance imaging (MRI) measures alignment of magnetic fields to generate images of anatomical brain features. Positron emission tomography (PET) measures brain function by displaying metabolic changes in different brain regions. Functional magnetic resonance imaging (fMRI) measures neural function by showing differences in the magnetic properties of oxygenated and deoxygenated blood. These and other brain imaging modalities have been used primarily to diagnose and monitor neuropsychiatric diseases. They may also be used to identify biomarkers that could predict their onset and development.

Structural and functional neuroimaging can detect biomarkers that "have the potential to explain effects of psychiatric disorders on the brain" (Calhoun and Arbabshirani 2013, p. 207; Singh et al. 2013). Abnormalities in the default mode network displayed by functional imaging, and reductions in gray matter volume displayed by structural imaging, are biomarkers for schizophrenia (Calhoun and Arbabshirani 2013, p. 210). Changes in the anterior cingulate cortex (ACC) detected by fMRI, diffusion tensor imaging (DTI) and voxel-based morphometry (VBM) can enable researchers to distinguish the first signs of unipolar depression from those of bipolar depression (Almeida and Phillips 2013). Yet even as researchers identify an increasing number of biomarkers for neuropsychiatric disorders, they have limited predictive value on their own. This is because of the multifactorial etiology of these disorders and the heterogeneity of symptoms in people affected by them. It is unlikely that a single biomarker will have a significant impact on diagnosis and treatment in psychiatry (Boksa 2013, p. 76). Neurobiological signatures associated with disease will not replace but combine with other investigative methods to confirm or predict its development, severity, responses to treatment and possible prevention.

Neuropsychiatric disease risk assessment based on neural biomarkers is probabilistic rather than deterministic. Researchers cannot draw direct inferences about future disease or behavior from populations to individuals within them. It could not be known whether an individual with a biomarker for depression, for example, would in fact develop this disorder. Informing a research subject in a neuroimaging clinical trial that she had a biomarker suggestive of a neuropathology without explaining the uncertainty surrounding it could unduly cause anxiety and harm her. Michael Rutter comments that, to be beneficial to patients, "inferences based on biomarkers must be shown to be robust and valuable at the individual level" (Rutter 2013, p. 189; Baum 2016). This underscores researchers' obligation to explain to subjects the uncertainty in probabilistic risk assessment using biomarkers in psychiatry in obtaining informed consent before enrolling them in imaging clinical trials. Researchers have the same obligations of respect for subjects' autonomy in providing all relevant information when obtaining consent and of nonmaleficence in protecting subjects from harm in imaging studies of early biological signatures associated with neurodegenerative diseases like Parkinson's and Alzheimer's (Emanuel et al. 2008, Part 8; Beauchamp and Childress 2019, Chaps. 4, 5).

The tendency to overgeneralize or oversimplify neuropsychiatric biomarker findings can have broader ethical and social implications over the course of a person's life. Ilina Singh and Nikolas Rose point out that, among children, "biomarker information might reshape the beliefs, practices and decision-making of the people in the child's environment, including parents, teachers and health providers" (Singh and Rose 2009, p. 204). This could influence how they interpret the child's behavior and their interactions with the child. The idea that a child might be predisposed to mental illness based on a biomarker could lead others to limit opportunities for him and the development of his autonomous agency. It could lead to different forms of discrimination. Insurance companies with access to biomarker information could deny medical insurance to those deemed at risk of developing a costly mental illness. For adults with an illness, companies could deny medical coverage based on a biomarker as evidence of a pre-existing condition. They could deny employment for similar reasons. More disturbing is the possible use of brain biomarkers to construct profiles of future antisocial or criminal behavior in children and adolescents. This information could be used to discriminate against people on scientifically questionable and ethically unacceptable grounds.

Biomarkers identified by brain imaging will likely be an important component in more accurate diagnosis and effective treatment of neuropsychiatric disorders. But more longitudinal studies are needed to clarify their explanatory significance and predictive value. They are only one component among other biological, psychological and environmental components in the development and progression of diseases of the brain and mind. Researchers should be especially circumspect in or avoid discussing the possible use of biomarkers to predict future behavior with research subjects. Singh and Rose emphasize that "prospective research on these issues is needed to inform policies and practices that will maximize the positive potential of biomarker information and protect individuals and families from harm" (Singh and Rose 2009, p. 204). The ethical and social considerations that I have discussed should inform how the research is conducted and translated into clinical practice.

Neuroimaging can also reveal the neural correlates of reasoning and decision-making. They are associated with the orbitofrontal cortex (OFC), ventromedial prefrontal cortex (vmPFC) and the ACC (Bechara et al. 2000; Pessoa 2013). But the ability of imaging to explain or predict behavior, specifically criminal behavior, has been questioned (Jones 2013; Freeman 2010; Morse 2010). It remains an experimental tool in forensic psychiatry. Neuroimaging could be combined with biomarkers of DNA methylation in the blood correlating with aggressive behavior (van Dongen et al. 2021). But this combination would not be enough to establish a causal connection between these epigenetic signatures and criminal acts because of differences between peripheral blood and the brain and variability in how these signatures are expressed in behavior (Morse and Roskies 2013).

There are epistemic gaps between functional neuroimaging and behavior that underscore its limitations as a technique for explaining it (Roskies 2008). The blood-oxygenation level-dependent (BOLD) signal in fMRI lags several seconds behind the brain activity it records and thus is not a direct measure of this activity. The signal-to-noise ratio in fMRI is another complicating factor. This measures the amount of relevant information (signal) that is corrupted by junk information (noise). The ratio is too low in a single scan of one brain for it to have any neurophysiological value. Because of this, images from fMRI have to be averaged over scans of many brains to be statistically significant (Roskies 2013a, 2013b). In addition, the fMRI signal cannot distinguish between excitatory and inhibitory neural activity. This makes it difficult to know whether brain activity displayed on a scan has an enabling or disabling effect on a person's mental capacities.

One meta-analysis showed a false-positive rate of up to 70% among researchers interpreting data from fMRI results. The authors pointed out that spatial autocorrelation in statistical analyses of the data may lead investigators to make misleading claims about brain activity (Eklund et al. 2016). Also, the content of the intention and other mental states of a criminal offender at the time of an action are not reducible or identical to the brain activity that generates and sustains it. Identifying the neural correlates of mental states is not equivalent to identifying these states themselves. While they have a neural underpinning, the content of an agent's mental states is inferred from his actions. All these factors show that there is considerable inferential distance between imaging data and a person's behavior. This highlights the explanatory limitations of brain imaging.

Unless imaging showed significant structural and functional abnormalities in the OFC, vmPFC, ACC and other brain regions mediating reasoning and decision-making, a brain scan or series of scans showing some abnormalities would not demonstrate that a person was unable to control his behavior, had difficulty controlling it, or was able but failed to control it when committing a criminal act (Glannon 2010, 2014). Also, in both healthy and diseased brains, neural activity can change between an earlier time when an individual committed a criminal act and a later time then his brain is scanned. Differences in neural activity over time raise questions about using neuroimaging to determine whether an individual had or lacked cognitive control when he acted. These differences confound making judgments of responsibility, mitigation or excuse (Glannon 2015).

In cases where the behavioral evidence was ambiguous in determining whether a person had enough cognitive control of an action to be responsible for it, information from structural and functional neuroimaging could clarify this question. But imaging alone would not provide a conclusive answer to it because control involves more than processes in one's brain. It also involves the content of one's mental states as representations of external events and how one perceives and responds to social cues when acting. Neuroimaging cannot determine that a defendant was capable or incapable of appreciating the wrongfulness of his actions. This capacity depends on normal cortical function but is not identical to it. Moral reasoning cannot be measured by images of cerebral blood flow or glucose metabolism because it is not located in the brain.

Some researchers argue that brain abnormalities detected by imaging may serve as biomarkers to predict future criminal behavior. These biomarkers could be used to establish recidivism rates in predicting which criminal offenders would re-offend, and which might be rehabilitated. In one study involving 96 male prisoners asked to perform a computer task, those with lower activity in the ACC detected by fMRI were 2-to-3 times more likely to be re-arrested after being released from prison (Aharoni et al., 2013; Slobogin 2013). This study suggests that functional imaging could monitor the activity of offenders and serve as part of a harm reduction program in reducing the probability of repeat offences.

Predicting uncontrollable criminal behavior presumably would depend on a certain degree of neural dysfunction. This in turn would depend on what neuroscientists deemed the necessary degree of dysfunction that would be predictive. Consensus among neuroscientists on what this degree would be may be difficult to reach (Jones et al. 2013). Again, data-based behavioral predictions would be probabilistic rather than deterministic. A brain abnormality will not necessarily manifest in criminal behavior. It may be a risk factor for crime. But many people may not re-offend despite a neural or mental abnormality. Among those who re-offend, an abnormality alone would not answer the question of whether they had or lacked the capacity to refrain from performing criminal acts (Morse 2010, 2015). In these respects, functional neuroimaging would have questionable predictive value. Neuropsychological testing could be used in combination with behavioral evidence and brain imaging to refine prediction of future criminal behavior. This could indicate an increased statistical probability of an offender re-offending. But this increase may not be enough to establish a causal connection between these measures of the person's brain and mind and future criminal acts.

Artificial intelligence (AI) could also be used to predict recidivism. An algorithm incorporating information about an offender's past behavior could generate a profile indicative of future behavior. However, like the other predictive techniques I have discussed, AI would be probabilistic rather than deterministic and would not definitively show that an offender would perform specific actions. More fundamentally, algorithms used for this purpose are imperfect and may create an illusion of accuracy and fairness in assessing defenders' future behavior (Dressel and Farid 2018).

Longitudinal studies have shown associations between structural brain abnormalities and persistent life-course antisocial behavior (Carlisi et al. 2020). Some might claim that imaging identifying these abnormalities in children or adolescents with behavioral problems could be used as one component of a program to modify the social environment and have a salutary effect on their developing brains. If imaging were ethically justified for this purpose, then competent adolescents and the parents of noncompetent children would have to give informed consent or proxy consent to have their brains scanned. They could refuse to do this as an expression of their individual and parental autonomy (Emanuel et al. 2008, Part 8). This could limit the use of neuroimaging as part of early behavioral intervention.

The problem of inferential distance between imaging data and criminal behavior raises questions about the probative value of imaging in the criminal law. This distance casts doubt on the idea that this data could be a reliable basis for normative assessments of persons' behavior. It goes some way to explaining why neuroscience has not resulted in a move away from behavioral and mental criteria in judgments of moral and criminal responsibility. While the most significant advances in brain imaging have been primarily in functional imaging, structural imaging has been more influential in criminal cases. But imaging alone will not provide conclusive answers to questions about control and responsibility.

More advanced imaging techniques may show stronger correlations between neural processes and the mental capacities necessary for reasoning, decision-making and action. This may give them a greater role in forensic psychiatry research and applications. These techniques may include DTI, which can detect changes in white-matter volume and connectivity, VBM, which can detect gray-matter changes, diffusion functional MRI (dfMRI) and diffusion spectrum imaging (DSI) (Roskies 2013a). Still, correlation is not causation. These or other techniques may not provide a conclusive answer to the question of whether or to what extent structural and functional brain abnormalities impair rational and moral agency. What matters in normative assessments of behavior is not brain processes as such. Rather, what matters is whether or to what extent these processes enable or disable the mental capacities necessary for behavior control.

The inferential distance between brain activity and brain images, on the one hand, and between neural and mental processes, on the other, shows the limitations of neuroimaging in confirming or predicting criminal behavior and justifying normative responses to it. Brain scans can show the neurobiological underpinning of reasoning and decision-making. But they cannot determine whether a person lacked the capacity to control her behavior at a specific time, had difficulty controlling it, or had this capacity but failed to exercise it. In light of these limitations, in the foreseeable future neuroimaging will likely supplement rather than supplant behavioral criteria in judgments about responsibility, mitigation and excuse in the criminal law.

In clinical trials conducted to answer questions about the relation between the brain and behavior, brain scans of subjects whose behavior is under investigation are compared with scans of controls. A potentially serious ethical, social and legal problem is unauthorized third-party access to neuroimaging data. It can be a challenge for this information to remain within the research setting. When subjects consent to participate in these studies, they assume that the information will be accessible only to the researchers conducting them. They may also be accessible to medical specialists if incidental findings warrant referral to them. Other parties should be prohibited from having access to this information, unless the subject consents to this as well. Unauthorized access should be prohibited based on autonomy and the negative right to non-interference in one's body and brain. The scope of this right includes not only a prohibition against altering the brain but also accessing and using information about it. As noted in the discussion of brain biomarkers, some parties could make invalid inferences from neuroimaging data to a person's future behavior. Prospective employers could prejudicially use data associated with clinically ambiguous or benign brain features to make hiring decisions based on the questionable risk of the person developing a neurological or psychiatric disorder. Prospective insurers could also draw invalid inferences about risk from the data to deny persons life or medical insurance.

There has been at least one actual example of this consequence of research participation. In a letter to *Nature*, an individual who volunteered for an MRI brain study reported having to contest insurance eligibility after a tumor was incidentally discovered in his carotid artery to the left of his brainstem. He emphasized the obligation of investigators to point out this and other potential consequences of undergoing brain scans in the process of obtaining informed consent for those who volunteer for this research (Anonymous 2005). The obligation to protect research subjects from the use of brain information that violates individual autonomy and privacy applies both to those in the experimental group and those in the control group. Just as the 2008 US Genetic Information Nondiscrimination Act (GINA) prevents discrimination based on a person's genetic profile, laws should be implemented and enforced to prevent discrimination based on information about a person's brain.

8.3 Neurostimulation for Prolonged Disorders of Consciousness

Millions of people globally have long-term motor and cognitive disabilities from different types of brain injury. These include traumatic brain injury (TBI) and anoxia/hypoxia of the brain. They often progress from coma as the immediate effect of injury. Coma involves extensive injury to the cortex, axonal connections, underlying white matter tracts and the thalamus. It is a state of complete unresponsiveness (Giacino et al. 2014). Some comatose patients eventually lose all integrated brain functions and are declared brain-dead. Others regain full consciousness, usually within 2–4 weeks of injury. Still others progress from coma to the vegetative state, (VS), in which they show arousal and have sleep–wake cycles but are unaware of themselves and their surroundings. The vegetative state has also been described as "unresponsive wakefulness syndrome" (Laureys et al. 2010). Most neurologists consider a persistent VS to become a permanent VS 3 months after an anoxic injury or 12 months after a TBI.

Some patients in a persistent VS progress to a minimally conscious state (MCS). This is "a condition of severely altered consciousness characterized by minimal but definite behavioral evidence of self or environmental awareness" (Giacino et al. 2014, p. 100). Emergence from the MCS is defined as: "the re-emergence of a functional communication system or restoration of the ability to use objects in a functional manner" (Giacino et al. 2014, p. 101). Emergence from this state is unpredictable and may or may not occur two to five years after brain injury (Fins 2015, p. 163).

The VS and MCS have been described as prolonged disorders of consciousness (DOCs) (Giacino et al. 2014). Deep brain stimulation of the central thalamus may promote recovery from these disorders because of its ability to directly target neural pathways mediating awareness, cognition and motor functions. DBS might induce axonal regeneration in brain regions that have been damaged and dysfunctional for an extended period. It might ameliorate motor and cognitive functions that are impaired following TBI. It might not have this potential in patients with anoxic or hypoxic injury and extensive axonal damage. In an FDA-approved clinical trial of the first use of DBS for a patient in the MCS, Greg Pearson, who had been in this state for 6 years following an assault, recovered some motor and cognitive capacities from this technique. DBS was administered over a 12-month period between 2006 and 2007 (Schiff et al. 2007). This trial tested the hypothesis that electrical stimulation of the central thalamus could modulate the meso-circuit consisting of the thalamus, basal ganglia and frontal cortex and restore some degree of functional recovery. The degree of recovery in Pearson (who died in 2011) showed the therapeutic potential of DBS for this patient population (Fins 2015, pp. 233 ff.).

Yet DBS remains an experimental neuromodulating technique with limited efficacy in restoring cognitive and motor functions in severely brain-injured patients. In the 14 years since the first human trial, studies have not shown significant improvement from the technique for most of these patients. The results of a study published in 2017 showed that thalamic stimulation of vegetative and minimally conscious patients "did not induce persistent, clinically evident conscious behavior in the patients" (Magrassi et al. 2016). There are important neurological differences between vegetative and minimally conscious patients, and these differences can determine whether neurostimulation has therapeutic potential for them Bur even those with more preserved brain function may not experience meaningful recovery from this technique.

In a more recent study involving 14 patients in the VS or MCS, only two regained consciousness and the ability to live independently (Chudy et al. 2018). Responding to the outcome of this study, Emad Eskandar points out that "a small number of patients with VS or MCS will spontaneously recover. It is unclear whether the number of patients that recovered during this study is significantly different from what would have been expected based on the rate of spontaneous recovery" (Eskandar 2018, p. 1187). Neuroimaging identifying areas of preserved brain function as potential targets of stimulation could enable researchers to identify patients with these brain features as more likely to respond to stimulation.

Still, outcomes with a fairly high success rate (above 30%) would be necessary to establish that it had a favorable benefit-risk ratio regarding patient recovery (Emanuel et al. 2008, Part 6). DTI has been used in one study to accurately predict the extent of chronic neurodegeneration in people with traumatic brain injury (Graham et al. 2020). This would allow researchers to exclude patients with diffuse axonal injury from clinical trials testing DBS to restore cognitive and motor functions. Yet among those with less diffuse injury who are included in these trials, predicting recovery from intermittent to full awareness and functional independence from neurostimulation remains fraught with uncertainty.

Commenting on the application of DBS to patients with TBI, Samuel Shin and co-authors state: "Major challenges in developing these approaches include the identification of specific phenotypes of TBI patients whose symptoms are amenable to therapeutic DBS... Given the expanding view of the potential to modulate functional brain networks with spatially precise DBS delivery, structural and functional neuroimaging studies should be used in future TBI trials to target biomarkers of functional deficits in a patient-specific fashion" (Shin et al., 2014, p. 1226). DTI and algorithmic EEG, for example, could identify biomarkers of neural deficits and intact neural functions to identify patients who would be more likely to respond to neurostimulation.

Recently, DBS resulted in significant functional recovery for a patient. In 2019, researchers reported that an implant delivering electrical stimulation to an area of the brain of a woman with a TBI from an automobile accident 18 years earlier restored near-normal levels of neural function. She is now able to live independently (Thibaut et al. 2019). More careful selection of patients with certain neural signatures and advanced brain stimulation may restore near-normal, or even normal, levels of cognitive and motor capacities and functional independence in more of these patients. Currently, though, the number of cases with these positive outcomes is low.

More placebo-controlled clinical trials are needed to determine whether these outcomes can be improved. There are challenges in conducting this research. The number of patients with enough preserved thalamocortical and corticocortical connections to be candidates for clinical trials may be small. This can affect the statistical significance of trial outcomes. Also, the design of a DBS clinical trial for post-coma DOCs may not be sensitive to the fact that some patients with these disorders may recover spontaneously without neurostimulation (Eskander 2018). It could be especially problematic if the trial involves subjects in the first two or three years after

TBI. Spontaneous recovery is more likely during this period than later. Neurostimulation during the early post-TBI period may be more likely to produce positive effects as well. This could confound interpreting outcomes and establishing the efficacy of DBS for these research subjects. These are among the "logistical and methodological difficulties of conducting placebo-controlled trials in this population" (Giacino et al. 2014, p. 107).

These trials have ethical implications. If a minimally conscious patient lacked the capacity to consent, his family could give proxy consent to allow him to participate as a subject in a study assessing the ability of DBS to induce recovery (Emanuel et al. 2008, Part 8; Appelbaum 2007). As their substitute decision-makers, families generally act in the best interests of noncompetent patients. Participation in research yielding scientific knowledge about DBS that led to its therapeutic use for this and other prolonged disorders of consciousness would be in the patient's best interests. Placebo-controlled trials would be the most scientifically robust way to demonstrate whether DBS could enable patients to emerge from the MCS and recover cognitive and motor functions.

Some families may be motivated by a therapeutic misconception based on hope of positive outcomes (Mathews et al. 2018). This misconception may be stronger if the trial begins not long after a TBI and the family believes that the therapeutic potential of the technique is greater if it is applied sooner rather than later. They may have unrealistic expectations and give proxy consent believing that the research will directly benefit the patient. Despite being informed by researchers, they may not duly consider that the primary goal of a clinical trial is to generate knowledge about the safety and efficacy of the technique. The lack of any therapeutic alternatives may make some families seize the opportunity to have a parent, child or sibling as a research subject in such a trial.

Yet knowing that the patient may be assigned to the placebo arm of the trial may disincline some families from giving proxy consent. This could affect the number of subjects enrolled in a trial and the statistical significance of the results. Processing information about the goals and technical details of a trial could be difficult for substitute decision-makers if they are still adjusting to the neurological and emotional fallout from having a loved one with a severe brain injury. This may be especially difficult for them if trial enrolment begins not long after a TBI. Proxy consent for a patient to be in a clinical trial years after the injury may avoid this problem. It may raise a different problem, however. Families may consent motivated by physical and emotional fatigue or exhaustion from caring for the patient for an extended period. It is also possible that, at this point, a patient may no longer meet inclusionary criteria for research participation because of extensive neurodegeneration from the injury. These are some of the research challenges that must be resolved in determining whether DBS is a safe and effective treatment for the MCS and other prolonged disorders of consciousness.

8.4 Brain-Computer Interfaces for Movement and Communication

A brain-computer interface (BCI), or brain-machine interface (BMI) "is a computer-based system that acquires brain signals, analyzes them, and translates thein into commands that are relayed to an output device to carry out a desired action" (Shih et al. 2012, p. 268; Wolpaw and Wolpaw 2012a, 2012b; Ramsey and Millan 2020). BCIs utilize wired or wireless systems to record electrical signals in neural ensembles in premotor, motor and supplementary motor areas, the prefrontal and posterior parietal cortex and other cortical regions. They then transmit these signals as sensorimotor input to an external device in producing motor output. This includes moving a computer cursor, prosthetic limb, robotic arm, or selecting letters from a wheelchair-mounted computer tablet to communicate. The interface can enable a person to overcome motor impairment from traumatic brain injury, stroke, limb loss or neurodegenerative diseases. It can re-establish sensorimotor capacities and some degree of functional independence (Lebedev and Nicolelis 2017). Many people with different nervous system disorders could benefit from this technology with greater functional independence and improved quality of life.

BCIs are still at an early stage of development in experimental neuroscience. Some of these systems consist of scalp-based EEG to record and transmit brain signals to a computer. Others consist of electrodes implanted epidurally or subdurally. A third type consists of microelectrode arrays implanted in the cortex. They have been tested in studies involving a small number of patients and are some way from being translated into standard therapy for paralysis and other forms of motor impairment. Not all paralyzed patients are appropriate subjects to use BCIs in research settings. Depending on the type and extent of brain injury, some may not be capable of the goal-directed thinking or responding to operant conditioning from investigators. They may be too cognitively impaired for the planning, sustained attention and task execution to successfully use a BCI. The main ethical question in these cases is not whether they could give informed consent. Rather, the question is whether they would be vulnerable to harm from unreasonable or unrealizable expectations of restoring motor control by using the interface. This suggests that researchers have an obligation of nonmaleficence to exclude patients who lack the requisite cognitive capacities from BCI research on the grounds that it may be more likely to harm than benefit them (Beauchamp and Childress 2019, Chap. 5).

Many hours of training are necessary to successfully operate a BCI. There is considerable variation among users in the time required to use it (Wolpaw and McFarland 2004). Depending on their cognitive, emotional and volitional capacities, some subjects may be more successful in using it than others. Success or failure in moving objects or communicating with a BCI is influenced by the extent of brain injury or disease. It is also influenced by the mental effort of the subject in selecting a goal and achieving it, as well as how the user interacts with the investigator.

Factors in addition to the inability to sustain attention, focus and patience may explain the failure of some subjects in trying to use BCIs (Birbaumer et al. 2014). The patient's experience of losing motor control from paralysis or limb loss can weaken or undermine the motivation to learn to use them. The intensity of the learning process can contribute to mental fatigue. In the case of a robotic arm or prosthetic limb, failure to move them could impair proprioceptive and somatosensory feedback from the arm or limb to the brain and mind and influence the subject's perception of these objects. Instead of perceiving them as forms of extended embodiment, they might perceive them as foreign objects that interfere with rather than promote restoration of motor function. Investigators must take all these factors into account in conducting studies to test the safety and efficacy of BCIs.

One possible unwanted intervention in neural prosthetics such as BCIs is brain-jacking. Hackers could gain unauthorized access to these systems through physiological attack vectors (Pycroft et al. 2016; Pugh et al. 2018). They could disrupt electrical signals in the brain and prevent subjects from performing intended motor tasks or force them to perform unintended actions. In addition, hackers could violate neural implant security and the subject's neural and mental privacy by accessing information from intracortical implants. Device-makers could design systems to prevent this intervention, or at least reduce the risk of this from occurring. Indeed, because of the potential harm to BCI users, manufacturers would have an ethical obligation to design and produce systems that would avoid these violations. But more sophisticated hackers may find ways of subverting implant security mechanisms.

Alternatively, BCI users could wear a separate security device that could detect and neutralize any disruption in the transmission of electrical signals in the brain to the computer for motor output. Although such a device may not require any mental effort from the subject, it could add to their mental burden and anxiety about success or failure in operating a neural interface Combined with the challenges of using a BCI, an additional device could be burdensome for the user and have a negative effect on their experience. This would provide an ethical obligation for device-makers to design BCIs that would be immune to unauthorized third-party access. It would also provide an ethical obligation for investigators to work closely with manufacturers to prevent interference in the operation of these systems. This would pertain both to protecting research subjects from harm and ensuring the scientific integrity of a BCI study and the reliability of its outcomes.

8.5 Access to Neuroscience Research

In the last section, I claimed that investigators conducting studies involving BCIs had an obligation of nonmaleficence to prevent harm to research subjects using these systems. This obligation could mean not selecting potential subjects deemed cognitively incapable of learning to use a BCI to move objects or communicate. It may seem unfair to exclude some patients from and include other patients in experimental uses of the technology when they all have severe motor impairment and the same need. Moreover, it may seem unfair to exclude some people with paralysis or other forms of motor impairment from access to a BCI when it is the only system that would restore some degree of motor control and functional independence.

Fairness consists in meeting claims in proportion to their strength. It is a measure of how claims to a good are adjudicated and how the good is distributed. A fair distribution of a scarce resource in health care is one that gives priority to meeting stronger claims of need (Daniels 2008, Chap. 4; Segall 2010, Part II). Patients with the most severe brain injuries have a greater need for potentially restorative therapy than those with less severe injuries. But if they lack the cognitive capacity to be research subjects or complete tasks in a study, then they could not contribute to scientific knowledge of an intervention in the brain and its therapeutic potential. It would not be unfair to exclude a brain-injured patient from a BCI study on these grounds. It could be ethically justified to treat some patients in this group unequally by denying them access to the technology in a research setting. The critical factor in this inequality would be the extent of brain injury and its effects on cognitive and motor functions.

This same reasoning applies to other areas of neuroscience research requiring subjects to complete cognitive and motor tasks. There are two ethically relevant issues here: the cognitive capacity to give informed consent to participate in a proof-of-concept study or clinical trial; and the cognitive capacity to perform tasks required of the research. Access to research would depend on these two capacities. Researchers would not be harming or unfairly treating patients who lacked these capacities by not allowing them to participate in research. On the contrary, they would be discharging their duty of nonmaleficence by acting in this way.

For patients enrolled in a clinical trial testing a device that provided them with the only therapy for their condition, denying them access to the device beyond the conclusion of the trial may seem unfair. In cases where the trial showed a clear benefit, patients could lose access to a therapeutic intervention if the device-maker decided to discontinue providing it. This question depends on the financial interests of the manufacturer and whether the outcome of a trial indicates a good return on their investment in it. Patients may also lose access because of the costs of continued neuromodulation or neural interfacing (Underwood 2015, 2017). This could be especially problematic in a market-based health care system like that in the US. It could be problematic as well if not all costs were covered by a universal single-payer system. Research subjects have a direct relationship with investigators but only an indirect relationship with the device-maker. Thu second relationship may be too weak to claim that a manufacturer such as Medtronic or St. Jude Medical had an ethical obligation to provide a device to patients indefinitely. Ideally, device makers could lower the costs of their systems. But this by itself would not solve all problems surrounding access.

In studies and trials of neural prosthetics showing no clear benefit and some risk, there would be no obligation of device-makers and investigators to continue providing them to patients. Instead, there would be an obligation to discontinue a trial and access to the technology. Consider the BROdmann Area 25 Deep brain Neuromodulation (BROADEN) trial. This was a study of deep brain stimulation for treatment-resistant depression. St. Jude Medical, the implant manufacturer, stopped the trial before its intended completion because the probability of a successful outcome at a critical stage of the trial was very low. There was no statistically significant response between active and sham control groups, and 29 of the 90 participants experienced adverse events (Holtzheimer et al. 2017).

Despite these outcomes, 44 participants wanted to keep their implants (Underwood 2017). The discrepancy between the actual response and subjects' claim that they responded to neurostimulation could be attributed to their belief that it was their last hope of controlling symptoms. But this would not generate an obligation for the manufacturer and investigators to continue providing the implant. This study shows that claims of need must be weighed against whether or to what extent these claims can be met. Both factors must be components of a fair and just policy about access to neurotechnology.

8.6 Conclusion

I have examined the main ethical, legal and social issues in three areas of neuroscience research. Emerging and future neurotechnologies will also generate questions about how research should be developed and conducted and how it should be translated into clinical applications for people affected by nervous system disorders. This includes more advanced neuroimaging with high spatial and temporal resolution and closed-loop neuromodulation with real-time responses to changes in the brain. It also includes BCIs with interfacing modalities that can augment motor and cognitive functions in people with brain injury or disease, as well as in healthy people. As an example of brain-body interaction, identifying biomarkers in maternal plasma reactivity could be used to more accurately predict neurodevelopmental disorders in children. This could allow for more effective behavioral interventions (Ramirez-Celis et al. 2021). Like other areas of medical research, investigators testing different techniques to measure or modify brain function have an obligation to obtain informed consent from patients they enroll as research subjects, or proxy consent from others acting in their best interests. They also have an obligation to protect research subjects from harm associated with untoward effects of experimental interventions in the brain. Fair selection criteria of patients with brain injuries or neurodegenerative diseases depend on the nature of the injury or disease and the extent to which they retain the cognitive and motor capacities necessary to perform experimental tasks.

Investigators conducting imaging studies designed to predict future neuropsychiatric disorders or diagnose criminal behavior must also protect subjects from the harm of discriminatory use of information about their brains. They must be circumspect in evaluating biomarkers that are probabilistic rather than deterministic about disease and behavior. The same protections given to patients with brain injuries should be given to asymptomatic research subjects undergoing neuroimaging. Protections should also include systems that reduce or eliminate the risk of brainjacking in DBS and BCIs.

Ethical discussions of neuroscience are often a reaction to new developments in this field. This can limit the impact of these discussions and their goal of pointing out how this research can affect subjects. Yet while anticipating ethical, legal and social issues may promote fruitful discussion when then arise, it can also lead to speculation about possible scenarios that never occur. Some developments in neuroscience are not foreseeable, and we can only consider their normative implications when they develop. In general, though, ethical questions should inform and be informed by science. This interaction between empirical and normative domains will appropriately guide and promote the translation of neuroscience research into therapeutic interventions in the brain.

Acknowledgments I am grateful to three anonymous reviewers for helpful comments on an earlier version of this chapter.

References

Aharoni, E., G. Vincent, C. Harenski, V. Calhoun, W. Sinnott-Armstrong, M. Gazzaniga, et al. 2013. Neuroprediction of future arrest. *Proceedings of the National Academy of Sciences* 110: 6223–6228.

Almeida, J., and M. Phillips. 2013. Distinguishing between unipolar depression and bipolar depression: Current and future clinical and neuroimaging perspectives. *Biological Psychiatry* 73: 111–118.

Anonymous. 2005. How volunteering for an MRI scan changed my life. *Nature* 434: 17.

Appelbaum, P. 2007. Assessment of patients' competence to consent to treatment. *New England Journal of Medicine* 357: 1834–1840.

Baum, M. 2016. *The neuroethics of biomarkers: What the development of bioprediction means for moral responsibility, justice and the nature of mental disorder*. Oxford: Oxford University Press.

Beauchamp, T., and J. Childress. 2019. *Principles of biomedical ethics*. New York: Oxford University Press.

Bechara, A., A. Damasio, and H. Damasio. 2000. Emotion, decision-making and the orbitofrontal cortex. *Cerebral Cortex* 10: 295–307.

Birbaumer, N., G. Gallegos-Ayala, M. Wildgruber, S. Silvoni, and S. Soekadar. 2014. Direct brain control and communication in paralysis. *Brain Topography* 27: 4–11.

Boksa, P. 2013. A way forward for research on biomarkers for psychiatric disorders. *Journal of Psychiatry and Neuroscience* 38: 75–77.

Calhoun, V., and M. Arbabshirani. 2013. *Neuroimaging based automatic classification of schizophrenia*, 206–230. Singh, Sinnott-Armstrong and Savulescu.

Carlisi, C., T. Moffitt, A. Knodt, H. Harrington, D. Ireland, T. Melzer, et al. 2020. Associations between life-course persistent antisocial behaviour and brain structure in a population-representative longitudinal birth cohort. *The Lancet Psychiatry* 7: 245–253.

Chudy, D., V. Deletis, F. Almahariq, P. Marcinkovic, J. Skrlin, and V. Paradzik. 2018. Deep brain stimulation for the early treatment of the minimally conscious state and vegetative state: Experience in 14 patients. *Journal of Neurosurgery* 128: 1189–1198.

Daniels, N. 2008. *Just health: Meeting health needs fairly*. New York: Cambridge University Press.

Dressel, J., and H. Farid. 2018. The accuracy, fairness, and limits of predicting recidivism. *Science Advances* 4: eaao5580. https://doi.org/10.1126/sciadv.aa05580.

Eklund, A., T. Nichols, and H. Knutsson. 2016. Cluster failure: Why fMRI inferences for spatial extent have inflated false positive rates. *Proceedings of the National Academy of Sciences* 113: 7900–7905.

Emanuel, E., C. Grady, R. Crouch, R. Lie, F. Miller, and D. Wendler, eds. 2008. *The Oxford textbook of clinical research ethics.* New York: Oxford University Press.

Eskandar, E. 2018. Thalamic stimulation in vegetative or minimally conscious patients. *Journal of Neurosurgery* 128: 1187–1188.

Fins, J. 2015. *Rights come to mind: Brain injury, ethics and the struggle for consciousness.* New York: Cambridge University Press.

Freeman, M., ed. 2010. *Law and neuroscience: Current legal issues.* Oxford: Oxford University Press.

Giacino, J., J. Fins, S. Laureys, and N. Schiff. 2014. Disorders of consciousness after acquired brain injury: The state of the science. *Nature Reviews Neurology* 10: 99–114.

Glannon, W. 2010. *What neuroscience can (and cannot) tell us about criminal responsibility*, 13–28. Freeman.

Glannon, W. 2014. The limitations and potential of neuroimaging in the criminal law. *Journal of Ethics* 18: 153–170.

Glannon, W., ed. 2015. *Free will and the brain: Neuroscientific, philosophical and legal perspectives.* Cambridge, UK: Cambridge University Press.

Graham, N., A. Jolly, K. Zimmerman, N. Bourke, G. Scott, J. Cole, et al. 2020. Diffuse axonal injury predicts neurodegeneration after moderate-severe traumatic brain injury. *Brain* 143: 3685–3698.

Holtzheimer, P., M. Husain, S. Lisanby, S. Taylor, L. Whitworth, S. McClintock, et al. 2017. Subcallosal cingulate deep brain stimulation for treatment-resistant depression: A multisite, randomized, sham-controlled trial. *The Lancet Psychiatry* 11: 839–849.

Jones, O., A. Wagner, D. Faigman, and M. Raichle. 2013. Neuroscientists in court. *Nature Reviews Neuroscience* 14: 730–736.

Laureys, S., G. Celesia, F. Cohadon, J. Lavrijsen, J. Leon-Carrion, W. Sannita, et al. 2010. Unresponsive wakefulness syndrome: A new name for the vegetative state or apallic syndrome. *BMC Medicine* 8: 65. https://doi.org/10.1186/1741-015-8-68.

Lebedev, M., and M. Nicolelis. 2017. Brain-computer interfaces: From basic science to neuroprostheses and neurorehabilitation. *Physiological Reviews* 97: 769–837.

Magrassi, L., G. Maggioni, C. Pistarini, C. Di Perri, S. Bastianello, A. Zippo, et al. 2016. Results of a prospective study (CATS) on the effects of thalamic stimulation in minimally conscious and vegetative patients. *Journal of Neurosurgery* 125: 972–981.

Mathews, D., J. Fins, and E. Racine. 2018. The therapeutic misconception: An examination of its normative assumptions and a call for its revision. *Cambridge Quarterly of Healthcare Ethics* 27: 154–162.

Morse, S., and A. Roskies, eds. 2013. *A primer on criminal law and neuroscience.* New York: Oxford University Press.

Morse, S. 2010. *Lost in translation: an essay on law and neuroscience*, 529–562. Freeman.

Morse, S. 2015. *Neuroscience, free will and criminal responsibility*, 251–286. Glannon.

Pessoa, L. 2013. *The cognitive-emotional brain: From interactions to integration.* Cambridge, MA: MIT Press.

Phillips, M., ed. 2015. *Ethical issues in neuroscience research: Integrative approaches and paths to progress.* New York: Nova Publishers.

Pugh, J., L. Pycroft, A. Sandberg, T. Aziz, and J. Savulescu. 2018. Brainjacking in deep brain stimulation and autonomy. *Ethics and Information Technology* 20: 219–232.

Pycroft, L., S. Boccard, S. Owen, J. Stein, J. Fitzgerald, A. Greer, et al. 2016. Brainjacking: Implant security issues in invasive neuromodulation. *World Neurosurgery* 92: 454–462.

Ramirez-Celis, A., M. Becker, M. Nuno, J. Schauer, N. Aghaeepour, and J. Van de Water. et al. (2021). Risk assessment analysis for maternal autoantibody-related autism (MAR-ASD): a subtype of autism. *Molecular Psychiatry* 26. https://doi.org/10.1038/s41380-020-00998-8.

Ramsey, N., and J. Millan, eds. 2020. *Handbook of clinical neurology: Brain-computer interfaces.* Amsterdam: Elsevier.

Roskies, A. 2008. Neuroimaging and inferential distance. *Neuroethics* 1: 19–30.

Roskies, A. 2013a. *Brain imaging techniques*, 37–74. Morse and Roskies.

Roskies, A. 2013b. *Other neuroscientific techniques*, 75–88. In Morse and Roskies.

Rutter, M. 2013. *Biomarkers: potential and challenges*, 188–205. In Singh, Sinnott-Armstrong and Savulescu.

Schiff, N., J. Giacino, K. Kalmar, J. Victor, K. Baker, M. Gerber, et al. 2007. Behavioural improvements with thalamic stimulation after severe traumatic brain injury. *Nature* 448: 600–603.

Segall, S. 2010. *Health, luck and justice.* Princeton: Princeton University Press.

Shih, J., D. Krusienski, and J. Wolpaw. 2012. Brain-computer interfaces in medicine. *Mayo Clinic Proceedings* 87: 268–279.

Shin, S., E. Dixon, D. Okonkwo, and M. Richardson. 2014. Neurostimulation for traumatic brain injury: A review. *Journal of Neurosurgery* 121: 1219–1231.

Singh, I., W. Sinnott-Armstrong, and J. Savulescu, eds. 2013. *Bioprediction, biomarkers and bad behavior: Scientific, legal and ethical challenges.* Oxford: Oxford University Press.

Singh, I., and N. Rose. (2009). Biomarkers in psychiatry. *Nature* 202–207.

Slobogin, C. (2013). *Bioprediction in criminal cases*, 77–90. Singh, Sinnott-Armstrong and Savulescu.

Thibaut, A., N. Schiff, J. Giacino, S. Laureys, and O. Gosseries. 2019. Therapeutic interventions in patients with prolonged disorders of consciousness. *Lancet Neurology* 18: 600–614.

Underwood, E. 2015. Brain implant trials raise ethical concerns. *Science* 348: 1186–1187.

Underwood, E. 2017. Brain implant trials spur ethical discussions. *Science* 358: 710.

Van Dongen, J., F. Hagenbeek, M. Suderman, P. Roetman, K. Sugden, A. Chiocchetti, et al. 2021. DNA methylation signatures of aggression and closely related constructs: A meta-analysis of epigenome-wide studies across the lifespan. *Molecular Psychiatry.* https://doi.org/10.10381/s41380-020-0087-x.

Wolpaw, J., and D. McFarland. 2004. Control of a two-dimensional movement signal by a non-invasive brain-computer interface in humans. *Proceedings of the American Academy of Sciences* 101: 17849–17854.

Wolpaw, J, and E. Wolpaw. 2012a. *Brain-computer interfaces: Something new under the sun*, 3–12. Oxford University Press.

Wolpaw, J., and E. Wolpaw, eds. 2012b. *Brain-computer interfaces: Principles and practice.* New York: Oxford University Press.

Walter Glannon is Professor Emeritus of Philosophy at the University of Calgary. Previously, he held academic appointments at McGill University and the University of British Columbia. He is the author or editor of 12 books and has published more than 170 articles. He was Canada Research Chair in Medical Bioethics and Ethical Theory at the University of Calgary from 2006 to 2010 and a recipient of a grant from the Templeton Foundation in 2012–2013. He is a fellow of the Hastings Center.

Chapter 9
Should Prisoners' Participation in Neuroscientific Research Always Be Disregarded When Making Decisions About Early Release?

Elizabeth Shaw

9.1 Introduction

On 21st May 1924, fourteen-year-old Bobby Franks was walking home from school, when his cousin, nineteen-year-old Nathan Leopold and Leopold's close friend, eighteen-year-old Richard Loeb, pulled up in a car alongside him. The older teenagers lured Franks into the car and killed him with a chisel. At their trial, Leopold and Loeb pled guilty to premeditated murder. The prosecution argued that the teenagers should be hanged, but their defence lawyer, Clarence Darrow, ultimately convinced the judge to spare them the death penalty.[1] However, even Darrow was doubtful about whether his clients could ever be reformed and returned to society, saying, "I know that these boys are not fit to be at large. I believe they will not be until they pass through the next stage of life, at forty-five or fifty. Whether they will be then, I cannot tell...I would not tell this court that I do not hope that some time, when life and age has changed their bodies, as it does, and has changed their emotions, as it does, that they may once more return to life"(McKernan 1989). Loeb spent 12 years in prison before a fellow inmate killed him with a razor. Leopold spent 34 years in prison, during which time he volunteered to participate in medical research and was injected with an experimental malaria vaccine. The prosecutor who had advocated hanging the teenagers was so impressed by Leopold's conduct in prison (which also included working in the prison hospital, library and school) that he offered to write a letter to the parole board supporting Leopold's release. After release, Leopold led a law-abiding life and worked in hospitals and church missions and taught mathematics.

---

[1] *People v Leopold and Loeb*, Cook County Crim Ct III [1924]. Darrow's speech is reprinted in McKernan (1989).

E. Shaw (✉)
University of Aberdeen, Aberdeen, Scotland
e-mail: eshaw@abdn.ac.uk

© The Author(s), under exclusive license to Springer Nature Switzerland AG 2023
T. Zima and D. N. Weisstub (eds.), *Medical Research Ethics: Challenges in the 21st Century*, Philosophy and Medicine 132, https://doi.org/10.1007/978-3-031-12692-5_9

He also willed his body to be used for medical purposes after death and his corneas were posthumously donated to two patients.

After very serious crimes, such as the brutal murder of Bobby Franks, it can be hard to imagine what kind of evidence could convince society that the perpetrator is suitable to be released. Arguably, a perpetrator's willing participation in medical research might qualify as an important part of such evidence, from the perspective of certain mainstream theories of punishment, given that the offender's reasons for volunteering to be a research subject might include an altruistic desire to contribute to society and help make amends for his crime. Participation in medical research was routinely taken into account by parole boards in the United States until the 1980s when much tighter legal restrictions resulted in medical research on prisoners becoming far less common—in the mid-1970s, 85% of all phase 1 trials in the United States were carried out on prisoners, dropping to 15% by the 1980s (Charles et al. 2016: 246; Arboleda-Flórez and Weisstub 2013: 115). While some penal theories might regard participation in medical research as relevant to parole decisions, from the perspective of medical ethics, there are strong objections against allowing prisoners to participate in medical research, due to factors such as the vulnerability of the prisoner population and the coerciveness of the prison environment. Medical ethics and some penal theories seem to concur on the importance of certain factors, e.g. consent. Valid consent is crucial from the standpoint of medical ethics, which seeks to safeguard patient autonomy; and, from a penal theoretic perspective, an offender's participation in medical research would only be a strong indicator that he has reformed if he has autonomously chosen to participate. However, medical ethics arguably diverges from penal theory regarding other factors. For example, many medical ethicists argue that research on prisoners should be restricted to research with negligible risks and/or which directly benefits the research participants, e.g. by providing them with access to new treatments for diseases from which they suffer (see, e.g. Appleman 2020). However, the lower the risks and the greater the personal benefits the research presents to participants themselves, the less valuable participation may seem as an indicator of remorse/reform. This is because, firstly, it is sometimes thought that an act cannot appropriately signify remorse for a crime unless that act is burdensome (e.g. Duff 2001), and if participation in an experiment is very low risk, the act of participating might not be considered sufficiently burdensome. Secondly, if participants benefit significantly from the experiment, this raises doubts about whether they were motivated to take part by these benefits rather than by genuine remorse. This chapter will discuss these areas of convergence and divergence between medical ethics and penal theories on this topic and will focus specifically on whether prisoners' participation in neuroscientific research should always be disregarded when making decisions about early release. While there is a significant medical ethical literature on medical research on prisoners, very little has been written on this topic from a penal theoretic perspective. This topic is of current interest, because of the growing literature in support of somewhat loosening restrictions on medical research on prisoners (see, e.g. Gostin 2007; Gostin et al. 2006; Arboleda-Flórez and Weisstub 2013; Charles et al. 2016).

This chapter will begin, in Sect. 1, by briefly outlining the prevalence and types of medical research that are currently being carried out on prisoners, or which might be carried out in the future. This section will end by highlighting *neuroscientific* medical research, because research into "treating" criminal behaviour is particularly relevant to prisoner populations and this kind of research typically involves the application of neuroscientific techniques or insights. Section 2 will briefly outline the laws and guidelines relevant to medical research on prisoners. Section 3 will consider suggestions that neuroscientific research on prisoners should be prohibited altogether; will summarise general medical ethical concerns with medical (and specifically neuroscientific) research on prisoners and will contrast these concerns with perspectives on this topic from selected mainstream penal theories. In the light of these considerations, Sect. 4 will discuss whether, if prisoners are permitted to participate in neuroscientific research, their participation should provide grounds for early release. The purpose of this chapter is not to settle these questions conclusively. Rather, its aim is to identify some restrictions that might plausibly be placed on the practice of prisoner experimentation and to map out different responses that penal theorists might make to the question of whether participation in neuroscientific research should be considered when making decisions about early release—a topic that has previously primarily been explored from the standpoint of medical ethics, rather than penal theory.[2]

9.2 Types and Prevalence of Medical Research on Prisoners

When discussing medical research on prisoners, it is important to distinguish between research involving *clinical interventions* and those that do not; between *beneficial* versus *non-beneficial* research; and between *prison-related* and *non-prison-related* research. A review of current practices in the UK found that 100 applications were made to the Research Ethics Service (RES) to conduct research on prisoners over the 2-year period studied and that the majority (53%) of this research involved questionnaires (Charles et al. 2016). Only seven out of the 100 studies planned to use *clinical interventions*—three of these involved mental health therapies and four of them tested diagnostic procedures (one of these involved imaging, the other two involved diagnostic devices). Most of the studies were "*non-beneficial*" in the sense that the individuals who took part in the studies did not, in the short-term, derive a direct clinical benefit from participating. However, the majority of the research could potentially help improve the long-term health of prisoners generally, as most of the studies were on mental health and infection—"areas that reflect the primary

---

[2] There are areas of overlap between medical ethics and penal theory, arising from the fact that certain moral principles are relevant to both fields, such as respect for autonomy and the prohibition against inflicting unjustified harm. However, the medical ethics *literature* on this topic can be distinguished from the penal theoretic *literature,* as the former focuses primarily on issues such as the doctors' duties to their patients, whereas an examination of the topic through the lens of penal theory would focus on the purposes of punishment.

health needs of prisoners", whose rates of mental illness and communicable disease are much higher than that of the general population (Charles et al. 2013: 249). Most of this research was *prison-related*, i.e. it aimed to investigate issues in the specific context of the prison environment and participants were selected because they were prisoners; as opposed to non-prison-related research, which includes prisoners on the same basis as non-prisoners. The commonest kind of non-prison-related research just involved the collection of data or tissue samples (although even this kind of research is not risk-free, given confidentiality concerns). To summarise: most studies that are officially classed as medical research in the UK involve low physical risks and the majority do not provide immediate clinical benefits to the participants themselves, but rather help to expand the knowledge base about conditions that can affect the general population and are particularly relevant to prison populations. Officially, the situation seems similar in the U.S., where "most North American prison institutions respect a functional ban on [non-beneficial] research except for…situations involving negligible risks" (Arboleda-Flórez and Weisstub 2013: 115) and beneficial research is also strictly limited, to the extent that arguably prisoners are being problematically denied access to certain experimental interventions which might effectively treat diseases from which they suffer (Gostin 2006, 2007).

However, the picture becomes more complicated if one questions what counts as "medical research" and who counts as a "prisoner". Arguably, some of the "treatments" that are currently provided to offenders, although not officially classed as medical "research", raise similar ethical concerns to medical experimentation. There can be a fine line between "therapy, innovative therapy and therapeutic experimentation" (Arboleda-Flórez and Weisstub 2013: 109). For example, sex offenders can currently be offered (in Scotland) or compelled (in some US jurisdictions) to receive testosterone-lowering drugs to reduce sexual thoughts and behaviour. However, these programmes started being implemented well before the long-term safety and effectiveness of these interventions were known and there is still uncertainty about their safety and effectiveness (Greely 2006; Greely and Farahany 2019). Within the US regulatory regime (described below) there are two main loopholes which enable medical research to be carried out on detained persons. Firstly, the Office for Human Research Protections' (OHRP) definition of "prisoner" is narrow, excluding those detained in drug rehabilitation facilities or any programme run by a private contractor (Reiter 2009, Gostin 2006). Secondly, the OHRP only regulates federally funded research, meaning that research carried out by drug companies without federal funding is not subject to OHRP oversight. In addition to the exploitation of loopholes, some studies have been carried out on prisoners that violate the letter, not just the spirit, of laws and regulations (for discussion see Appleman 2020 and Reiter 2009). In 2009, Reiter observed that in practice "experimentation on prisoners regularly occurs… despite strict federal regulations, which have now been in place for forty years, and which were intended to limit severely such experimentation, if not to eliminate it entirely" (Reiter 2009: 526) and there has been growing pressure since then to expand experimentation further—a trend discussed by Appleman (2020). It is not the aim of this chapter to argue that the volume of research carried out on prisoners should either increase or decrease, but rather to discuss moral considerations (including consent, harm and the motivation behind/aims of the research) that should inform decisions about what kind of research (if any) should be carried out on prisoners and under what constraints. Specifically, this chapter will examine these issues from the angle of penal theory—a perspective that has received insufficient attention. This kind of examination of the topic is one necessary step in the process of ensuring that medical research on prisoners is guided by morally relevant considerations, rather than less relevant/irrelevant factors, such as who is funding the research and whether the place where the research subjects are involuntarily detained is labelled a "prison" or not.

Neuroscientific research is a type of research that is particularly important for the study of prisoners. Given the connection between the brain and behaviour, the application of neuroscientific techniques and insights may seem a promising method of reducing reoffending. Although neuroscientific research into criminal behaviour has historically been associated with extremely damaging and distressing interventions, such as prefrontal lobotomies, and *Clockwork Orange*-style aversive conditioning (Ryberg 2020, Chap. 6), it should be stressed that modern neuroscientific interventions ("neurointerventions") to treat criminal behaviour do not have to be particularly invasive or high-risk. For example, there has been some research on the use of fish oil supplements to reduce aggressive behaviour via the effect of omega-3 fatty acids on the brain (e.g. Firestone et al. 2005). Researchers are also investigating the use of tDCS (transcranial direct current stimulation) to reduce risk-taking (e.g. Fecteau et al. 2007). TDCS involves placing a device outside the skull that transmits weak electrical stimulation to the brain and is generally regarded as safe. Neurofeedback involves learning to "retrain" thought processes by watching real-life imagery of how one's brain responds to stimuli. Neurofeedback has been studied as a potential way of reducing impulsivity and enhancing self-control (for discussion see Focquaert 2014). Neuroscientific insights and techniques could potentially also guide the use of psychological therapies, by helping to diagnose different conditions and identify which individuals might be most responsive to particular psychological interventions (see e.g. Cornet et al. 2014). On the riskier end of the spectrum of interventions are testosterone-lowering drugs (which, as mentioned above, might arguably still be regarded as an experimental treatment). Testosterone has a significant effect on how the brain functions (and thereby affects psychological processes and behaviour), and researchers have suggested that the effect of testosterone-reduction on the brain may play an even more important role in sex offender rehabilitation than the other effects of this treatment on the body (Greely and Farahany 2019). As Grubin explains, "In the male brain, testosterone receptors are most dense in hypothalamic nuclei, the amygdala and other areas of the limbic system, the prefrontal cortex and the temporal cortex, all parts of the brain known to be involved in processing sexual stimuli or initiating or maintaining sexual behavior…testosterone has effects on the responsiveness of both general and specific neurological arousal mechanisms, it influences the processing of sexual sensory stimuli, it impacts on motivation, attention, and mood, and it is associated with aggression and dominance, all of which are potentially relevant to sexually problematic behaviour…Therefore treatments aimed at moderating the activity of testosterone...can...weaken the foundation on which sex offending sits." (Grubin 2018: 711–712).

9.3 Laws and Guidelines

The current framework of laws and guidelines regulating experimentation on prisoners (and other vulnerable groups) began to be introduced after a vast number of atrocities perpetrated in the twentieth century in the name of medical research. In Europe, these legal developments were a response to the Nazi war crimes. The judges who tried the Nazi doctors formulated the Nuremberg Code (1947) as part of their judgement.[3] This Code prioritises consent as a fundamental requirement of medical experiments, listing consent as the first of ten principles—the other nine focusing on the need to protect participants from various harms and to attend to the "humanitarian importance" of the research.[4] The Declaration of Helsinki (1962) incorporated many of the Nuremberg Code's requirements and later revisions of the Declaration introduced new provisions, such as the requirement for independent ethical review committees (2000), but the provisions concerning consent are less stringent than the Nuremberg Code, containing exceptions to the consent requirement where obtaining consent would be impossible, or impractical or would undermine the experiment's validity. However, the International Ethical Guidelines for Biomedical Research Involving Human Subjects (1982) provide that, where research subjects are incapable of consenting, or have limited capacity to consent, the research must either benefit the participants or involve negligible risks.[5] The commentary to these guidelines identifies prisoners as a group with limited capacity to consent, due to limited autonomy.[6] The European Prison Rules (2006) prohibit any experiments on prisoners conducted without their consent or which may result in physical harm, mental distress or damage to health. The Nuremberg Code, Helsinki Declaration, International Ethical Guidelines and European Prison Rules are not legally binding but have influenced various legally binding provisions. Article 7 of the International Covenant on Civil and Political Rights (1966), which is a binding covenant, provides that no-one shall be subjected to scientific experiments without their free consent,[7] as does the legally binding Convention on Human Rights and Biomedicine (1997).[8] The

---

[3] *US v Brandt* 2 T.W.C. 171.

[4] Ibid.

[5] These protections are intended to apply to all prisoners as a class (seemingly on the basis that imprisonment always places some limitation on autonomy and hence on the capacity to consent). They are not intended to apply to prisoners on a case-by-case basis commensurate with the extent of the particular individual's capacity to consent.

[6] Henceforth: "International Ethical Guidelines (1982)".

[7] International Covenant on Civil and Political Rights (adopted 16 December 1966, entered into force 23 March 1976) 999 United Nations Treaty Series 171 (ICCPR).

[8] Council of Europe. Convention for the Protection of Human Rights and Dignity of the Human Being with Regard to the Application of Biology and Medicine: Convention on Human Rights and

Additional Protocol to the Convention on Human Rights and Biomedicine concerning Biomedical Research (2005) explicitly addresses experiments on prisoners, permitting them to be involved in beneficial research (in accordance with the principle that prisoners should receive an equivalent standard of healthcare to non-prisoners) and permitting them to participate in research which does not benefit the specific participants, provided that the research would be impossible without using prisoners, that the research benefits prisoners generally, and that risks are minimal.[9]

In the UK, the Royal College of Physicians' guidelines (2007: para 8.47) provide that, "research that can be conducted on patients or healthy volunteers who are not in prison should not be conducted on prisoners. Incarceration in prison creates a constraint which could affect the ability of prisoners to make truly voluntary decisions without coercion to participate in research".

Despite the plethora of laws and guidelines in Europe adopting a very restrictive approach to prisoner experimentation, as noted above, there is a surprising willingness to provide offenders with biomedical treatments to reduce reoffending, such as testosterone-lowering drugs, that might arguably be classed as experimental given the relative shortage of evidence about their efficacy (at least for certain types of offending) and potential long-term side-effects, such as loss of bone density, permanent reduction in fertility, and weight gain (Lewis et al. 2017).[10] Even irreversible surgical castration of sex offenders has been practiced in Germany (until 2017)[11] and the Czech Republic (ongoing), despite uncertainty about its efficacy for reducing certain kinds of sex offending. This practice has prompted criticism from the European Committee for the Prevention of Torture and Inhuman or Degrading Treatment or Punishment (2009, 2012).[12]

The United States was slower than European jurisdictions to introduce strict regulations in this area, with unethical medical procedures being performed on prisoners on a large scale until the mid-1970s and (to a lesser extent) even in more recent years. The experiments which prompted the U.S. reforms were characterised by serious harm or risk to participants, "a lack of transparency, an absence of a therapeutic aim… a significant knowledge and power imbalance between the subjects and the experimenters… [and] serious questions regarding consent" (Reiter 2009: 507). The

---

Biomedicine. (Opened for signature 4 April 1997, entered into force 1 December 1999) Strasbourg: Council of Europe Publishing. ETS No.164.

[9] Council of Europe. Additional Protocol to the Convention on Human Rights and Biomedicine, concerning Biomedical Research. (Opened for signature 25 January 2005, entered into force 1 December 2007) Strasbourg: Council of Europe Publishing. CETS No.195. Principles 48.1 and 48.2 respectively.

[10] The side effects noted in the text can persist after the treatment has stopped. Other side effects, such as hot flushes, depression and feminisation of the body may be limited to the duration of the treatment (Lewis et al. 2017). However, it should be noted that some offenders receive treatment for many years (Sifferd 2020).

[11] In Germany this practice has now effectively ceased (Council of Europe 2017). However, it remains legal: Law on Voluntary Castration and Other Methods of Treatment of 1969, ss2 and 3.

[12] For a discussion of whether providing such biomedical interventions to offenders is compatible with the European Convention on Human Rights see Shaw (2017).

malaria study in which Nathan Leopold participated (see Introduction) suffered from some of these unethical practices. Other examples include experiments on prisoners in Washington and Oregon in 1971 examining the effects of exposing people to radiation, with many survivors of this experiment suffering from illnesses likely caused by the radiation, including prostate cancer, vision loss and vascular disease. An experiment in Pennsylvania between 1965 and 1966 involved exposing prisoners' skin to the main poisonous ingredient of Agent Orange (used in the Vietnam war) at doses 468 times higher than the chemical manufacturer's original protocol for the experiment, which left participants suffering, even decades later, from severe blistering, scars, cysts and rashes. These scandals prompted a Department of Health, Education and Welfare report (1976), leading in 1978 to the introduction of title 45, part C of the Code of Federal Regulations, which state that prisoner experimentation should be minimally risky and non-intrusive and confined to the following four categories: "(1) research about the effects of incarceration, (2) research about prisons as institutions, (3) research about conditions particularly affecting prisoners, and (4) research about practices expected to improve the health of individual subjects".[13] However, these strict regulations (which if interpreted literally are arguably over restrictive—Gostin 2006, 2007) have in practice provided insufficient protection to research subjects, due to weak or inconsistent enforcement. As noted above, there are loopholes that can allow research to escape OHRP oversight. One area of current concern is the ongoing use of new/experimental drug addiction therapies whose safety (and relative efficacy compared to existing treatments) is yet to be established (Appleman 2020). A 2006 report by a committee of the Institute of Medicine recommended improvements to the oversight of prisoner experimentation, including widening the definition of "prisoner" and creating a publicly accessible database of all research conducted on prisoners, but these recommendations have not yet been implemented (Gostin 2006, 2007).

9.4 Neuroscientific Research on Offenders: Perspectives from Medical Ethics and Penal Theory

The disturbing history of repeated unethical experimentation on offenders raises the question whether such experimentation should be completely or largely prohibited. For example, McTernan (2018) argues that this history creates a "defeasible presumption" against using (experimental) neurointerventions to "treat" crime—a presumption which, she claims, is difficult to rebut, given that key factors that led to the historical abuses are still present today, e.g. prisoners' vulnerability and stigmatised status and scientists' limited knowledge of how the brain works. McTernan's conclusion about neurointerventions rests on the principle that "if a course of action has been historically ethically terrible we have reason to doubt that it is now the

---

[13] Additional DHHS Protections Pertaining to Biomedical and Behavioural Research Involving Prisoners as Subjects. 45 C.F.R. § 46.306(a)(2)(i)-(iv)—summarised in Reiter (2007).

correct course of action to take" (McTernan 2018: 278). To support this principle, she imagines a surgeon who, believing they have discovered an improved way of performing a procedure, tries the procedure with devastating results for the patient. Then the surgeon tries the procedure a second time, again with appalling results. Intuitively, the second use of the procedure is even more morally problematic than the first. Unless the surgeon can point to good reasons for thinking it would be different the second time, their failure to learn from their previous mistake displays "moral hubris", thereby *wronging*, as well as harming, the patient by unjustifiably exposing them to known risks (McTernan 2018). McTernan's analysis raises concerns that are clearly relevant from the perspective of medical ethics, as harming patients through moral hubris violates the doctor's duty of care. From the perspective of certain penal theories, harming offenders through moral hubris not only wrongs offenders, but also arguably undermines the criminal justice system's moral authority to punish, given that the system would have failed to uphold and embody the community's core moral values, which is its fundamental role according to mainstream retributive and communication theories (e.g. Duff 2001). From a consequentialist perspective, the likelihood of harm seems more directly relevant than the culpability of the experimenter, but consequentialists would consider the experimenter's culpable moral hubris to be indirectly problematic, if it undermined trust in doctors and the criminal justice system, potentially leading to increased ill health and crime.

Ryberg (2020) challenges McTernan's formulation of the "lessons from history" argument. McTernan's presumption against repeating a type of action that has previously caused harm relies on the ability to categorise a particular action as being a certain "type" of behaviour, which may prove problematic, given that actions can be described in different ways. On what basis does one decide that the act of providing a modern neurointervention to an offender in the context of an experiment that complies with current regulations is the same "type" of action as the historical abuse of a neurointervention? Ryberg suggests that the fact that both acts involve similar technologies is insufficient to categorise them as belonging to the same type. He asks us to imagine someone who in the past used a mobile phone to detonate a bomb and later has the opportunity to use a mobile phone to call an ambulance to save an injured person. The fact that this person previously used the same technology to cause harm provides no reason to hesitate to use it to do good. McTernan might reply that there are more similarities between the historical and modern uses of neurointerventions, besides the use of similar technologies—similarities such as the purposes for which the interventions are/were used (e.g. modification of socially undesirable behaviour) and the vulnerability of the recipients. Hence there are more reasons to categorise modern and historical uses of neurointerventions as belonging to the same type of action than to categorise the two actions involving the mobile phones as belonging to the same type of action. Nevertheless, this does not settle the question of whether modern neurointerventions should be prohibited all things considered.

Although McTernan and Ryberg differ about the precise formulation of the "lessons from history argument", they agree that it is vital to examine the precise ways in which past uses of neurointerventions have gone wrong, to avoid repeating these mistakes. Some of the main failings of historical prisoner experiments were:

(a) failure to obtain consent from participants that is genuinely (i) informed and (ii) voluntary; (b) mental and physical harm to participants caused by the intended effects and side-effects of the experimental interventions; and (c) experimenters' questionable motives.[14] The remainder of this section will discuss aspects of each of these issues in turn, focusing specifically on experimental neurointerventions designed to reduce criminal behaviour (known as "neurocorrectives"—Douglas 2014), while emphasising areas of convergence and divergence between medical ethics and penal theories.

9.4.1 Consent

(i) *Informed*

Prisoners typically have a lower standard of education and are more likely to suffer from mental disorders and learning disabilities than the general population, which can potentially reduce their capacity to give informed consent. Even less vulnerable populations struggle to understand consent forms. One study revealed that at least 80% of participants (who were not prisoners and half of whom had some level of higher education) were unable to answer correctly basic questions about the experimental nature, safety and relative efficacy of interventions, after having read and signed forms asking for their consent to receive these interventions (Schumacher et al. 2017). Obtaining informed consent from prisoners to receive *experimental neurocorrectives* is particularly challenging. It may be especially difficult to imagine what the effects of certain neurointerventions would be like. For example, a neurointervention that enhanced empathy in offenders with psychopathic tendencies might seem like the holy grail of neurorehabilitation research, since studies suggest that this group of offenders commit a disproportionate number of serious crimes (Kiehl 2011). However, someone with little or no empathy might find it impossible clearly to imagine what it would be like to feel compassion for others, as this intervention might produce a radical change in the kind of mental experiences they have, not just a difference in degree. An inadequate understanding of the psychological effects of such neurointerventions may be even more problematic than a failure to understand the physical effects of treatments, given that the mental capacities and dispositions potentially targeted by neurointerventions may be central to a person's sense of self. Furthermore, due to the nature of characteristics like insufficient empathy, high impulsivity and recklessness, which neurocorrectives seek to modify, participants with these characteristics might struggle to appreciate or give sufficient weight to possible *long-term* consequences of experimental interventions. It has been suggested that low-empathy individuals, who feel a lack of connection to others, may also fail

---

[14] Other historical problems included: failure to ensure that the participants' medical information was kept confidential; and conducting experiments that from the outset were unlikely to produce significant benefits for participants/wider society or whose benefits were not sufficient to justify the risks. However, these issues will not be discussed in the present chapter.

to feel a connection with their own future selves (Levy 2014). Impulsive and reckless individuals may dismiss future risks (which they may later regret), due to being excessively influenced by incentives, such as the possibility of early release, a desire to escape boredom or by short-term pressures within the prison environment (discussed below).

From the standpoint of medical ethics, the purpose of obtaining informed consent is to protect the patient's autonomy. However, refusing to allow an offender to receive an experimental neurointervention might also interfere with their autonomy. Some offenders suffer from conditions linked to criminality (such as addictions, sexual disorders, and conditions characterised by impulsivity), which they find intensely distressing, and certain prisoners have mounted lawsuits in order to gain access to potential treatments for such conditions (Fischer 2006: 2–3). If an offender feels tormented by a condition linked to antisocial behaviour and is willing to face the risks attaching to an experimental treatment, is the state justified in withholding such a treatment in order to "protect" the offender from the consequences of his choice? It might be worried that we cannot be sure that a prisoner's apparent desire to receive an intervention is genuinely autonomous. However, a blanket presumption that *no prisoner* can *ever* make up their own mind about whether to receive such an intervention also seems like a failure to respect their autonomy. Some of the laws and guidelines discussed in the previous section take into account the need to balance similar considerations. For example, the International Ethical Guidelines (1982) provide that research that benefits participants may sometimes be legitimate, even when the usual requirement for informed, voluntary consent cannot be adequately met.[15] Arguably, even if there is some uncertainty about the validity of the offender's consent, if the treatment has an appropriately high chance of benefitting the offender (balanced against the harm of leaving him untreated) and all reasonable steps have been taken to seek valid consent, it seems ethically permissible to provide the intervention.

From the perspective of consequentialist penal theories, the interests of prisoners need to be balanced against the interests of others including potential victims. If an experimental neurocorrective has a significant chance of preventing large numbers of people from becoming victims of serious crimes, this could, in theory, outweigh the interests of prisoners who might be harmed in such experiments. Indeed, it might be thought that this kind of consequentialist reasoning implies that the consent requirement should be abandoned for experiments that have a sufficiently high likelihood of preventing serious harm. Following this line of reasoning, no form of draconian treatment of offenders could be excluded, as long as it maximised overall good consequences. Such unpalatable implications are considered by many to be grounds for rejecting (pure) consequentialist theories of punishment. Consequentialists could respond, that, in practice, it would almost never promote good consequences overall to treat prisoners in such inhumane ways, as this could, among other things, undermine

---

[15] The clearest example of an experiment that would be beneficial for the offender, is where the offender suffers from a serious disease, such as cancer, for which existing treatments have proven ineffective and participation in the trial of an experimental treatment provides the only hope for recovery. However, the decision to participate in such a trial would be irrelevant to considerations of reform/repentance as grounds for early release, so will not be discussed here.

faith in the authorities and cause social chaos. However, this type of response may not seem wholly satisfactory as it relies on contingent empirical claims—for example, claims about society's reaction to the maltreatment of prisoners and about the inability of the authorities to conceal this maltreatment. Furthermore, this response arguably fails to capture our intuitions about *why* non-consensual prisoner experimentation is objectionable—intuitions which seem to be based on the wrong done to the prisoners themselves, rather than the risks to wider society.

Traditional retributivism faces difficulties explaining why the physical/mental distress caused by experimental neurocorrectives could not constitute the "suffering" which retributivists hold that offenders deserve, given the vagueness of many retributive theories about exactly what this suffering should consist in (Caruso 2020). Ryberg (2020), without endorsing retributivism, has argued that, in principle, neurocorrectives could be imposed as a form of retributive punishment. Given that punishment need not be imposed with the offender's explicit consent, this line of reasoning would also (problematically) imply that the consent requirement for prisoner experimentation could be abandoned. Some retributivists might oppose experimental neurointerventions on the ground that retributivists cannot accept dual-purpose "punishments" whose goal is not merely to inflict deserved suffering, but also to produce some future good outcome—see Shaw (2018), noting that the influential retributivist, Moore (1997), insists that retributivism should focus solely on punishing past wrongdoing and "cannot share the stage" with forward-looking penal theories. While this retributive argument might preclude non-consensual prisoner experimentation, it fails, like the above-mentioned consequentialist reasoning, to provide an intuitive explanation of *why* this practice would be wrong. Another retributive argument against the idea that experimental neurocorrectives could constitute a form of punishment is that there is a risk that different participants would experience different levels of adverse side-effects (Steffen 2020: 157). This would violate the principle that offenders who have committed the same crime should suffer the same severity of punishment. However, since offenders differ markedly in how aversive they find traditional punishments, e.g. an offender's physical and psychological characteristics can make a large difference to how much he suffers in prison (Kolber 2009), the logic of this argument could make it hard for retributivists to justify any form of punishment. Furthermore, even if it were possible to guarantee that individuals who were forced to receive harmful neurocorrectives suffered identical adverse side-effects, the practice still seems intuitively objectionable and thus the "unequal-level-of-suffering" argument does not seem to capture our intuitions about what makes this practice wrongful.

In contrast, communication theories of punishment (e.g. Duff 2001) arguably contain conceptual resources that both help to explain what is wrong about non-consensual prisoner experimentation and which suggest how genuine informed consent might be obtained.[16] Communication theory emphasises the importance of ongoing dialogue with the offender and building respectful relationships between the offender and other members of the moral community including criminal justice

---

[16] Although there are reasons discussed below why communication theorists might oppose prisoner experimentation, even if consensual.

actors who are working with the offender. I have previously argued that communication theory precludes the non-consensual use of neurocorrectives, because this practice would objectify offenders, by failing to engage them in the kind of dialogue required by the theory, by refusing to listen to them, excluding them from the moral community, and by portraying them as radically defective and inferior to others with regard to a fundamental aspect of their agency (Shaw 2014). For the purposes of the present chapter, it is important to highlight that communication theory's emphasis on dialogue could increase the chance that consent to neurocorrectives would be genuinely informed. (Appleman 2020) reports that participants' comprehension of relevant issues markedly increases when those administering the medical intervention took the time to provide a detailed oral explanation to participants and to answer their queries, rather than just giving them a consent form. Appleman notes that this way of obtaining consent would be time-consuming and expensive. However, communication theory provides an additional justification for investing these resources, given that ongoing dialogue with offenders should be happening anyway throughout their punishment, according to this theory. (Furthermore, taking these measures to obtain genuine, informed consent might save resources in the long-term, because the interventions are more likely to be successful if they are consensual and the offender has a respectful relationship with those administering them.) However, the people administering any neurocorrectives should be different individuals from the other criminal justice actors who are working with the offender, to prevent undue influence. There is some convergence between the communication theory of punishment's emphasis on dialogue and certain themes from the medical ethics literature. Just as communication theory stresses that dialogue is a two-way process and that the authorities should acknowledge their own fallibility; medical ethicists have highlighted the need for forensic researchers to reflect on their own moral commitments, past experiences and shortcomings, in order to avoid these factors having a negative impact on recipients of interventions. Arboleda-Flórez and Weisstub (2013: 113, summarising the views of Candilis et al. 2007) observe that, "it is essential that forensic experts find vehicles to explore and articulate their own personal values, identifications, and life histories in order to do justice in a particular instance… [and that] every effort be made to hear the voices…of parties in conflict, actually or potentially".

(ii) *Voluntary*

While it is common for medical ethicists to assert that prisoners' consent to neurocorrectives could never be voluntary, due to the coerciveness of the prison environment, criminal justice theorists have debated whether we should assume that prisoners are incapable of making such choices and even whether their consent should be required at all, given that "traditional" interventions like imprisonment are imposed on offenders without their consent (see, e.g., the discussion in McMillan 2014; Douglas 2014, 2019). Assuming that consent should be required before neurocorrectives can legitimately be given to offenders (in defence of this view see, e.g. Shaw 2014, 2019), the question arises whether the chance of early release as a result of receiving experimental neurointerventions would prevent the voluntariness requirement for valid consent from being met. The answer to this question seems crucially to depend on whether the prison conditions are excessively harsh. In addition, to the loss of liberty and interference with relationships with family and friends that imprisonment inevitably involves, prisoners may be subject to violence from other prisoners or staff, live in overcrowded or unsanitary conditions, be at higher risk of catching diseases and have limited access to medical treatment (Appleman 2020). In contrast, voluntariness of consent may be less problematic in regimes that are more humane and less restrictive of the offenders' freedoms, such as certain Norwegian prisons, which strive to uphold the principle of normality which states that "life inside [prisons] will resemble life outside as much as possible".[17]

9.4.2 Harm

As stated above, offenders may have conditions associated with an increased risk of criminality that also cause harm to their mental and physical wellbeing. If existing therapies have not proved successful, receiving an experimental neurocorrective might be in the offender's best interests overall, even if this neurocorrective carries some risk of harm. I have argued elsewhere (Shaw 2018) that individuals with psychopathic personality disorder suffer as a *direct* consequence of their condition (rather than only suffering due to society's punitive response to their criminal behaviour). Psychopathy is linked to much higher rates of physical illness, addiction, injury and early death, due to characteristics of the disorder such as impulsivity, dismissal of long-term considerations and tendency to enter into conflicts (Hare 1993). It seems plausible that psychopaths regard these harms to themselves as aversive, even if they typically lack insight into their causes. Given that many psychopaths lack this insight there may be particular challenges in obtaining *informed* consent from psychopaths (see above and Shaw 2018). However, since psychopathy exists on a spectrum (Hare 1993), some psychopaths may have sufficient insight to give informed consent, and some of those prisoners who suffer from other conditions (e.g. addictions and sexual disorders) might also have enough insight to be able to give informed consent or might be brought to gain such insight after, e.g., talking therapies. Overall, if valid consent can be obtained and the balance of harms suggests that the intervention is likely to be beneficial for the offender, then it seems ethically acceptable to offer the intervention.

This leaves the question of whether prisoners should ever be permitted to consent to neurocorrectives that are all-things-considered harmful to them. From the standpoint of mainstream medical ethics, the answer seems to be "no". However, various penal theories might permit such interventions. As mentioned above, consequentialist theories might permit offenders to participate in experiments that harm them but promote the welfare of a greater number of other people, and retributive theories might imply that the harm caused by such interventions is sometimes deserved, e.g. if the harm the offender would experience is in proportion to the harm he inflicted on

---

[17] http://www.kriminalomsorgen.no/informationin-english.265199.no.html.

the victim (although many consequentialists and retributivists attempt to resist these conclusions). Penal theorists who stress the importance of moral reform might argue that society should not interfere with a prisoner's choice to suffer harm through participating in a medical experiment, if the prisoner is motivated by a desire to "give something back to society, to redeem, atone, and reconcile" (Garnett 1995: 481). Furthermore, some penal theorists, unlike mainstream medical ethicists, might prioritise offenders' moral welfare (Morris 1981) over their physical welfare. Such a conception of moral "welfare" and moral "harm" might imply that permitting offenders to atone for their crime in this way would not necessarily harm them overall. However, the communication theory might preclude allowing offenders to participate in medical experiments that seriously physically harm them overall, because allowing this would send out confusing messages to the public about the values embodied in the criminal justice system. Communicative punishment is meant to send a clear message condemning the offender's crime as morally wrong, because it violated fundamental moral norms, such as prohibitions against causing serious physical harm. If punishment itself seriously violates the offender's physical integrity, then the criminal justice system seems to be communicating the problematic message: "do what I say, not what I do." To theorists who advocate respecting the autonomous choice of a prisoner to allow himself to be seriously harmed in a medical experiment, communication theorists might reply that the prisoner should be free to make such choices after release, but doing so under the auspices of the criminal justice system would taint that system. The offender, according to communication theory, should not be permitted to choose a harsher regime for himself than that which would send out the appropriate message about his offense, any more than he should be permitted to select a more lenient one.[18]

9.4.3 The Motives Behind/Purposes of Experimenting on Prisoners

Historically, experimenters have sometimes inappropriately prioritised the socially beneficial aims of experiments over the welfare of participants. Moreover, some experimenters neglected *both* participants' welfare *and* the beneficial purposes

---

[18] Medical ethicists and penal theorists, regardless of which penal theory they subscribe to, might also oppose allowing a prisoner to consent to neurocorrectives that would seriously harm the prisoner overall, on the basis that (a) if the intervention really is so harmful, then there are reasons to suspect that that the recipient's apparent consent is not genuine and (b) given the lessons from history, the state cannot be trusted to restrict their use of such harmful interventions only to prisoners who validly consent, so the state should be prohibited from ever using seriously harmful interventions. However, even if it could be shown that (a) and (b) were not the case, allowing prisoners to consent to seriously harmful interventions still arguably seems problematic, so the communication theorist's explanation (outlined in the text) of why it would be problematic may seem attractive (at least to those who reject an alternative explanation, based simply on a paternalistic prohibition against seriously harming oneself in order to benefit others).

of experiments, because the experimenters cared more about ethically problem-atic considerations, such as profit and reputational enhancement (Appleman 2020). Furthermore, if it can be argued that an experiment promotes some worthy purpose, this may make it psychologically easier for experimenters to rationalise inflicting unjustified harm on participants, when the experimenters may also be motivated, at some level, by less worthy considerations (Haslam et al. 2016). It might somewhat help to reduce these problems (without completely solving the problem of conflicts of interests) if research on prisoners were limited to experiments aimed at benefit-ting the prisoners themselves either as a group or individually (see the Laws and Guidelines section above). Without such a constraint, there is a danger that prisoners might be viewed simply as a resource. For example, Albert Kligman, the researcher behind the Agent Orange experiments said that, when he got access to prisoners to test chemicals on, he was excited to have so many "acres of skin" at his disposal (quoted in Reiter 2009: 502). Bomann-Larsen (2013) argues that the only kind of intervention that should be given to offenders, as a condition of early release, are interventions that aim to prevent them from repeating the specific offences for which they were convicted. She claims that being found guilty of a particular crime does not give the state a general permission to do just anything to an offender that might be useful to the state. Rather, a finding of guilt only gives the state the right to impose measures on an offender that are directly related to the crime they were proved to have committed. However, other theorists have argued that this constraint is too narrow, that the state has a legitimate interest in crime prevention *generally,* and that, in prin-ciple, it might be permissible to use interventions that aim to prevent crimes other than those of which the recipients were convicted (Ryberg and Petersen 2013). Other penal theorists might only require that the intervention is in some way "appropriately connected" to the offenders' crime. However, the idea of an "appropriate connec-tion" is rather vague, and, at the most abstract level, it might be argued that almost any benefit conferred on society could be seen as helping to make up for the wrong done to society. There have been some attempts to place limits on the idea of an "appropriate connection" between the crime and the kind of actions that would make amends for the crime (see, e.g., Lee 2016). For example, someone who vandalised a building, might be required to help repair it and someone who drove recklessly, might be compelled to visit hospital wards for car crash victims. Whether/how one might flesh out the idea of an appropriate connection between committing a crime and receiving an experimental intervention is beyond the scope of this chapter.[19]

---

[19] Determining with precision which measures are appropriately connected to which crimes is also challenging in the context of traditional punishments—it is not a challenge that is unique to prisoner experimentation.

9.5 Should Participation in Neuroscientific Research Provide Grounds for Early Release?

In the light of the appalling history of prisoner experimentation, the considerations discussed above, particularly those stemming from the communication theory of punishment, suggest that neuroscientific research in exchange for early release should only be permitted (if at all)[20] if very strict conditions were met. These should include (at least) the following requirements. Firstly, in order to obtain *informed* consent, researchers should engage in extended dialogue with offenders about the nature of the experiment and possible risks, and researchers must reflect on and be open about any of their values, life experiences, vested interests etc. that might adversely impact on the participants. The use of traditional consent forms is not sufficient, given the barriers to informed consent that particularly affect prisoners. Secondly, given such barriers to consent, some degree of uncertainty almost always surrounds the validity of prisoners' consent. To counterbalance this problem, in addition to taking all reasonable steps to obtain consent, it should be required that participants receive some benefit from the results of the experiment (at least if the experiment carries any significant risks). Thirdly, if prison conditions are excessively harsh, then the *voluntariness* of consent is dubious and the state lacks the right to enrol such prisoners in experiments, as well lacking the right to expose them to such prison conditions (Focquaert 2014). Fourthly, prisoners should not be permitted to receive interventions that are likely seriously to harm them overall, even if it is clear that they desire to receive such interventions. Finally, there may be some reasons for thinking that if participation in neuroscientific research were to provide grounds for early release, such research would need to be appropriately related to the offender's crime. Such "appropriately related" interventions would most obviously include treatments to prevent the same sort of crime being repeated. However, there is not scope within this chapter to explore this issue in detail.

The remaining question is whether releasing offenders *early* would be problematic from a penal theoretical perspective. From a retributive standpoint, anything that allows the offender to evade (all or part of) his deserved punishment would be problematic. For example, Kant wrote: "What, therefore, should one think of the proposal to preserve the life of a criminal sentenced to death if he agrees to let

---

[20] Ryberg (2020) persuasively argues, that what is permissible in principle might be very different from what is permissible in practice under non-ideal conditions. Thus, although Ryberg is a strong proponent of the view that neurointerventions are permissible in principle, he casts doubt on whether it would be permissible to use them under the conditions that currently obtain in the US criminal justice system, due to the "irrationality" of many practices within that system (Ryberg 2020: 205). Similarly, the five prerequisites that I outline in this section for the permissibility of neurointerventions might need to be supplemented with further requirements depending on different practices in different jurisdictions. Indeed, in some jurisdictions whose current practices are particularly "irrational", it might be impossible to implement enough reliable safeguards to make the use of experimental neurointerventions on prisoners ethically acceptable. This article has aimed to identify some of the most important *necessary* conditions for neuroscientific prisoner experimentation to be justifiable, rather than setting out sufficient conditions suitable for all jurisdictions.

dangerous experiments be made on him and is lucky enough to survive them, so that in this way physicians learn something new of benefit to the commonwealth? A court would reject with contempt such a proposal from a medical college, for justice ceases to be justice if it can be bought for any price whatsoever." (Kant 1785: 141). However, more recently there have been attempts to justify parole and early/delayed release in terms of communicative retributivism (see e.g. O'Hear 2011; Dagan and Sergev 2015). Nevertheless, Bülow (2018: 13) rejects the idea that early release is compatible with communicative retributive theories, arguing that, "from a retributivist perspective, indeterminate sentencing is deemed problematic because of proportionality and because punishment should reflect crime severity and not the risks of future delinquency or behavior in prison." He considers the alternative possibility that mercy (a doctrine independent from the primary justification of punishment) might justify early release for good behaviour, but ultimately suggests that, "the most promising rationale for indeterminate sentencing stems from mixed theories of punishment such that retributivist concerns set the minimum and maximum time in prison and forward-looking consequentialist concerns decide the actual date of release" (Bülow 2018: 14). Those who still seek to reconcile early release with a communicative theory of punishment (rather than with mixed theories) might claim that the "quality" of one's conduct during one's sentence might justify a reduced "quantity". It might be justifiable, on this view, to release offenders early if, through good conduct during part of their sentence, offenders manage to make "precisely the amends they ought to make to apologise for their crimes" before the whole sentence is complete (Lee 2016: 19[21]). Perhaps willingness to receive experimental neurocorrectives might constitute such amends. One might worry, however, that this kind of indeterminate sentence could create too much uncertainty and inconsistency between cases. An alternative might be to offer the offender a choice at the outset between different punishments which were considered (roughly) equally burdensome: either a shorter prison term plus receiving an experimental neurocorrective, or a longer prison term without the neurocorrective. The purpose of this chapter is not to settle these questions. Its aim was to identify some restrictions that might plausibly be placed on the practice of prisoner experimentation (summarised at the start of this section) and to map out different responses that penal theorists might make to the question of whether participation in neuroscientific research should be considered when making decisions about early release—a topic that has previously primarily been explored from the standpoint of medical ethics, rather than penal theory.

---

[21] Lee was referring to amends made *before* the sentence was imposed (and did not commit himself to the view that this justified a reduced sentence), but the logic of this argument could also be applied to good behaviour *during* one's sentence.

9.6 Conclusion

This chapter began by briefly outlining the prevalence and types of medical research that are currently being carried out on prisoners, or which might be carried out in the future. This section ended by highlighting *neuroscientific* medical research, because research into "treating" criminal behaviour is particularly relevant to prisoner populations and this kind of research typically involves the application of neuroscientific techniques or insights. Section 2 briefly outlined the laws and guidelines relevant to medical research on prisoners. Section 3 considered suggestions that neuroscientific research on prisoners should be prohibited altogether and summarised general medical ethical concerns with medical (and specifically neuroscientific) research on prisoners and contrasted these concerns with perspectives on this topic from selected mainstream penal theories. In the light of these considerations, Sect. 4 set out some requirements that should be met before neuroscientific research in exchange for early release should be permitted (if at all) and mapped out different responses that penal theorists might make to the question of whether participation in neuroscientific research should be considered when making decisions about early release—a topic that had previously primarily been explored from the standpoint of medical ethics, rather than penal theory.

References

Appleman, L. 2020. The captive lab rat: Human medical experimentation in the carceral state 61 B.C.L. Rev. 61: 1.

Arboleda-Flórez, J., and D. Weisstub, 2013. Forensic research with the mentally disordered offender. In *Ethical issues in prison psychiatry*, ed. N. Konrad, B. Völlm, and D. Weisstub. Dordrecht: Springer.

Bomann-Larsen, L. 2013. Voluntary rehabilitation? On neurotechnological behavioural treatment, valid consent and (in)appropriate offers. *Neuroethics* 6: 65.

Bülow, William. 2018. Deserved delayed release? The communicative theory of punishment and indeterminate prison sentences. *Criminal Justice Ethics* 37 (2): 164.

Caruso, G. 2020. *Rejecting retributivism: free will, punishment, criminal justice*. New York: Cambridge University Press (in press).

Charles, A., A. Rid, H. Davies, and H. Draper. 2016. Prisoners as research participants: Current practice and attitudes in the UK. *Journal of Medical Ethics* 42 (4): 246.

Cornet, L., C. Kogel, H. Nijman, A. Raine, and P. Laan. 2014. Neurobiological changes after intervention in individuals with antisocial behaviour: A literature review. *Criminal Behaviour and Mental Health* 25 (1): 10.

Council of Europe. 2006. *Recommendation of the Committee of Ministers to Member states on the European Prison Rules*. Strasbourg: Council of Europe Publishing. Rec(2006)2.

Council of Europe. 2017. News. Germany: Uneven progress in treatment of detained persons and detention conditions, says anti-torture committee. (1st June 2017) https://www.coe.int/en/web/cpt/-/germany-uneven-progress-in-treatment-of-detained-persons-and-detention-conditions-says-anti-torture-committee Accessed October 26, 2020.

Dagan, N., and D. Sergev. 2015. Retributive whisper: Communicative elements in Parole. *Law and Social Inquiry* 3: 611.

Douglas, T. 2014. Criminal rehabilitation through medical intervention: Moral liability and the right to bodily integrity. *Journal of Ethics* 18 (2): 101.

Douglas, T. 2019. Nonconsensual neurocorrectives and bodily integrity: A reply to Shaw and Barn. *Neuroethics* 12 (1): 107.

Duff, A. 2001. *Punishment, communication and community.* Oxford: Oxford University Press.

European Committee for the Prevention of Torture and Inhuman or Degrading Treatment or Punishment. 2009. *Report to the Czech Government on the visit to the Czech Republic carried out by the European Committee for the prevention of torture and inhuman or degrading treatment or punishment.* Strasbourg: Council of Europe Publishing. CPT/Inf (2010) 22.

European Committee for the Prevention of Torture and Inhuman or Degrading Treatment or Punishment. 2012. *Report to the German Government on the visit to the Germany carried out by the European Committee for the prevention of torture and inhuman or degrading treatment or punishment.* Strasbourg: Council of Europe Publishing. CPT/Inf (2012) 6.

Fecteau, S., D. Knoch, F. Fregni, N. Sultani, P. Boggio, and A. Pascual-Leone, 2007. Diminishing risk-taking behavior by modulating activity in the prefrontal cortex: A direct current stimulation study. *The Journal of Neuroscience* 27 (46): 12500.

Firestone, P., Kevin L. Nunes, H. Moulden, I. Broom, and J. Bradford. 2005. Hostility and recidivism in sexual offenders. *Archives of Sexual Behavior* 34 (3): 277–283.

Fischer, J. 2006. *My way: Essays on moral responsibility.* Oxford: Oxford University Press.

Focquaert, F. 2014. Mandatory neurotechnological treatment: Ethical issues. *Theoretical Medicine and Bioethics* 35: 59.

Garnett, R. 1995. Why informed consent? Human experimentation and the ethics of autonomy. *The Catholic Lawyer* 36: 455.

Gostin, L. 2007. Biomedical research involving prisoners: Ethical values and legal regulation. *JAMA* 297: 737.

Gostin, L., C. Vanchieri, and A. Pope, The Institute of Medicine Committee on Ethical Considerations for Revisions to the Department of Health and Human Services Regulations for Protection of Prisoners Involved in Research. 2006. *Ethical considerations for research involving prisoners.* Washington, DC: The National Academies Press.

Greely, H., and N. Farahany. 2019. Neuroscience and the criminal justice system. *Annual Review of Criminology* 2: 451.

Greely, H. 2006. Neuroscience and criminal justice: Not responsibility but treatment. *Kansas Law Review* 1104.

Grubin, D. 2018. The pharmacological treatment of sex offenders. In *The Wiley Blackwell Handbook of Forensic Neuroscience,* ed. A. Beech A. Carter, R. Mann, and P. Rotshtein. Chichester: Wiley-Blackwell.

Hare, R. 1993. *Without conscience: The disturbing world of the psychopaths among us.* New York: The Guilford Press.

Haslam, S., S. Reicher, and M. Birney. 2016. Questioning authority: New perspectives on Milgram's 'obedience' research and its implications for intergroup relations. *Current Opinion in Psychology* 11: 6.

Kolber, A. 2009. The subjective experience of punishment. *Columbia Law Review* 109: 192.

Lee, A. 2016. Defending a communicative theory of punishment: The relationship between hard treatment and amends. *Oxford Journal of Legal Studies* 37 (1): 217.

Levy, N. 2014. Psychopaths and blame: The argument from content. *Philosophical Psychology* 27 (3): 351.

Lewis, A., D. Grubin, C. Ross, and M. Das. 2017. Gonadotrophin-releasing hormone agonist treatment for sexual offenders: A systematic review. *Journal of Psychopharmacology* 31 (10): 1281.

McKernan, M. 1989. *The amazing crime and trial of Leopold and Loeb.* Notable Trials Library, Reprint Edition.

McMillan, J. 2014. The kindest cut? Surgical castration, sex offenders and coercive offers. *Journal of Medical Ethics* 40: 583.

McTernan, E. 2018. Those who forget the past: An ethical challenge from the history of treating deviance. In *Treatment for crime: Philosophical essays on neurointerventions in criminal justice*, ed. D. Birks, and T. Douglas. Oxford: Oxford University Press.

Moore, M. 1997. *Placing blame*. Oxford: Oxford University Press.

Morris, H. 1981. A paternalistic theory of punishment. *Philosophical Quarterly* 18 (4): 263.

O'Hear, M. 2011. Beyond rehabilitation: A new theory of indeterminate sentencing. *American Criminal Law Review* 48 (2): 1247.

Reiter, K. 2009. Experimentation on prisoners: Persistent dilemmas in rights and regulations. *California Law Review* 97: 501.

Royal College of Physicians. 2007. Guidelines on the practice of ethics committees in medical research with human participants. Available at: https://shop.rcplondon.ac.uk/products/guidel ines-on-the-practice-of-ethics-committees-in-medical-research-with-human-participants?var iant=6364998469. Accessed September 2020

Ryberg, J. 2020. *Neurointerventions, crime, and punishment*. Oxford: Oxford University Press.

Ryberg, J., and T. Petersen. 2013. Neurotechnological behavioural treatment of criminal offenders: A comment on Bomann-Larsen. *Neuroethics* 6: 79–83.

Schumacher, A., W. Sikov, M. Quesenberry, H. Safran, H. Khurshid, K. Mitchell, and A. Olszewski. 2017. Informed consent in oncology clinical trials: A brown university oncology research group prospective cross-sectional pilot study. *PLoS ONE* 12 (2): e0172957.

Shaw, E. 2014. Direct brain interventions and responsibility enhancement. *Criminal Law and Philosophy* 8 (1): 1.

Shaw, E. 2019. The right to bodily integrity and the rehabilitation of offenders through medical interventions: A reply to Thomas Douglas. *Neuroethics* 12: 97.

Shaw, E. 2017. Retributivism and the biomedical moral enhancement of offenders through brain interventions. In *Moral enhancement: Critical perspectives*, ed. M. Hauskeller, and L. Coyne. Cambridge: Cambridge University Press.

Shaw, E. (2018) The treatment of psychopathy: Conceptual and ethical Issues. In *Neurolaw and responsibility for action*, ed. B. Donnelly-Lazarov, P. Raynor, and D. Patterson. Cambridge: Cambridge University Press.

Sifferd, K. 2020. Chemical castration as punishment. In *Regulating human mental capacity*, ed. N. Vincent, T. Nadelhoffer, and A. McCay. Oxford: Oxford University Press.

Steffen, J. 2020. Moral cognition in criminal punishment. *British Journal of American Legal Studies* 9 (1): 143.

World Health Organization and the Council of International Organization of Medical Societies. 1982. *Commentary to the international ethical guidelines for biomedical research involving human subjects*. Guidelines 9, 12, and 13.

World Medical Association. 1962. Declaration of Helsinki—Ethical principles for medical research involving human subjects.

Elizabeth Shaw is a senior lecturer in the School of Law at the University of Aberdeen. Her research interests include: free will and theories of punishment, criminal responsibility, neuroethics and neurolaw, psychopathy, mental incapacity defences, the rights to bodily and mental integrity, and the use of medical interventions to rehabilitate criminals. She is a co-director of the Justice Without Retribution Network.

Chapter 10
Applying Neuroscience Research: The Bioethical Problems of Predicting and Explaining Behavior

David Freedman

Neuroscience research has already changed the expectations of how it can explain and predict behavior, but it does so in ways that raise significant ethical and scientific questions. Some of the current research recruits prison inmates as the subjects of studies of behavior, using brain imaging to offer supposedly predictive and character defining models. Most of the current research is crosssectional, yet broad conclusions are offered about behavior without having established ecological validity or acknowledging how social and environmental conditions shape behavior. Moreover, while neuroscience on the one hand demonstrates that the brain is not fixed, that the neurodevelopmental trajectory defines a changing structure and function across the lifespan, the use of the research in courts typically fails to address these changes, opting instead for static labeling approaches which ignore the very science on which it is based. In discussing these bioethical issues, this chapter will address the importance of considering the social context of behavior; the ways in which cross-sectional studies (e.g., most neuroimaging studies) are used to make unsupported, broad claims about behavior; how neuroscience has become another tool used to label, categorize and exercise control over some people; and whether neuroscience research adequately details its assumptions and limitations when applied.

10.1 Introduction to the Neuroscience of the Brain-Behavior Relationship

Neuroscience research, and clinical applications of that research, have reshaped nearly every aspect of medicine in the last few decades. Structural and functional

D. Freedman (✉)
International Academy of Law and Mental Health, Montreal, QC, Canada
e-mail: df2379@gmail.com

© The Author(s), under exclusive license to Springer Nature Switzerland AG 2023
T. Zima and D. N. Weisstub (eds.), *Medical Research Ethics: Challenges in the 21st Century*, Philosophy and Medicine 132,
https://doi.org/10.1007/978-3-031-12692-5_10

brain alterations, often identifiable during the premorbid and prodromal periods of illness, are well-established in many psychiatric and neurologic conditions. Such deviations from typical structure and function, observed across the lifespan for those who have or will develop psychiatric and neurological conditions, have wrought a reconceptualization of disease onset and course (Allison et al. 2019; Jalbrzikowski et al. 2019). At its core, neuroscience research has changed the ways in which the relationship between brain and behavior are studied and conceptualized, increasing the focus on biological and reductionist approaches to illness. Yet, at its simplest, being able to "see" the brain while a person is alive, whether by structural or functional imaging, as well as gaining more insight into the processes of neurotransmission, whether electrical or chemical, have made less opaque the processes by which the brain is related to behavior.

Neuroscience research incorporates many medical sciences and types of research, but is now synonymous with neuroimaging and technological approaches to medicine. It includes studies of the brain, how neurons grow and connect, what typical and atypical development and functioning look like, how disease and injury might change brain function, adaptability and plasticity. In turn, this broad definition of the research agenda supports efforts to apply the research towards understanding the processes that underlie perception, cognition, functioning, and behavior and the complex interplay of genetic risks and life experience which shape decision-making. Understanding and explaining this complexity of functioning and behavior, how cognition and experience shape the ways in which people make sense of the moment, the range of options perceived as available to respond and act, and the executive functioning tools to weigh, determine, initiate, and carry out a course of action, are all within neuroscience's self-defined purview.

Non-invasive access to the brain has changed the conception of development, shifting notions of when adolescence transitions to adulthood, and thereby changing ideas regarding the culpability of youth in criminal justice settings (Steinberg 2013). In the last decades, the concepts of developmental trajectory and life-course changes in brain structure and function have been demonstrated through neuroscience techniques. The complexity of networked connections, integrated neuronal pathways, which mature over time and continue to adapt throughout life, has become an accepted baseline principle (Bassett & Sporns 2017). These advances are significant and contain the prospect of advancing the ontology of the human experience.

Neuroimaging, a key tool of neuroscience, has also re-defined the conception of the physical brain from lobe-based localization to network-based integration. Lobe-based views have historically led to horrific experimentation on people with mental illnesses: for instance, removing portions of the frontal lobe to control behavior (El-Hai 2005). Although still used to describe behavior and as a shorthand way to reference brain functions, lobe-based approaches are largely out of step with current evidence. Localizing functions to regions of the brain most responsible for them, especially when diagnosing or treating an acquired injury such as stroke or lesion, remains an important tool for neuroscience research, but neuroscience research makes clear that most brain functions are networked and occur through processes which maintain equilibrium over time. For instance, visual stimuli which may at first appear threatening activates a survival response but also an evaluative process to ascertain the scope of the threat. Damage or malformation anywhere within this network may result in aberrant or out of context behavioral responses.

However, while such advances are important and suggest the possibility of new approaches to preventing, ameliorating and treating neurological and psychiatric illnesses, they also bring with them substantial risks. The application of neuroscience advances without acknowledgment of the limits and assumptions which underlie the research, poses an ethical crisis for medicine and risks of dehumanizing and stigmatizing people with neurologic and psychiatric illnesses. Research on stigma over the past few decades has found that as more people adopt the neurobiological explanations for mental illnesses, such explanations have increased, rather than decreased, social distancing and rejection of the mentally ill (Pescosolido et al. 2010).

In the medical treatment and research arenas, neuroscience advances have changed the core understandings of mental illness, re-shaped the research paradigms, and altered the way funding flows from both government agencies, private foundations and corporate investors. It has prioritized technological advance and biologically driven explanations for behavior back to the forefront. It risks further separating human behavior from the contextual reality of lived experience, and threatens a dystopian world in which characteristics and biological markers are used to define future intentions and actions before an individual actor even conceives of the possibility of them.

10.2   Assumptions and Limits of Neuroscience Research

A recent article in the American Bar Association Journal, which reaches lawyers throughout the U.S., begins with a criminal case in which the defendant's intent to commit the offense, and therefore his culpability for the offense, were put at issue based, in part, on functional MRI (fMRI) scans which purported to differentiate between true and false memories (Davis 2020). The article goes on to discuss funding from the MacArthur Foundation Research Network on Law and Neuroscience and private industry for these sorts of efforts to use neuroscience to answer legal questions of intent, culpability, lying and memory. By combining neuroimaging technology and machine learning, these efforts have been trying to develop technology capable of both determining and predicting criminal behavior and the suspect's mental state at the time of the offense. Similar efforts have attempted to use brain scans as lie detectors. Although those efforts have been critiqued as not yet ready for use (Farah et al. 2014), the implication of such reviews is that neuroscience can and will be ready soon.

Ironically, this sort of research, trying to differentiate true from false, intentionality, or real versus false memory, relies on the identification of specific brain regions claimed to be involved. Ironic because one of the most significant advances enabled by neuroimaging is better understanding the networked processes of complex cognitive behaviors. This research attempts to localize the mind to a specific lobe or region of interest claimed to be involved in social behavior. These researchers often publish findings which localize responses to stimuli to a specific region, for instance the dorsolateral prefrontal cortex which Farah et al. reported in a meta-analysis to be associated with lying. Yet, that location is also associated with nearly every serious mental illness, as well as executive functioning, working memory, cognitive flexibility, social cognition, deductive reasoning, language processing and motor function. Any or all of these functions may be activated by a feigned dissimulation task, making the interpretation of the behavior and its prediction implausible. Moreover, none of these "behaviors" are performed by the dorsolateral prefrontal cortex in isolation. Rather, they occur or are produced through networked, integrated communication through the brain. And finally, the claims obscure the differences between mind and brain, arguing for a reductionist view that consciousness for a specific type of behavior can be located in a tiny region of the brain without any acknowledgement that the field of neuroscience, and neuroimaging in particular, have already disproven such reductionist views of the brain and mind.

An additional irony is that one of neuroscience's most important contributions has been assisting in understanding the developmental and degenerative processes of brain function across lifespan, yet the vast majority of studies are cross-sectional, ignoring one of neuroscience's most important contributions. Normal variation at any given age, among healthy people, is wide. During developmental periods, the pace of maturation of brain structure and function is also quite varied within typical development, with some youth developing earlier than others, and the association of brain structure with behavior fluctuating widely within the range of typical development (Walhovd et al. 2014).

Although more recent efforts have been initiated to conduct life-course or longitudinal neuroimaging research (Telzer et al. 2018), most research has had to rely on cross-sectional, mixed-method, aggregated or group averaged data, typically across short time frames, rather than following people longitudinally (Caspi et al. 2020; Cope et al. 2020; Modabbernia et al. 2021; Walhovd et al. 2018). Cross-sectional research increases the likelihood of misinterpretation, overinterpretation, bias and confounding (Rothman et al. 2008).

So, what should be concluded from neuroimaging that shows differences in a specific region or lobe when trying to predict a behavior? Which of the many functions and networks should be given priority in the interpretation? What conclusions might be drawn about the relationship between brain and mind from such studies, if any? And when would it be reasonable to say that the predictive power of differential activation in that region is sufficient to judge the cause or intent or truthfulness or meaning of social cognition behaviors?

The appeal of answering predictive behavioral questions, as well as hoping to identify the mind within the brain, have been largely too seductive to be tempered by the limits of the evidence, the lack of scientific rigor, or the moral and ethical doubts which should attach to the pursuit (Choudhury and Slaby 2012; Gkotsi and Gasser 2016). Most of the studies on which neuroscience makes claims to explain the mind suffer from selection bias, uncontrolled confounding, insufficient sample sizes, a lack of representativeness, a lack of norms, a lack of understanding of the range of typical and atypical brain functions and structure, lack of replication, and a host of other science-based problems (Button et al. 2013; Eack et al. 2012; Orem et al. 2019; Peprah et al. 2015; Poldrack et al. 2017; Wahlund and Kristiansson, 2009).

These methodological limitations are nowhere more clear than in the recruitment of prison inmates as subjects for neuroimaging studies. Any discussion of prisoner research must first be clear that who goes to prison, meaning how the criminal justice system selects, prosecutes and sentences, is inherently unequal on the bases of race and class (Alexander 2012). This means that any prisoner based research will replicate the inherent biases and inequalities of the criminal justice system. Even if replicating, and likely worsening inequality by asserting a biologically defined difference in those who are imprisoned, is not bad enough, prisoner based studies will also never be methodologically sound because of the selection bias, confounding and lack of generalizability. In short, these studies will never develop reliable and valid evidence-based observations which have any application to the general population.

Nevertheless, this research has pursued two ideas: first, that the criminal brain is different from the non-criminal brain, an age-old concept which has repeatedly failed but persists (Pustilnik 2009); and second, that neuroscience can predict dangerousness and future behavior based on supposed brain defects or differences from non-criminal brains (Kiehl et al. 2001; Raine et al. 2000), another hypothesis which repeatedly fails (DeMatteo et al. 2020; Edens et al. 2015; Poldrack et al. 2018). As discussed below, these two efforts are based on fundamental and core misconceptions of social cognitive behaviors, and misconstrue every part of the relationship between brain and mind: they assume the mind is wholly encompassed by the physical structure of the brain alone, and that the physical structure is sufficient to explain the complexity of lived experience, adaptability, and the interaction of individual experience with social and structural forces.

Despite the consistent failure of this enterprise, the unethical nature of the research goes largely unchallenged. Prison inmates, at least in the U.S., whose participation may be the result of coercion, in that they feel they have no choice to refuse or will suffer consequences for refusal; because they are bored and have limited options for being out of their cells; because they hope for some treatment or remediation of medical difficulties, especially in the face of long waits for medical care and poor quality prison health care; or from altruism, wanting to believe that advances in science will benefit others (Christopher et al. 2011). This last reason is the most disturbing in the face of the decades of failures of this type of research to identify any meaningful tools to predict future behavior or violence, and because the agenda of these studies is fundamentally at odds with such unfounded hopefulness.

In addition, methodological issues beyond the recruitment of prisoners continue to go unaddressed in this research which seeks to predict behavior. First, as noted above, research which localizes brain defects is not specific, meaning that such deficits are widely observed in many people with psychiatric and neurological conditions, as well as healthy people. Even if they were shown to consistently differentiate healthy and ill, or violent and non-violent, people on average, the lack of specificity would make the predictive potential quite poor. In fact, clinical predictions of future violence are inaccurate more than two-thirds of the time and with a positive predictive value (the proportion of people predicted to be violent who later are) estimated to be 0.41 (Fazel et al. 2012), a prediction rate substantially worse than chance. One often cited study of prisoner future violence, which sought to use brain imaging-based deficits to predict violent and non-violent re-arrest was unable to produce an estimate of future violent re-arrest because there were simply too few violent behaviors and re-arrests in the subjects followed for 4 years, despite the prediction and expectation based on the neuroimaging model that there would be (Aharoni et al. 2013).

Neuroimaging methodology too often fails to address the poor sensitivity and specificity of studies (Noble et al. 2020), and fails to address the lack of reliability (Elliott et al. 2020). The meta-analysis conducted by Elliott et al. calculated test–retest reliability of fMRI studies, finding poor reliability (with a mean intraclass correlation coefficient of 0.397), well below the threshold for clinical use or applying to an individual. In a second analysis, Elliott et al. reported on whether reliability for regions of interest specific to the task (i.e., target regions, the regions of the brain expected to be in use to complete the assigned task while being imaged) compared to regions not expected to be involved (non-target) would have higher test–retest reliability. They reported that intraclass correlation coefficients were exceedingly poor for both (mean for target = 0.27; mean for non-target = 0.23), with no statistical difference between region types. As those researchers note, it is not the tool (fMRI) that is the problem, but the strategy for applying and interpreting the meaning of the results. Such poor task focused test–retest reliability poses substantial questions for cross-sectional imaging, suggesting that brain activation on the same task differs over time or between sessions.

The question of the reliability and validity of findings is also one of generalizability: to whom might these findings apply? Much of the research suffers from selection bias and a lack of representativeness because the samples are not community-drawn and population representative. Similarly, the lack of population norms and understanding of the range of typical parameters, so that atypical can be a meaningful marker, render many of the neuroimaging study findings of little value beyond the study subjects themselves. And the lack of consideration of confounding, those variables which may influence both brain and behavior, variables which are often unmeasured or uncontrolled, should further limit the breadth of conclusions drawn, but more often go unaddressed.

Additionally, the belief that neuroimaging could uncover the truth of social cognitive behaviors, the mind beneath the brain, was based on the belief that brain activation differs between such tasks as trying to deceive someone and not trying to deceive them. These studies almost always rely on "known groups," that is, people told to lie about a specific thing that usually has nothing to do with them. Researchers teach or assign people to "play" liars and then treat the findings as though they are liars (Farah et al. 2014). Is it lying to remember the instructions to lie? or to engage motor functions to respond to the prompts while being imaged? or to mobilize areas of the brain which respond to novel settings or those that might heighten anxiety? anxiety which might be caused by the imaging machine or the social pressure to perform the task assigned? or hoping to benefit the research by performing well? or the healthy subject bias of who volunteers to participate in such studies? or relying on working memory to remember the instructions or attend to the stimuli? Of course not, but the research methods are confounded by all these simultaneously operating cognitive processes. Thus, even in the laboratory, which artificially limits, as best it can, external stimuli and extraneous interference, the results are not precise or specific.

Nor does neuroimaging research account for such confounders as placebo effect, although it has been documented. Reporting on pain research and neuroimaging of placebo, Wager et al. found that placebo analgesia reduced activity in the areas of the brain related to pain (Wager et al. 2004). Changes in response to placebo treatments of depression have also been widely reported (Mayberg et al. 2002; Peciña et al. 2015). The implications of these studies are that simply being neuroimaged may alter brain activity. With no agreed upon referents, no estimation of error, let alone a gold standard for identifying reliability, these research approaches are seriously flawed. Yet, it is the interpretation and application of techniques, not the technology, that raise doubts as whether they should be considered science based, appearing to follow closer to the tradition of phrenology (Parker Jones et al. 2018). This is all the more so when neuroscience is used for behavioral analysis and prediction.

10.3 Social Neuroscience

A tenet of scientific research is to isolate the process under investigation and control for all other variables; yet, social behavior is not an isolated process. Interpersonal interactions are dynamic, reciprocal, and context-dependent events in which the behavior of one individual is influenced by the other. Research on neural systems of social behavior must account for and/or examine the influence of these variables

[(Hooker 2015) p. 124].

It would be revolution enough if neuroscience was able to explain the thoughts of a person isolated in a laboratory, even one assigned a narrow task. Perhaps Heisenberg's uncertainty principle from quantum physics, that we are unable to accurately measure both momentum and position equally well at the same time, could be instructive to neuroscience: suggesting that researchers can either observe the brain isolated in a laboratory accurately or observe behavior as people move through the world accurately, but not both at once with reasonable precision. Although the analogy underestimates the problem because the behavior in the world is not a constant and adapts to, and interacts with, its surroundings.

Behavior occurs in social context, not solely inside a person's head. To begin to make sense of cognitive social behavior, at least three issues are relevant: (1) developmental and life-course trajectory; (2) interaction and social cognition; and (3) reciprocity. Each of these, discussed below, are relevant to understanding how typical people behave in the world. For people with psychiatric and neurological conditions, multiple layers of complexity must be considered for each.

10.3.1   Developmental and Life-Course Trajectory

As mentioned, one of neuroscience's achievements has been to identify how and when the brain changes over the course of development and degeneration. To explain social cognitive behavior, it is necessary to consider the place along that trajectory that the brain is engaging in the world. Recently, a group of mental health and advocacy organizations, clinical and research practitioners and academics, filed a Amici brief in the U.S. Supreme Court in support of stopping the execution of Billy Joe Wardlow by the state of Texas (Amici Curiae re Wardlow v. Texas 2020). In order to sentence someone to death in Texas, a jury must find that the person is more likely than not to commit future acts of violence. Seeking intervention by the U.S. Supreme Court to stay the execution of Mr. Wardlow, the Amici presented evidence to the court which set out why it is scientifically impossible to reliably predict the future behavior of people still in their developmental period (identified in this brief as 21 years old or younger, although the brief also cites literature which indicates the developmental period lasts until the mid-20s). Reviewing the neuroscience evidence, the Amici describe how risk taking and impulsivity are correlated with brain development, subsiding as the brain more fully matures.

This is born out by longitudinal research of youthful offenders, the vast majority of whom desist from recidivist behaviors as they age into adulthood (Laub and Sampson 2009). That same life-course research, which followed 500 juveniles convicted of delinquency into their 70s, found that about 3% of subjects were chronic offenders, and even those 3% essentially stopped offending in their early 50s. In short, neuroscience has identified changes in the brain over the life course, but this does not predict behavior because people, and their circumstances, change over the life course. The reductionist predictive neuroscience approach ignores the developmental and degenerative evidence, suggesting in its place that while the brain changes, the characteristics of the person, and therefore their behaviors, are static (Siegel and Victoroff 2009).

Even assuming that the conclusion is valid and reliable, that an observed difference in a specific part of the brain was generalizable and predictive, this static labeling based on such a finding has devastating consequences for the person so labeled, and is a tool of social control and punishment far beyond the evidence. Once labeled as disordered or damaged, or antisocial or psychopathic, or a future danger, social institutions restrict and control a person's life options, movement and self-efficacy. In the U.S., this type of social control has a long history of being unequally applied to people of color and the poor, acting as a means of social control and reinforcement (Alexander 2012). Where neuroscience research is used to define the static characteristics of an individual, ignoring its own evidence of dynamic change over time, it becomes a tool of social control and no longer a scientific pursuit.

10.3.2 Interaction and Social Cognition

In adulthood, typically developed, healthy people have the perceptual ability to assess another person's intentions and react to them, to perceive and interpret the environment in which they interact, and have developed the communicative and narrative capacity to make sense of other's actions and their own in relation to the other (Gallagher 2020). The iterative, evaluative, context specific nature of human interaction defines social cognitive behavior: in a specific context, in relation to that context and other people, based on the constant stream of stimuli which require evaluation, re-assessment and adaptability.

Neuroscience research, due to technological limitations in part, and in part to the dominant individualistic paradigm of the fields that comprise it, fails to adequately address the difference between social interaction and laboratory studies. This limitation is not unknown to researchers, but is not taken at all seriously enough, with many researchers suggesting that despite the limits, they can predict behavior and intentions from the isolated, acontextual, artificial laboratory experiments. Even at best, this research would falsely reduce social interaction and behavior to a localized cognitive process of a single, isolated individual.

10.3.3 Reciprocity

People act in, and shape, social context, but they are also shaped by the environment. In a large, community based study of youth (ages 8–21), both low socio-economic status and more traumatic stressful events, increased psychiatric and neurological symptoms, as well as being found to alter the structure and function of the brain (Gur et al. 2019). In other research, adverse neighborhood characteristics and victimization has been reported to increase the risk of psychosis symptoms in youth 4.8 times, after controlling for family socioeconomic status, family psychiatric history, and adolescent substance problems (Newbury et al. 2017). The social determinants of health and mental illness have long been established, with a wide array of factors having been shown to adversely impact development and functioning across the lifespan (Crossley et al. 2019).

Environments which shape behavior, symptoms, and brain functional and structural development, range widely from parenting style (hostile parenting is associated with reduced limbic system connectivity and impaired affective processing) (Kopala-Sibley et al. 2020) to neighborhood characteristics (urbanicity, racial segregation and density of poverty are associated with later life mental illness and altered brain structure and function) (Solmi et al. 2020) to experiences of racial discrimination (which increase symptoms of mental illness and worsen physical health) (Van Dyke et al. 2020; Williams et al. 1997).

This is, perhaps, most clear when considering childhood trauma and the long-term consequences of such experiences. Complex trauma is well-established to have long-term physical and psychiatric consequences, as well as biological, genetic and neurological ones: shortening telomere length (the tip of DNA strands associated with length of life and the pace of aging); disrupting and dysregulating hormonal and endocrine systems; increasing inflammation and inflammatory responses, while also interfering with immune functioning; reducing brain volumes in, and connectivity between, the amygdala, hippocampus and prefrontal cortex; as well as impairing neurological and cognitive development and increasing the risk of later life psychiatric illnesses (Cassiers et al. 2018; Danese and McEwen 2012; Odgers and Jaffee 2013; Sipahi et al. 2014; Varese et al. 2012). Trauma, and exposure to an accumulation of adverse childhood experiences, bears consequences across life-span, including worse economic status, lower school attainment, and higher levels of poverty (Bunting et al. 2018; Metzler et al. 2017).

Trauma affects behavior and social cognitive processes over the life-course because the perception of social cues, perceptual reasoning, threat perception, cognitive flexibility, social cognition, inhibitory systems of self-regulation, theory of mind and conceptions of self and self-worth are all disrupted and dysregulated (Chu et al. 2019; De Bellis and Zisk 2014; Su et al. 2019; Vasterling and Arditte Hall 2018). In this way, early childhood experiences shape the development and life course of that child, and those experiences are carried throughout life, continuing to shape and define how people experience future events. The social context shapes the individual and the individual's response, based on all that the participants bring into that moment, and childhood trauma usually continues to influence social interaction and perception throughout life.

10.4   Social Neuroscience and Criminal Law

From this perspective of social neuroscience, it makes sense that the science would become more integrated into approaches to criminal law, especially in death penalty sentencing cases, in which many courts have routinely held that defense lawyers can and should present evidence of structural or functional brain impairment to fact-finders (see e.g., *Fairchild v. Workman*, 579 F.3d 1134, 1150–51 (10th Cir. 2009) (suggesting that if the defense lawyer had presented the jury with evidence of brain damage as a partial explanation for the defendant's crime, it "could have provided an important explanation for the jury" regarding the sentence); *United States v. Montgomery*, 635 F.3d 1074 (8th Cir. 2011) (expert testimony that Montgomery's PET scan showed abnormalities in the limbic and somatomotor regions of her brain was sufficiently reliable to be admitted"); *Johnson v. United States*, 860 F.Supp.2d 663, 893 (N.D. Iowa 2012) (failure to investigate and present MRI, PET, and EEG evidence at trial was prejudicial ineffective assistance of counsel); *United States v. McCluskey*, (D.N.M. 2013, No. 10-2734, Doc. 1481, Verdict Form) (in a non-unanimous split, John McCluskey was sentenced to life instead of death when one or more jurors found as mitigating that "John's brain is damaged in the regions regulating behaviors and emotions, social cognition, reward, and conflict resolution")).

These cases come as no surprise as they fit squarely within the U.S. Supreme Court's requirements that defense lawyers have an obligation to investigate and present evidence regarding their client's unique life experiences and the human frailties which make a person [*Lockett v. Ohio*, 438 U.S. 586 (1978)]. A defendant's cognitive functioning is considered as mitigating in capital cases, even if the impairment bears no direct link with the homicidal behavior [*Tennard v. Dretke*, 542 U.S. 274 (2004)]. Thus, in capital sentencing, neuroscience evidence does not need to causally explain a specific behavior, it needs to help explain a person, his/her qualities and characteristics, and how he/she is more than the offending behavior.

Yet, neuroscience is relevant not only in the death penalty sentencing setting. Farahany recently conducted a systematic review of U.S. court opinions and reported that neuroscience evidence was introduced in an increasing number of criminal cases, not limited to capital sentencing, between 2005 and 2012 (Farahany 2016). Moreover, at least by 1975, in a non-capital armed robbery case, the 9th Circuit Court of Appeals held that a defendant was entitled to obtain funds for an EEG because the defense turned on the question of the defendant's mental condition [*U.S. v Hartfield* 513 F.2d 254 (1975)]. Most famously, in the prosecution of John Hinckley, who shot President Reagan and his Press Secretary James Brady, the defense introduced CT scan images to support their contention that Hinckley suffered from brain atrophy and ventricular enlargement related to psychosis [*U.S. v John Hinckley*, 672 F.2d 115 (1982)].

From the defense perspective, a perspective which requires heightened reliability and validity because it must counter to overwhelming resources and power of prosecutorial agencies, neuroscience evidence must always be presented within the context of the multi-generational, bio-psychosocial history that situates a defendant and explains the development and life-course of the defendant (Freedman and Woods 2018; Freedman and Zaami 2019). For instance, demonstrating structural and functional changes in post-traumatic brain injury is compelling and comprehensible when the before and after social cognitive behaviors are also presented and explained. That is, demonstrating the before and after behaviors through descriptions and records can bring to life the neuroimaging of current damage. Neuroscience evidence should always be presented within that bio-psychosocial framework, such that examples and descriptions from lay witnesses and institutional records from across the defendant's life trajectory can tie the neuroscience evidence to the lived experience of the defendant. This is equally as true for neurodevelopmental conditions as it is for acquired injuries.

Most of this is ignored in the ongoing debates about neuroscience and law, which continue to focus on theoretical questions related to free will, determinism, culpability and mental state, and most importantly, blame and retribution (Bennett 2016; Slobogin 2017). A great deal of attention has been paid to these philosophical questions (Jones 2017; Morse 2013, 2016; Zeki et al. 2004). The framework of the questions in these debates, even among those who seek to redefine the focus, works to reinforce the status quo, ignoring or denying the central role of structural inequality in criminal justice and prosecutorial enforcement of mores and norms. Although silent about it, these debates are fundamentally about the maintenance of social control and power, about the ways in which the criminal justice system is used to regulate behavior, but to regulate specific behaviors of the dispossessed.

Morse most explicitly argues for the status quo by advocating a wholesale rejection of the relevance of neuroscience, saying that neuroscience changes nothing in criminal law because law must and should regulate behavior regardless of the cause of the behavior. He puts forward a perspective that we can simply observe the behavior and know whether it should be adjudged immoral or improper, and take action to control it and punish the offender without resorting to neuroscience (Morse, 2013, 2016). To support his rejectionist approach, Morse [(Morse 2016) (at p. 338)] mischaracterizes a quote from a plurality opinion written by Justice Marshall in a 1968 Supreme Court case, *Powell v. Texas* [392 U.S. 514 (1968)] in which Justice Marshall says that the dissent "goes much too far on the basis of too little knowledge" (at p. 521). Morse implies that Justice Marshall is referring to the medical knowledge of the defendant's alcoholism, but he is referring to the poor record made in the lower court as to the bases for its findings. Justice Marshall is commenting on the evidence presented in the case as insufficient, not the capacity of medical science to answer the question posed.

Contrary to Morse's implication, Justice Marshall goes on to say that one reason for his opinion is that criminal law sentencing has a determinate jail period, whereas the alternative sentencing suggested, therapeutic civil commitment, could subject the defendant to an indefinite period of detention. He continues:

> Faced with this unpleasant reality, we are unable to assert that the use of the criminal process as a means of dealing with the public aspects of problem drinking can never be defended as rational. The picture of the penniless drunk propelled aimlessly and endlessly through the law's 'revolving door' of arrest, incarceration, release and re-arrest is not a pretty one. But before we condemn the present practice across-the-board, perhaps we ought to be able to point to some clear promise of a better world for these unfortunate people. Unfortunately, no such promise has yet been forthcoming. If, in addition to the absence of a coherent approach to the problem of treatment, we consider the almost complete absence of facilities and manpower for the implementation of a rehabilitation program, it is difficult to say in the present context that the criminal process is utterly lacking in social value (p. 531).

Morse misunderstands the human rights basis for Justice Marshall's opinion. And yet, what Justice Marshall seeks to avoid is precisely what Morse's rejectionist approach would do: use the law to excessively punish those who have a medical condition, not because we lack knowledge of the condition, but because, in Morse's view, such knowledge is irrelevant. Morse's rejectionist approach refuses to acknowledge advances in medicine and science in favor of punishing the observed behavior, in the case cited, public drunkenness. When applied to law, Morse's rejectionist approach would not only limit neuroscience's use, it would reject all context, historical and immediate, mitigation and any explanation of behavior, because what he argues is that nothing should be considered beyond the specific behavior. To argue, as Morse does, that "we" can simply observe unwanted behavior and regulate it, intentionally obscures the more than 400 years of oppressive usage of criminal law to maintain power and inequality. It reaffirms the status quo of unequally and unjustly applied prosecutorial power, the power of the state which regulates some but not others, based on race, class and gender. It pretends that what we "see" is outside the implicit and explicit biases that shape us.

Moreover, Morse's rejectionist approach, in which he argues that the cause of a behavior, or the understanding of an action, or the context of it, are irrelevant because criminal law only cares what the action is, is wrong on the law. Since its common law foundations, criminal adjudication has considered the intentions and reasoning for an act, and the causal basis for offending, in determining liability and culpability. Morse's rejectionist idea, although focused on rejecting the relevance of neuroscience, is in reality a call to disregard all evidence other than the prosecutorial depiction of the offense itself.

A counter to Morse, which attempts to reframe the debate about the role of neuroscience in criminal law, suggests a more nuanced use of the science, focused on how the evidence is being used to determine criminal liability and punishment (Slobogin 2017). Slobogin suggests five uses of neuroscience evidence and the relative meaning and strengths of each: first, evidence of brain abnormality, which he describes as weakly relevant to criminal liability; second, where the abnormality is common in those who offend, which is the arena of prediction neuroscience, that damage to a specific part of the brain could be argued to cause criminal behavior; third, where the behavior is the outcome of an abnormality, such that people with a type of brain impairment are more likely to offend; fourth, individualized behavioral assessment, in his example through neuropsychology rather than brain imaging, which then ties the neuroscience to lab based, group-normed behavioral assessment; and fifth, individualized assessment compared to legal standards, such that a person whose brain functions like a twelve year old would be considered in law as a juvenile, regardless of the person's chronological age.

Even this nuanced and thoughtful approach to weighing how neuroscience could be used in various ways does not adequately conceptualize behavior as historically situated and contextualized, nor does it adequately come to grips with how a person is more than a specific act committed. Criminal enforcement is not an ahistorical or apolitical equal application of communal policy:

> Thus, as policing established itself in new roles of monitoring and regulating public space, policing schools, participating in gentrification projects, carrying out the vast expansion of criminal law, and generating revenue for cash-strapped municipal budgets in a time of fiscal austerity, citizen exposure to police oversight was normalized and institutionalized, crowding out other imperatives and evolving a "policing state" that is historically unprecedented and breathtaking in its scope and limited accountability compared to other local bureaucracies. That this occurred even as police have been the primary target of two of the nation's largest social movements in the last century—the black freedom struggle in the 1960s and Black Lives Matter today—speaks to how established and autonomous police are as a political force. (p. 195) (Weaver and Geller 2019)

Further, these social control policies produce the very behavior they seek to regulate, worsening psychological functioning and increasing criminal behaviors that bring youth into contact with policing agencies (Del Toro et al. 2019). In this way, the status quo not only continues to reinforce inequality and injustice, but causes a downward cycle of replication, increasing the injury and destruction of structural oppression.

Behavior, and its regulation by law, cannot be considered without addressing structural inequality. Behavior is not free, it is not the choice to act without regard to setting or historical moment, it is specifically and identifiably bounded by history and context. Neuroscience's potential to change how social control is enforced through prosecutorial power, used for the benefit of the already powerful, requires it to establish how and why cognitive behavior is bounded and shaped not by a localized neuronal response, but social cognitive processes in historical context. Instead of debating straw-man theories about free will and determinism, legal theorists could pay more attention to how neuroscience might be a tool used to help fix structural inequalities and oppression. Morse's reductionist view, that we already know simply by the action itself, is ahistorical and acontextual, and serves to maintain the status quo, and it is inaccurate and misguided.

10.5 Research Ethics of Applied Neuroscience

Neuroscience research should consider structural inequality, developmental and life-course trajectory, interaction and social cognition, and reciprocity. Failing to do so, in addition to being poor science, poses a substantial ethical risk: by reducing complexity, interaction and adaptability to localized points in the brain, it provides a scientific sheen to social inequalities and reinforces stigma. Neuroscience's effort to predict violence by neuroimaging specifically ignores a wealth of research on social context and does just this.

First, it ignores the cultural, community and cross-cultural meaning of behavior and interaction (Gallagher 2020). It reinforces social inequality and disparities by assuming that the dominant cultural experience is the only one that exists, or at best is normative and all others are non-normative. Particularly in cross-cultural social interactions, the perception and re-enactment if inequality should be accounted for if one's research agenda is the accurate prediction of behavior and intentionality.

For example, consider the research on jury decision-making. In a study, mock jurors were assigned to all white jury panels or multiracial panels. The mock jurors watched a video of trial evidence and deliberated. Those jurors in mixed-race jury groups were less likely to presume guilt before deliberations, deliberated longer, and were more thorough in their evaluation of the evidence presented (Sommers 2006). The findings show that white jurors act differently when in multiracial groups, meaning that social context is crucial to understanding behavior and social cognitive behaviors.

Second, the research on neighborhood effects and violence demonstrates convincingly that context matters. Concentrated disadvantage, segregation based on race, and low collective efficacy combine to increase the risk of violence regardless of the individual characteristics of the people in the neighborhood (Morenoff et al. 2001; Sampson et al. 2005). Further, modifiable built environment changes can reduce violence in neighborhood (Culyba et al. 2016), but can also improve health more generally (Duchowny et al. 2020; Morenoff and Lynch 2004).

The consequence of neuroscience's approach to explaining behavior, such as violence, as an individualized, brain location specific occurrence, is that ecological validity and structural inequality are ignored. These validity problems are sometimes referred to as the individualistic fallacy, which assumes that individual-level outcomes, such as behavior and functioning, should be attributed solely to the individual, fundamentally denying the structural, social, institutional, contextual and environmental frame within which behaviors and functioning occur (Silver 2000). In the neuroscience research setting, the individualistic fallacy is compounded by an unfounded hope that mean region of interest measurement differences observed in a laboratory can be applied to an individual interacting in the world.

This reductionist approach also worsens stigma. By asserting that the brain structural or functional deficits are the cause of behaviors which should be controlled or from which people should distance themselves, rather than a social setting or social interactions in which the person is acting, neuroscience effectively reinforces the stereotypes that people with psychiatric or neurological conditions are dangerous, to be feared, or to be kept away. Neuroscience risks being a tool of reinforcing and recreating structural inequality, raising serious doubts as to the ethical basis for its research agenda.

10.6 Towards an Ethical Neuroscience Research Agenda

Neuroscience research has made enormous strides in understanding psychiatric and neurological conditions. Increasingly, however, the (mis)application of neuroscience research, especially into legal settings (i.e., "neurolaw"), poses a host of problems. Neuroscience has become a primary tool for, or at least is proposed as a means of, explaining, predicting and controlling behavior, primarily through criminal and civil legal systems without regard for the structural inequality maintained by those systems. As the demands for the application of neuroscience research to real world settings, researchers have an enhanced obligation to articulate the assumptions, limits, validity and appropriate interpretation and applications of their findings and to maintain the integrity of the research.

The Bradford Hill criteria of causal relationships might be useful to consider for this purpose: strength of the association, consistency of the observation, specificity, temporality, dose-response or exposure-response, scientific plausibility of the association, coherence, experimentally supported, and similarity with other evidence (Hill 1965). More recently, concepts such as confounding, interaction and bias have been added to this approach for assessing causal relationships (Rothman et al. 2008). As discussed, neuroscience's effort to explain, predict and control social cognitive behavior fails this test of causal inference. The application of neuroscience research currently rests on a number of assumptions which too infrequently acknowledge the limitations of the findings.

The desire to explain and predict behavior based on brain structure and function may be understandable, but when the consequences could include prior restraint or incarceration, and the worsening of structural inequality beneath a veneer of science, the pursuit becomes unethical. Reductionist approaches to explaining social cognitive behavior are, at core, unethical. To engage in research related to social cognitive behavior in a manner consistent with ethical principles, neuroscientists should consider how their research on social cognitive behavior might be used. Professional societies should establish use-based criteria which can be considered by Institutional Review Boards when they evaluate research proposals. Such criteria should explicitly consider the relationship between the proposed study and structural inequality, stigma and socio-cultural setting.

Second, researchers should consider whether consulting with, or providing support to, prosecutorial agencies meets ethical standards because of the specific ways in which these agencies reproduce and maintain structural inequality. While many scientists seek to maintain separation from the political nature of criminal justice by holding themselves out as independent and not beholden, this is pretense and fails to live up to obligations to act ethically. This is not a blanket condemnation of prosecutorial functions. Clearly there is an ethical difference between prosecuting civil rights violations and enforcing stop-and-frisk policing. It is, however, a suggestion that the consequence of those functions be considered and addressed directly by neuroscience researchers.

Third, the limitations of the science should be more clearly and more frequently articulated: the reliability (e.g., test–retest) of the technology; the poor replication between laboratories; the conceptual problems of applying group mean data to individuals; the non-representativeness of research subjects, including the use of prisoners; the lack of established normative data on the range of typical brain structure and function, and such normative data across lifespan; the reductionist conception of social cognitive behavior; the lack of ecological validity of the tasks used and lack of understanding of what is being measured during those tasks; all raise significant ethical issues about how neuroscience is being used.

Fourth, neuroscience as a field, and its many subsidiary branches, still needs to increase the number of people of color and women recruited into and retained as researchers. As the research cited above on multiracial jury panels suggests, perhaps multiracial research teams will be less likely to engage in studies or to apply them in ways that replicate inequality. Similarly, recruitment of broadly population representative subjects into the research studies, studies which should be longitudinal, require additional effort.

Finally, ethical neuroscience should draw bright line distinctions between science which seeks to advance knowledge and science which seeks to be applied. It is not sufficient to defer such questions of the use of neuroscience to the courts, hoping that they will correctly apply their gatekeeper functions to the admissibility [*Daubert v. Merrell Dow Pharmaceuticals*, 516 U.S. 869 (1993)]. Neuroscience needs to self-regulate, to establish rules to govern the approaches to applying neuroscience, rules which recognize the social and contextual. The consequences to individual people of getting this wrong can be severe and neuroscience researchers have an obligation to guard against the misuse and misapplication of the science.

References

Aharoni, E., G.M. Vincent, C.L. Harenski, V.D. Calhoun, W. Sinnott-Armstrong, M.S. Gazzaniga, and K.A. Kiehl. 2013. Neuroprediction of future rearrest. *Proceedings of the National Academy of Sciences* 110 (15): 6223–6228. https://doi.org/10.1073/pnas.1219302110.

Alexander, M. 2012. *The new Jim Crow: Mass incarceration in the age of colorblindness.* New York: The New Press.

Allison, S.L., R.L. Koscik, R.P. Cary, E.M. Jonaitis, H.A. Rowley, N.A. Chin, H. Zetterberg, K. Blennow, C.M. Carlsson, S. Asthana, B.B. Bendlin, and S.C. Johnson. 2019. Comparison of different MRI-based morphometric estimates for defining neurodegeneration across the Alzheimer's disease continuum. *NeuroImage Clinical* 23: 101895 https://doi.org/10.1016/j.nicl.2019.101895

Bassett, D.S., and O. Sporns. 2017. Network neuroscience. *Nature Neuroscience* 20 (3): 353–364. https://doi.org/10.1038/nn.4502.

Bennett, E. 2016. Neuroscience and criminal law: Have we been getting it wrong for centuries and where do we go from here? *Fordham Law Review* 85: 437–451.

Bunting, L., G. Davidson, C. McCartan, J. Hanratty, P. Bywaters, W. Mason, and N. Steils. 2018. The association between child maltreatment and adult poverty—A systematic review of longitudinal research. *Child Abuse & Neglect* 77: 121–133. https://doi.org/10.1016/j.chiabu.2017.12.022.

Button, K.S., J.P.A. Ioannidis, C. Mokrysz, B.A. Nosek, J. Flint, E.S.J. Robinson, and M.R. Munafò. 2013. Power failure: Why small sample size undermines the reliability of neuroscience. *Nature Reviews Neuroscience* 14: 365. https://doi.org/10.1038/nrn3475.

Caspi, A., R.M. Houts, A. Ambler, A. Danese, M.L. Elliott, A. Hariri, H. Harrington, S. Hogan, R. Poulton, S. Ramrakha, L.J. Rasmussen, and T.E. Moffitt. 2020. Longitudinal assessment of mental health disorders and comorbidities across 4 decades among participants in the dunedin birth cohort study. *JAMA Network Open* 3(4). https://doi.org/10.1001/jamanetworkopen.2020.3221

Cassiers, L.L.M., B.G.C. Sabbe, L. Schmaal, D.J. Veltman, B. Penninx, and F. Van den Eede. 2018. Structural and functional brain abnormalities associated with exposure to different childhood trauma subtypes: A systematic review of neuroimaging findings. *Front Psychiatry* 9: 329. https://doi.org/10.3389/fpsyt.2018.00329.

Choudhury, S., and J. Slaby, eds. 2012. *Critical neuroscience a handbook of the social and cultural contexts of neuroscience.* Chichester, Sussex: Wiley.

Christopher, P.P., P.J. Candilis, J.D. Rich, and C.W. Lidz. 2011. An empirical ethics agenda for psychiatric research involving prisoners. *AJOB Primary Research* 2 (4): 18–25. https://doi.org/10.1080/21507716.2011.627082.

Chu, D.A., R.A. Bryant, J.M. Gatt, and A.W.F. Harris. 2019. Cumulative childhood interpersonal trauma is associated with reduced cortical differentiation between threat and non-threat faces in posttraumatic stress disorder adults. *Australian and New Zealand Journal of Psychiatry* 53 (1): 48–58. https://doi.org/10.1177/0004867418761578.

Cope, L.M., J.E. Hardee, M.E. Martz, R.A. Zucker, T.E. Nichols, and M.M. Heitzeg. 2020. Developmental maturation of inhibitory control circuitry in a high-risk sample: A longitudinal fMRI study. *Developmental Cognitive Neuroscience* 43: 100781. https://doi.org/10.1016/j.dcn.2020.100781.

Crossley, N.A., L.M. Alliende, T. Ossandon, C.P. Castaneda, A. Gonzalez-Valderrama, J. Undurraga, M. Castro, S. Guinjoan, A.M. Díaz-Zuluaga, J.A. Pineda-Zapata, C. López-Jaramillo, and R.

Bressan. 2019. Imaging social and environmental factors as modulators of brain dysfunction: Time to focus on developing non-western societies. *Biological Psychiatry Cognitive Neuroscience Neuroimaging,* 4 (1): 8–15. https://doi.org/10.1016/j.bpsc.2018.09.005

Culyba, A.J., S.F. Jacoby, T.S. Richmond, J.A. Fein, B.C. Hohl, and C.C. Branas. 2016. Modifiable neighborhood features associated with adolescent homicide. *JAMA Pediatrics* 170 (5): 473–480. https://doi.org/10.1001/jamapediatrics.2015.4697.

Danese, A., and B.S. McEwen. 2012. Adverse childhood experiences, allostasis, allostatic load, and age-related disease. *Physiology & Behavior* 106 (1): 29–39. https://doi.org/10.1016/j.physbeh.2011.08.019.

Davis, K. 2020, June/July. Millions have been invested in the emerging field of neurolaw. Where is it leading? *ABA Journal.* Retrieved from https://www.abajournal.com/magazine/article/millions-have-been-invested-in-the-emerging-field-of-neurolaw.-where-is-it-leading

De Bellis, M. D., and A. Zisk, 2014. The biological effects of childhood trauma. *Child and Adolescent Psychiatric Clinics of North America,* 23 (2): 185–222. https://doi.org/10.1016/j.chc.2014.01.002

Del Toro, J., T. Lloyd, K.S. Buchanan, S.J. Robins, L.Z. Bencharit, M.G. Smiedt, K.S. Reddy, E.R. Pouget, E.M. Kerrison, and P.A. Goff. 2019. The criminogenic and psychological effects of police stops on adolescent black and Latino boys. *Proceedings of National Academic Science United States of America* 116 (17): 8261–8268. https://doi.org/10.1073/pnas.1808976116

DeMatteo, D., S.D. Hart, K. Heilbrun, M.T. Boccaccini, M.D. Cunningham, K.S. Douglas, J.A. Dvoskin, J.F. Edens, L.S. Guy, D.C. Murrie, R.K. Otto, and T.J. Reidy. 2020. Statement of concerned experts on the use of the Hare Psychopathy Checklist—Revised in capital sentencing to assess risk for institutional violence. *Psychology, Public Policy, and Law* 26 (2): 133–144. https://doi.org/10.1037/law0000223.supp (Supplemental)

Duchowny, K.A., M.M. Glymour, and P.M. Cawthon. 2020. Is perceived neighbourhood physical disorder associated with muscle strength in middle aged and older men and women? Findings from the US health and retirement study. *Journal of Epidemiology and Community Health* 74 (3): 240–247. https://doi.org/10.1136/jech-2019-213192.

Eack, S.M., A.L. Bahorik, C.E. Newhill, H.W. Neighbors, and L.E. Davis. 2012. Interviewer-perceived honesty as a mediator of racial disparities in the diagnosis of schizophrenia. *Psychiatric Services (washington, D.c.)* 63 (9): 875–880. https://doi.org/10.1176/appi.ps.201100388.

Edens, J.F., S.E. Kelley, S.O. Lilienfeld, J.L. Skeem, and K.S. Douglas. 2015. DSM-5 antisocial personality disorder: Predictive validity in a prison sample. *Law and Human Behavior* 39 (2): 123–129. https://doi.org/10.1037/lhb0000105.

El-Hai, J. 2005. *The lobotomist: A maverick medical genius and his tragic quest to rid the world of mental illness.* Hoboken, NJ: Wiley.

Elliott, M.L., A.R. Knodt, D. Ireland, M.L. Morris, and R. Poulton. 2020. What is the test-retest reliability of common task-functional MRI measures? New empirical evidence and a meta-analysis. *Psychological Science* 31 (7): 792–806. https://doi.org/10.1177/0956797620916786.

Farah, M.J., J.B. Hutchinson, E.A. Phelps, and A.D. Wagner. 2014. Functional MRI-based lie detection: Scientific and societal challenges. *Nature Reviews Neuroscience, 15,* 123. https://doi.org/10.1038/nrn3665. https://www.nature.com/articles/nrn3665#supplementary-information

Farahany, N.A. 2016. Neuroscience and behavioral genetics in US criminal law: An empirical analysis. *Journal of Law and the Biosciences* 2 (3): 485–509. https://doi.org/10.1093/jlb/lsv059.

Fazel, S., J.P. Singh, H. Doll, and M. Grann. 2012. Use of risk assessment instruments to predict violence and antisocial behaviour in 73 samples involving 24,827 people: Systematic review and meta-analysis. *BMJ* 345: e4692. https://doi.org/10.1136/bmj.e4692.

Freedman, D., and G.W. Woods. 2018. The developing significance of context and function: Neuroscience and law. *Behavioral Sciences & the Law* 36 (4): 411–425. https://doi.org/10.1002/bsl.2351.

Freedman, D., and S. Zaami. 2019. Neuroscience and mental state issues in forensic assessment. *International Journal of Law and Psychiatry* 65: 101437. https://doi.org/10.1016/j.ijlp.2019.03.006.

Gallagher, S. 2020. *Action and interaction*. New York: Oxford University Press.

Gkotsi, G.M., and J. Gasser. 2016. Neuroscience in forensic psychiatry: From responsibility to dangerousness. Ethical and legal implications of using neuroscience for dangerousness assessments. *International Journal of Law and Psychiatry* 46: 58–67. https://doi.org/10.1016/j.ijlp.2016.02.030.

Gur, R.E., T.M. Moore, A.F.G. Rosen, R. Barzilay, D.R. Roalf, M.E. Calkins, K. Ruparel, J.C. Scott, L. Almasy, T.D. Satterthwaite, R.T. Shinohara, R.C. Gur. 2019. Burden of environmental adversity associated with psychopathology, maturation, and brain behavior parameters in youths. *JAMA Psychiatry*. https://doi.org/10.1001/jamapsychiatry.2019.0943

Hill, A.B. 1965. Environment and disease—Association or causation. *Proceedings of the Royal Society of Medicine-London* 58 (5): 295–300. https://doi.org/10.1177/003591576505800503

Hooker, C.I. 2015. Social neuroscience and psychopathology: Identifying the relationship between neural function, social cognition, and social behavior. In *Social neuroscience: Brain, mind, and society*, ed. R.K. Schutt, L.J. Seidman, and M.S. Keshavan, 123–145. Cambridge, MA: Harvard University Press.

Jalbrzikowski, M., D. Freedman, C.E. Hegarty, E. Mennigen, K.H. Karlsgodt, L.M. Olde Loohuis, R.A. Ophoff, R.E. Gur, and C.E. Bearden. 2019. Structural brain alterations in youth with psychosis and bipolar spectrum symptoms. *Journal of the American Academy of Child and Adolescent Psychiatry*. https://doi.org/10.1016/j.jaac.2018.11.012

Jones, O.D. 2017. Keynote: Law and the brain—Past, present, and future. *Arizona State Law Journal* 48: 917–933.

Kiehl, K.A., A.M. Smith, R.D. Hare, A. Mendrek, B.B. Forster, J. Brink, and P.F. Liddle. 2001. Limbic abnormalities in affective processing by criminal psychopaths as revealed by functional magnetic resonance imaging. *Biological Psychiatry* 50 (9): 677–684.

Kopala-Sibley, D.C., M. Cyr, M.C. Finsaas, J. Orawe, A. Huang, N. Tottenham, and D.N. Klein. 2020. Early childhood parenting predicts late childhood brain functional connectivity during emotion perception and reward processing. *Child Development* 91 (1): 110–128. https://doi.org/10.1111/cdev.13126.

Laub, J.H., and R.J. Sampson. 2009. *Shared beginnings*. Divergent Lives: Harvard University Press.

Mayberg, H.S., J.A. Silva, S.K. Brannan, J.L. Tekell, R.K. Mahurin, S. McGinnis, and P.A. Jerabek. 2002. The functional neuroanatomy of the placebo effect. *American Journal of Psychiatry* 159 (5): 728–737. https://doi.org/10.1176/appi.ajp.159.5.728.

Metzler, M., M.T. Merrick, J. Klevens, K.A. Ports, and D.C. Ford. 2017. Adverse childhood experiences and life opportunities: Shifting the narrative. *Children and Youth Services Review* 72: 141–149. https://doi.org/10.1016/j.childyouth.2016.10.021.

Modabbernia, A., A. Reichenberg, A. Ing, D.A. Moser, G.E. Doucet, E. Artiges, T. Banaschewski, G.J. Barker, A. Becker, A.L. Bokde E.B. Quinlan, S. Desrivières, and I. Consortium. 2021. Linked patterns of biological and environmental covariation with brain structure in adolescence: a population-based longitudinal study. *Molecular Psychiatry*. https://doi.org/10.1038/s41380-020-0757-x

Morenoff, J.D., and J.W. Lynch. 2004. *What makes a place healthy? Neighborhood influences on racial/ethnic disparities in health over life course*. Washington, DC: The National Academies Press.

Morenoff, J.D., R.J. Sampson, and S.W. Raudenbush. 2001. Neighborhood inequality, collective efficacy, and the spatial dynamics of urban violence. *Criminology* 39 (3): 517–559. https://doi.org/10.1111/j.1745-9125.2001.tb00932.x.

Morse, S.J. 2013. Brain overclaim redux. *Law and Inequality* 31: 509.

Morse, S.J. 2016. Actions speak louder than images: The use of neuroscientific evidence in criminal cases. *Journal of Law and the Biosciences* 3 (2): 336–342. https://doi.org/10.1093/jlb/lsw025.

Newbury, J., L. Arseneault, A. Caspi, T.E. Moffitt, C.L. Odgers, and H.L. Fisher. 2017. Cumulative effects of neighborhood social adversity and personal crime victimization on adolescent psychotic experiences. *Schizophrenia Bulletin*. https://doi.org/10.1093/schbul/sbx060.

Noble, S., D. Scheinost, and R.T. Constable. 2020. Cluster failure or power failure? Evaluating sensitivity in cluster-level inference. *NeuroImage* 209: 116468. https://doi.org/10.1016/j.neuroimage.2019.116468.

Odgers, C.L., and S.R. Jaffee. 2013. Routine versus catastrophic influences on the developing child. *Annual Review of Public Health* 34: 29–48. https://doi.org/10.1146/annurev-publhealth-031912-114447.

Orem, T.R., M.D. Wheelock, A.M. Goodman, N.G. Harnett, K.H. Wood, E.W. Gossett, D.A. Granger, S. Mrug, and D.C. Knight. 2019. Amygdala and prefrontal cortex activity varies with individual differences in the emotional response to psychosocial stress. *Behavioral Neuroscience* 133 (2): 203–211. https://doi.org/10.1037/bne0000305

Parker Jones, O., F. Alfaro-Almagro, and S. Jbabdi. 2018. An empirical, 21st century evaluation of phrenology. *Cortex* 106: 26–35. https://doi.org/10.1016/j.cortex.2018.04.011.

Peciña, M., A.S. Bohnert, M. Sikora, E.T. Avery, S.A. Langenecker, B.J. Mickey, and J.K. Zubieta. 2015. Association between placebo-activated neural systems and antidepressant responses: neurochemistry of placebo effects in major depression. *JAMA Psychiatry* 72 (11): 1087–1094. https://doi.org/10.1001/jamapsychiatry.2015.1335.

Peprah, E., H. Xu, F. Tekola-Ayele, and C.D. Royal. 2015. Genome-wide association studies in Africans and African Americans: Expanding the framework of the genomics of human traits and disease. *Public Health Genomics* 18 (1): 40–51. https://doi.org/10.1159/000367962.

Pescosolido, B.A., J.K. Martin, J.S. Long, T.R. Medina, J.C. Phelan, and B.G. Link. 2010. "A disease like any other"? A decade of change in public reactions to schizophrenia, depression, and alcohol dependence. *American Journal of Psychiatry* 167 (11): 1321–1330. https://doi.org/10.1176/appi.ajp.2010.09121743.

Poldrack, R.A., C.I. Baker, J. Durnez, K.J. Gorgolewski, P.M. Matthews, M.R. Munafo, T.E. Nichols, J.B. Poline, E. Vul, and T. Yarkoni. 2017. Scanning the horizon: Towards transparent and reproducible neuroimaging research. *Nature Reviews Neuroscience*, 18 (2): 115–126. https://doi.org/10.1038/nrn.2016.167

Poldrack, R.A., J. Monahan, P.B. Imrey, V. Reyna, M.E. Raichle, D. Faigman, and J.W. Buckholtz. 2018. Predicting violent behavior: what can neuroscience add? *Trends in Cognitive Sciences* 22 (2): 111–123. https://doi.org/10.1016/j.tics.2017.11.003.

Pustilnik, A.C. 2009. Violence on the brain: A critique of neuroscience in criminal law. *Wake Forest Law Review* 44: 183.

Raine, A., T. Lencz, S. Bihrle, L. LaCasse, and P. Colletti, P. 2000. Reduced prefrontal gray matter volume and reduced autonomic activity in antisocial personality disorder. *Arch Gen Psychiatry* 57 (2): 119–127; discussion 128–119. Retrieved from http://www.ncbi.nlm.nih.gov/pubmed/10665614

Rothman, K.J., S. Greenland, and T.L. Lash. 2008. *Modern epidemiology*, 3rd ed. Philadelphia: Wolters Kluwer Health/Lippincott Williams & Wilkins.

Sampson, R.J., J.D. Morenoff, and S. Raudenbush. 2005. Social anatomy of racial and ethnic disparities in violence. *American Journal of Public Health* 95 (2): 224–232. https://doi.org/10.2105/AJPH.2004.037705.

Siegel, A., and J. Victoroff. 2009. Understanding human aggression: New insights from neuroscience. *International Journal of Law and Psychiatry* 32 (4): 209–215. https://doi.org/10.1016/j.ijlp.2009.06.001.

Silver, E. 2000. Race, neighborhood disadvantage, and violence among persons with mental disorders: The importance of contextual measurement. *Law and Human Behavior* 24 (4): 449–456. https://doi.org/10.1023/a:1005544330132.

Sipahi, L., D.E. Wildman, A.E. Aiello, K.C. Koenen, S. Galea, A. Abbas, and M. Uddin. 2014. Longitudinal epigenetic variation of DNA methyltransferase genes is associated with vulnerability to post-traumatic stress disorder. *Psychological Medicine* 44 (15): 3165–3179. https://doi.org/10.1017/s0033291714000968.

Slobogin, C. 2017. Neuroscience nuance: dissecting the relevance of neuroscience in adjudicating criminal culpability. *Journal of Law and the Biosciences*, lsx033-lsx033. https://doi.org/10.1093/jlb/lsx033

Solmi, F., G. Lewis, S. Zammit, and J.B. Kirkbride. 2020. Neighborhood characteristics at birth and positive and negative psychotic symptoms in adolescence: Findings from the ALSPAC birth cohort. *Schizophrenia Bulletin* 46 (3): 581–591. https://doi.org/10.1093/schbul/sbz049.

Sommers, S.R. 2006. On racial diversity and group decision making: Identifying multiple effects of racial composition on jury deliberations. *Journal of Personality and Social Psychology* 90 (4): 597–612. https://doi.org/10.1037/0022-3514.90.4.597.

Steinberg, L. 2013. The influence of neuroscience on US Supreme Court decisions about adolescents' criminal culpability. *Nature Reviews Neuroscience* 14 (7): 513–518. https://doi.org/10.1038/nrn3509.

Su, Y., C. D'Arcy, S. Yuan, and X. Meng. 2019. How does childhood maltreatment influence ensuing cognitive functioning among people with the exposure of childhood maltreatment? A systematic review of prospective cohort studies. *Journal of Affective Disorders* 252: 278–293. https://doi.org/10.1016/j.jad.2019.04.026.

Telzer, E.H., E.M. McCormick, S. Peters, D. Cosme, J.H. Pfeifer, and A.C.K. van Duijvenvoorde. 2018. Methodological considerations for developmental longitudinal fMRI research. *Developmental Cognitive Neuroscience* 33: 149–160. https://doi.org/10.1016/j.dcn.2018.02.004.

Van Dyke, M.E., N.K. Baumhofer, N. Slopen, M.S. Mujahid, C.R. Clark, D.R. Williams, and T.T. Lewis. 2020. Pervasive discrimination and allostatic load in African American and white adults. *Psychosomatic Medicine* 82 (3): 316–323. https://doi.org/10.1097/psy.0000000000000788.

Varese F., F. Smeets, M. Drukker, R. Lieverse, T. Lataster, W. Viechtbauer, J. Read, J. Van Os, R.P. Bentall. 2012. Childhood adversities increase the risk of psychosis: A meta-analysis of patient-control, prospective- and cross-sectional cohort studies. *Schizophrenia Bulletin* 38 (4): 661–671. https://doi.org/10.1093/schbul/sbs050

Vasterling, J.J., K.A. Arditte Hall. 2018. Neurocognitive and information processing biases in posttraumatic stress disorder. *Current Psychiatry Reports* 20 (11): 99. https://doi.org/10.1007/s11920-018-0964-1

Wager, T.D., J.K. Rilling, E.E. Smith, A. Sokolik, K.L. Casey, R.J. Davidson, S.M. Kosslyn, R.M. Rose, and J.D. Cohen. 2004. Placebo-induced changes in FMRI in the anticipation and experience of pain. *Science* 303 (5661): 1162–1167. https://doi.org/10.1126/science.1093065

Wahlund, K., and M. Kristiansson. 2009. Aggression, psychopathy and brain imaging—Review and future recommendations. *International Journal of Law and Psychiatry* 32 (4): 266–271. https://doi.org/10.1016/j.ijlp.2009.04.007.

Walhovd, K.B., A.M. Fjell, R. Westerhausen, L. Nyberg, K.P. Ebmeier, U. Lindenberger, D. Bartrés-Faz, W.F. Baaré, H.R. Siebner, R. Henson, C.A. Drevon, C. Lifebrain. 2018. Healthy minds 0–100 years: Optimising the use of European brain imaging cohorts ("Lifebrain"). *European Psychiatry* 50 (47): 56. https://doi.org/10.1016/j.eurpsy.2017.12.006

Walhovd, K.B., C.K. Tamnes, and A.M. Fjell. 2014. Brain structural maturation and the foundations of cognitive behavioral development. *Current Opinion in Neurology* 27 (2): 176–184. https://doi.org/10.1097/wco.0000000000000074.

Weaver, V.M., and A. Geller. 2019. De-policing America's youth: disrupting criminal justice policy feedbacks that distort power and derail prospects. *The ANNALS of the American Academy of Political and Social Science* 685 (1): 190–226. https://doi.org/10.1177/0002716219871899.

Williams, D.R., Y. Yan, J.S. Jackson, and N.B. Anderson. 1997. Racial differences in physical and mental health: socio-economic status stress and discrimination. *Journal of Health Psychology* 2 (3): 335–351. https://doi.org/10.1177/135910539700200305.

Zeki, S., O.R. Goodenough, J. Greene, and J. Cohen. 2004. For the law, neuroscience changes nothing and everything. *Philosophical Transactions of the Royal Society of London. Series B: Biological Sciences* 359 (1451): 1775–1785. https://doi.org/10.1098/rstb.2004.1546

Case Citations

Amici re Wardlow v. Texas, Brief of Amici Curiae Professional Organizations, Practitioners, and
    Academics in the Fields of Neuroscience, Neuropsychology, and other Related Fields in Support
    of Petitioner, No. 19–8712 (US Supreme Court 2020).
Daubert v. Merrell Dow Pharmaceuticals, 516 U.S. 869 (1993)
Fairchild v. Workman, 579 F.3d 1134 (10th Cir. 2009)
Johnson v. United States, 860 F.Supp.2d 663 (N.D.Iowa, 2012)
Lockett v. Ohio, 438 U.S. 586 (1978)
Powell v. Texas, 392 U.S. 514 (1968)
Tennard v. Dretke, 542 U.S. 274 (2004)
U.S. v Hartfield 513 F.2d 254 (1975)
U.S. v John Hinckley, 672 F.2d 115 (1982)
U.S. v. Montgomery, 635 F.3d 1074 (8th Cir. 2011)
U.S. v. McCluskey, (D.N.M. 2013, No. 10–2734, Doc. 1481, Verdict Form)

David Freedman is a psychiatric epidemiologist whose research primarily focuses on neurodevelopmental disorders and cognition. He works as an expert consultant in capital cases. He
has published numerous academic papers on prenatal and perinatal risks for psychotic disorders, cognitive functioning and behavior, forensic assessment and issues related to mental illness
and intellectual disability, indigent defense and death penalty litigation. Having been awarded
an NIMH Fellowship in Psychiatric Epidemiology, he worked as a Postdoctoral Scholar in the
Departments of Psychiatry and Biobehavioral Sciences, at the Semel Institute for Neuroscience
and Human Behavior, University of California, Los Angeles; and obtained an Sc.M. from the
Harvard School of Public Health. David has taught for some years a course at Columbia Law
School entitled Mental Health and Criminal Defense. He is a Senior Research Consultant with
the International Academy of Law and Mental Health.

Chapter 11
Direct Benefit, Equipoise, and Research on the Non-consenting

Stephen Napier

11.1 Introduction

Research on human subjects aims to obtain knowledge of vital importance for human health and functioning. Most guidelines or regulations for pediatric research (whether in the U.S. or elsewhere) require that if a research intervention exposes pediatric subjects to more than minimal risk, a prospect of direct benefit is required—along with some other important criteria. It is also required that the methods and design will yield scientifically valid knowledge (Emanuel et al. 2000). Obtaining valid knowledge is typically met via randomized-controlled trials, a sub-set of which are no-treatment or placebo-controlled trials. One ethical criterion for assessing the permissibility of randomized controlled trials (RCT) is referred to as clinical equipoise. The researchers should be in a state of aporia as to whether one or the other arm of the study is overall better. I present an argument to the effect that more than minimal risk research on pediatric subjects cannot meet both ethical requirements: the prospect of direct benefit, and clinical equipoise. The basic idea is that if there is in fact a *prospect* of direct benefit, either the subjects in the experimental or control arm will receive a benefit, *but not both*. Clearly, the no-treatment group cannot benefit. So, clinical equipoise is threatened. Conversely, if clinical equipoise is met, then it cannot be known or anticipated that the subjects in either arm of the study will benefit by the research. So, a prospect of direct benefit cannot be met.[1]

---

[1] The problem I highlight in this paper should not be confused with the problem of justifying a placebo control in the setting of active treatments. Whether such studies can satisfy clinical equipoise is a different question from the one I address here. My question is this: *if clinical equipoise is satisfied, can a study promise a prospect of direct benefit?*

S. Napier (✉)
Villanova University, Villanova, PA, USA
e-mail: stephen.napier@villanova.edu

© The Author(s), under exclusive license to Springer Nature Switzerland AG 2023
T. Zima and D. N. Weisstub (eds.), *Medical Research Ethics: Challenges in the 21st Century*, Philosophy and Medicine 132,
https://doi.org/10.1007/978-3-031-12692-5_11

But this argument is outlined merely in the service of motivating my response to it which recapitulates a permutation of component analysis (Weijer 2000) according to which different ethical principles apply to different activities of a research study. After arguing that this can be the only solution to the problem posed by satisfying both direct benefit and clinical equipoise, I argue that certain classes of neuroscientific research presents complications that are irresolvable by component analysis or by the focus on commutative justice. To avoid any misunderstanding, the structure of the chapter is (1) presentation of a problem—(2) outline the only solution to that problem,—and then (3) pose a problem for the solution.

In the next section, I explain the two primary ethical principles around which my argument is constructed: the idea of a prospect of direct benefit and clinical equipoise. The third section explains and defends the argument just adumbrated according to which randomized no-treatment controlled trials in pediatric subjects cannot be ethically permitted. I call this the moratorium argument because, if correct, the argument would justify not doing any such research. This is the initiating problem and the fourth section responds to it by considering a version of component analysis. I then turn to review considerations specific to neuroscientific research that complicate the assessment of both the moratorium argument and the component analysis rejoinder. The fifth section concludes with a comment on policy implications.

11.2   Direct Benefit and Clinical Equipoise

This section defines the key terms and ethical principles for the discussion that follows. It is important to understand that the problem and its solution pertain to research on pediatric subjects, more generally subjects who cannot give consent. I shift to the language of nonconsenting subject instead of pediatric subject to bring into relief the moral ideals that justify pediatric research. The ideal is this: additional protections afforded pediatric subjects are motivated by a concern to protect those subjects who cannot protect themselves—because they cannot consent or dissent. They cannot protect their own interests by, for example, consenting to risks that are greater than those encountered in daily life for the purposes of obtaining scientific knowledge. So, the moral justification for pediatric research applies to all subjects who cannot consent.[2]

One means by which additional protections are afforded nonconsenting subjects is to require that more than minimal risk research offers a prospect of direct benefit.[3] Minimal risk is a technical term referring to the daily life risk of harm that may be experienced by the subject population specifically (the so-called relative standard)

[2] The applicability of the moral justification is, in this case, broader than its regulatory applicability since sub-part -D governs pediatric subjects alone, not psychiatric subjects, adults suffering from neurodegenerative diseases, or mentally disabled adults, etc.

[3] If the evidence on the therapeutic misconception is correct, neither do adults understand the complexities of RCTs. But that fact would only argue for added protections for adults, not for taking away the subpart D protections for pediatric subjects.

or in age matched healthy children (the so-called absolute standard). 45 CFR Part 46 subpart D outlines the general parameters for approving this research on pediatric subjects. I will assume the argument valid and sound for this chapter, that more than minimal risk research on nonconsenting subjects requires promising a favorable benefit to harm ratio (Jonas 1969). Specifically, 45 CFR part 46.405 says,

> HHS will conduct or fund research in which the IRB finds that more than minimal risk to children is presented by an intervention or procedure that holds out the prospect of direct benefit for the individual subject,…only if the IRB finds that: (a) The risk is justified by the anticipated benefit to the subjects; (b) The relation of the anticipated benefit to the risk is at least as favorable to the subjects as that presented by available alternative approaches…. (U.S. Code of Federal Regulations, cf. Title 21 §50.52)

I shall refer to the requirement that 'more than minimal risk research on nonconsenting subjects is permissible only if there is a prospect of direct benefit' as the direct benefit requirement. There are, of course, other ethical criteria that a study must satisfy (Emauel et al. 2000). But if the subject cannot consent to risks she or he would otherwise not experience outside the research setting, exposing subjects to those risks anyway is unjust and disrespectful without some promise of benefit from the research.

The notion of a "direct" benefit is a benefit that accrues to the subjects in virtue of the research interventions. The benefit a subject might get by, for example, participating in the study such as a gift card, or a good feeling of altruism are considered collateral benefits (King 2000, 333). The notion of "in virtue of" is causal. One must have justification for thinking that the intervention being tested will cause a therapeutic outcome. "Direct benefit… is properly defined as benefit arising from receiving the intervention being studied" (King 2000, 333). And finally, the notion of a "prospect" does not mean hopeful, but is epistemic in that there must be a reason for thinking that the experimental drug, device, or surgical procedure will benefit the subjects receiving that drug, device, or surgical procedure. The notion of a "prospect" should not be interpreted strictly in probabilistic terms because there might be no frequency upon which to base such a judgment. This might be the case in phase I, first in-human trials for example. Clinical experience or biochemical and physiological knowledge can count as reasons for thinking that there might be a direct benefit.

Before discussing clinical equipoise, it is important to understand that the phrase 'justification for research' means at least two different things. It means both giving a *reason for* the research study, and it can mean giving reasons that there is *no ethical barrier* to doing the research study. There might be no ethical barrier to doing an anonymous survey of people's pizza preferences, but we can easily imagine no scientifically valid reason for doing so. The direct benefit requirement is meant to provide *reasons for* doing the research. If one is exposing someone else to more than minimal risks, one needs a reason for doing so.[4]

To summarize, the direct benefit requirement is an ethical requirement for justifying more than minimal risk research on nonconsenting subjects. Though rarely stated, the reason is that it would be *unjust* to expose those subjects to such risks.

---

[4] I thank an anonymous reviewer for prompting me to address these terms.

The notion of justice relevant here is commutative justice. Commutative justice is the virtue that perfects the relationships between individual persons such that if person A causes person B to suffer, A avoids doing an injustice if there is some compensation to B. We can now ask: How would a commutatively just person behave towards a nonconsenting individual whom she exposes to risks of harm? The least the just person would do is expose the person to risks only if there is a prospect of benefiting the person by such exposure. An exemplar of commutative justice would not use one nonconsenting person S for the good of others without any compensating benefit to S.

The second notion is clinical equipoise which is an ethical standard that any RCT must meet, whether no-treatment, placebo, or active controlled. It is the requirement that, given the state of knowledge when the study is proposed, the "arms" of the study must be clinically equivalent in terms of the anticipated benefits and harms. This does not mean that the arms of the study must promise *the* same benefits or harms, rather what is required is that the ratio of harm-to-benefit is axiologically symmetrical across the arms of the study. The basic moral ideal that grounds requiring clinical equipoise is also commutative justice. A just person would not let a patient be randomly assigned to an arm of a study that is *knowingly* inferior.

The focus on commutative justice as the root justification for clinical equipoise is an anomaly (cf. Field and Behrman 2004), so some word in its defense is warranted. A reason for thinking that equipoise should apply to research on nonconsenting subjects is that children are more vulnerable to being disadvantaged. It is unclear, however, what counts as being disadvantaged, even some adult subject populations can be considered disadvantaged. Even if conceptual clarity can be reached, the reason why disadvantaging (however it is construed) someone is wrong is because it violates the respect due to human persons, i.e. it violates commutative justice.

Another justification for clinical equipoise is that it functions as a "bridge" between the physician's therapeutic obligation to an individual patient, and the demands of research practice. But this only tells us what the principle does (i.e. its function) but does not explain why there needs to be a "bridge" in the first place. What is the ethical concern in wanting to cleave the value of beneficence and scientifically valid research? The reason for requiring clinical equipoise is so that nonconsenting subjects (the focus of this chapter) are not exposed to unjust treatment in research; that is the ethical concern. A marker for unjust treatment is unequal access to known benefits or unequal exposure to additional risks. And well accepted notions of justice require treating equals equally. Clinical equipoise is violated precisely when there is knowingly unequal access to benefits or unequal exposure to harms. Consequently, violating clinical equipoise violates our common conception of commutative justice.

I have explained equipoise with reference to the given state of knowledge at the time the study begins. But commentators differ on the criteria for satisfying clinical equipoise. Some require *expert disagreement* about which of the control or experimental cohort is superior (Freedman 1987). Some require only *no consensus* about which of the control or experimental cohort is superior.[5] Yet another criterion, much more objective, is that the *total evidence* available does not favor one or the other arm of the study (as suggested by Levine 1986, 187). Though these distinctions are fine-grained, they can yield widely divergent judgments on controlled trials.

To see the divergence, consider the childhood adeno-tonsillectomy trial (CHAT) for pediatric obstructive sleep apnea (OSA) (Marcus et al. 2013).[6] The design of the trial was to randomize children suffering from OSA to a no treatment group (referred to as watchful waiting) and a standard therapy group which received an adenotonsillectomy (A-T). When the trial was first proposed, there had been no RCT to test whether an A-T was effective in resolving a child's OSA and the symptoms thereof. Children who suffer from OSA suffer from the typical sequelae of hypoxemia and hypercapnia: cardiac problems, neuropsychiatric deficiencies, mood impairments, and impaired school performance. Because OSA is typically caused by anatomical changes, namely, disproportionate adenotonsillar enlargement, A-T is standard therapy. Redline et al. (2011, 1509) state, "standard practice usually involves adenotonsillectomy as the primary treatment for childhood OSA." Inclusion criteria for the study required that an otolaryngologist certify that a subject is an appropriate surgical candidate (2011, 1511) and it was and is the most common surgical procedure performed on children (Mahant et al. 2015).

On the expert disagreement or no consensus interpretations, the CHAT study does not satisfy clinical equipoise. If a child has moderate to severe OSA, one should perform an A-T. What seemed to justify equipoise was the objective state of the evidence.[7] There was no previous controlled trial that compared the risks of surgery and neuropsychiatric development with the risks of waiting and neuropsychiatric development. The state of the evidence was what justified equipoise in this study, not what the experts agreed or disagreed upon.

---

[5] Freedman (1987, 120) seems to think that disagreement and no consensus are synonymous as he mentions both as characterizing clinical equipoise. They are different. As I am using the term, non-consensus occurs if there is an *absence of evidence* to determine one way or the other, which arm of a study is better or worse overall. This can occur if the trial is a first-in-human trial or the preliminary data is based on poorly controlled pilot studies. Expert disagreement occurs if *evidence exists* to support differing opinions on the overall merits of the respective arms of the study. The distinction can be understood as tracking the distinction between having no evidence at all, and having ambiguous evidence.

[6] Potential conflict of interest disclosure: I worked for an IRB at the time it approved this study, though I was not a voting member.

[7] Neither the authors of the study (Marcus et al. 2013) nor commentators on it (Witmans and Shafazand 2013) mention the fact that A-T can deleteriously affect one's immune system since the adenoids and tonsils are lymphatic organs. A-Ts are associated with a 17% increase risk for infectious diseases, and a 2–threefold increased risk for respiratory infections (Byars et al. 2018). In fairness to the CHAT researchers, these numbers were not published until 2018. However, it was known that the adenoids and tonsils make up the lymphatic organ system. It should have been anticipated that the immune system might be compromised. In any case, the facts of the Byars study place added burdens on the A-T side and brings into equilibrium the relative benefit-harm ratio between A-T and waiting.

Another lesson that can be gleaned from the OSA CHAT study is that satisfying clinical equipoise is consistent with the study being a placebo/no treatment-controlled trial in the setting of active treatments (Emanuel and Miller 2001, 917). If those active treatments are risky themselves—as A-T is since it involves surgery with anesthesia—such that they cause other health issues or compromise patient compliance, then a no treatment control is permissible. The reason why a no treatment control is justified when the available treatment has deleterious side effects is because the benefit-to-harm *ratio* in the no treatment arm is axiologically symmetrical to the expected benefit-to-harm ratio in the experimental arm. For example, the benefit of being in the A-T arm of the CHAT trial is having one's OSA resolved; and the harms are the risks of surgery and a diminution in one's immune system. The benefits of being in the waiting arm are retaining one's lymphatic organs and avoiding the risks of surgery; but the harm—or deprivation of benefit—is having one's OSA remain when it could be resolved. A reasonable person could evaluate the benefit-to-harm ratio in these arms as roughly equivalent.

There are detractors to the clinical equipoise requirement. Brody (2012), for example, argues that clinical equipoise is an inapposite moral principle governing research because the purpose of research is to obtain knowledge about what an intervention does to groups of patients. Once one understands the goals of research, concern about the welfare of individual subjects is misplaced. "Clinical equipoise and all other forms of equipoise make sense as a normative requirement for clinical trials *only* on the assumption that investigators have a *therapeutic obligation* to the research participants" (Brody 2012, 210, emphasis mine). Brody denies that there is such a therapeutic obligation in clinical research. If by "therapeutic obligation" one means that one should be concerned about the welfare of this patient/subject then the purpose of research on Brody's understanding is immoral by its intentions alone. It is morally wrong that the justification for research activities should preclude or ignore their welfare; and this is the case especially for research on non-consenting subjects. In point of fact, §46. 405 (b) endorses clinical equipoise when it requires the following for more than minimal risk research: "The relation of the anticipated benefit to the risk is *at least as favorable* to the subjects as that presented by available alternative approaches" (emphasis mine).

Another detractor to my understanding of equipoise is King (2000). She notes that "equipoise is a reasonable difference of opinion about what *will be* the better *treatment* for future patients—not about what *is* better for current subjects" (2000, 337). Two points are noteworthy. First, equipoise is not identical to a difference of opinion; a difference of opinion is the evidence or reason for thinking that equipoise is satisfied. Equipoise *is*, on my understanding, the axiological symmetry between the different arms of a study vis-à-vis the current state of knowledge. Second, it is hard to pinpoint what kind of evidence would justify an opinion about what *will* be better if not the evidence provided by the proposed clinical trial. To be fair, King's point is that when clinicians disagree prior to the conduct of an RCT, they think in terms of what *will be* the more effective therapy. Psychologically speaking, King is correct. Epistemically speaking, that understanding is not correct because judgments of what "will be" effective treatment *in the absence of a trial whose purpose is to provide the missing evidence*, is mere conjecture. The only justification for having an opinion about what *will be* the better treatment for future patients is the evidence provided by the proposed clinical trial. Such evidence does not exist without the trial.

To summarize, clinical equipoise is an ethical criterion which also follows upon our conception of commutative justice. One should not expose a patient to a risk of being randomized to a knowingly inferior form of clinical care. The key term in this explanation is knowingly. Thus, clinical equipoise is satisfied if the researchers are in a state of epistemic aporia about which arm of the study holds out the better benefit-to-harm ratio. There are at least three different epistemic standards which justify aporia: expert peer disagreement, expert peer no consensus, and the total available evidence criterion. Which standard one chooses can affect whether an individual study meets clinical equipoise. The choice of standard, however, does not affect the moratorium argument since that argument depends merely upon the researcher having justification (however measured) that equipoise is satisfied.

11.3   The Moratorium Argument

To understand the moratorium argument better, I focus on a placebo-controlled RCT even though I think the argument applies to all no-treatment controlled studies such as studies on significant risk devices or surgical procedures (research where a placebo *pill* is not used as a control). Furthermore, I will understand clinical equipoise as understood from the perspective of the researcher doing the research. From that standpoint, the researcher must be justified in thinking that, so far as we know at the start of the study, the arms are clinically equivalent. Finally, the phrase "arms of the study being clinically equivalent" is shorthand for the following claim: the researcher is justified in believing that the benefit-to-harm ratio between the different arms of the study are roughly symmetrical. And symmetry is understood in terms of how a reasonable person would *evaluate* the relative benefit-to-harm ratio. It might be better to think of clinical equipoise as requiring an axiological symmetry—see my commentary on the CHAT trial above.

Consider the set of evidence which if believed, would support the belief that an experimental drug holds out a prospect of direct benefit, e.g., reduction in tumor size. Assume also that the study is a placebo-controlled trial. It follows that subjects in the placebo arm will not derive the direct benefit from the experimental drug. Of course, there is other relevant evidence such as the evidence that the experimental drug might cause certain toxicities. This is evidence that the placebo-controlled subjects avoid certain harms. One might think at this point that the direct benefit and clinical equipoise requirements are both satisfied. One might reasonably judge that (A) the direct benefit plus the toxicities experienced in the experimental arm are reasonable in relation to (B) having no prospect of direct benefit plus no toxicities either.

The problem is this. Assume that (A) and (B) are axiological symmetrical—as is required by clinical equipoise. But (B) is axiologically equivalent to standard care or minimal risk, i.e., the value of being in the placebo arm of the study is equivalent to not being in the study. This has two important implications. First, if the value of being in the study is equivalent to not being in the study, that value gives one no *reason for* being in the study. Second, if the value of being in the study is equivalent to not being in the study, that value need not require *additional protection* for research subjects. Since (A) and (B) are axiologically symmetrical and the overall value in (B) provides no *reason for* doing the study and no additional protections, it follows that the symmetrical value in (A) provides no reason for the study and no additional protections. Exposing subjects to greater than minimal risk lacks a reason for, and the value accruing to the subjects does not provide *additional* protections. If the researcher justifiably believes that reduction in tumor size is *axiologically symmetrical*—all things considered—to receiving *no treatment*, that value cannot provide the *additional* protections required when one exposes another to greater than minimal risk. The direct benefit is not meaningful if it is offset by a harm that makes it axiologically symmetrical to the minimal risk arm. Such a direct benefit fails to provide a value that justifies additional protections for nonconsenting subjects exposed to greater than minimal risk. To see this, consider the following example.

Suppose I am out hiking with my daughter and we happen upon what looks like a deep creek. I want to dive in for a swim but I am uncertain of its depth. Whereas I might be justified in risking it myself, to ask my daughter to jump in first is risible. The point is that delegating risks to others looks *prima facie* haughty, arrogant, or cowardly. The example illustrates the relationship between delegating risks to others, and the strength of justification required to do so. The researcher of course does not bear the risk of being wrong that a drug does not have serious side effects, the subjects do. Moreover, exposing others who are nonconsenting to greater than minimal risks requires added justification for such exposure. Whereas no one would lock a parent up for allowing their child to be exposed to daily life risks, I would be locked up if I threw my four-year-old into the watering hole to see how deep it is. My intuitions do not change even supposing that taking a dip is considered a benefit by all parties.

Though the example tolerates various permutations the point for now is only to illustrate that the level of justification one needs to take on certain risks oneself can tolerate being weaker than the level of justification required for exposing *another* person to risks, and weaker still to the level of justification required for exposing a *nonconsenting* person to those same risks.

The additional protections afforded nonconsenting subjects is because they cannot consent. This fact increases the epistemic pressure on exposing them to greater than minimal risks in the setting of axiologically equivalent study arms. And my point is that when the higher risk arm is axiologically equivalent to not being in the study, there is yet no justification for exposing nonconsenting subjects to the greater risk arm. One needs a stronger justification than mere axiological equivalence to not being in the study.

If one is not convinced yet, there is one more moral intuition relevant for defending the moratorium argument. That intuition pertains to the difference between tolerating risks for which we cannot control and those that have sources in another person's freely chosen actions. For illustration, consider a hypothetical example simply to illustrate the ethical point: suppose that in each year of normal driving, one risks a

1% chance of hospitalization. Intuitively, this is a tolerable risk. Suppose *a researcher* injects me with a live COVID-19 virus that carries a 1% chance of hospitalization. These might not be the actual numbers, but suppose they are. Intuitively the latter is not a tolerable risk—at least, not without significant contextualization. It is when the risk of harm is *due to someone's agency* that concerns about commutative justice are raised. To offset *that* increase risk of harm requires a more favorable benefit-to-harm ratio. The point is that it is not enough to satisfy commutative justice that the ratio between the minimal risk arm and the greater than minimal risk arm are symmetrical.

This conclusion becomes stronger once we consider that for numerous first in-human trials and many types of neuroscientific research, the risks are *unanticipated*. To articulate this ethical worry, I must resort to the first-person and the reader is invited to do likewise.[8] If I, qua researcher, expose someone to anticipated risks, I can make a measured judgment about whether the study is worth it. If I am testing an intervention with a notable subset of unanticipated risks,[9] I cannot measure whether the test intervention is *worth* studying. To test the intervention anyway is exculpating only if I get morally lucky. A commutatively just person would not *just so happen* to avoid harming others. And the wrong would be a function of an imbalance of harms over benefits which may actually occur. Exposing subjects who cannot consent to those risks strikes me as treating them unjustly.

Once we consider that many research interventions involve *agential* sources of *unanticipated* risks on *nonconsenting* subjects, we can see that we need a greater justification than mere axiological equivalence with those in the minimal risk arm of the trial. The most important idea is that if the researcher is truly in aporia about the merits of a drug D versus placebo, then the researcher does not have evidence for believing that D is overall better than no treatment. *And if she is justified in believing that, she cannot be justified in believing that D will provide any meaningful direct benefit.* By meaningful direct benefit, I mean a benefit that justifies the researcher exposing nonconsenting subjects to unanticipated risks. If one strengthens the direct benefit requirement to be meaningful in the way explained, equipoise is destabilized. The greater risk is justified, but overall, more favorable than being in the placebo/no-treatment arm of the study. Consequently, if clinical equipoise is satisfied, the direct benefit requirement is not—at least not understood as providing a reason for the study and offering additional protections. If the direct benefit requirement is meaningfully satisfied, then clinical equipoise is not.

---

[8] The reason is that moral discourse cannot proceed all the way "down" to assessing concrete actions unless the interlocutors *see* or *feel* for themselves the moral values at stake with those actions.

[9] The reader might find oxymoronic the idea that a researcher can expect unanticipated risks. The tension is resolved once we consider how an intellectually humble researcher would approach, for example, doing a first in-human trial.

11.4   Rebutting the Moratorium Argument

11.4.1   Component Analysis

The key premise in the moratorium argument is that axiological symmetry with a no-treatment arm of a trial gives one no reason for exposing nonconsenting subjects in the experimental arm of the study to more serious and unanticipated risks. Agreed, the greater risks are balanced by the greater probability of benefit in the experimental arm, but when that 'balance' is symmetrical with no-treatment, i.e., not being in the study, the overall value fails to justify the greater risks to non-consenting subjects.

Responding to the moratorium argument requires discerning a pattern on how similar problems affecting pediatric research are responded to by other authors. One of those problems is the problem of doing diagnostic testing on subjects in a more than minimal risk study (Friedman et al. 2012). In many cases, such diagnostic testing is more than minimal risk itself, such as biopsy procedures. On the one hand such testing offers no prospect of direct benefit to the subjects. On the other hand, such testing is required to make the study scientifically valid. For research to be scientifically valid, one must know whether the drug, device, or surgical procedure has worked. Diagnostic testing is an example of a component in the study that holds out no prospect of direct benefit, but if one does not do it, one violates the ethical requirement of scientific validity. Call this the diagnostic case.

The strategy in resolving the diagnostic case is to understand that different ethical principles apply to different components of a study, hence the phrase 'component analysis'. Weijer articulates the idea this way: "[Research] activities involve procedures administered with different intents. Some interventions may be administered for therapeutic purposes, while others solely to answer a scientific question" (Weijer 2000, 352). In pediatric research, the ethical requirement to promise a direct benefit applies to the researchers' justification for thinking that *the test intervention*, and the test intervention alone, will have therapeutic effects. The requirement to realize scientific validity, however, does not apply to *what* effects the test intervention might cause but to *our knowledge* of those effects.

One may think about the diagnostic case in this way. The prospect of direct benefit is justified by evidence that P is *potentially* effective. More than minimal risk diagnostics are justified when those diagnostics are necessary for knowing whether P is *actually* effective. In a sense, what justifies more than minimal risk diagnostic procedures is the ethical requirement of there being a prospect of direct benefit. What good are entirely unknown direct benefits to the subject or the researcher? Component analysis is simply taking on two parallaxes of the same ethical requirement. It is the same ethical requirement looked at as potentially being satisfied and then as actually being satisfied.

The argument for the superiority of component analysis, though not explicitly mentioned by Weijer (2000), is that there is no concrete action that answers to the label "research activity." Rather, research actions include this research coordinator procuring informed consent from this subject, or this nurse researcher administering the test article to this subject. Because concrete actions are open to moral justification, it is apposite that different moral principles will "apply" because one is performing different actions with different intents. The justification for component analysis follows simply from the categorical difference in actions performed as part of the research project. Viewed in this way, component analysis is the only approach to assess research projects, but my argument below does not depend on this stronger claim.[10]

In response to the moratorium argument, component analysis yields the following results. The prospect of direct benefit is meant to offset whatever risks above minimal that the subjects are exposed to *due to the research intervention*. Subjects receiving standard clinical care with non-invasive follow-up testing as part of a study do not *need* to be promised a prospect of direct benefit to treat them justly. The research interventions in the no treatment arm of the study are not morally significant in principle. So, if we keep in view the principled justification for the direct benefit requirement, we see a principled difference between how the subjects may be treated in the different arms of a study. I may not want to describe this as satisfying clinical equipoise, but as satisfying a moral equipoise. Moral equipoise involves treating the subjects justly in their respective cohorts, since the cohorts are exposed to different anticipated and unanticipated risks.

What justifies the agential source of risk (hereafter understood also as unanticipated risk unless otherwise noted) on the nonconsenting? If we keep in view the principled justification for clinical equipoise and direct benefit, namely, both are animated by a concern for commutative justice, it cannot be inconsistent with either requirement that subjects in the experimental arm who experience an agential source of risk enjoy a slightly more favorable benefit to harm ratio. Satisfying a meaningful direct benefit treats those in the experimental arm justly, and those in the placebo arm are not deprived of just treatment since they are not owed untested therapies.

---

[10] In a very careful and informative study, Rid and Wendler (2011) critique component analysis. They state, "The fundamental problem with different ethical requirements for therapeutic and nontherapeutic interventions is that they introduce different thresholds of acceptable net risk in biomedical research and thereby render risk-benefit evaluations incoherent. For example, "component analysis" allows competent participants to consent to sometimes significant risks without any compensating potential clinical benefits, as long as the risks result from a nontherapeutic intervention…" (2011, 148). If this is what component analysis allows, then Rid and Wendler are correct to abandon it. However, I do not think that component analysis has such consequences. As explained in the text, component analysis is measuring the various action types that take place within a research protocol against their *respective* moral principles. By respective moral principles I mean that certain action types are best analyzed only against certain ethical principles. For example, a person's arrogance cannot be assessed with reference to the principle of autonomy. We think arrogance is bad, yet if someone harbors supremacist views about oneself but never shares them or acts on them, no one's autonomy is violated. The more fitting principle to assess the moral deficiency of arrogance is the principle of justice or better yet, the virtue of humility. Responding more directly to Rid and Wendler's concerns, component analysis does not justify participants taking on significant risks without any compensating benefit. The reason is that component analysis doesn't tell you which principles to select. It just tells you that you must select the most fitting ones. If one selects the virtue of commutative justice, component analysis does not allow competent participants to take on significant risks without any benefit (see Napier, 2019, chapter 11 for further discussion).

But this forces a slight shift in understanding equipoise. Equipoise now should be understood not as treating the values in the respective arms as quantities of value that have to be equal, but as requiring just treatment *proportionate to* the subjects' respective contexts. Notably, to the extent that the reader finds the introduction of "proportionality" problematic, to that extent, the moratorium argument gains plausibility. The only other option to avoid the moratorium argument is to challenge the arguments for requiring a meaningful direct benefit.

11.4.2   Case-Based Complications

The key ideas in the foregoing analysis can be summarized as follows. The researcher can be viewed as a commutatively just person who aims not to mistreat other persons especially those who cannot consent to greater than minimal risk. Equipoise and direct benefit are two distinct ethical principles whose petals flower from the same virtue, i.e., commutative justice. They are meant to justify different aspects of a study. Hence, their satisfaction is justified based on *distinct* evidence sets. Hence, both can be met in any given research study.

There are significant complications to this analysis for neuroscientific research in relation to other categories of disease processes and organ dysfunction. The human brain and nervous system are very complex organ systems that involve integrated processing and higher cognitive ability that holds few analogs to animal models. Consequently, testing the safety of a drug or biologic on animal models will not necessarily tell researchers what the risks might be for human beings (Mathews et al. 2008). Furthermore, there is the complexity of the brain and nervous system itself in relation to drug or biologic based therapeutic interventions. Kimmelman explains, "[s]trategies that deliver therapeutic agents to the entire brain risk deranging delicately balanced functions in off-target brain structures" (Kimmelman 2013, 252–253). Even if targeted therapeutics could be developed, those therapeutics would likely be in the form of biologics or implanted neural interfaces (INI) to bypass the blood–brain barrier. Biologics, however, are a more complex cocktail of molecular entities, additives, vectoring agents, and may require supportive drugs for their administration (for example, immunosuppressants) (Kimmelman 2013, 255). In addition to these sources of unanticipated risks, there are known risks associated with the administration of biologics, namely, neuro-surgery and its risks (Selden et al. 2013), and some INI's involve percutaneous implants which increase the risk of infection (Lew 2017).

There are two issues in this regard. First, a placebo-controlled trial using biologics is likely to require neurosurgery to by-pass the blood–brain barrier as noted. As such, it would be a sham control as well. Given the ethical concerns over sham surgeries, research on diseases of the neurological system such as Batten—discussed below—should rarely be placebo controlled. On the one hand, making comparative inferences using historical controls is dubious, and likely requires a large cohort population—which, in the case of rare diseases is not likely. On the other, without a control group at all, it is harder to determine safety and efficacy. For example, Worgall et al. (2008) discovered that 1 (out of 10) subjects died after suffering a single seizure 14 days after infusion. Notably, this was an expected event since the authors note that there were "no unexpected serious adverse events" (2008, 463). Was this event caused by the gene therapy or the disease since seizures are associated with Batten? How can a non-controlled trial distinguish adverse events that are caused by the therapy when those events are also associated with the disease pathology? The ethical risk of miscoding adverse events as "unrelated" is high in non-controlled studies (FDA 2001, 27).

In the setting of unanticipated and known risks, there is also a track record of limited benefit in addressing certain neurological disorders (I am thinking specifically of Huntington's disease, Alzheimer's disease, but also Batten disease). Evidence of limited benefit stems both from the limited inferences we can draw from animal models (Descotes 2005), and the limited success of previous first in-human trials (see Kimmelman 2013, 252–54 for review). Related to these concerns is the fact that researchers' perception of acceptable risk is predominantly affected by whether the research subject population has clinically available options. Deakin and colleagues surveyed gene therapy researchers and they noted that "if the clinical context was dire, then it was more acceptable for the risk–benefit ratio to be skewed in favor of risk" (Deakin et al. 2013, 813). A similar increase in risk tolerance in the setting of clinical paucity is prevalent among parents and the FDA (Unguru 2015).[11]

Finally, Kimmelman (2013, 255) explains the following dilemma in terms of enrollment issues and trial design for research on many neurological disorders. If the study enrolls patients advanced in their disease, it is more difficult to detect efficacy because brain tissue damage is more difficult to reverse given the lack of cellular plasticity of the human brain. At this stage of medical advancement, one cannot expect reversal but rather only stymieing the disease process. If the study enrolls patients early in their disease course, certain patient populations would not be tested given the lack of validated early detection diagnostics—such as that pertaining to neuronal ceroid lipofuscinoses (Augustine et al. 2013, 3), or clinical equipoise might not be met since the patients at the time of the study might still enjoy a favorable quality of life and symptom management—such as those in early stage dementia or Alzheimer's disease.

Consider the following case study for illustration. Neuronal ceroid lipofuscinoses (NCL) is a family of diseases that affect the central nervous system and brain. Today there are no proven therapies for many of these diseases. A species of NCL is Batten disease. A nice layperson description of the pathophysiology of Batten is provided by Sardiello (2017). Most cities have garbage management infrastructure. Garbage is taken out of the city for if it were to remain in the city the functionality of the city—living conditions, the spread of diseases etc.—would quickly deteriorate. A

---

[11] Observations such as these might be where the notion of disadvantage might be useful as a marker or indicator for unjust treatment. Children who, no fault of their own, suffer from diseases with few clinical options are exposed to research projects that tolerate a higher risk than research projects on subjects with clinical options. Of course, the reason why this treatment is wrong is because it fails to treat *all* children as persons, and instead treats them in light of accidental features.

similar thing happens with Batten disease. The lysosome clears out the garbage from the "city" of the cell. In Batten disease the lysosome does not function well or at all. This leads to cellular "garbage" that builds up and manifests in the following progressive symptoms: blindness, dementia, psychosis, paraplegia, seizures, and premature death.

Suppose an RCT is proposed to test immune suppressant therapy on Batten patients (Augustine et al. 2018). Because there is no previously available treatment, those in the placebo control arm receive standard care. (It is more common for research on rare diseases with no approved treatment modality that they are historically or concurrent-patient controlled. In the case of historical controls, the principle of clinical equipoise is not apposite since, given the electronic medical record, they may be deceased or not proximate to the research location). Those in the experimental arm receive the therapy. Because Batten is a rare disease the subject population is expected to be small and thus harder to discern safety and tolerability. Furthermore, as is common with rare diseases, the therapy is being tested for the first time in the subject population.

Here is a study where the placebo control group will experience the typical pathophysiology and progression of the disease. Those in the experimental group will receive the therapy. To satisfy the requirement that there must be a prospect of direct benefit, the researcher must have evidence or justification for thinking that the drug will stymie[12] further neurodegeneration caused by NCL dysfunction. To satisfy the clinical equipoise requirement, the researcher must have evidence or justification for thinking that the relative benefit to harm ratio between the two arms of the study are axiologically symmetrical—in other words, being in the placebo arm of the study is axiologically comparable to being in the experimental arm of the study. What justifies the agential source of risk on the nonconsenting is the contextual feature of there being desolate treatment options for a serious and irreversible disease with death often occurring prior to when standards of subject consent could be satisfied. The direct benefit is made meaningful by these contextual features.[13]

Instead of arguing that the agential source of risk needs to be compensated for one might argue instead that having a placebo control is depriving those subjects in the placebo arm of a benefit. Therefore, such studies should be single arm trials. At least for commentators on NCL, the argument is that there is no other option for treating NCL and it is a particularly devastating degenerative disease. We should try anything, the argument goes, that has physiological reasons supporting its therapeutic efficacy. Augustine states, "In a rapidly progressing fatal disease, there is perhaps greater urgency on the part of parents to ensure their child is exposed to an active treatment condition, before the possible window of therapeutic opportunity is lost" (Augustine et al. 2013, 4).

---

[12] Currently, whatever therapeutic modalities are being tested none of them can reverse the damage done by the progression of the disease at the time they received the modality.

[13] I harbor the opinion, however, that this claim is wrongheaded—which is why I title these "case based complications." I do not think children who are at a clinical dead end can be subjects of research with a risk–benefit ratio "skewed in favor of risk" (Deakin op. cit.). That would treat children differently depending on accidental features, i.e., whether they have clinical options or not.

In response, it is correct to observe the ethical importance of NCL being a rare disease, with no therapeutic option, and a very devastating pathophysiology. But those features are ethically relevant in a way that justifies exposing subjects *in the experimental arm* to unanticipated risks of harm. The reason is because the unanticipated risks of harm are likely not to exceed in magnitude or probability the magnitude or probability of harms caused by the disease without any experimental treatment. The reason those features do not justify culling the placebo-control arm of the study and conducting it with a single arm is because patients are not owed untested therapies with unanticipated harms.[14] It is not a violation of commutative justice to withhold giving them putative therapies with no evidence of safety or tolerability. It *is* a violation of commutative justice to give a putative therapy with no evidence of safety or tolerability to first in-human subjects *unless* those subjects have no other therapeutic option and the side effect profile of the test drug is not expected to exceed the probability and magnitude of harms incurred by subjects who do not receive the test drug. Finally, it is notable that given the evidence highlighted by Kimmelman (2018), subjects in the experimental arm of trials on neurodegenerative diseases typically do not fare as well as no-treatment cohort subjects.

So, there must be an added justification for exposing nonconsenting subjects in the experimental arm to unknown risks of harm. But that justification is supplied by contextual features that include the nature of the disease, and the absence of alternatives. Consequently, studies of this sort satisfy both a prospect of direct benefit requirement and clinical equipoise. These ethical criteria are met because they involve believing two different things about the study. For direct benefit, the researcher needs justification for believing that the added risks caused by the research intervention (in this case, gene therapy) is compensated for by the research intervention causing the remaining lysosomes to function correctly. For clinical equipoise, the researcher needs justification for believing that the different arms of the study are axiologically symmetrical. There are no prospects of direct benefit being in the placebo arm, but likewise there is no prospect of suffering unanticipated side effects from a first in-human experimental drug. There is a prospect of direct benefit to being in the experimental arm, but likewise a prospect of suffering unanticipated side effects from a first in-human experimental drug. A reasonable person could evaluate these options as axiologically equivalent.

In relation to the difficulties mentioned so far, it might be helpful to review a study that is ethically more manageable. Consider a placebo-controlled trial using trofinetide for fragile X syndrome (Berry-Kravis et al. 2020). Fragile X syndrome is a genetic disorder that causes intellectual disability, autism spectrum disorder, social phobias, and various morphological abnormalities (e.g. flat feet). Patients with the fragile X deficiency usually live a normal lifespan and can lead a relatively healthy and productive life. At the time the study is proposed, there was, and is, no

---

[14] The argument in this paragraph is only to the point that desolate treatment options for a small subject population is not a reason for not controlling the proposed study. There are *other* reasons noted above that do justify not controlling the study, namely, if the route of administration requires neurosurgery, and thus a sham control.

established treatment for fragile X syndrome. Trofinetide is a neuropeptide that has anti-inflammatory effects and was originally used to manage symptoms of stroke (Cartagena et al. 2013). This was not a first in-human trial, but close to it. It had been tested in pediatric subjects with Rett syndrome only a few years prior (Glaze et al. 2017). Going into the study for fragile X, the researchers knew that trofinetide can increase the incidence of diarrhea and vomiting. It is otherwise well-tolerated. It was also known that trofinetide can reduce the symptoms of Rett syndrome. The goals of the fragile X study were to measure efficacy of the drug in terms of reduced intellectual and emotional symptomology; and to test safety and tolerability.

Was there a reason for thinking that the different arms of the study can be viewed as axiological symmetrical? Evidence from the Rett studies indicated improvement in various symptoms at higher doses, but uncomfortable gastrointestinal side effects—diarrhea and vomiting. Furthermore, efficacy was expected to be more pronounced at the younger ages, but diarrhea and vomiting might be viewed by younger patients with intellectual disabilities as more burdensome. A reasonable person can view the benefit-to-harm ratio in the respective arms as axiologically symmetrical. By receiving placebo, the child may forgo a possible diminution in his symptoms, but he also avoids possible gastrointestinal disruptions. By receiving trofinetide, the child may experience diminution in his symptoms, but may experience gastrointestinal disruptions.

Is there a prospect of direct benefit for the subjects who receive trofinetide? Theoretically, a neuropeptide with anti-inflammatory effects may be expected to reduce the deleterious effects of a fragile X deficiency. Furthermore, promising preclinical data on a related disease, i.e., Rett syndrome, indicate that trofinetide would reduce the intellectual and emotional symptoms of fragile X. Lastly, the agential source of risk on the nonconsenting is offset in this case by the fact that (1) the diarrhea or vomiting would resolve quickly—there were minor withdraw effects from the drug. (2) The diminution of the intellectual and emotional symptoms can be viewed as important enough to tolerate a risk of reversible diarrhea or vomiting. (3) There are no treatment options and few options for symptom management of fragile X syndrome. Together, these reasons offset the agential source of risk on the nonconsenting posed by being in the experimental arm of the study.

11.5   Policy Implications

Following the advice of Klaus Rose, I wish here to outline how we can avoid protecting children *against* research, and instead protect their health and welfare *through* research (Rose 2009, 129). My first recommendation is educational. IRB members (in the United States), and REB members (in Canada) are typically not aware of the complexities facing researchers for neurodegenerative diseases—particularly those utilizing biologics or INI's. Many of the concerns outlined in the previous section might be worthwhile primers on assessing any individual research project. Of particular importance for educating such members is the Deakin et al. (2013) study indicating a higher risk tolerance for disease populations who suffered from serious diseases and clinically limited options. Knowing this is an unconscious tendency in one's risk perception might better dispose such members to caution against research projects that are objectively too risky to proceed.

A second recommendation is that research in this particular category, namely phase 1 studies on neurodegenerative diseases, cannot be approved under category 405 (of the Common Rule) and Title 21 §50. 52 (of the FDA Rule). It is arguable that such phase 1 research projects do not satisfy the prospect of direct benefit requirement—which requires that the risk is justified by the anticipated benefit. The smaller the chance of benefit, and the greater the expectation of risk, the less chance one has to justify such research. I do not argue, however, that such research should not take place. A better procedure for reviewing such research and its accompanying approval category should be category 407 (for the Common Rule) or Title 21 §50. 54 (for the FDA rules). Though such categories require the review by a centralized review mechanism (e.g. the Secretary for the Health and Human Services), given my first recommendation, I do not think this centralization is required. In any case, centralization of a review mechanism is better able to see patterns when all relevant knowledge is collated and reviewed. Furthermore, with a centralized review mechanism, those reviewers will acquire more experience reviewing the most difficult research projects. Individual IRB's (or REB's) might see a phase 1 neurodegenerative research project too infrequently to make expert judgments on their permissibility.

In conclusion, I have reviewed an important argument against more than minimal risk research on the nonconsenting. Obviating that argument required exploring different implications of commutative justice. I also noted that these implications are harder to "apply" to risky research on neurodegenerative diseases, namely, those affecting children. The ethical complexities of doing such research present conundrums and dilemmas at various decision points, the resolution of which is likely not enveloped by applying a few principles. Thinking through such cases as a commutatively just person would, may be a better moral guide than a checklist approach.[15]

References

Augustine, E.F., H.R. Adams, and J.W. Mink. 2013. Clinical trials in rare disease: Challenges and opportunities. *Journal of Child Neurology* 28 (9): 1142–1150.

Augustine, E.F., C.A. Beck, H.R. Adams, S. Defendorf, A. Vierhile, D. Timm, J.M. Weimer, J.W. Mink, F.J. Marshall. 2018. Short-term administration of mycophenolate is well-tolerated in CLN3 disease (juvenile neuronal ceroid lipofuscinosis). *JIMD Reports* 43: 117–124. Springer, Berlin, Heidelberg.

Berry-Kravis, Elizabeth, et al. 2020. A double-blind, randomized, placebo-controlled clinical study of trofinetide in the treatment of fragile X syndrome. *Pediatric Neurology* 110: 30–41.

Brody, Howard. 2012. 'A critique of clinical equipoise: Therapeutic misconception in the ethics of clinical trials. In *The ethical challenges of human research: Selected Essays*, ed. Miller, Franklin

---

[15] I thank two anonymous referees for comments on a previous draft.

G. New York: Oxford Scholarship Online (Published online May 2015). https://doi.org/10.1093/acprof:osobl/9780199896202.003.0015. Accessed Sept 02, 2020.

Byars, S.G., S.C. Stearns, and J.J. Boomsma. 2018. Association of long-term risk of respiratory, allergic, and infectious diseases with removal of adenoids and tonsils in childhood. *JAMA Otolaryngology-Head & Neck Surgery.* 144 (7): 594–603.

Cartagena, C.M., K.L. Phillips, G.L. Williams, et al. 2013. Mechanism of action for NNZ-2566 anti-inflammatory effects following PBBI involves upregulation of immunomodulator ATF3. *Neuromolecular Medicine* 15: 504–514. https://doi.org/10.1007/s12017-013-8236-z.

Deakin, C.T., I.E. Alexander, C.A. Hooker, and I.H. Kerridge. 2013. Gene therapy researchers' assessments of risks and perceptions of risk acceptability in clinical trials. *Molecular Therapy* 21 (4): 806–815.

Descotes, J. 2005. Immunotoxicology: Role in the safety assessment of drugs. *Drug Safety.* 28 (2): 127–136.

Emanuel, E.J., D. Wendler, and C. Grady. 2000. What makes clinical research ethical? *JAMA* 283 (20): 2701–2711.

Emanuel, E.J., and F.G. Miller. 2001. The ethics of placebo-controlled trials—A middle ground. *New England Journal of Medicine* 345 (12): 915–918.

Field, M.J., and R.W. Behrman, eds. 2004. *Ethical Conduct of clinical research involving children.* Washington DC: Institute of Medicine of the National Academies, The National Academies Press.

Friedman, A., E. Robbins, and D. Wendler. 2012. Which benefits of research participation count as 'direct'? *Bioethics* 26 (2): 60–67.

Freedman, B. 1987. Equipoise and the ethics of clinical research. *New England Journal of Medicine* 317: 141–145.

Glaze, Daniel G., et al. 2017. A double-blind, randomized, placebo-controlled clinical study of trofinetide in the treatment of Rett syndrome. *Pediatric Neurology* 76: 37–46.

Jonas, H. 1969. Philosophical reflections on experimenting with human subjects. *Daedalus* 219–247.

Kimmelman, J. 2013. Biologics, ethics and the human brain. In *Neuroethics in practice*, ed. M. Farah and A. Chatterjee, 251–264. New York: Oxford University Press.

Kimmelman J. 2018. Better to be in the placebo arm for trials of neurological therapies? *Cell Transplantation* April: 677–681. https://doi.org/10.1177/0963689718755708

King, N.M. 2000. Defining and describing benefit appropriately in clinical trials. *The Journal of Law, Medicine & Ethics* 28 (4): 332–343.

Levine, Robert J. 1986. *Ethics and regulation of clinical research*, 2nd ed. New Haven: Yale University Press.

Lew, C.D. 2017. What do new neuroscience discoveries in children mean for their open future? In *Neuroethics: Anticipating the future*, ed. J. Illes, 159–179. New York: Oxford University Press.

Mahant, Sanjay, et al. 2015. Association of national guidelines with tonsillectomy perioperative care and outcomes. *Pediatrics* 136 (1): 53–60.

Marcus, C.L., R.H. Moore, C.L. Rosen, B. Giordani, S.L. Garetz, H.G. Taylor, R.B. Mitchell, R. Amin, E.S. Katz, R. Arens, S. Paruthi. 2013. Childhood adenotonsillectomy trial (CHAT). A randomized trial of adenotonsillectomy for childhood sleep apnea. *New England Journal of Medicine.* 368 (25): 2366–2376.

Mathews, D.J., J. Sugarman, H. Bok, D.M. Blass, J.T. Coyle, P. Duggan, et al. 2008. Cell-based interventions for neurologic conditions: Ethical challenges for early human trials. *Neurology* 71 (4): 288–293.

Napier, Stephen. 2019. *Uncertain bioethics: Moral risk and human dignity.* New York: Routledge.

Redline, S., R. Amin, D. Beebe, R.D. Chervin, S.L. Garetz, B. Giordani, C.L. Marcus, R.H. Moore, C.L. Rosen, R. Arens, and D. Gozal. 2011. The Childhood adenotonsillectomy trial (CHAT): Rationale, design, and challenges of a randomized controlled trial evaluating a standard surgical procedure in a pediatric population. *Sleep* 34 (11): 1509–1517.

Rid, A., and D. Wendler. 2011. A framework for risk-benefit evaluations in biomedical research. *Kennedy Institute of Ethics Journal* 21 (2): 141–179.

Rose, K. (2009). Ethical principles of pediatric research and drug development: A guide through national and international frameworks and applications to a worldwide perspective. In *Pediatric drug development: Concepts and Applications*, eds. A. E. Mulberg, S. A. Silber and J. N. van den Anker, 115–132. Hoboken, N.J.: John Wiley & Sons.

Sardiello, Marco. 2017. Research reveals strategy to potentially treat juvenile Batten disease. https://www.youtube.com/watch?v=qPzIqtKXCPo. Accessed September 9, 2020.

Selden, N.R., A. Al-Uzri, S.L. Huhn, et al. 2013. Central nervous system stem cell transplantation for children with neuronal ceroid lipofuscinosis. *Journal of Neurosurgery Pediatrics* 11 (6): 643–652. https://doi.org/10.3171/2013.3.PEDS12397.

Unguru, Y. 2015. Ethical challenges in early-phase pediatric research for life-limiting illness. *Seminars in Pediatric Neurology* 22: 177–186.

United States Food and Drug Administration. 2001 Guidance for industry E 10 choice of control group and related issues in clinical trials. https://www.fda.gov/media/71349/download. Accessed November 9, 2020.

Weijer, C. 2000. The ethical analysis of risk. *Journal of Law, Medicine & Ethics* 28: 344–361.

Witmans, M., and S. Shafazand. 2013. Adenotonsillectomy or watchful waiting in the management of childhood obstructive sleep apnea. *Journal of Clinical Sleep Medicine* 9 (11): 1225–1227.

Worgall, S., D. Sondhi, N.R. Hackett, et al. 2008. Treatment of late infantile neuronal ceroid lipofuscinosis by CNS administration of a serotype 2 adeno-associated virus expressing CLN2 cDNA. *Human Gene Therapy* 19 (5): 463–474. https://doi.org/10.1089/hum.2008.022.

Dr. Stephen Napier is an associate professor of philosophy at Villanova University. He specializes in clinical bioethics and theories of knowledge. His previous employment includes a clinical ethics fellowship at St. Thomas Hospital in Nashville, he was a research protections analyst at Cincinnati Children's Hospital, and an ethicist with the National Catholic Bioethics Center. His publications include two books, *Virtue Epistemology: Motivation and Knowledge* (Bloomsbury, 2008), and most recently Uncertain Bioethics: Moral Risk and Human Dignity (Routledge, 2019). He has published various articles that address issues in the philosophy of religion, beginning of life issues, end-of-life issues, and research ethics.

Part V
Surgery

Chapter 12
The Ethics of Surgical Research and Innovation

Wendy A. Rogers and Katrina Hutchison

12.1   Introduction

Surgery is a powerful tool in the contemporary medical armamentarium. Surgeons can remove body parts that are diseased or cancerous; implant prostheses to replace failing joints; and transplant tissues and organs. Surgery can save many lives and improve the quality of others. But surgical advances come at a cost. Surgery is invasive, not easily reversible and generally expensive. The successes have at times been matched by failures of new techniques, tools and devices that cause ongoing or lethal harm to patients. As surgery is such a high stakes intervention, there is a strong prima facie ethical imperative for research to establish the safety and effectiveness of new treatments. Surgical research is, however, challenging for a number of reasons. The boundaries of surgical innovation blur with those of research at one extreme, and with variations in usual practice at the other, raising questions about the necessity and timing of formal investigation. Surgical interventions are hard to test in clinical trials as they do not occur in a pre-specified and static format such as a chemical formula but evolve rapidly as techniques change and new iterations of devices emerge. The safety and efficacy of surgery depends on the skills and experience of the surgeon involved, creating uncertainty regarding the generalizability of surgical research. Unlike taking a medication, which can usually be ceased if there are adverse events, surgery is irreversible. This creates an extra level of risk in research, as patients may be left considerably worse off. Placebo controlled trials, while possible are controversial and expensive to run. Finally, a complex web of potentially competing interests surrounds surgical research as surgeons occupy multiple roles including clinician, researcher and sometimes device developer.

W. A. Rogers (✉) · K. Hutchison
Macquarie University, North Ryde, NSW, Australia
e-mail: wendy.rogers@mq.edu.au

© The Author(s), under exclusive license to Springer Nature Switzerland AG 2023   217
T. Zima and D. N. Weisstub (eds.), *Medical Research Ethics: Challenges in the 21st Century*, Philosophy and Medicine 132,
https://doi.org/10.1007/978-3-031-12692-5_12

In this chapter we investigate the ethics of surgical research and innovation. Section 12.1 is a case study of a surgical innovation used to treat vaginal prolapse, which illustrates the potential harms of, and challenges posed by introducing surgical innovations into practice. Section 12.2 explores four of these challenges in detail. These are: the lack of clear boundaries between surgical practice, research and innovation; the irreversible nature of surgery; methodological issues in surgical research; and conflicts of interest. In Sect. 12.3 we argue for the value of a structured approach to introducing surgical innovations, such as that developed by the IDEAL Collaboration, which proposes specific research methods for five stages across the development and deployment of surgical innovations. We identify ethical issues arising across these stages, ranging from consent through to broader issues such as justice in access to new interventions. Given the complex nature of surgical interventions, it is argued that a structured approach such as IDEAL is essential to ensure that the surgical interventions offered to patients are as safe and effective as possible. This approach must be supplemented with additional scrutiny to ensure that the risks and benefits are not distributed unfairly across social groups.

12.2 Case Study: Treatment Innovation for Pelvic Organ Prolapse (POP)

In 2002 the FDA approved the first of a class of devices known as "mesh kits" to treat pelvic organ prolapse (POP) (Food and Drug Administration 2011). These kits contained pre-cut synthetic polypropylene tissue repair mesh, along with placement tools and fixings for a trans-vaginal approach. This approval elevated synthetic mesh from an off-label option of last resort—something to try when conservative management and natural tissue surgical repairs had failed—to a front-line treatment for women suffering POP.

The kits promised to solve a real need in the context of POP. POP is a chronic condition affecting quality of life for millions of women, with symptoms including discomfort, incontinence and sexual dysfunction. Conservative treatments such as pelvic floor muscle training, or the use of rubber device inserted into the vagina to hold the prolapsed tissue in place, can improve symptoms but are not curative (Hagen and Stark 2011; Dällenbach 2015). Moreover, they require ongoing vigilance to maintain. The surgical alternative, natural tissue repair surgery, often does not produce a lasting solution. Between 30 and 50% of women who have primary POP surgery experience further prolapses (American College of Obstetricians and Gynaecologists 2019).[1] Synthetic mesh kits promised to address this problem—a fix-and-forget solution sold in a neat package that any surgeon or gynaecologist could be trained to implant.

---

[1] Note, however, that the American College of Obstetricians and Gynaecologists believe these figures were probably inaccurate, and that a more accurate figure is between 6 and 30% (American College of Obstetricians and Gynaecologists 2019: 397).

The mesh kits were approved via the U.S.A.'s Food and Drug Administration (FDA) 510 k regulatory process. This premarket approval process involves manufacturers claiming that the new product is substantially equivalent to, and therefore equally safe and effective as, an already legally marketed device. The 510k for the mesh kits was based on 'similarity' to mesh kits used to repair inguinal hernias, predominantly in men. Consequently, they came to market without high quality clinical trials of the mesh kits for POP– that is, without robust evidence of effectiveness or safety for this new application (Hutchison and Rogers 2017). Between 2002 and 2008 scores of different mesh kits came to market from multiple device manufacturers. Their convenience, combined with enthusiastic marketing from manufacturers, led to rapid, widespread uptake. As Wall and Brown (2010) noted, the mesh kits created a perception of the operation as straightforward and standardised. The kits promised higher surgical volumes and appeared to mutually benefit surgeons, device companies, and women with prolapses. The kits had the advantage of including "everything you need to operate (except skill and good clinical judgment) right in the box" (Wall and Brown 2010:30.e1). At the same time, the number of different kits on the market meant that there was no single, standardised procedure or set of tools.

Training was typically organised by manufacturer representatives and delivered by proctorship from "thought leader" clinicians on the manufacturers' payroll. S. Abbas Shobeiri describes the perfunctory and inadequate nature of at least some training offered by proctors and device reps:

> I went to vaginal mesh kit training courses and as an educator found the level of training lacking. I asked the trainer who was an obstetrician/gynecologist colleague and the company business director if they had ever "failed" any surgeons during their years of training, and they said no. At one training session, the company representatives were given the course completion certificates for surgeons who failed to show up so they could hand deliver them to the absent surgeons (Rostaminia and Shobeiri 2019: 6).

Despite the initial promise of mesh kits as a solution to POP, the product did not live up to the hype. There were high complications rates, which were not detected for some time due to lack of rigorous trials, registries or other forms of postmarket surveillance. Many kits have now been recalled or withdrawn from the market.

The most common adverse events associated with mesh kits for POP reported to the FDA's Manufacturer and User Facility Device Experience (MAUDE) database between 2005 and 2010, were: mesh erosion (mesh moves from the implant site and can break through into e.g. the vaginal canal, rectum or bladder), pain, infection, bleeding, dyspareunia (pain during sexual intercourse), organ perforation, urinary problems, vaginal scarring/shrinkage, neuromuscular problems, and recurrent incontinence (Brill 2012; Abbott et al. 2014; O'Connor and Madden 2020). In addition to dyspareunia affecting the implanted woman, the partner can also experience pain or injury during intercourse due to protrusion of mesh into the vagina.

These adverse events are not necessarily reversible. Mesh can be difficult to remove or to remove completely because it becomes incorporated into scar tissue. Removal sometimes requires multiple surgeries (Margulies et al. 2008), while complete removal risks recurrence of the prolapse (Kisby and Linder 2020). Pain symptoms, including dyspareunia, are not always resolved by removing the mesh

(Kisby and Linder 2020; Margulies et al. 2008). This may be due to nerve damage (Rostaminia et al. 2019).

Women's risk of pain and need for revision/excision of mesh varies according to the volume of POP repairs undertaken by the surgeon involved (e.g. Rogo-Gupta and Castellanos 2016). One study found that low volume surgeons (who performed as little as one case per year) had significantly higher reoperation rates than surgeons who performed 2 or more cases, and that more than half of POP repairs were performed by low volume surgeons (Eilber et al. 2017).

12.3   Challenges in Surgical Research

Rigorous research is required to avoid the introduction of potentially harmful surgical innovations, such as the POP mesh kits. However, as we describe here, several features of surgical innovation make research challenging.

12.3.1   Lack of Clarity Between Clinical Practice and Research

A fundamental prerequisite for triggering research is a clear distinction between variations in ordinary clinical practice and the proposed introduction of novel interventions. The distinction should track changes that are sufficiently novel so as to raise the risk that safety and efficacy may be impacted, justifying research evaluation before introduction into practice. For pharmaceutical interventions, the boundary is relatively clear. Any change to the molecular structure of a drug creates a new agent that requires its own evidence base to establish safety and efficacy prior to approvals. This clarity is lacking in surgery, for a number of reasons.

First, surgery has to be responsive to the anatomy and circumstances of each individual patient. Standard techniques for well-known procedures may be unsuitable for some patients. For example, a neurosurgical patient may have unusual bony protuberances or past scarring that require the use of additional tools or an alternative technique. Surgeons respond to these daily challenges by varying the techniques and tools they use, the types of prostheses they implant and the ancillary care they provide before and after surgery. These variations are not investigated in formal research as they are responses to the needs of individuals rather than potentially widespread changes to practice. The use of mesh for POP initially fitted this pattern of responding to individual variations. Surgeons cut it to shape the anatomy of the women involved who had specific indications for using mesh rather than any other alternatives. Mesh kits were later presented as a harmless variation on these custom uses, rather than an innovation.

A second reason for the blurry boundary between routine practice and novel interventions requiring research is that surgeons themselves hold divergent views on when a change is novel enough to warrant research before introducing it into practice (Rogers et al. 2014). Part of this confusion may stem from the widespread and inconsistent use of the term 'surgical innovation'. Surgical innovations occupy a middle ground between variations in practice and major changes that clearly warrant research, such as a completely novel procedure like the first heart or uterus transplantation. Some surgeons use 'innovation' to refer to the first time that they perform a particular operation or use a new tool. For others, an innovation is an invention or major change warranting research (Rogers et al. 2014). There is some evidence that surgeons may use the term opportunistically, as labelling an activity research triggers mandatory ethical and regulatory oversights that do not apply equally to innovation (Reitsma and Moreno 2005). As well as the lack of distinction between innovation and research, the relationship between them is poorly conceptualised. Some surgeons describe the relationship as an iterative process in which an innovation is tried in practice and if promising, is then subject to testing in research, while others describe a more linear approach in which research informs (or should inform) the development of the innovation (Rogers et al. 2014). The former approach of 'trial and error' is problematic for several reasons, including the lack of reporting of failed innovations.

Surgical interventions are not static, providing a third reason as to why it is difficult to determine when surgical research should occur. Surgical interventions evolve during their development, and continue to evolve in practice. For example, the innovation of minimally invasive oesophagectomy underwent modification as it was being evaluated in research. Early results revealed adverse events associated with the initial procedure leading to adaptation of the technique that was tested in subsequent research (Blazeby et al. 2011). This rapid evolution is not limited to surgical techniques; tools and devices can also change quickly in response to surgeon feedback, such that the results of a trial on one device may be obsolete by the time the results are available as the original iteration of that device has been superseded.

12.3.2   Irreversible Nature of Surgery

A second challenge for surgical research concerns the nature of surgery itself. Many surgical procedures involve cutting the skin and subcutaneous tissue to remove material and/or implant a device. While every care is taken, even the most skilled of surgeons will leave a scar, and there may be unavoidable damage to nerves and other tissues specific to particular procedures. Major interventions such as liver transplantation or radical prostatectomy can change the anatomy and physiology of patients in far-reaching ways. As the mesh case study shows, removal of implants may not be possible, and even complete removal did not resolve the ongoing pain caused by nerve damage of some patients. The irreversible and potentially high-impact nature of surgery provides strong justification for performing research, to ensure that only safe and effective interventions are offered to patients. But the very same features of invasiveness and irreversibility mean that participation in surgical research is correspondingly more risky than participation in a drug trial. While it is usually possible to withdraw from a pharmaceutical trial, cease the medication and return to one's pre-trial state, this is impossible with surgical trials. The operation cannot be undone in a way that restores the pre-trial status quo. At best, a device may be removed or a surgical procedure revised, but this is not always possible, as demonstrated by examples including the POP mesh case and the metal on metal Articular Surface Replacement (ASR) hips.

The ASR hip, marketed by the Johnson and Johnson subsidiary, De Puy, was widely promoted as a long-lasting prosthesis with novel metal-on-metal technology that promised reduced risk of fracture and dislocation (Johnson and Rogers 2014). It was introduced without clinical research, on the basis of substantial equivalence to existing prostheses (the 510 K pathway). Within a few years, it became obvious that the ASR hip was not performing as expected. Patients experienced numerous side effects linked to metal toxicity, and there was a high failure rate related to adjacent tissue necrosis from the leakage of metal ions from the prosthesis. Many patients underwent revisions involving removal of the ASR hip and replacement with non-metal alternatives. Removal of the ASR prostheses led to lowered levels of cobalt and chromium for most of these patients, but some had persistent elevated levels and ongoing symptoms including pain (Lainiala et al. 2015). Like the vaginal mesh, it was not possible to 'undo' the surgery and in many instances, patients were left worse off than if they had not undergone the surgeries in the first place. These harms could have been minimised with rigorous research prior to widespread introduction into practice, but with both the ASR hip and the vagina mesh, the prostheses were approved based on their similarity to existing ones.

12.3.3 Methodological Challenges in Surgical Research Including Placebo-Controlled Trials

There is a well-established research pathway for producing evidence about the effectiveness and safety of medical interventions. This evidence-based approach prizes systematic reviews of randomised controlled trials (RCTs) (1a) and individual RCTs with narrow confidence intervals (1b) as the highest (i.e. most reliable) levels of evidence for therapeutic interventions (Oxford Centre for Evidence Based Medicine 2020). However, there are specific reasons why it is difficult for surgical trials to conform to this type of research. High quality RCTs aim to minimise bias by 'double blinding', which involves concealing information from both participants and doctors about which intervention (the trial intervention or usual practice/placebo) the participant has received. But surgeons performing procedures in a surgical research trial cannot be blinded to the procedure they are delivering—i.e. a double-blind study design is not possible. Second, surgeons may not be in equipoise regarding the relative benefits of the comparator treatments (even if they should be in equipoise given the state of the evidence). In the context of vaginal mesh kits for POP, differing views were noted in the clinical community (e.g. Cooper 2016).[2] The outcomes of surgical procedures depend upon the surgeon's skills, skewing equipoise and potentially influencing surgical performance and/or observation of outcomes. Next, the composition of medicines used in clinical trials can be set prospectively, and medicines are typically packaged in a standard form and delivered by a standard method (for example, swallowed or injected). In contrast, individual surgeons are likely to do aspects of a procedure in idiosyncratic ways, and to refine their technique over time. Even if efforts are made to specify exactly how the procedure is to be done in the context of a trial, some variations are inevitable. New procedures that differ significantly from anything that has been done before pose the additional challenge of learning curves for surgeons. The learning curve refers to the time it takes for a qualified surgeon to become competent at a new procedure. Learning curves raise challenging questions about the surgeons involved in research; if those in trials are already expert in the procedure, the results of that trial will be better than if the research involves surgeons who have never before performed the procedure (Cook et al. 2004). In head-to head trials, outcomes in the intervention arm may be affected by learning curve effects that do not arise in the comparator arm.

The influence of surgeon factors on surgical trials is broader than learning curve effects. Where a trial compares two or more surgical interventions, the research team must decide whether any given surgeon will perform only one procedure, or will perform all the procedures. Both options can have problems. Where the same surgeon performs both/all procedures, they may have significantly more experience with one, potentially affecting outcomes. Where different surgeons perform the procedures on different arms of the trial, it may be challenging to match the groups for skill and experience. Broadly, some differences in outcome can be attributed to the surgeon rather than the procedure, but it may be difficult to ascertain the extent of this effect. A similar set of issues can arise between centres in multicentre surgical trials. Insofar as aspects of care delivery vary between centres—for example, pre and post-operative care protocols, discharge and follow-up—part of the variation in outcomes may be due to centre effects rather than differences in effectiveness between the techniques. Together, these challenges are called 'cluster effects'—clusters of patients treated by the same surgeon or at the same centre may have systematically different outcomes to other patients due to surgeon or centre effects rather than the difference in technique that the trial purports to investigate (Cook et al. 2012).

Surgery is known to have a strong placebo effect (Probst et al. 2016). This presents a barrier to comparing surgical interventions fairly with conservative or non-surgical interventions. Moreover, it might mean that the benefits attributed to some procedures believed to be effective, are due to the placebo effect rather than the presumed mechanism of action of the procedure. Where there is no safe, established comparator

---

[2] See also the variety expressed in submissions from clinical groups to the Australian Senate Inquiry (Senate Community Affairs References Committee, 2018).

surgical intervention, the potential for bias due to the placebo effect when comparing a surgical procedure against a non-surgical control can render the results unreliable. In such instances, although contentious, placebo-controlled RCTs may be ethically justified (Rogers et al. 2014).

The main ethical concern raised by placebo-controlled RCTs is that of intentionally exposing participants to harm (Macklin 1999). To be convincing, placebo interventions must sufficiently mimic the trial intervention such that participants cannot tell which they have received. This necessarily involves some kind of anaesthesia, incision and manipulation of tissue, which are actual rather than potential harms. In addition there are the risks of any surgery, such as those related to anaesthetic misadventure, infection, allergic reactions or accidents associated with hospital admission. Knowingly exposing trial participants to harm breaches the ethical obligation of researchers to minimise risk of harm to participants. On some views this is an unacceptable breach that cannot be outweighed by the potential gains in knowledge (Macklin 1999). Nonetheless, placebo-controlled trials may be the only research method that can provide evidence about the efficacy of new interventions, especially when there is no existing surgical procedure for a comparator, or the intervention is for an outcome such as pain, which is susceptible to a placebo response. Reliable evidence prior to the introduction of surgical innovations is necessary to prevent the exposure of large numbers of patients to interventions that are potentially ineffective and/or harmful. Proponents of placebo trials in surgery argue that the harms and risks can be minimised and that these are outweighed by the value of the knowledge gained. Future patients are thereby spared unsafe or ineffective procedures while society avoids the costs of ineffective surgery (Rogers et al. 2014).

12.3.4 Competing Interests and the Multiple Roles Occupied by Surgeons

To conclude this section, we identify a further challenge for surgical research—the multiple roles occupied by surgeons. Rogers and Johnson argue that in surgery, unlike many other areas of medicine, "the surgeon is both diagnostician and the instrument of therapy" (2013: 220). As a consequence of this dual role, surgeons are obliged not only to ensure the diagnosis is correct, but also to ensure that any ensuing surgical treatment is as effective as possible. In addition, innovation is widely regarded as an essential part of surgery. Surgery is a practical discipline that has and continues to evolve rapidly in response to the development of new materials and new tools. This rapid evolution creates its own pressure for surgeons to offer the latest technique or prosthesis, and to embrace innovation as part of their practice.

In many cases, the primary role-related interest of surgeons to act in the best interests of their individual patients does not come apart from other financial or non-financial interests associated with innovating. For example, spur of the moment innovations, such as using a titanium pin to stem uncontrolled presacral bleeding

(Qinyao et al. 1985), provide benefits to patients in otherwise desperate situations and pose few risks of conflict of interest. Likewise, variations in technique to adapt to unusual anatomical presentations are inevitable on a case-by-case basis, but do not give rise to conflicts of interest or within-role conflicts.

However, in other cases there are tensions between the financial or non-financial interests associated with research and innovation, and those associated with patient care (Johnson and Rogers 2014). The financial interests associated with innovating can include direct payments of various sorts—such as royalties, or engagement by device manufacturers to proctor other surgeons. In the vaginal mesh case study above, Rostaminia and Shobeiri (2019) describe the conflicted roles that thought leaders and industry sponsored proctors of vaginal mesh kits came to occupy in the context of rapid uptake and aggressive industry marketing of these kits. In other examples, surgeons have received royalties for their contributions to the invention of devices, making those surgeons less likely to speak out when the device in question led to poor outcomes—such as the ASR hip (Johnson and Rogers 2014). Financial conflicts can also be less direct. In their analysis of the use of mesh kits, Wall and Brown (2010) noted that these kits promised higher surgical volumes (and thus greater income) due to presenting the procedure as standard and straightforward.

Non-financial conflicts of interest arising from innovation can include the desire for esteem of peers, promotion or publication of findings. In addition, surgeons can have unsubstantiated optimism about the advantages of a new approach (optimism bias), or about their own skill in performing the innovation (Rogers and Johnson 2013).

In summary, there are various challenges that arise in attempting to guarantee that new surgical interventions are safe and effective for patients. These range from lack of clarity as to when research should occur through to methodological challenges inherent to the nature of surgery itself and the multiple roles of surgeons in the innovation-research nexus.

12.4   A Structured Approach to the Ethics of Surgical Research

Due to the challenges identified in Sect. 12.2, surgery has lagged behind other areas of medicine in developing a robust evidence-base (McCulloch et al. 2009). This situation is however changing in response to evidence of accumulating harms to patients caused by innovations such as the vaginal mesh for POP and the ASR hip, and increasing recognition that accepted surgical interventions, such as arthroscopic treatment of meniscal tears are not effective and are therefore wasteful (Sihvonen et al. 2013). In response there have been international efforts to develop a research pathway capable of creating a surgical evidence-base to guarantee patient benefits and avoid harms. The most prominent of these is the IDEAL Framework, first proposed in 2009 by an international collaboration (The IDEAL Collaboration, n.d). The initial iteration of IDEAL was based on a pathway that involved evaluating five stages that reflected the development and stabilisation of new surgical interventions. The stages are (1) Idea, (2) Development, (3) Exploration, (4) Assessment and (5) Long term study (McCulloch et al. 2009). Subsequent iterations of the IDEAL Framework have collapsed Development and Exploration into stages 2a and 2b, linked specific research questions and methodologies to each stage and recognised a pre-IDEAL stage in which the feasibility and definition of a new procedure occur (Hirst et al. 2019). The 2019 Framework includes for the first time explicit discussion of ethical issues raised at each stage of IDEAL (see Table 12.1) and was published with a companion paper analysing ethical issues in detail (Rogers et al. 2019).

The structured approach of the IDEAL Framework has the capacity to address several of the challenges of surgical identified above, and in so doing, allay some of the ethical concerns and/or provide mechanisms for addressing these (Rogers et al. 2019). The staged pathway of the IDEAL Framework recognises the lack of a clear distinction between innovation and research and the evolving nature of surgical innovations. It is designed to capture data collection from the very first iteration of a new intervention, which minimises harm to future patients and helps to ensure that consent in future stages is as informed as possible. The Framework stipulates research ethics review (Institutional Review Board) for Stages 2a–3 while recognising that Stage 1 innovations may not fit these processes yet require some form of oversight. The requirement for formal ethics approval helps to build the research culture in surgery, with the aim of minimising opportunities for untested innovations to spread in practice.

Regarding methodological challenges, the IDEAL Framework proposes specific research methods for each stage, with explicit recognition of issues such as learning curves and surgical equipoise. The incremental accumulation of evidence at each stage can inform subsequent stages, thereby minimising harms to patients and strengthening informed consent. Potential conflicts of interest occur across all stages of the Framework. While the proposed solutions are yet to be tested in practice, strategies such as minimising the influence of industry funders and balancing the optimism bias of surgeon innovators with independent oversight have prima facie appeal (Rogers and Johnson 2013).

The structured approach of IDEAL makes explicit and provides strategies for addressing ethical issues in surgical innovation and research. However, the process of bringing new surgical interventions into practice raises issues beyond those directly related to ensuring safety and efficacy.

12.4.1   Justice and Access

A distinct set of ethical considerations arise in relation to the fair and safe access of patients to the benefits of surgical research and innovation, as well as the fair distribution of the associated burden of risks and harms. Although research ethics approval processes typically protect vulnerable people from exploitation as participants in

Table 12.1 Stages of IDEAL and associated research and ethical issues (adapted from Rogers et al. 2019; Hirst et al. 2019)

IDEAL stage	Purpose	Research methods	Major ethical aspects
Pre-IDEAL	Feasibility and definition of the intervention	Various including simulator, modelling and animal	Animal welfare Bias and unfair assumptions Managing conflicts of interest (CoI)
Stage 1: Idea	Proof of concept	First in human: Structured case reports	Potential harm to patients Informed consent Team communication Managing CoI Reporting results External oversight and/or formal research ethics review
Stage 2a: Development	Development and refining the intervention	Prospective development studies	Potential harm to patients Informed consent (including Stage 1 outcomes) Managing CoI Reporting results Formal research ethics review Independent oversight
Stage 2b: Exploration	Achieving consensus on final details of the intervention	Prospective multi-centre cohort studies	Potential harm to patients (managing learning curve) Informed consent (including Stage 2a outcomes) Managing CoI Reporting results Formal research ethics review Independent oversight

(continued)

Table 12.1 (continued)

IDEAL stage	Purpose	Research methods	Major ethical aspects
Stage 3: Assessment	Comparative effectiveness testing	RCTs ± modifications: cluster, preference RCTs, stepped wedge, adaptive designs, placebo-controlled	Patient reported outcomes Fair recruitment Potential harm to patients (managing learning curve) Informed consent (including Stage 2 outcomes) Placebo surgery Managing CoI Reporting results Formal research ethics review Independent oversight
Stage 4: long term monitoring	Surveillance	Registries, rare case reports	Patient consent for data use Privacy Access and equity Managing CoI

research, they are silent on the eventual patterns of uptake and use. For example, it is beyond the scope of research ethics committees to ensure that findings of surgical research will, in future, deliver benefits fairly across different social groups.

David Hunter argues that in the context of innovative technologies it is important to distinguish injustices that occur at a point in time, from those that are ongoing and can become entrenched (Hunter 2013). With any new technology or procedure there will always be a first centre and first patients selected. Early injustices might arise only for a limited period due to such factors and may be ethically acceptable if they do not involve the unfair exploitation of vulnerable groups. We therefore focus here on injustices related to the introduction of surgical innovations that are systematic, entrenched, or that reinforce existing patterns of privilege and oppression.

Injustice can occur when the risks associated with new procedures disproportionately affect patients from one group due to gender, ethnic or geographical factors. The mesh case presented above affected women rather than men. While this is due to the condition—vaginal prolapse—there is evidence that women are disproportionately harmed by medical devices overall. The International Consortium of Investigative Journalists (ICIJ) found that women comprised 67% of the 340, 000 injury and death cases from medical devices recorded in the FDA's MAUDE database (Guevara 2019). Hutchison (2019) has argued that disproportionate harm to women from medical devices is the result of systemic factors in processes of design, regulation, and use.

Expensive devices such as surgical robots can be unfairly available to those who are better off or who live in favourable geographical locations, such as major centres with large teaching hospitals (Hutchison et al. 2016). For example, research into robotic prostatectomy in the USA shows that African-American and Hispanic patients, and patients of lower socio-economic status are less likely to access robotic prostatectomy than wealthier Caucasian patients (Kim et al. 2013). Geographical factors, such as living in a regional or remote area can increase the chance of patients suffering harm due to learning curves or low surgical volumes of a new procedure. The mesh case illustrates the differences in outcome for patients of low volume surgeons. Submissions to the Australian Senate Inquiry into trans-vaginal use of mesh products recommended that surgeons disclose their experience with the procedure to patients during consent processes, as well as details about outcomes at the hospital (Senate Community Affairs References Committee 2018). However, inevitably there are differences in volume for clinicians and hospitals in regional locations.

To summarise, it is important that ethical issues associated with justice and access are considered in the context of surgical innovation and research. Most considerations relating to access and justice are beyond the scope of research ethics committees and are also outside the scope of the IDEAL Framework, which focuses on evaluating the safety and efficacy of new interventions. It is therefore important that innovators, regulators, and funders remain vigilant about these issues after new procedures are introduced into practice, with particular focus on the risk that new treatment options will entrench or amplify existing social disadvantage.

12.5 Conclusion

This chapter has provided an overview of ethical issues arising in surgical research and innovation. Several of the issues identified—the lack of clear boundaries between clinical practice, innovation and research in surgery; the irreversibility of surgery; methodological challenges associated with surgical research; and conflicts of interest—can be addressed by recommendations of the IDEAL Framework. This Framework provides detailed guidance for the production and dissemination of high-quality evidence regarding the safety and efficacy of surgical innovations. However, issues associated with justice and access to new treatments are beyond the scope of the IDEAL Framework, and require ongoing vigilance on the part of all stakeholders, with particular attention to the risk new treatments will entrench existing patterns of privilege and disadvantage.

References

Abbott, S., C.A. Unger, J.M. Evans, K. Jallad, K. Mishra, M.M. Karram, C.B. Iglesia, C.R. Rardin, and M.D. Barber. 2014. Evaluation and management of complications from synthetic mesh after pelvic reconstructive surgery: A multicenter study. *American Journal of Obstetrics and Gynecology* 210 (2): 163.e1-163.e8. https://doi.org/10.1016/j.ajog.2013.10.012.

American College of Obstetricians and Gynecologists. 2019. Pelvic organ prolapse. *Female Pelvic Medicine & Reconstructive Surgery* 25 (6): 397–408.

Blazeby, J.M., N.S. Blencowe, D.R. Titcomb, C. Metcalfe, A.D. Hollowood, and C.P. Barham. 2011. Demonstration of the IDEAL recommendations for evaluating and reporting surgical innovation in minimally invasive oesophagectomy. *The British Journal of Surgery* 98 (4): 544.

Brill, A.I. 2012. The hoopla over mesh: What it means for practice. *OB GYN News* 47 (1): 14.

Cook, J.A., C.R. Ramsaya, and P. Fayers. 2004. Statistical evaluation of learning curve effects in surgical trials. *Clinical Trials* 1 (5): 421–427.

Cook, J.A., T. Bruckner, G.S. MacLennan, and C.M. Seiler. 2012 Clustering in surgical trials - database of intracluster correlations. Trials 13:2. https://doi.org/10.1186/1745-6215-13-2

Cooper, J.C. Why I use mesh: a personal perspective. International Urogynecology Journal 23 (8):971-3. Hutchison K., J. Johnson, and D. Carter. 2016. Justice and surgical innovation: the case of robotic surgery. Bioethics 30 (7): 536-546.

Dällenbach, P. 2015. To mesh or not to mesh: A review of pelvic organ reconstructive surgery. *International Journal of Women's Health* 7: 331–343.

Eilber, K.S., M.L. Alperin, A. Khan, N.T. Wu, C.T. Pashos, J.Q. Clemens, and J.T. Anger. 2017. The role of the surgeon on outcomes of vaginal prolapse surgery with mesh. *Female Pelvic Medicine & Reconstructive Surgery* 23 (5): 293–296.

Food and Drug Administration. 2011. *Urogynecologic surgical mesh: Update on the safety and effectiveness of transvaginal placement for pelvic organ prolapse.* https://www.fda.gov/media/81123/download

Guevara, M.W. 2019, November 25. We used AI to identify the sex of 340, 000 people harmed by medical devices. *International Consortium of Investigative Journalists.* https://www.icij.org/investigations/implant-files/we-used-ai-to-identify-the-sex-of-340000-people-harmed-by-medical-devices/

Hagen, S., and D. Stark. 2011. Conservative prevention and management of pelvic organ prolapse in women. *The Cochrane Database of Systematic Reviews*, *12*, CD003882–CD003882.

Hirst, A.G., Y.G. Philippou, J.G. Blazeby, B.G. Campbell, M.G. Campbell, J.G. Feinberg, M.G. Rovers, N.G. Blencowe, C.G. Pennell, T.G. Quinn, W.G. Rogers, J.G. Cook, A.G. Kolias, R.G. Agha, P.G. Dahm, A.G. Sedrakyan, and P.G. Mcculloch. 2019. No surgical innovation without evaluation: Evolution and further development of the IDEAL Framework and recommendations. *Annals of Surgery* 269 (2): 211–220.

Hunter, D. 2013. How to object to radically new technologies on the basis of justice: The case of synthetic biology. *Bioethics* 27 (8): 426–434.

Hutchison, K. 2019. Gender bias in medical implant design and use: A type of moral aggregation problem? *Hypatia* 34 (3): 570–591.

Hutchison, K., and W. Rogers. 2017. Hips, knees, and hernia mesh: When does gender matter in surgery? *IJFAB: International Journal of Feminist Approaches to Bioethics* 10 (1): 148–174.

Johnson, J., and W. Rogers. 2014. Joint issues—Conflicts of interest, the ASR hip and suggestions for managing surgical conflicts of interest. *BMC Medical Ethics* 15 (1): 63.

Kim, S.P., S.A. Boorjian, N.D. Shah, C.J. Weight, J.C. Tilburt, L.C. Han, R.H. Thompson, Q.-D. Trinh, M. Sun, J.P. Moriarty, and R.J. Karnes. 2013. Disparities in access to hospitals with robotic surgery for patients with prostate cancer undergoing radical prostatectomy. *The Journal of Urology* 189 (2): 514–520.

Kisby, C.K., and B.J. Linder. 2020. Management of vaginal mesh exposures following female pelvic reconstructive surgery. *Current Urology Reports* 21 (12): 1–9.

Lainiala, O., A. Reito, P. Elo, J. Pajamäki, T. Puolakka, and A. Eskelinen. 2015. Revision of metal-on-metal hip prostheses results in marked reduction of blood cobalt and chromium ion concentrations. *Clinical Orthopaedics and Related Research* 473 (7): 2305–2313.

Macklin, R. 1999. The ethical problems with sham surgery in clinical research. *The New England Journal of Medicine* 341 (13): 992.

Margulies, R.U., C. Lewicky-Gaupp, D.E. Fenner, E.J. Mcguire, J.Q. Clemens, and J.O.L. Delancey. 2008. Complications requiring reoperation following vaginal mesh kit procedures for prolapse. *American Journal of Obstetrics and Gynecology* 199 (6): 678.e1-678.e4.

McCulloch, P., D.G. Altman, W.B. Campbell, D.R. Flum, P. Glasziou, J.C. Marshall, and J. Nicholl. 2009. No surgical innovation without evaluation: The IDEAL recommendations. *The Lancet* 374 (9695): 1105–1112.

O'Connor, M., and B. Madden. 2020. Vaginal dialogues: The trials and tribulations of mesh in the repair of prolapse. *Journal of Law and Medicine* 27 (3): 618–633.

Oxford Centre for Evidence Based Medicine. 2020. *Levels of Evidence* (March 2009). https://www.cebm.ox.ac.uk/resources/levels-of-evidence/oxford-centre-for-evidence-based-medicine-levels-of-evidence-march-2009

Probst, P.C., K.J. Grummich, J.W. Harnoss, F.K. Hüttner, K.K. Jensen, S.K. Braun, M.K. Kieser, A.K. Ulrich, M.K. Büchler, and M.K. Diener. 2016. Placebo-controlled trials in surgery: A systematic review and meta-analysis. *Medicine* 95 (17): e3516–e3516.

Qinyao, W., S. Weijin, Z. Youren, Z. Wenqing, and H. Zhengrui. 1985. New concepts in severe presacral hemorrhage during proctectomy. *Archives of Surgery* 120 (9): 1013–1020.

Reitsma, A.M., and J.D. Moreno. 2005. Ethics of innovative surgery: US surgeons' definitions, knowledge, and attitudes. *Journal of the American College of Surgeons* 200 (1): 103–110.

Rogers, W., K.A. Hutchison, and A.A. Mcnair. 2019. Ethical issues across the IDEAL stages of surgical innovation. *Annals of Surgery* 269 (2): 229–233.

Rogers, W.A., and J. Johnson. 2013. Addressing within-role conflicts of interest in surgery. *Journal of Bioethical Inquiry* 10 (2): 219–225.

Rogers, W.A., M.A. Lotz, K.A. Hutchison, A.A. Pourmoslemi, and A.A. Eyers. 2014. Identifying surgical innovation: A qualitative study of surgeons' views. *Annals of Surgery* 259 (2): 273–278.

Rogers, W., K. Hutchison, Z.C. Skea, and M.K. Campbell. 2014. Strengthening the ethical assessment of placebo-controlled surgical trials: Three proposals. *BMC Medical Ethics* 15 (1): 78. https://doi.org/10.1186/1472-6939-15-78.

Rogo-Gupta, L., and M. Castellanos. 2016. When and how to excise vaginal mesh. *Current Opinion in Obstetrics and Gynecology* 28 (4): 311–315.

Rostaminia, Ferzandi T., and S.A. Shobeiri. 2019. Vaginal mesh and pain complications. *The Innovation and Evolution of Medical Devices*, 263–279. ed. Shobieri, S.A. Springer.

Senate Community Affairs References Committee. 2018. *Report: Number of women in Australia who have had transvaginal implants and related matters.* Commonwealth of Australia. https://www.aph.gov.au/Parliamentary_Business/Committees/Senate/Community_Affairs/MeshImplants/Report

Shobeiri, S.A. 2018. *The innovation and evolution of medical devices: Vaginal mesh kits.* Springer.

Sihvonen, R., M. Paavola, A. Malmivaara, A. Itälä, A. Joukainen, H. Nurmi, J. Kalske, and T.L.N. Järvinen. 2013. Arthroscopic partial meniscectomy versus sham surgery for a degenerative meniscal tear. *The New England Journal of Medicine* 369 (26): 2515–2524.

The IDEAL Collaboration. n.d. *Idea, development, exploration, assessment, long-term follow-up, improving the quality of research in surgery.* http://www.ideal-collaboration.net/framework/

Wall, L.L., and D. Brown. 2010. The perils of commercially driven surgical innovation. *American Journal of Obstetrics and Gynecology* 202 (1): 30.e1-30.e4.

Wendy A. Rogers is Distinguished Professor of Clinical Ethics at Macquarie University. Her research interests include the ethics of surgical research and innovation, ethical issues raised by AI in healthcare, the ethics of synthetic biology, and transplant abuse in China. Prof Rogers has served two terms on the Australian Health Ethics Committee and been involved in revisions to the current and previous versions of the *National Statement on Ethical Conduct in Human Research.* She was named in 2019 by *Nature* as one of ten people who matter in science for her work exposing the publication of unethical Chinese transplant research.

Katrina Hutchison is Lecturer in the Philosophy Department at Macquarie University. She currently holds a three year Australian Research Council DECRA fellowship. She works mainly on topics in surgical ethics, including the ethics of surgical innovation. Dr Hutchison was involved in developing the Macquarie Surgical Innovation Identification Tool (MSIIT) which has informed policy on safely introducing surgical innovations. A major strand of her research focuses on gender biases in healthcare and philosophy, and she is co-editor of the book *Women in Philosophy: What Needs to Change.*

Part VI
Palliative Care

Chapter 13
Opening Death's Door: Psilocybin and Existential Suffering in Palliative Care

Duff R. Waring

13.1 Introduction

As his first wife lay dying of cancer, Aldous Huxley wanted her to attain the spiritual solace she had sought for so long. With her son and sister at her bedside, Huxley whispered in his wife's ear, urging her to "Go forward into the light" of "peace" and "pure being" (Dunaway 1989, pp. 318–320). Huxley wanted to see that light himself and it is easy to imagine that his last hours of conscious life were anything but dim. After ten years of self-experimentation with psychedelic drugs, Huxley requested that his second wife inject him with 100 µg of lysergic acid diethylamide when he lay on his own deathbed.[1] She then read aloud from the *Tibetan Book of the Dead*, a Buddhist "manual for dying well" that aims to guide an "everlasting self across the threshold of consciousness into a room full of light" (Dunaway 1989, pp. 375–376; Huxley 1971, pp. 278–279, 307–308). He too, she told him, was moving toward that radiance (Huxley 1971, pp. 286–287). Based on his previous writings (Huxley 1954), we can surmise the following: Huxley thought that LSD was a key that would facilitate his passing from earthly life by unlocking the door to that numinous room. He took the results of that experiment to his grave but new directions in clinical research will enable dying cancer patients to report their responses to psychedelic drugs before they die. A renascent approach to assuaging existential suffering in palliative care employs psychedelic drugs to induce a transformative spiritual experience. According to some first-person accounts, this experience can reduce existential suffering and facilitate an acceptance of impending death.

---

[1] By her own account, she gave him another 100 ug an hour later (Huxley 1971, 282, 284).

D. R. Waring (✉)
York University, Toronto, Canada
e-mail: dwaring@yorku.ca

© The Author(s), under exclusive license to Springer Nature Switzerland AG 2023   235
T. Zima and D. N. Weisstub (eds.), *Medical Research Ethics: Challenges in the 21st Century*, Philosophy and Medicine 132,
https://doi.org/10.1007/978-3-031-12692-5_13

There is a thriving revival of research on classic psychedelic drugs (hereinafter PDs)[2] that has rich historical precedent in the history of psychiatry. The formal study of the effect range of PDs and their clinical potential was initiated in 1951 and lasted until the mid-70s. Between 1951 and 1961, over 1000 articles on LSD alone had been published in medical journals (Dyck 2012, 15). PDs were introduced as potential treatments for existential anxiety in terminal cancer patients in the late 50s at the Chicago Medical School and at UCLA. Pivotal work with dying patients was conducted from the late 60s until 1976 at the Maryland Psychiatric Research Center at Spring Grove State Hospital (Grob et al. 2013, 294). The recreational association of PDs with "hippie counterculture," their alleged incompatibility with a blinded, placebo-controlled trial design, and resulting political and regulatory pressure eventually shut PD research down (Oram 2018). This earlier research was focused mainly on LSD, the classic psychedelic most often associated with the "inebriant mania" of 60s—early seventies counterculture (Hoffman 2019, 28). It also has a nefarious and well-documented association with a CIA program of covert, non-consensual research on mind control that left its own human casualties (Lee and Shlain 2001). These popular cultural narratives have overshadowed the counter-story; that PDs have therapeutic promise and should be scientifically studied. The emerging, contemporary research on psychedelic healing is shining new light on that counter-story and may yet enable PDs to emerge from the clinical shadows.

Many of the contemporary investigations are US and European studies on psilocybin. Some PD studies aim to better understand their neurophysiologic, biochemical mechanisms of action. These studies would exemplify what Freedman termed "science in service to understanding," e.g., they might aim to present a physiologic analysis of a drug's specific effects, or delineate the neural correlates of altered conscious states (cf. Freedman 1990, 5).[3] My focus is on a subset of studies that

---

[2] Classic serotonergic psychedelics are substances that exert their effects mainly by an agonist (or partial agonist) action on the receptors for brain serotonin 5-hydroxytryptamine (5-HT) 2A See Nichols (2016), at 266. These substances are psilocybin, mescaline, LSD, and dimethyltryptamine (DMT). Definitions of "psychedelic" can be expansive, and include other mind-altering compounds that induce a different type of experience than that associated with one of the classics. These include the empathogens (entactogens), e.g., MDMA, MDA, some NMDA-antagonist dissociative anesthetics, e.g., ketamine and phencyclidine (PCP), and a host of plant-based shamanic medicines, e.g., ayahuasca. See Ben Sessa (2012), 27. Dr. Humphrey Osmond coined the term "psychedelic" in 1957 when studying LSD as a potential treatment for alcoholism. On his account, classic psychedelics have "a mind-manifesting" capability that induces beneficial "enlargements, burgeonings of reality." By "enriching the mind and enlarging [its] vision," psychedelics help us to "explore and fathom our own nature." See Humphrey Osmond (1957), at 428–429. This term has been popular in the master cultural narrative of mind-altering drugs for more than fifty years, but has not been generally endorsed "by the scientific community because it implies that these substances have useful properties" (Nichols 2016, 266). A key point is that the term "psychedelic" is also meant to denote a certain type of qualitative subjective experience with noetic dimensions. These range from psychodynamic insight to mystical-type revelation. This is one reason it has not found widespread acceptance in the psychiatric community: These experiences allegedly eschew medical science for the vagaries of spirituality and artifactual insight. As I will argue, this betrays a limited model of psychiatry.

[3] Freedman (1990) saw such studies as attempting a "clean" analysis of a drug's "specific" effects that is "uncontaminated by patient, investigator, or mechanism artifacts." As I hope to show, it is

exemplify "science in service to healing" (Freedman 1990, 4), i.e., studies that aim to assess the therapeutic potential of PDs for psychiatric disorders such as treatment-resistant depression, obsessive compulsive disorder, and substance use disorders.[4] A signal challenge of twenty-first century psychiatry is the effective treatment of existential/spiritual suffering in palliative care. The latter is an umbrella term that encompasses anxiety and depression about impending death and a felt loss of meaning in life. To date, three double-blind, placebo-controlled trials have been completed that assessed the potential of psilocybin to relieve anxiety and depression in cancer patients. The published scientific evidence does not yet support the use of any PD "for patient care by clinical practitioners outside the research setting" (Reiff et al. 2020, 404). This chapter will concentrate on research to assess the therapeutic potential of psilocybin in palliative care psychiatry, with reference to the studies on LSD which preceded it.[5] If a PD-induced experience (hereinafter PDE) can facilitate an acceptance of impending death, then it could be a valuable therapeutic anodyne.

The therapeutic use of PDs raises numerous questions about research trial design, the model of palliative care psychiatry that might accommodate it, and the kind of experience it induces. All of them implicate ongoing controversies about the dividing line between the freedom to cultivate spiritual experience and the protective constraints of public policy. I focus on four questions: (1) what is the ethical justification for this research on patients? (2) What types of research trial design are best used to assess the purported benefits of a PDE? (3) What is the epistemic warrant of the "mystical" and "noetic" facets of this experience? (4) What models of (a) psychiatry and (b) drug action could accommodate the therapeutic use of psilocybin in palliative care?

In response to each, I argue the following. (1) The expert clinical community is in a state of clinical equipoise regarding preferred treatments for existential suffering in palliative care. Under carefully controlled circumstances and with appropriate supervision, the risks of PDs are minimal and the benefits are at least roughly equivalent to those attained through other available treatments. (2) The crucial importance of set and setting to the PDE indicates a close relationship with psychological factors that drive placebo effects. Some of the early proponents of PDs argued that the standard, placebo-controlled trial design was thoroughly inappropriate for an assessment of their therapeutic benefits. The concern was that the powerful effects of PDs would easily enable patients to break the blind. I argue instead for a two-arm, unblinded

---

debatable whether that conception of a study trial is apposite to psychedelic drugs, as all three artifacts are inseparable from, and essential to, the range of experiences they induce.

[4] In 2018, the FDA declared it a "breakthrough therapy" in refractory depression and prioritized its consideration in the regulatory process (Reiff et al. 2020, 395).

[5] By contrast, current clinical research with LSD has a much smaller database that has been generated by observational studies. Circa 2020, Swiss and German researchers have conducted one randomized controlled Phase 2 trial to gage the safety and efficacy of LSD-assisted psychotherapy in patients with anxiety associated with life-threatening disease (Reiff et al. 2020, 392–397). The latter was the first authorized clinical study of LSD in over forty years (Gasser et al 2014). Two prospective clinical studies in Switzerland will examine LSD and LSD-assisted psychotherapy as treatments for patients suffering from illness-related anxiety (Reiff et al. 2020, 396–397).

trial design that understands the PDE as a total drug effect (hereinafter TDE). The latter notion best captures all of the factors which would drive a therapeutic response to PDs. I advocate the immediate commencement of Phase 4 clinical effectiveness studies if PDs are approved for clinical use. (3) Much depends on how we understand the nature of the PDE. Although it has inspired claims of metaphysical revelation, it might illustrate nothing more than the mind's ability to have it. Even if the noetic insights PDs induce are illusory, they can generate a transformative, "spiritual" kind of meaning response that persons outside the trial setting are free to cultivate with non-drug techniques. The PDE's truth value as a treatment is not just conceptual. In a science of healing, it is also pragmatic, i.e., how well it relieves existential suffering in palliative care. (4) The emerging model of narrative psychiatry could accommodate PD therapy (hereinafter PDT) for patients, as could a drug-based model of pharmacologic action.

13.2 The Three Trials

The first trial was a Phase 1 pilot study that aimed to determine whether psilocybin "might be effective in reducing anxiety, depression and physical pain, and therefore improving…quality of life" in 12 terminal cancer patients whose cancer-related anxiety met criteria for a DSM IV anxiety-related disorder (Grob et al. 2011; Clinical Trials.gov Identifier: NCT00302744). The protocol specified a "modest" 0.2-mg/kg dose of psilocybin that was not thought capable of inducing the "mystical experience" documented in earlier studies. It was thought "capable of inducing an alteration in consciousness with potential therapeutic benefit" (Grob et al. 2011, 76). The placebo control was 250 mg of niacin. The psilocybin experience was tolerated well by all patients (Grob et al. 2013, 294) and there were no clinically significant adverse events (Grob et al. 2011, 71). Repeated administration of quantitative rating scales showed a significant reduction in cancer-related anxiety at one and three months after treatment. Improved mood reached significance at six months post-treatment (Grob et al. 2011, 71, 77). "Overall, patients reported their participation … as having been a very valuable experience, allowing them to improve their quality of life and augmenting their capacity to withstand the psychological stresses of their medical condition" (Grob et al. 2013, 294–295). The blind was not maintained. "Although we used a within-subject, double-blind, placebo-controlled design, the drug order was almost always apparent to subjects and investigators whether the treatment was psilocybin or placebo. In fact, one consistent subject critique of the study was that the placebo sessions were perceived as far less worthwhile than those with psilocybin" (Grob et al. 2011, 77).

The second was a Phase 2 trial that used a randomized, double-blind crossover design. All 51 patients were told they would receive either a high or a low dose of psilocybin. The study aimed to determine whether the higher dose of psilocybin (22 or 30 mg/70 kg) can "produce personally and spiritually meaningful experiences in 51 patients with a life-threatening cancer diagnosis and symptoms of anxiety and/or depression. This could be important because spirituality has been associated with increased psychological coping and decreased depression in serious illness." The lower dose of psilocybin (1 or 3 mg/70 kg) was used as a control on the assumption that it would be inactive, although the investigators conceded that "some pharmacological activity of this dose cannot be ruled out entirely" (Griffiths et al. 2016, 1181, 1184, 1195. Clinical Trials.gov Identifier: NCT00957359).

It built upon a previous study conducted by the same principal investigator which examined the use of psilocybin in a nontherapeutic context. Griffiths et al. concluded after a 2-month follow-up that "When administered under supportive conditions, psilocybin occasioned experiences similar to spontaneously occurring mystical experiences" in 22 of 36 [healthy] volunteers. They reported this PDE as having "substantial and sustained personal meaning and spiritual significance" (2006, 282). These conclusions were confirmed after a subsequent 14-month follow-up study (Griffiths et al. 2008, 621).

The results of the second trial were promising. When administered under psychologically supportive conditions, a single high dose of psilocybin produced substantial and enduring reductions in depressed mood and anxiety, increased quality of life, increased meaning of life, death acceptance, optimism, and decreases in death anxiety. Ratings by the patients, clinicians, and community observers suggested that these effects endured at least 6 months for approximately 80% of the 51 patients. Patients related improvements in attitudes about "life/self, mood, relationships, and spirituality to the high-dose experience, with > 80% endorsing moderately or greater increased well-being/life satisfaction" (Griffiths et al. 2016, 1181). The overall response at 6 months on clinician-rated depression and anxiety was 78% and 83%, respectively. The "mystical experience" induced by the high dosage was found to have a mediating role in the positive therapeutic response (Griffiths et al. 2016, 1195).

The third was an Early phase 1 trial that also used a double-blind, placebo-controlled (using niacin) crossover design and involved 29 patients with cancer-related anxiety and depressive symptoms. A single "moderate" dose of psilocybin (0.3 mg/kg) was used in conjunction with psychotherapy and generated "mystical-type experiences" in the cohort of patients studied. The intensity of that experience "significantly mediated" clinical benefit (e.g. reduction in anxiety and depression symptoms) in the medium term of 6 weeks after the first dose (Ross et al. 2016a, 1165, 1177; Clinical Trials.gov Identifier: NCT00957359). Psilocybin was associated with substantial, immediate, and sustained improvements in depression and anxiety and led to decreased cancer-related demoralization and hopelessness. Patients reported improved spiritual wellbeing, increased quality of life, and improved attitudes towards death. Psilocybin was associated at the 6.5-month follow-up with enduring anxiolytic and anti-depressant effects: approximately 60–80% of patients continued to evince clinically significant reductions in depression, anxiety, and existential distress, and continued to report improved quality of life. This result "matches descriptive historical data from the open-label LSD-assisted psychotherapy trials for psycho-spiritual distress" that were conducted in the late 60s/early seventies in which the mystical experience was also suggestive of causality (Ross et al. 2016a, 1165, 1177; cf. Grof 2001, 27).

Like the first trial, the blind was broken, demonstrating again that niacin is not an effective placebo for psilocybin. The published study referred obliquely to "the use of a control with limited blinding" (Ross et al. 2016a, 1176). An online only Supplementary Appendix confirmed that investigators could distinguish the patients who received niacin from the patients who received psilocybin in 28/29 cases. "Of note, the participants were not asked to record their guesses as to which drug they received on dosing session days" Ross et al. 2016b).

The last two studies resulted in "surprisingly large and lasting antidepressant and anxiolytic effects. They provide complementary support for the efficacy of high-dose psilocybin for cancer-related psychological distress" (Johnson et al. 2019, 91). All three studies provide support for the therapeutic use of psilocybin with patients in palliative care. Aside from the broken blind in the first and third studies, there are significant limitations to this evidence, restricted as it is to three studies with small samples and short duration (Dos Santos et al. 2018, 11). Another limitation is "the use of nonrepresentative samples (relative to the general population) through self-selection of individuals into clinical trials who may be biased toward expecting beneficial effects, including mystical experience related to ingestion of psychedelics" (Reiff et al. 2020, 406).

As with most PD studies, the ingestion of psilocybin was combined with pre- and post-trial counselling sessions. The pre-trial sessions included the informed consent process, and explorations of the patients' existential distress. The goal of these sessions was to prepare the patients for the PDE by inculcating a receptive mindset, which unavoidably involved giving them some notion of what to expect. The goal of the post-trial sessions was to explore the patients' PDE and enable them to integrate it into their awareness of impending death. These sessions reflect the historical connection between PDs and psychotherapy. PDT has traditionally revolved around two components: an intense PDE and an intensive psychotherapy regimen in which it is embedded. Each component is thought essential to facilitating a reorientation of the "patients' perceptions of themselves and their place in the world, in ways that could [hopefully]lead to lasting positive changes in their values, attitudes, and behavior" (Oram 2018, 3) At the very least, they would combine to alleviate the desperation and denial that frequently accompanies a terminal illness (cf. Oram 2018, 39).

In sum, PDT is not exclusively pharmacologic. It is instead a drug-assisted psychotherapy. This begs the question as to whether a trial design that aims to estab-lish the efficacy of a pure drug effect is suitable for an intervention that involves more than drugs. Put another way, the three trials were as much about the psychotherapy surrounding the PDE as they were about the psilocybin that induced it. Whether this cardinal point is sufficiently stressed by contemporary researchers is open to question.[6] Indeed, some media reports have presented PDT as an "almost magic one-shot [pharmaceutical] cure." This alone could inflate expectations of successful

---

[6] Patients in the second psilocybin trial had approximately eight hours of preparation, and patients in the third trial had six. Even so, the central role of psychotherapy is not highlighted in the second study. There is a brief description of the preparatory session, but the term "psychotherapy" is not employed. The third study specifies that psilocybin is administered with "targeted psychotherapy,"

trcatment. But a diminished emphasis on psychotherapy might disconnect PDT from a component that is traditionally thought key to its therapeutic benefits (Oram 2018, 220–221). As I suggest below, this component also grounds a suitable trial design that accommodates the extra-pharmacologic, but no less essential, characteristic factors of PDT.

13.3   The Psychedelic Drug Experience (PDE)

A current theory about the mechanisms of action in PDT posits an induced "mystical-type experience" as a mediator for positive treatment outcomes. As defined by the *Pahnke-Richards Mystical Experience Questionnaire* (Pahnke 1969; Richards 1975), a "complete mystical experience" has been thought by some to be the most important mediator. A recent qualitative study "suggest[s] a more complex topography," wherein a single mediating factor might be insufficient to account for the multi-level phenomena of a PDE. That experience is not limited to mystical revelation and has a range of dimensions that carry "cognitive, emotional, behavioral, psychodynamic, spiritual, existential, and/or experiential components of significance" (Belser et al. 2017, 377). These can range from exalted feelings of joy, bliss, and love, to a movement from feelings of separateness to interconnectedness with others or the universe, revised life priorities, transient psychological distress that dissipates quickly in a supportive treatment context, lasting changes to one's sense of identity, and the gaining of "transpersonal insights into the nature of the universe or existence" (Belser et al. 2017, 362, 367–369). Indeed, "psycholytic" therapy avoids the mystical-type experience altogether. It employs smaller doses of PDs as means of facilitating a patient's access to their psyche in combination with insight-oriented psychotherapy.[7] The mystical experience induced in high-dose PDT is allegedly a conscious state that transcends the limitations of personal-identity to a participatory sense of interconnectedness with the universe. It has the following components: a sense of pure consciousness or a sense that all things are alive and interconnected, a sense of reverence, awe, or holiness, a sense of encountering "ultimate reality," a sense of transcending boundaries of time and space in an infinite present, a sense of universal love, peace, and tranquility, and "a sense that the experience cannot be

---

but the psychotherapeutic procedures employed were "relegated to an online-only supplementary appendix and could easily be overlooked by many readers" (Oram 2018, 220).

[7] Evidence from the late 60s suggested that anxiety neuroses respond best to psycholytic treatment and that psycholytic treatment is most effective when supplemented by group therapy between LSD sessions. See Alan Grieco & Robert Bloom, "Psychotherapy with Hallucinogenic Adjuncts from a Learning Perspective," *International Journal of the Addictions* 16, no. 5, 801–827, at 805. Research conducted in Switzerland between 1988 and 1993 indicated that the 85.1% of [121] patients [with personality, adjustment, and affective disorders] "considered themselves to have experienced good improvement or slight improvement during their psycholytic treatment.… After treatment, that percentage climbed to 90.9%" See Peter Gasser, "Psycholytic Therapy with MDMA and LSD in Switzerland." *Newsletter of the Multidisciplinary Association for Psychedelic Studies* (MAPS) 5, no. 3 (1994–95): 3–7.

adequately described in words—a sense of the reconciliation of paradoxes" (Grob et al. 2013, 300). There was no intention to induce mystical-type experiences in the first psilocybin trial, while some patients in the second and third reported having them.

(1) An Ethical Justification for PD Research on Dying Human Subjects

The therapeutic aim of using PDs in palliative care is to assuage "existential suffering" as precipitated by "death anxiety." By 2011, the notion of existential suffering was ascendant among palliative care physicians and nurses, although subject to multiple (i.e., at least fifty-six) definitions, some of which distinguished it from spirituality and some of which did not. Thematic concerns across definitions included, but were not limited to, lack of meaning or purpose, loss of connectedness to and a sense of isolation from others, and loss of identity, hope, and autonomy (Boston et al. 2011). A broad and provisional definition has emerged: Existential suffering is a patient's personal distress about fundamental questions that problematize the human condition, e.g., questions about whether one's life has meaning, purpose, and value, and questions about how to face unavoidable pains, afflictions, losses, and failures, and the experiences of guilt, futility, and isolation that every life involves. Existential suffering is precipitated in palliative care by death anxiety, i.e., the patient's powerful realization of impending annihilation and their sense that they are separating from life and ebbing toward death (Boston et al. 2011, 607). This anxiety can precipitate a rupture of personal identity that disconnects patients from the assumptive values, roles, and projects that formerly defined them, leaving in its wake a pervasive feeling of life's futility. Existential suffering in palliative care thus captures a key element of what Jerome Frank termed "demoralization," i.e., a persistent inability to cope that is conjoined with feelings of helplessness, hopelessness, subjective incompetence, and loss of meaning and purpose in life (Clarke and Kissane 2002, 733–734; Frank 1974).

The formalized interventions for treating existential suffering in palliative care are largely psychotherapeutic. They include Cognitive Behavioural Therapy, Meaning Centered Psychotherapy, Dignity Therapy, and Managing Cancer and Living Meaningfully Therapy, among others (Bates 2016; Breitbart and Poppito 2014; Chochinov 2012; Nissim et al. 2012). Antidepressants have also been used, as existential anxiety has been clinically associated with depression (Bates 2016). There is considerable debate in the expert clinical community about the best way to treat existential suffering. It is acknowledged as one of the most debilitating conditions that occur in dying patients "and yet the way we attend to this suffering in their last days is not well understood" (Boston et al. 2011, 604). A recent study concludes that "More high-quality studies with tailored outcome measures are required to fully evaluate the most effective interventions for death anxiety in patients with advanced cancer" (Grossman et al. 2018). It thus remains an open question as to whether PDT is better than, worse than, or as good as a manualized short-term psychotherapy in palliative care.

This debate indicates a state of clinical equipoise, which is confirmed when a state of honest, professional disagreement exists, or may soon exist, among members of the relevant expert community as to the publicly available evidence on the relative merits of non-validated treatment A and the standard, validated treatment B. It only takes evidence adduced by a respectable minority of expert clinicians to bring a non-validated treatment within the standard of care required by clinical equipoise. A "significant" or "respectable" minority of expert clinicians who favor the non-validated treatment is sufficient to establish this disagreement. If A and B are in clinical equipoise, then trial subjects are not randomized to an intervention that is known to be inferior. When non-validated treatment A and validated treatment B are in clinical equipoise, then the risks and benefits to research subjects of taking A are required to be approximately equal to those accepted by patients outside the trial who take B for the condition under study. The therapeutic risks of A would thus be acceptable (Miller and Weijer 2003).

There are certainly enough members of the expert PD clinical community to constitute a respectable minority, and there is more than enough pretrial evidence to support a null hypothesis about the relative risks and benefits of PDT for patients.

While the lasting benefits, if any, of the PDE are still open to question, there is consensus in the expert clinical community that PDs are "very physiologically safe" and can be used under medical supervision without serious adverse effects (Nichols 2016, 273). Transient, moderate increases in systolic and diastolic blood pressure, headaches, nausea, or vomiting have been reported (Dos Santos et al 2018, 10). PDs do not lead to addiction or dependence and no deaths are known to have occurred "after ingestion of typical doses of LSD, psilocybin, or mescaline" (Nichols 2016, 275). The main safety concerns about psychedelics are psychologic, e.g., strong dysphoria and anxiety. In the case of psilocybin, these concerns can be managed through interpersonal support without pharmacological intervention. Transient dysphoria does not appear to diminish the positive evaluation of the PD experience by patients. The risks of exacerbating psychotic disorders can be managed by screening out vulnerable patients with pre-existing disorders or apparent vulnerabilities to same. The risk of instigating novel and prolonged psychotic episodes in screened and supervised research subjects is thought to be low (Johnson et al. 2019, 85). Guidelines for the management of acute adverse reactions have been established for high dose PD trials that maximize patient safety (Johnson et al. 2008). With these procedures in place, the administration of psychedelics carries acceptable, if not minimal risk (Nichols 2016, 275–276). In sum, the administration of classic PDS in clinical research shows "a good safety and tolerability profile, with few transient and moderate adverse reactions." The induction of prolonged psychotic symptoms was not observed in any of the controlled studies described in this chapter. This can be attributed to the cautious and rigorous screening process in these studies (Dos Santos et al. 2018, 11). We can thus argue for the use of PDs by patients because their condition is dire and the alternative treatments are not known to be any better. Indeed, a standard effective treatment for existential suffering in palliative care has yet to determined.

The ethical justification for clinical PD research is not limited to clinical equipoise. We can argue on beneficent grounds for the compassionate use of PDs by patients who have "exhausted all proven treatment options." For this cohort, their condition is dire, the risks are low and manageable, and the alternatives to not receiving PDs can be construed as "expectedly worse." Even highly risky interventions could be "expectedly beneficial" for them (Greif and Surkala 2020, 490). I would construe one worse alternative as the demoralizing, potentially "devastating" impact of untreated existential suffering (cf. Greif and Surkala 2020, 490).[8] PDs can be construed as worth the gamble for such patients because "the potential risks seem to be low and the potential gains high" (Greif and Surkala 2020, 490). We can look at this through a wider lens. The worldwide death rate is expected to reach 70 million deaths per year by 2030, with some researchers predicting a 40% increase in the number of patients seeking palliative care (Dyck 2019, 104). We might anticipate that the need for enhanced means of treating end-of-life existential suffering will grow ever more acute. The long-term clinical potential of PDs is presently unclear, but the current research cautiously indicates a therapeutic promise worth further study.

There is still an ethical objection that can be made to the PDE, and thus to any therapy that would induce it. One might claim that using PDs to induce a mystical-type experience might change one's core values and beliefs, and seriously disrupt the continuity and genuineness of one's identity. This disruption can be seen as a loss of self that is too great to be offset by benefits. We can think of palliative care patients who want to continue feeling connected to the lives their disease has disrupted. The concern, simply put, is that PDs pose a threat to the continuity of one's self and the stable coherence of one's identity. Similar concerns were raised in the early 2000s about the risks of deep brain stimulation, and similar responses can be made with reference to PDs.

Radical changes to one's fundamental values and beliefs, and ways of living are not, by definition, incompatible with one's capacity as an autonomous decision-maker (Pugh 2020, 1675). Radical changes can be integrated successfully into a patient's autobiography, as addiction recovery narratives well illustrate. Patients can be quite capable articulating such changes with an intelligible account of a motivated, developing life story. As Baylis contends, this articulation can be plausibly expressed through informed consent to treatment that might bring about such change (Baylis 2013, 522). We should also be clear on what would constitute evidence that changes to one's personality or sense of identity are so radically different as to be patho-logical. We can imagine someone becoming permanently psychotic, or hypersexual. Fortunately, the available empirical evidence does not support the occurrence of such changes after ingestion of PDs. But the current state of that evidence does not obviate moral concern about the harm a fundamental change to one's person might inflict. Let's assume that PDs can invoke deep changes to one's values and metaphysical beliefs. These might be construed as a harmful disruption of the narrative continuity of one's self. We can imagine someone who was an avowed atheist before the PDE

---

[8] Greif and Surkala include "the virtually certain devastating outcomes stemming from their condi-tion, including long-term disability or death." PDs will not stave off disability or death, but they can apparently ameliorate the spiritual distress that accompanies a life-threatening condition. Other patients eligible for compassionate PDT should include "the profoundly depressed, the severely addicted, … and people wishing to end their lives" (2020, 290).

and a devout Catholic after. But for whom would this change be problematic? If not the post-PDE Catholic, then presumably for their significant others.

Absent mental incapacity, we generally privilege the first-person perspective in gauging the authenticity of one's core values and beliefs. To shift that privilege to third parties ignores the agency of a mentally capable person. If it is true that third parties have "considerable scope" in establishing a person's authentic values, then "the autonomy of a patient's decision is to be determined by its content (in this case, whether the content is compatible with third-party views on what is valuable)," rather than the decisional process that is based on what the patient values (cf. Pugh 2020, 1676). The possibility that persons might disagree with third parties about what is good captures much of what is morally significant about autonomy. PDs are compatible with the "Millian thought" that there is value in performing experiments in living, and in conducting "one's own mode of existence" (Pugh 2020, 1676). An ethical requirement for narrative coherence over a lifetime can amount to a paternalistic straitjacket that confines one to substantive notions of authentic identity imposed by others.

There might also be concern that PDs could hijack one's critical faculties or disenable one's capacity to reflectively step back and assess the PDE, i.e., that you cannot but think from the new perspective the PDE has induced (cf. Baylis 2013, 523). That concern would apply to any dramatic life-changing events, from the birth of a child to the death of a loved one, and thus be "trivially true" (cf. Baylis 2013, 523, 525). There is nothing to suggest that the PDE negates or lessens one's ability to reflect critically on one's values and beliefs. The three trials challenge us to think about patients who are looking to connect with a greater sense of meaning when the core values and beliefs they have developed have been threatened or diminished by illness. Indeed, most of those patients indicate that the PDE afforded them wider contexts of meaning with which to evaluate their lives. This theme runs through the narratives of many patients from the earlier decades of PD research. Far from threatening one's identity, PDs might expand it by enhancing the capacity for self-reflection and afford a deeper connection with values that patients already have, or a new appreciation of values they might have overlooked.

(2) **PDs and Research Trial Design**

Two of the three psilocybin studies noted above have used inactive placebo controls, e.g., niacin. It is questionable whether the placebo-controlled, double blind trial design is apposite for psychedelics. I argue that it isn't for at least two reasons. First, the PDE is not a pure drug effect from which psychological contaminants have been purged. It is a synergistic interaction of the drug, the patient's mindset, and the environmental setting in which the drug is ingested. As such, the PDE is driven significantly by what the patient expects, as well as an environment that supports those expectations. Pre-trial briefing and informed consent procedures, not to mention the popular cultural lore around PDs, will unavoidably contribute to the PDE. Indeed, the careful preparation of patients is intended to facilitate the PDE's therapeutic benefits. Second, patients in the drug arm of a high dose PD trial will inevitably break the blind, as will many patients in the control arm since it is difficult to concoct a reliable placebo control that mimics a PDE without actually being one.

The double-blind random control trial design has a contentious history in PD research. It was becoming the institutionalized, methodological norm at the same time as LSD and psilocybin were emerging as therapeutic agents in psychiatry (Hendy 2018, 154). As early pioneers of PD research, Humphrey Osmond and Abram Hoffer objected to its imposition in the mid-50s. They published the results of qualitative, unblinded trial designs on the efficacy of LSD as a treatment for alcoholism. From their perspective, the placebo-controlled, blinded trial design would not capture the deeper, spiritual meanings of the patients' subjective experience. It was a methodology designed to enhance objective, empirical, and statistical measures and reduce subjective interference. By contrast, Osmond and Hoffer's PD research emphasized an appreciation of subjective perception as a means of guiding the therapeutic process. They saw the growing reliance on double blinds as generating impersonal results that were predetermined by pharmaceutical companies and necessary for any drug's regulatory approval. Osmond and Hoffer were much more interested in clinical effectiveness than pharmacologic efficacy (Dyck 2012, 29–30, 48–50). Their study designs were basically open label, uncontrolled studies which depended greatly on the patients' self-reports of the meaning that the LSD experience afforded them. They saw the double blind as methodologically redundant because "the profundity of responses was so significant that neither the observer nor the subject had any doubts about whether the placebo or LSD had been administered" (Dyck 2012, 50–51).

PD research in the 60s was thwarted partly by its association with recreational drug use in the counterculture. It was also thwarted by its "inability to fit into the new institutional norms for pharmaceutical research." Those norms were legally mandated by the US Food and Drug Administration in 1963 after the thalidomide scandal (Hendy 2018, 155–156). From the institutionalized, regulatory science perspective, the double-blind was crucial in establishing a drug's efficacy. The placebo response is a "universal confound" that a controlled research design should screen out. We can supposedly subtract it from the response level in the active drug arm to estimate a pure, psychologically uncontaminated drug effect. If the blind is broken, then the results of the efficacy trial are compromised since we will not know whether, and if so the extent to which, the patient's response is fueled by the drug as opposed to the confounding psychological factors which drive the placebo response.

This manner of thinking is arguably irrelevant to PD research if the notion of a purely physiologic drug effect is not apposite to the PDE. That notion posits the body as distinct from the mind, whereas PD therapists posit the mind and body as synergistically conjoined, with the PDE driven significantly by the mind-manifesting power of the PD and other artifactual factors. Given their ability to enhance suggestibility and expectancy effects, PDs can be understood as mobilizing the placebo response through the parallel phenomena of set and setting (cf. Hendy 2018, 160, 156). The response to PDs is properly understood as extra-pharmacologic, i.e., as shaped significantly by the set and setting in which PDs are ingested. Set refers to the psychological mindset of the patient, e.g., personality, preparation, expectation, and intention. This mindset is shaped by the pre-trial briefing and informed consent process with an aim to making the patient receptive to the PDE. Setting is the physical environment in which the experience is cultivated. It includes the physical, social, and cultural surroundings in which the PDE occurs. Patients undergo their PDE in a physical setting designed to support a positive response. Hence the use of music, and aesthetically appealing objects to enhance the environment and "charge it with positive cues." "Thorough preparation" aims to enhance expectation and intention. Some of the variables of set and setting are also understood as contributing factors to the placebo response, e.g., expectation, social interaction [with therapists], and cultural milieu (Hartogsohn 2016, 1260–1261). In a standard drug efficacy trial, set and setting would be design artifacts that bias the expectancies of research subjects toward a type of psychological response that obscures a pure drug effect. In a PD trial in service to healing, set and setting are characteristic factors of the trial design that aim to facilitate, if not induce, the PDE. A biologist's artifact is, to a clinician who studies psychedelic therapy, a mechanism of action (cf. Freedman 1990, 5).[9] From this perspective, a pure drug experience that is somehow distinguished from psychologic and artifactual factors misconstrues what the PDE is all about. Artifactual factors that a standard efficacy trial is designed to screen out are precisely the extra-pharmacologic factors that a PD trial in service to healing aims to screen in. PD research is thus congruent with the notion of a placebo response as a medically interesting phenomenon that can have beneficial effects (cf. Hendy 2018, 160).

As set and setting operate synergistically with the PD, the PDE is best understood as a total drug effect (hereinafter TDE). According to Claridge (1970), the pharmacology of a specific drug does not completely explain its full effect. Four extra-pharmacological factors contribute as well. These are: (1) the attributes of the drug ingested, e.g., the shape, color, taste, smell, and other physical properties of the drug. This factor is illustrated by placebo effect studies which show that patients report a greater response to placebo pills after taking two rather than one, and that patients receiving blue pills report sedative effects while patients receiving red pills report stimulating effects. (2) The attributes of the person (patient) ingesting the drug, e.g., their mood, mindset, expectations, and any personal preparations they have made to ready themselves for a therapeutic psychedelic session. (3) The attributes of the person (physician/researcher) providing the drug, e.g., their expectations of the patient and their attitude toward the experience the drug should provide. (4) The physical environment in which the drug is ingested, e.g., a comfortable, safe, and aesthetically appealing setting. Helman has subsequently added a fifth factor, which he terms the "macro-context," e.g., "the wider social, cultural and economic milieu in which prescribing, and ingestion, take place" (2001, 5). Healing expectancy may thus be influenced by the psychospiritual narratives that have made the "psychedelic" or "mind-expanding" experience part of our popular cultural lore. These six factors can be seen as contributing to the TDE of a particular substance (Feeney 2017). The TDE is thus more than a physiologic reaction to the isolated ingredients of a chemical compound. It is an emergent, interactive response that is greater than the

---

[9] Freedman sees "no reason to control for a mechanism artifact that is only partly (and quite possibly synergistically) responsible for benefit" (1990, 5).

sum of psychologic and chemical constituents. There is no point in designing a blind that cannot be maintained. Absent an effective active placebo, we will not accurately capture the TDE in the drug arm of a PD trial unless patients know what they are taking.

If understood as a TDE, the response to PDs is driven by some of the same extra-pharmacologic factors that drive the placebo response. It is not, however, felicitous to construe PDs as placebo treatments for existential anxiety, if "placebo" is understood as a confound to be eliminated from a trial. This would confuse the differences between the use of placebos in regulatory science and the use of set and setting in PDT. Placebo research has traditionally measured non-drug factors that influence response to pharmaceuticals and other therapeutic interventions. Set and setting have traditionally been specific, characteristic factors of PDT which play an important role in shaping the PDE. Where placebo research usually measures effects on health indices, e.g., symptom reduction, set and setting contribute to a broader range of multidimensional phenomena, e.g., psychological insights, expressions of creativity, and peak spiritual experiences (Hartogsohn 2016, 1262).

Regulatory authorities and researchers alike might better promote the science of healing by employing a trial design that accepts set and setting as characteristic factors of a beneficial TDE. A suitable trial design for PDT in palliative care would acknowledge the characteristic factors of drug and psychotherapy. An apposite design would acknowledge the intrinsic coupling of PDs with psychotherapy. A basic two-arm trial design would compare PDs with pre- and post-trial psychotherapy to psychotherapy alone. One such design has been proposed by Rick Doblin, the Executive Director of the Multidisciplinary Association for Psychedelic Studies, a leading US non-profit organization that funds PD research. This design would not exclude the use of placebos for regulatory compliance, and would conform to trial designs that have previously received regulatory approval. Psychedelic psychotherapy would be compared with the identical psychotherapy without a PD, either with or without an active or inactive placebo. In this two-arm trial design, PDT would need to be "statistically more effective than the identical psychotherapy without the use of a psychedelic drug in order for a psychedelic to be approved as a prescription medicine" (Doblin 2001, 255).

This basic trial design was used in the NIMH Collaborative Study of Depression (Elkin et al. 1989), which compared responses across four treatment groups: "interpersonal psychotherapy, cognitive behavior therapy, pharmacotherapy (imipramine) with clinical management, and placebo with clinical management." Clinical management involved providing guidelines to manage medications and side effects, review the patient's clinical status, and to provide patients with supportive encouragement and direct advice if necessary. Although specific psychotherapeutic interventions were prohibited to avoid an overlap with the two psychotherapies, the clinical management component "approximated a 'minimally supportive therapy' condition" (Doblin 2001, 247–248).

No attempt was made to create a double-blind between the two psychotherapy treatment arms and the pharmacotherapy arms. Even so, "the groups/treatments were successfully and repeatedly compared with each other in a succession of studies.

This indicates that the lack of a double-blind between the psychotherapy-only group and the psychedelic psychotherapy group is not necessarily a fundamental design flaw when attempting to compare the efficacy of different treatments" (Doblin 2001, 253).[10]

A variant of this basic design is used currently in the MAPS-sponsored Phase 3 Study of MDMA-assisted psychotherapy for severe posttraumatic stress disorder. Patients will be randomized to receive three sessions of either MDMA or placebo along with psychotherapy over a 12-week treatment period. They will also receive three sessions each of non-drug preparatory and integration sessions. Therapy team members are trained in a manualized "MDMA-assisted psychotherapy." The placebos are "comprised entirely of lactose" (Multidisciplinary Association for Psychedelic Studies 2020). No patient is thus exposed to a form of treatment that is known to be inferior, or denied treatment by receiving a placebo.

Debate continues as to what might constitute an active PD placebo. Some propose that sub-threshold doses of the PD under study could function effectively as active placebos that might maintain a blind. Recent research on LSD indicates that the minimal dose at which subjective and performance effects were notable is 5 mcg (Hutten et al. 2020). This option merits further study, as typical sub-threshold doses are not well established. A dose that is too low may not maintain a blind in the control arm any better than a sugar pill, but a dose with minimal mind-manifesting effects might enhance the placebo response. This would set a higher benchmark for the drug arm to exceed, but it might also be too close to the high dose PD response to provide an effective contrast. It could reduce "between group effect sizes," since even very low doses of PDs "could have significant anti-depressant effects" (O'Donnell et al. 2019).

Broken blinds aside, there are pragmatic policy reasons for including inactive placebos in PD research. It is arguably crucial to be seen to be complying with regulatory requirements for licensed clinical use. To that end, contemporary PD researchers are not protesting the placebo control regime they way Osmond and Hoffer did in the 50s. Instead, they often concede the regulatory expediency of compliance. This move might come back to haunt them, as there is widespread acknowledgement that blinding is problematic.[11] If this trend continues, PD researchers could be required to explain why breaking the blind is *not* problematic. If regulatory authorities think that

---

[10] Doblin endorsed further measures for adding placebos to a trial design that guarantees psychotherapy for both groups (Doblin 2001).

[11] An ongoing multi-site Phase 2 study of the effects of psilocybin on major depression uses 100 mgs of niacin as the placebo control. Set and setting will be attended to by means of a pre-dose preparation session and an "aesthetically pleasing room." There will also be "three post-dose integration sessions during which participants are encouraged to discuss their intervention experience." Whether these sessions amount to psychotherapy is unclear. Regardless, skin flushing, dizziness, rapid heartbeat, and nausea are not likely to be confused with a full-blown PDE. Even if they are, those who do experience a PDE are even less likely to mistake it for the side effects of vitamin B. It remains to be seen whether researchers will check to see if the blind was maintained during the trial, and whether regulatory authorities will disavow the results if it was not. See *A Study of Psilocybin for Major Depressive Disorder (MDD).* ClinicalTrials.gov Identifier: NCT03866174. Available at: https://clinicaltrials.gov/ct2/show/NCT03866174. Accessed Nov. 20, 2020.

it is, then PD studies could be dismissed again as having failed to demonstrate a pure drug effect. Researchers could respond that breaking the blind is not a problem unique to PD trials, and cite studies showing that in trials of established antidepressants, "a significant majority" of patients and clinicians could accurately guess whether they were in the treatment or placebo group (O'Donnell et al. 2019). It remains to be seen whether regulatory authorities will use the broken blind as a reason to disavow PD research. History may not repeat itself, as that move could put regulators on the hook for explaining why the concern for unblinding would not apply to trials of other psychiatric drugs. The use of placebos in PD trials might reflect a prudent policy direction that better enables regulatory authorities to approve PD trials. Contemporary PD researchers recognize "political compromise" and "partnership" with the "drug policy bureaucracy" as being "crucial to bureaucratic change." Strategies for the regulatory approval of PDs should be aligned with "the current momentum of government," i.e., PD research aimed at treating drug addiction or alcoholism would stand a better chance of approval than studies targeting the benefits of recreational use (Goldsmith 2015, 341–342). Access to PDT may have to carefully limited after approval. Some proponents go the extra mile and argue for the immediate implementation of Stage 4 clinical effectiveness trials should PDs be approved for therapeutic use (Doblin 2015, 372).

13.4   The Epistemic Warrant of the PD Experience

The therapeutic use of PDs raises philosophic questions about the nature of the PDE, especially its allegedly noetic and "mystical" dimensions. While these "spiritual experiences" have been known to convince persons of their veridicality, the PDE might be nothing more than a manufactured response which set and setting are designed to invoke, and one that provides comforting illusions at the expense of veridical insights into the nature of the universe. Even so, PDs can be used to induce an experience that some patients have found to be deeply meaningful in a way that lessens existential suffering. That is a response that patients might want to settle for if administered openly and with careful preparation. If these spiritual responses can be reliably induced, then using PDs to cultivate an emollient subjective experience is arguably compatible with respecting patient autonomy and the goals of palliative care psychiatry.

It is easy to see why PDs invite the veridicality question. They instigate a wide panoply of experiences that have included communication with alien beings and mental telepathy. Some of the early researchers saw PDs as the gateway to paranormal science (Luke 2008). If truth counts for knowledge, then just how "real" and true is the PDE? If "real" is understood from the perspective of one conception of philosophical naturalism, then existence is limited to the natural world and excludes anything supernatural. As anti-supernaturalism, naturalism turns to the hard sciences, e.g., physics, chemistry, neurology, not so much to define what is real, but to arbitrate what is not real. The meanings of nature are thus limited to what these sciences claim to know. It is not uncommon to equate naturalism with materialism. By this account, naturalism reflects a scientistic ideology (Alexander 2013, 104). Mystical experiences of any kind would be considered illusory. This can be seen as ethically troubling: A promising treatment modality in mental health care "seems to work, at least in part, by inducing highly compelling metaphysical illusions" (Letheby 2016, 30–31). An ethics of self-searching honesty would presumably impel efforts to avoid being seduced by the fool's gold of comforting illusions. Even a well-known and cautious proponent of PD research asks whether it simply foists "a comforting delusion on the sick and dying" (Pollan 2015).

Naturalism may not be constrained to this position, as it lacks a "very precise" philosophical meaning (Papineau 2020). On the formulation above, one could claim that spiritual experiences, cosmic or otherwise, are illusory and illegitimate. That claim can also be seen as ethically troubling. To claim that the PDE is illusory is to deride as superstition a rich culture of indigenous spiritual practices that have been handed down through millennia. To be fair, naturalism is not constrained to this position, as it has also accommodated "possibilities for deep, qualitatively expressive meaning achieved through the interaction of human beings with each other and with nature" (Alexander 2013, 248). Naturalists like Santayana, Emmerson, and Dewey invoked a "human eros," or a spiritual aesthetics of living with nature (Alexander 2013, 247–249). The noetic insights of a PDE could be inaccurate and unreliable, but a naturalist might still accommodate them as meaningful responses "which help us to articulate our implicit assumptions when we are threatened with paradox and have difficulty finding a solution" (cf. Papineau 2020). Indeed, Letheby (2021) rigorously defends a naturalistic account of the PDE by which the epistemic profile of PDT involves less risks and more benefit than one might think.

Suppose we concede that the PDE has epistemic faults, i.e., it is illusory. It can apparently confer psychological benefits in terms of a reduction is existential suffering. It may also confer epistemic benefits. A psilocybin-induced PDE can lead to durable increases in the personality trait of openness to experience (Griffiths 2006), which likely means more engagement with the world, increased curiosity, and more new experiences, all of which are potential sources of knowledge (Letheby 2016, 31, 33–34, 35). If the epistemic and psychological benefits can be seen to outweigh the epistemic faults, it is arguably redundant to deride someone for believing that another mode of reality pertains after physical death, that love is an animating force of the universe, or that consciousness survives bodily death in the form of a transpersonal, unitive energy.

More to the point, how wrong is it to reconfigure a sense of meaning and purpose through what *might* be a comforting metaphysical illusion? We do not know that existence is limited to the natural world. We cannot demonstrate that non-naturalistic beliefs like "Reality is immaterial, [and] God exists" are false, as they are "not likely to be contradicted by empirical evidence or refuted by the best natural science" (Greif and Surkala 2020, 492). We allow persons considerable freedom to formulate their spiritual beliefs, and PDT need not function as a metaphysical corrective to beliefs that are incompatible with naturalism.

The spiritual experience can thus be seen as an understanding one's life in a wider, cosmological view of one's own. Suppose a patient comes to believe that the universe is a benevolent and welcoming place, especially after death. Suppose further that the universe is actually indifferent, i.e., the patient's belief in love as the animating principle of the universe is just flat-out wrong. We might be better off believing that it is, especially in our violet hours. The benefit of dying in peace can outweigh the alleged harm of believing that the universe is a kinder, gentler place than it is. PDs might enable troubled patients to die in a state of epistemic innocence about how things actually are, but if innocence is bliss, it can also be a therapeutic benefit that patients would be only too happy to have. If this benefit is denied on the basis of its questionable veridicality, then we need reasons for why our preferred criteria of truth should prevail (cf. Flanagan and Graham 2017). It is questionable whether there is anything inherently wrong in using a chemical to find more meaning in one's life, and a greater sense of connection to significant others or the universe. The mystical facets of the PDE are compatible with acceptable psychological transformations that persons have achieved by more time-consuming means, e.g., meditation. There is a limited opportunity for experienced benefits in palliative care. In a psychedelic science devoted to healing, a philosopher's "illusion of profoundness" is a clinician's successful treatment, and a patient's experience of healing (cf. Hartogsohn 2018, 4). It would be prima facie consistent with the goals of palliative care if existential suffering were relieved, and the patient felt better about death and dying.

The PDE can be credibly understood as a spiritual experience, or one that enables a person to become aware of a transcendent meaning and purpose that need not be connected to organized religious ritual. Spirituality in the context of palliative care has been defined as the "aspect of humanity that refers to the way individuals seek and express meaning and purpose and the way they experience their connectedness to the moment, to self, to others, to nature, and to the significance of the sacred" (Puchalski et al. 2009, 642). It has also been defined as finding a self-transcendent meaning and purpose in life and death (Vachon et al. 2009, 55). Enhanced meaning in life is thought to enable persons to move beyond their present with a greater sense of connection to an order of reality larger than themselves (Hill and Hood 1999). Transcendence includes an experience of an enduring self that is connected to something sacred and thus more than a dying body. Bossis points to a "growing body of clinical research" which shows that enhanced existential/spiritual well-being may buffer against hopelessness, loss of meaning, anxiety, and depression. Meaning and transcendence are seen as crucial factors in the development of spiritual well-being and improved quality of life for patients (Bossis 2014, 276).

The mystical experience does not exhaust the PDE if it occurs at all. Spirituality can have decidedly secular existential dimensions. Even naturalists can experience an enhanced sense of meaning and purpose in life, and a greater sense of connectedness to others (Letheby 2021). They might even revere nature and see intrinsic value in conserving it. As adjuncts to traditional forms of psychotherapy, PDs can facilitate changes in attitude toward personal problems, "and the psychodynamic experience of becoming aware of an unconscious conflict (Greif and Surlaka 2020, 492). The veridicality question kicks up more dust than naturalist critics and spiritual believers are able, or even need to settle. A more pressing concern is whether the PDE leads to lasting benefits. From a healing perspective, an experience alone does not guarantee a beneficial transformation; the PDE has to cash out in new and hopefully lasting patterns of thought, attitude, and behavior. These would be benefits that are best assessed in PDT by longer-term Stage 4 clinical effectiveness studies.

13.5 An Apposite Model for Psychedelic Psychiatry

Different theoretical frameworks can be proposed to accommodate the use of PDs in palliative care psychiatry. A spiritually supplemented biopsychosocial model has been proposed (Bossis 2014, 252; Sloshower 2018, 117). "Narrative psychiatry" could harbor their use, as the PDE has to be integrated via dialogue and articulation in the post-drug therapy sessions. Narrative psychiatry is an emerging therapeutic praxis that stresses the development of positive mental health. The concerns of narrative psychiatry in palliative care have a pronounced but not exclusive focus on a patient's reckoning with existential/spiritual suffering. Attention to patient narrative is central to clinical praxis in psychiatry, e.g., history-taking and diagnosis, engagement in a therapeutic alliance, and various facets of psychotherapy.

Some argue for the integration of narrative and evidence-based medicine in psychiatry and see no conflict between the two (Holmes 2000, 93, 96). Bracken and Thomas et al. make a less conciliatory argument for a conflict between a hegemonic, technological/biomedical model of psychiatry and "a more nuanced form of medical understanding and practice" that is especially sensitive to the "complex interplay" of biological, psychological, social, and cultural forces that underlie mental health problems. They cite growing evidence that a primarily biomedical approach to diagnosis and treatment is inadequate and that a "radical shift" is required for psychiatry to regain its therapeutic footing. To improve therapeutic outcomes for patients, psychiatry must focus primarily on contexts, relationships, and the promotion of dignity, respect, and engagement in a meaningful life. This shift would not eschew the "tools of empirical science" or reject medical techniques, but it would re-position the ethical, and hermeneutic facets of psychiatry as primary (Bracken et al. 2012, 432). This is precisely the kind of shift that narrative psychiatrists like Bradley Lewis endorse.

Lewis views psychiatry as "part medicine, part psychotherapy," and narrative psychiatry as part narrative medicine and part narrative psychotherapy integration. Like narrative medicine, narrative psychiatry is patient-centered; its first objective is for physicians and patients to achieve a shared understanding of how the latter experience their situation. Like narrative approaches to psychotherapy integration, narrative psychiatry employs a "metatheoretical frame" that accommodates a wide range of psychotherapeutic praxes. Narrative psychotherapy integration regards an informed patient's choice of therapy as an autonomous exercise in "self-fashioning" in line with their chosen goals and values. All therapies, regardless of their specific techniques and theoretical orientation, are but different ways of supplying therapeutic narratives by which patients acquire new perspectives on their situation and "new tools for coping with their problems"(Lewis 2011, 52–53). Different psychotherapies enable different ways of telling stories about psychic life and suffering. The aim is to fit a psychotherapy to the values, preferences, and theoretical sympathies of the patient instead of trying to fit a patient to a psychotherapy. Lewis distinguishes narrative psychiatry from narrative medicine by its "prominent consideration of psychiatric medications" (Lewis 2011, 67). Patients may choose to understand their illness stories through the lens of a "neurochemical self" narrative and use drugs to organize their "sense of self." Anti-depressants and psychotics are among the array of "possible tools that can be woven together into an eclectic narrative of treatment" (Lewis 2011, 71, 73). So, I would argue, are PDs.

The narrative psychiatrist's therapeutic concerns about existential suffering and the clinical relevance of illness stories are intensified in work with dying patients. Narratives have been crucial to palliative care from its contemporary inception. The core concepts of hospice care were originally formulated by Dame Cicely Saunders on the basis of over one thousand patient narratives that she collected herself. Saunders defined the "total pain" of hospice patients as encompassing all facets of their illness, e.g., physical pain, social problems, spiritual needs, and mental distress, the latter being "perhaps the most intractable pain of all" (Saunders 1963, 197). "Spiritual pain" evinces "bitter anger at the unfairness of what is happening (at the end of life) and above all a desolate felling of meaninglessness" (Saunders 1988, 29). The very notion of a "good death" is subject to multiple narratives promoted by interest groups that reflect different values about the sanctity of life (Frank 2009, 162, 168–172). An ascendant narrative approach relates a good death to giving dying patients an opportunity to review and recount their life stories in ways that resolve existential/spiritual concerns about personal fulfillment, and enable an acceptance of impending death (Das Gupta et al. 2009, 38). Palliative care associates spirituality with the human quest for purpose and meaning. Since this quest can become exigent toward the end of life, "good palliative care" is claimed by some to embrace "good spiritual care" by definition (Stanworth 2009, 210–2011).

The use of drugs in palliative care psychiatry has usually involved psychotropic medications for comorbid mental disorders like depression and dementia. PDs could address its existential/spiritual concerns. There is an established and growing literature that uses psycho-spiritual narratives to articulate the PDE (Dickens 2014, 365–380). Indeed, the protocol designs of much contemporary PDT research encourage spiritual narrative through qualitative, first-person accounts, questionnaires, and rating scales. Evidence thus far indicates that the PDE helps some patients to come to grips with what the end of their life means in a transpersonal scheme of things, and to accept their impending death.

Narrative psychiatry in palliative care might offer pharmacology a unique therapeutic niche. Given its explicit concerns with spirituality and positive psychological growth, it is arguably amenable to a treatment approach whereby a deeply meaningful spiritual experience is induced with PDs. While a narrative approach does not eschew pharmacologic treatments, their use need not be seen as limited to the control of pathologic symptoms, e.g., dulling anxiety about impending death. Some drugs might reliably induce an experience by which one gains a reconfigured understanding of death that facilitates its acceptance. This experience is plausibly described as spiritual. Palliative care is but one of the medical areas in which meaning affects perceptions of illness and healing, and PDs are one way to induce transformative meaning. The therapeutic praxis of narrative psychiatry in palliative care turns significantly on inducing meaning, with drugs being a ready option.

The use of PDs would also require a radical shift in the prevailing model of drug action. PDT is not easily accommodated in the regnant disease-centred model. This model claims that the therapeutic effects of drugs result from their corrective effects on the mechanisms of a particular, underlying disease state. Those drug effects "move the human organism from an abnormal physiological state towards more normal one" (Moncrieff 2008, 14, 18). PDT is better understood in terms of Moncrieff's drug-centred model of action. This model claims that the therapeutic value of a drug is determined by the particular quality of the altered state it produces. That experiential state is mediated by the patient's biological response to the drug, as well as the context in which the drug is taken, e.g., the social circumstances and emotional state of the patient. The drug-centred model claims that we can understand the effects of therapeutic drugs the same way we understand the effects of recreational drugs. The effects of both are sought after by patients, valued as desirable in themselves, and also as ways to help them deal with "current difficulties." These effects include "euphoria, stimulation, indifference, disinhibition," sedation and "psychedelic experiences" (Moncrieff 2008, 14–15, 17).

The biological model of psychiatry that promotes mental disorders as neurochemical imbalances and psychotropics as magic bullets that can equilibrate them has reached a stalemate, if not a dead end. There is now a much wider, and more sophisticated awareness of the limitations of the randomized, placebo-controlled trial design. There are reduced prospects for "new and potentially lucrative psychiatric drugs," with some major pharmaceutical companies actually ceasing research on psychotropics. Moreover, the psychiatric profession is now taking stock, and re-evaluating "the conceptual, theoretical and methodological limitations of the biochemical, brain imaging, and genetic work of previous decades" (Harrington 2019, 271, 275). Psychedelics will not likely revolutionize psychiatry, but they might be used in service to a discipline that "seeks to understand ways that human brain functioning, disordered or not, is sensitive to culture and context," as the recent acknowledgement of the therapeutic potential of the placebo response has shown (Harrington 2019, 275–276).

13.6 Conclusions: We Have All Been Here Before

The renewed zeal for PDT invokes a sense of déjà vu, but it is unclear whether history will repeat itself. Despite intriguing promise, the studies conducted during the initial era of psychiatric interest (1951–1976) did not survive the advent of the double-blind, placebo-controlled trial as a de facto requirement for funded research. Debate continues as to whether that trial design can accommodate the "non-specific" facets of PDT. The respectability of PD research was diminished further by its association with the hedonistic, and occasionally harmful drug use in the mid-60s countercul-ture. All of which is to say, the current enthusiasm is not an exercise in nostalgia. Contemporary investigators are aiming to surpass the debatable results of the earlier PD research. Doblin published a long-term follow up and methodological critique (1991) of Walter Pahnke's legendary Good Friday Experiment. The latter was one of the first attempts to assess the mystical-type dimensions of the PDE with a blinded trial design (Pahnke 1963). More recently, Roland Griffiths and his team (2006) made no secret of their attempt to update Pahnke's study with a rigorously designed, blinded trial of their own. The present-day research on psilocybin and end-of-life suffering is picking up where the equivocal results of the uncontrolled Spring Grove pilot studies left off. Unlike some of the original advocates, contemporary researchers display a greater willingness to work with de facto regulatory requirements.

The limitations of the current psychedelic studies should temper clinical enthu-siasm. Some of this research has been lauded for its rigor, but talk of an actual paradigm shift is premature. That shift would move away from a psychiatry that claims to rectify discrete neurochemical disorders with specific drugs to one that endorses the combination of PDs and psychotherapy as a means of inducing deeply meaningful, cathartic experiences, possibly of a mystical-type, with positive, hope-fully long-term mental health outcomes (Schenberg 2018, 5–6). It is question-able whether contemporary researchers will be able to successfully deliver PDs to approved clinical practice if they decouple the PDE from psychotherapy and mini-mize the problems of rigid conformity to the standard RCT design. That design assesses the efficacy of drugs that will supposedly have specific corrective effects on targeted pathological symptoms. By contrast, PDs are used to facilitate the cultivation of a unique subjective experience that is not exclusively dependent on their molecular composition. The expectations, beliefs, and intentions of the patient are key determi-nants of that experience, and its integration in the patient's life involves some degree of psychotherapy before and after the PDs are ingested. PDs are not magic bullets, and the variability of their effects means that the PDE is as much devised as induced. Where the standard, placebo-controlled trial design would attempt to screen expec-tations, beliefs, and intentions out, PDT therapy aims to incorporate them. A trial design that can accommodate those determinants might better assess the therapeutic potential of PDs. A trial design that cannot might consign them to another round of therapeutic oblivion (Oram 2018, 205–222). The lack of an effective placebo should not be a barrier to ongoing research, as concerns about blinding are not limited to PDs.

We should also consider the social, cultural, and economic milieu in which any pharmaceutical is prescribed (Helman 2001, 5). Drug fads have come and gone, and PDs are still just drugs. The perceived benefit from classical PDs might wax and wane if they enter approved clinical practice and gain widespread acceptance (Pollan 2018, 382). Indeed, a growing array of new PDs might make the resurgent interest in psilocybin as passé as William James' endorsement of nitrous oxide. PDs may never have the market share of antidepressants, but they could become another means of managing one's neurochemical self, and inform an understanding by which our desires, moods, and discontents are mapped on regions of the brain (Rose 2007, 187–188). Microdosing PDs is thus endorsed by some as an effective means of regulating "the vicissitudes of mood" and having "a really good day" (Waldman 2017, xix, xx). PDs may lose some of their spiritual appeal as chemical paths to the sacred. Indeed, they already inform a naturalistic "neurochemical reshaping of personhood" (Rose 2003, 59). In their study of "changes in brain network properties related to psilocybin-induced ego-dissolution," Lebedev et al. argue that the maintenance of "self" or "ego" is a "perceptual phenomenon" that rests on the "normal functioning" and "inter-hemispheric communication" of the medial temporal lobe circuitry and the "salience network" (Lebedev et al. 2015, 1, 10). Letheby and Gerrans argue that ego dissolution in the PDE shows that the self is nothing more than "a useful Cartesian fiction" concocted by the mind; a "false representation" of a "substance or entity" which has our "properties and experiences" (2017, 1–2, 9). By these accounts, PDs do not reveal an absolute reality within, just a mind that plays tricks on itself. This is a far cry from the theological spirituality of some proponents, who posit PDs as doorways to the divine. We do well to remember that a mystical-type experience is not the only PDE on offer. Ingested in lower doses and combined with insight-oriented psychotherapy, PDs can apparently induce meaningful, and possibly transformative responses in patients who are more interested in dealing with problems in living than merging with God (Grieco and Bloom 1981). Therapy aside, PDs can also inform wider research on brain-related functioning and the "biological bases of consciousness and self-awareness" (Letheby 2021, 2).

Recent developments in Canada auger well for the clinical acceptance of psilocybin in palliative care. In 2021, Health Canada authorized the compassionate use of psilocybe mushrooms for at least twenty-four terminal cancer patients suffering from "end-of-life distress," each of whom was given a one-year exemption from the *Controlled Drugs and Substances Act* (S.C. 1996, c. 19). Exemptions were also granted to nineteen health-care providers who can use and possess the mushrooms for "professional training purposes." These included family physicians, nurses, psychologists, psychiatrists, clinical counselors, and social workers (Canadian Healthcare Network.ca, Jan. 18/2021). As of January 5, 2022, Health Canada has superseded the exemption process by amending its Special Access Program (Health Canada 2021). Physicians are now permitted to request that their eligible patients have access to illegal psychoactive substances, like MDMA and psilocybin, for PDT. The case-by-case decisions are reserved for serious treatment-resistant or life-threatening conditions, in instances where other therapies have failed, or are unsuitable or unavailable in Canada. While there are no reports that these patients are reflecting on the *Tibetan Book of the Dead,* at least one of them has confirmed the crucial integration of psilocybin and psychotherapy in alleviating "unbearable" anxiety about impending death: "It's not like you take a pill and everything is fantastic. It doesn't work like that any more than regular therapy does. There is work to be done. There are challenges to face. There are issues that need to be worked through the same as any other session. The main difference is that with the psychedelic assisted therapy, it can get your ego out of the way so you can get at some things" (Canadian Healthcare Network.ca, Jan. 18/2021). All told, PD research in palliative care should continue given its initial promise. PDs have been used therapeutically for millennia, but only Stage 4 clinical effectiveness studies will tell us whether they are dependable resources for healing. Leading proponents concede this and advocate the inception of such studies if and when a sufficient number of Stage 3 trials are successfully completed and PDs are approved for clinical use.

Acknowledgements Thanks to Drs. David Weisstub, R. Steven Turner, and two anonymous referees for comments on previous drafts.

References

Alexander, Thomas M. 2013. *The human eros: Eco-ontology and the aesthetics of existence.* New York: Fordham University Press.
Bates, Alan T. 2016. Addressing existential suffering. *British Columbia Medical Journal* 58 (5): 268–273.
Baylis, Francoise. 2013. 'I am who I am': On the perceived threats to personal identity from deep brain stimulation. *Neuroethics* 6: 513–526.
Ben Sessa. 2012. *The psychedelic renaissance: Reassessing the role of psychedelic drugs in 21st century psychiatry and society.* London: Muswell Hill Press.
Belser, Alexander B., Gabrielle Agin-Liebes, T. Cody Swift, Sara Terrana, Neşe Devenot, Harris L. Friedman, Jeffrey Guss, Anthony Bossis, & Stephen Ross. Patient experiences of psilocybin-assisted psychotherapy: An integrative phenomenological analysis. *Journal of Humanistic Psychology* 57, no.4 (2017): 354–388.
Bossis, Anthony, P. 2014. Psilocybin and mystical experience: Implications for the alleviation of existential and psycho-spiritual distress at the end of life. In *Seeking the Sacred with Psychoactive Substances: Chemical Paths to Spirituality and to God*, vol. 2: Insights, arguments, and controversies, ed. by J. Harold Ellens, 251–284. Santa Barbara, California: Praeger.
Boston, Patricia, Anne Bruce, and Rita Schreiber. 2011. Existential suffering in the palliative care setting: An integrated literature review. *Journal of Pain Symptom Management* 41 (3): 604–618.
Bracken, Pat, Philip Thomas, Sami Timimi, Eia Asen, Graham Behr, Carl Beuster, Seth Bhunnoo, Ivor Browne, Navjyoat Chhina, Duncan Double, Simon Downer, Chris Evans, Suman Fernando, Malcolm R. Garland, William Hopkins, Rhodri Huws, Bob Johnson, Brian Martindale, Hugh Middleton, Daniel Moldavsky, Joanna Moncrieff, Simon Mullins, Julia Nelki, Matteo Pizzo, James Rodger, Marcellino Smyth, Derek Summerfield, Jeremy Wallace, and David Yeomans. 2012. Psychiatry beyond the current paradigm. *The British Journal of Psychiatry* 201: 430–434.
Breitbart, William S., and Shannon R. Poppito. 2014. *Individual meaning-centered psychotherapy for patients with advanced cancer: A treatment manual.* Oxford and New York: Oxford University Press.
Canadian HealthcareNetwok.ca. *Where Do 'Magic Mushrooms' Belong in Patient Treatment?* 18 Jan 2021. Available at: https://www.canadianhealthcarenetwork.ca/physicians/news/where-do-magic-mushrooms-belong-in-patient-treatment-61485. Accessed 16 Feb 2021.
Chochinov, Harvey Max. 2012. *Dignity therapy: Final words for final days.* Oxford and New York: Oxford University Press.

Clarke, David M., and David W. Kissane. 2002. Demoralization: Its phenomenology and importance. *Australian & New Zealand Journal of Psychiatry* 36 (6): 733–742.

Claridge, Gordon. 1970. *Drugs and human behaviour*. London: Allen Lane.

Das Gupta, Sayantani, Craig Irvine, and Maura Spiegal. 2009. The possibilities of narrative palliative care medicine: 'Giving sorrow words'. In *Narrative and stories in health care: Illness, dying, and bereavement*, ed. Yasmin Gunaratnum, and David Oliviere, 33–46. Oxford: Oxford University Press.

Dickens, Robert. 2014. Variety of religious paths in psychedelic literature. In *Seeking the sacred with psychoactive substances: Chemical paths to spirituality and to god, vol. 1: History and practices*, ed. J. Harold Ellens, 365–380. Santa Barbara, California: Praeger.

Doblin, Rick. 1991. Pahnke's 'good friday experiment': A long-term follow-up and methodological critique. *The Journal of Transpersonal Psychology* 23 (1): 1–28.

Doblin, Rick. 2001. *Regulation of the medical use of psychedelics and marijuana*. PH.D. dissertation in Public Policy from the Kennedy School of Government, Harvard University, June 2001. Available at: https://maps.org/news/332-maps-resources/research-papers/5402-dissertation-rick-doblin,-ph-d. Accessed 13 Feb 2021.

Doblin, Rick. 2015. Regulation of the prescription use of psychedelics. In *The psychedelic policy quagmire: Health, law, freedom, and society*, ed. J. Harold Ellens, and Thomas B. Roberts, 365–386. Santa Barbara, California: Praeger Inc.

Dos Santos, Rafael G., Jose Carlos Bouso, and Miguel Angel Alcazar-Corcoles. 2018. Efficacy, tolerability, and safety of serotonergic psychedelics for the management of mood, anxiety and substance use disorders: A systematic review of systematic reviews. *Expert Review of Clinical Pharmacology* 11, no. 9: 1–14.

Dunaway, David King. 1989. *Huxley in Hollywood*. New York: Anchor Books—The Knopf Doubleday Publishing Group.

Dyck, Erica. 2012. *Psychedelic psychiatry: LSD on the Canadian Prairies*. Winnipeg: University of Manitoba Press.

Dyck, Erica. 2019. Psychedelics and dying care: A historical look at the relationship between psychedelics and palliative care. *Journal of Psychoactive Drugs* 51 (2): 102–107.

Elkin Irene, M. Tracey Shea, Watkins J, Stanley Imber, Stuart Sotsky, Joseph Collins, David Glass, Paul Pilkonis, William Leber, & John Docherty. 1989. National Institute of mental health treatment of depression collaborative research program. General effectiveness of treatments. *Archives of General Psychiatry* 46, no. 11 (1989): 971–982.

Feeney, Kevin. *Beyond set and setting: A new understanding of psychedelics and healing*. Available at: https://chacruna.net/beyond-set-and-setting-psychedelic-healing/. Accessed 25 Nov 2020.

Flanagan, Owen, and George Graham. 2017. Truth and sanity: Positive illusions, spiritual delusions, and metaphysical hallucinations. In *Extraordinary science and psychiatry*, ed. Jeffrey Poland and Serife Tekin, 293–313. Cambridge: The MIT Press.

Frank, Arthur. 2009. The necessity and dangers of illness narratives, especially at the end of life. In *Narrative and stories in health care: Illness, dying, and bereavement*, ed. Yasmin Gunaratnam and David Oliviere, 161–176. Oxford: Oxford University Press.

Frank, Jerome D. 1974. *Persuasion and healing: A comparative study of psychotherapy*. New York: Schocken Books.

Freedman, Benjamin. 1990. Placebo controlled trials and the logic of clinical purpose. *IRB: A Review of Human Subjects Research* 12, no. 6 (1990): 1–5.

Gasser, Peter. Psycholytic therapy with MDMA and LSD in Switzerland. *Newsletter of the Multidisciplinary Association for Psychedelic Studies (MAPS)* 5, no. 3 (1994–95): 3–7.

Gasser, Peter, Dominic Holstein, Yvonne Michel, Rick Doblin, and Bera Yazar-Klosinski. 2014. Safety and efficacy of lysergic acid diethylamide-assisted psychotherapy for anxiety associated with life-threatening diseases. *The Journal of Nervous and Mental Disease* 202 (7): 513–520.

Goldsmith, Neil M. 2015. Changing psychedelic policy. In *The psychedelic policy quagmire: Health, law, freedom, and society*, ed. J. Harold Ellens, and Thomas B. Roberts, 335–352. Santa Barbara, California: Praeger Inc.

Grieco, Alan, and Robert Bloom. 1981. Psychotherapy with hallucinogenic adjuncts from a learning perspective. *International Journal of the Addictions* 16 (5): 801–827.

Greif, Adam, and Martin Surkala. 2020. Compassionate use of psychedelics. *Medicine, Health Care and Philosophy* 23: 485–496.

Griffiths, Roland R., William A. Richards, Una D. McCann, and Robert Jesse. 2006. Psilocybin can occasion mystical-type experiences having substantial and sustained personal meaning and spiritual significance. *Psychopharmacology (Berl)* 187, no. 3 (2006): 268–283.

Griffiths, Roland R., William A. Richards, Matthew W. Johnson, Una D. McCann, and Robert Jesse. 2008. Mystical-type experiences occasioned by psilocybin mediate the attribution of personal meaning and spiritual significance 14 months later. *Journal of Psychopharmacology* 22 (6): 621–632.

Griffiths, Roland R., Matthew W. Johnson, Michael A. Carducci, Annie Umbricht, William A. Richards, Brian D. Richards, Mary P. Cosimano, and Margaret A. Klinedins. 2016. Psilocybin produces substantial and sustained decreases in depression and anxiety in patients with life-threatening cancer: A randomized double-blind trial. *Journal of Psychopharmacology* 30 (12): 1181–1197.

Grob, Charles S., Alicia L. Danforth, Gurpreet S. Copra, Marycie Hagerty, Charles R. McKay, Adam L. Halberstadt, and George R. Greer. 2011. Pilot study of psilocybin treatment in patients with advance-stage cancer. *Archives of General Psychiatry* 68, no. 1 (2011): 71–78.

Grob, Charles S., Anthony P. Bossis, and Roland R. 2013. Griffiths. Use of the classic hallucinogen psilocybin for treatment of existential distress associated with cancer. In *Psychological aspects of cancer: A guide to emotional and psychological consequences of cancer, their causes and their management*, ed. B.I. Carr, and J. Steel, 291–308. New York: Springer US.

Grof, Stanislav. 2001. *LSD psychotherapy: Exploring the frontiers of the hidden mind*, 3rd ed. Sarasota Florida: Multidisciplinary Association for Psychedelic Studies.

Grossman, Christopher H., Joanne Brooker, Natasha Michael, and David Kissane. 2018. Death anxiety interventions in patients with advanced cancer: A systematic review. *Palliative Medicine* 32 (1): 172–184.

Harrington, Anne. 2019. Mind fixers: Psychiatry's troubled search for the biology of mental illness. New York and London: W.W. Norton and Company.

Hartogsohn, Ido. 2016. Set and setting, psychedelics and the placebo response: An extra-pharmacological perspective on psychopharmacology. *Journal of Psychopharmacology* 30 (12): 1259–1267.

Hartogsohn, Ido. 2018. The meaning-enhancing properties of psychedelics and their mediator role in psychedelic therapy, spirituality, and creativity. *Fronters in Neuroscience* 12 (129): 1–5.

Health Canada. *Regulations amending certain regulations relating to restricted drugs* (Special Access Program): SOR/2021-271.

Helman, Cecil G 2001. Placebos, and Nocebos: The cultural construction of belief. In *Understanding the placebo effect in complementary medicine: Theory, practice and research*, ed. D. Peters, 3–16. London: Churchill Livingstone.

Hendy, Katherine. 2018. Placebo problems: Boundary work in the psychedelic science renaissance. In *Plant medicines, healing and psychedelic science: Cultural perspectives*, ed. Beatriz Caiuby Labate, and Clancey Cavnar, 151–166. Switzerland: Springer International Publishing.

Hill, Peter C., and Ralph W. Hood Jr. (eds.). *Measures of religiosity*. Birmingham, AL: Religious Education Press.

Hoffman, Albert. 2019. *LSD: My problem child*. London and New York: Oxford University Press.

Holmes, Jeremy. 2000. Fitting the biopsychosocial jigsaw together. *British Journal of Psychiatry* 177: 93–94.

Hutten, Nadia R.P.W., Natasha L. Mason, Patrick C. Dolder, Eef L. Theunissen, Friederike Holze, Holze Matthias E. Liechti, Amanda Feilding, Johannes G. Ramaekers, and Kim P.C. Kuypers. 2020. Mood and cognition after administration of Low LSD doses in healthy volunteers: A placebo controlled dose-effect finding study. *European Neuropsychopharmacology*. https://doi.org/10.1016/j.euroneuro.2020.10.002 (in press).

Huxley, Aldous. 1954. *The doors of perception and heaven and hell*. London: Chatto & Windus.

Huxley, Laura. 1971. *This timeless moment: A personal view of Aldous Huxley*. New York: Ballantine Books.

Johnson, Matthew W., William A. Richards, and Roland R. Griffiths. 2008. Human hallucinogen research: Guidelines for safety. *Journal of Psychopharmacology* 22 (6): 603–620.

Johnson, Matthew W., Peter S. Hendricks, Frederick S. Barrett, and Roland Griffiths. 2019. Classic psychedelics: An integrative review of epidemiology, therapeutics, mystical experience, and brain network function. *Pharmacology & Therapeutics* 197 (3): 83–102.

Lebedev, Alexander, Martin Lovden, Gidon Rosenthal, Amanda Fielding, David J. Nutt, and Robin L. Carhart-Harris. 2015. Finding the self by losing the self: Neural correlates of ego-dissolution under psilocybin. *Human Brain Mapping* 36 (2015): 3137–3153.

Lee, Martin A., and Bruce Shlain. 2001. *Acid dreams: The complete social history of LSD: The CIA, the sixties, and beyond*. London: MacMillan UK.

Letheby, Chris. 2016. The epistemic innocence of psychedelic states. *Consciousness and Cognition* 39: 28–37.

Letheby, Chris, and Phillip Gerrans. 2017. Self unbound: Ego dissolution in psychedelic experience. *Neuroscience of Consciousness* 1: 1–11.

Letheby, Chris. 2021. *The philosophy of psychedelics*. Oxford: Oxford University Press.

Lewis, Bradley. 2011. *Narrative psychiatry: How stories can shape clinical practice*. Baltimore: The Johns Hopkins University Press.

Luke, David. 2008. Psychedelic substances and paranormal phenomena: A review of the research. *The Journal of Parapsychology* 72: 77–107.

Miller, Paul, and Charles Weijer. 2003. Rehabilitating equipoise. *Kennedy Institute of Ethics Journal* 13: 93–118.

Moncrieff, Joanna. 2008. *The myth of the chemical cure: A critique of psychiatric drug treatment*. Hampshire: Palgrave MacMillan.

Multidisciplinary Association for Psychedelic Studies (MAPS). 2020. *Phase 3 program of MDMA-assisted psychotherapy for the treatment of severe posttraumatic stress disorder (PTSD)*. ClinicalTrials.gov Identifier: NCT03537014. Available at: https://clinicaltrials.gov/ct2/show/NCT035 37014. Accessed Nov 21/2020.

Nichols, David E. 2016. Psychedelics. *Pharmacological Review* 68 (2016): 264–355.

Nissim, Rinat, Emily Freeman, Chris Lo, Camilla Zimmermann, Lucia Gagliese, Anne Rydal, Sarah Hales, and Gary Rodin. 2012. Managing Cancer and Living Meaningfully (CALM): A qualitative study of a brief individual psychotherapy for individuals with advanced cancer. *Palliative Medicine* 26 (5): 713–721.

O'Donnell, Kelley C., Sarah E. Mennenga, & Michael P. Bogenschutz. 2019. Psilocybin for depression: Considerations for clinical trial design. *Journal of Psychedelic Studies*. https://doi.org/10. 1556/2054.2019.026.

Oram, Matthew. 2018. *The trials of psychedelic therapy: LSD psychotherapy in America*. Baltimore: Johns-Hopkins University Press.

Osmond, Humphrey. 1957. A review of the clinical effects of psychotomimetic agents. *Annals of the New York Academy of Sciences* 66 (3): 418–434.

Pahnke, Walter N. 1963. *Drugs and mysticism: An analysis of the relationship between psychedelic drugs and the mystical consciousness*. Ph.D. Thesis. Boston, Massachusetts: Harvard University.

Pahnke, Walter N. 1969. Psychedelic drugs and mystical experience. *International Journal of Psychiatry in Clinical Practice* 5 (1969): 149–162.

Papineau, David. "Naturalism", *The stanford encyclopedia of philosophy* (Summer 2020 Edition), ed. Edward N. Zalta. https://plato.stanford.edu/archives/sum2020/entries/naturalism/.

Pollan, Michael. 2015. The trip treatment: Research into psychedelics, shut down for decades, is now yielding exciting results. *The New Yorker* 90, no. 47 (9 Feb 2015): 36.

Pollan, Michael. 2018. *How to change your mind: What the new science of psychedelics teaches us about consciousness, dying, addiction, depression, and transcendence*. New York: Penguin Random House.

Puchalski, Christina M., Robert Vitillo, Sharon K. Hull, and Nancy Reller. 2014. Improving the spiritual dimension of whole person care: Reaching national and international consensus. *Journal of Palliative Medicine* 17 (6): 642–656.

Pugh, Jonathan. 2020. Clarifying the normative significance of 'personality changes' following deep brain stimulation. *Social and Engineering Ethics* 26: 1655–1680.

Richards, William A. 1975. *Counseling, peak experiences and the human encounter with death: An empirical study of the efficacy of DPT-assisted counseling in enhancing the quality of life of persons with terminal cancer and their closest family members.* Ph.D. Thesis. Washington DC: Catholic University of America.

Reiff, Colin M., Elon E. Richman, Charles B. Niemeroff, Linda L. Carpenter, Alik S. Widge, Carolyn I. Rodriguez, Ned H. Kalin, William M. McDonald, & the Work Group on Biomarkers and Novel Treatments. 2020. A division of the American Psychiatric Association Council of Research. Psychedelics and psychedelic-assisted psychotherapy. *American Journal of Psychiatry* 177, no. 5 (2020): 391–410.

Rose, Nikolas. 2003. Neurochemical selves. *Society* 41, no. 1 (2003), 46–59.

Rose, Nikolas. 2007. *The politics of life itself.* New Jersey: Princeton University Press.

Ross, Stephen, Anthony Bossis, Jeffrey Guss, Gabrielle Agin-Liebes, Tara Malone, Barry Cohen, Sarah E. Mennenga, Alexander Belser, Krystallia Kalliontzi, James Babb, Zhe Su, Patricia Corby, and Brian L. Schmidt. 2016a. Rapid and Sustained symptom reduction following psilocybin treatment for anxiety and depression in patients with life-threatening cancer: A randomized controlled trial. *Journal of Psychopharmacology* 30, no. 12 (2016a): 1165–1180.

Ross, Stephen, Anthony Bossis, Jeffrey Guss, Gabrielle Agin-Liebes, Tara Malone, Barry Cohen, Sarah E. Mennenga, Alexander Belser, Krystallia Kalliontzi, James Babb, Zhe Su, Patricia Corby, & Brian L. Schmidt. 2016b. *Supplement to: Rapid and sustained symptom reduction following psilocybin treatment for anxiety and depression in patients with life-threatening cancer: A randomized controlled trial.* https://doi.org/10.1177/0269881116675512.

Saunders, Cicely. 1963. The treatment of intractable pain in terminal cancer. *Proceedings of the Royal Society of Medicine* 56: 195–197.

Saunders, Cicely. 1988. Spiritual pain. *Journal of Palliative Care* 4 (3): 29–32.

Schenberg, Eduardo Ekman. 2018. Psychedelic-assisted psychotherapy: A paradigm shift in psychiatric research and development. *Frontiers in Pharmacology* 9: 1–11.

Sessna, Ben. 2012. *The psychedelic renaissance: Reassessing the role of psychedelic drugs in 21st Century psychiatry and society.* London: Muswell Hill Press.

Sloshower, Jordan. 2018. Integrating psychedelic medicines and psychiatry: Theory and methods of a model clinic. In *Plant medicines, healing and psychedelic science: Cultural perspectives,* ed. Beatriz Caiuby Labate, and Clancey Cavnar, 113–132. Switzerland: Springer International Publishing.

Stanworth, Rachel. 2009. Spiritual care and attentiveness to narrative. In *Narrative and stories in health care: Illness, dying, and bereavement,* ed. Yasmin Gunaratnum and David Oliviere, 207–220. Oxford: Oxford University Press.

Vachon, Melanie, Lise Fillion, and Marie Achille. 2009. A conceptual analysis of spirituality at the end of life. *Journal of Palliative Medicine* 12 (1): 53–59.

Waldman, Ayelet. 2018. *A really good day: How microdosing made a mega difference in my mood, my marriage, and my life.* New York: Anchor Books.

Duff R. Waring is a philosopher/lawyer who has specialized in mental health law, psychiatric patient advocacy, bioethics, and philosophy of psychiatry. He joined the department of philosophy at York University in 2004 where he currently is a Full Professor.

Part VII
Diagnostics, Risk, and Prediction

Chapter 14
How Risky Can Biomedical Research Be?

On Setting an Upper Limit of Risk in Non-beneficial Research Involving Volunteers

Joanna Różyńska

14.1 Introduction

The main ethical concern regarding biomedical research involving human subjects is that it exposes participants to the risk of harm for the benefit of others. Therefore, the requirement for a favorable risk–benefit profile is a key ethical standard in research ethics and law (Emanuel et al. 2000; Rid and Wendler 2010; Rid 2012). Although the language slightly varies, all international guidelines and regulations on biomedical research impose on researchers and research ethics committees (RECs) an obligation to protect participants from unjustified and excessive risks, i.e., risks that are unnecessary, disproportionate to the social and scientific value of a study, and not properly minimized. There is wide agreement that where the research has no potential to produce results of direct clinical benefit to the subjects (so-called "non-beneficial research"), and the subjects are unable to give consent, the risks involved should not exceed a certain minimal threshold (Kopelman 2004). Yet, how the threshold should be defined and implemented still remains debatable (Freedman et al. 1993; Kopelman 2004; Resnik 2005; Wendler 2005, 2018; Wendler and Glantz 2007; Nelson and Ross 2005; Glass and Binik 2008; Rid and Wendler 2010; Rid 2012, 2014; Binik 2014; DeGrazia et al. 2017; Rossi and Nelson 2017). However, even more controversial are two other issues: (1) whether there should be an upper limit of allowable risk in non-beneficial research involving individuals able to give consent, particularly healthy volunteers, and if yes, (2) how it should be identified.

Some ideas presented in this essay have been already published in: Różyńska, J. 2016. Etyka i ryzyko w "nieterapeutycznych" badaniach biomedycznych (Ethics of risk in "non-therapeutic" biomedical research), Przegląd Filozoficzny. Nowa seria 98(2): 213–226 (published in Polish). However, they have been considerably developed and significantly revised in the present chapter.

J. Różyńska (✉)
University of Warsaw, Faculty of Philosophy, Warsaw, Poland
e-mail: j.rozynska@uw.edu.pl

© The Author(s), under exclusive license to Springer Nature Switzerland AG 2023   265
T. Zima and D. N. Weisstub (eds.), *Medical Research Ethics: Challenges in the 21st Century*, Philosophy and Medicine 132,
https://doi.org/10.1007/978-3-031-12692-5_14

This paper focuses on the second problem, one of the unmet challenges for risk–benefit assessment in biomedical research. The majority of international ethical and legal instruments governing biomedical research do not specify a maximal threshold of acceptable risk in non-beneficial research involving volunteers. There is also relatively little in-depth discussion on the subject in the bioethics literature. The problem has been analyzed only by a few authors, most comprehensively by London (2006), Miller and Joffe (2009), and Resnik (2012). Several others, notably Kimmelman (2009), Simonsen (2012), Rid (2012, 2014), and only recently Parquette and Shah (2020), have provided inspiring comments and suggestions.

The structure of this inquiry is as follows. First, to set the basis for the discussion, I briefly recall arguments against and for placing an upper limit on research risks to which competent participants can be exposed. I claim that there should be a maximum permissible risk threshold in non-beneficial research involving volunteers. Limiting research risks is necessary to protect research participants from excessive risks and maintain public trust in research practice. Then, two approaches to setting a risk ceiling, endorsed by international ethical guidelines and regulations on biomedical research—i.e., the no catastrophic harm/risk approach and the pure procedural approach—are analyzed. I argue that the pure procedural approach has significant advantages—it is context-sensitive and flexible, but it is not free from some serious problems. A remedy to these problems is to provide RECs with clear guidelines that will help them to determine ethically and socially acceptable limits of risk in high-risk biomedical studies. I examine whether the two main strategies for limiting research risks developed in the bioethics literature, the so-called numerical approach and the comparative approach, can provide RECs with such guidance. I conclude that they both suffer from severe conceptual and normative deficiencies, making them of little practical use. Ultimately, I propose a new approach, called the ELS procedural approach, and argue that this approach may be more promising than current regulatory and conceptual alternatives.

14.2 Limiting Research Risks—For and Against

There is no consensus on whether there should be an upper limit of risk to which a competent, informed, and consenting individual may be exposed for the sake of science and society. Some ethicists argue that the imposition of such a maximal risk threshold would constitute an unjustified paternalistic infringement of the potential participant's autonomy (Bergkamp 2004; Rajczi 2004; Edwards et al. 2004; Miller and Wertheimer 2007; Shaw 2014; Steel 2020). For instance, David Shaw claims that competent persons should have a right to participate even in extremely risky research projects, provided they fully understand the risks involved. If we allow competent adults to practice dangerous sports or hobbies (such as skydiving, potholing, or bungee jumping), which confer no benefit to others, it would be morally inconsistent—he argues—to disallow them to participate in scientifically valuable and socially beneficial research projects. Thus, neither legal regulations nor RECs should stop willing individuals from participating in non-beneficial high-risk research and deny them "a right to help their communities, patients worldwide, and future generations of patients" (Shaw 2014: 1009).

The majority of ethicists do not support this line of thinking. Two main arguments for setting maximal risk threshold have been developed in the literature: an argument from the protection of the research enterprise and an argument from the protection of the weaker party (cf. Miller 2003; Hope and McMillan 2004; Miller and Joffe 2009; Kimmelman 2009; Rid and Wendler 2011; Resnik 2012; Rid 2012; Różyńska 2015; Binik 2020; Shah et al. 2020; Paquette and Shah 2020).

The first argument stresses the need to protect the biomedical research 'enterprise'—viewed as a complex, collaborative social practice directed towards producing a common good (London et al. 2010)—from the loss of public confidence in the practice and willingness to support it. As David Resnik observes: "The death of a healthy volunteer in biomedical research can be a traumatic event, often leading to investigations and sanctions from oversight authorities as well as lawsuits. Additionally, negative publicity from the incident can have adverse impacts on the institution and the scientific community by eroding public trust in research." (Resnik 2012: 140). He recalls the tragic history of the death of Jesse Gelsinger in a gene therapy trial at the University of Pennsylvania in 1999, which not only led to the suspension of the study, federal investigations, and private lawsuits but also had a serious slowing down effect on the field of gene transfer research all over the world (Steinbrook 2008). Regrettably, there were other well-known cases of young and previously healthy subjects who died or were seriously harmed as a consequence of participation in a research study, for example, the 2001 death of Ellen Roche in a study of asthma at the Johns Hopkins Medical Institution (Steinbrook 2002), or the 2006 catastrophic multisystem failure in healthy volunteers in TeGenero's TGN1412 phase I trial of a novel monoclonal antibody, conducted at Parexel's clinical pharmacology research unit at Northwick Park Hospital in London (Emanuel and Miller 2007). All these cases have raised serious questions about the safety and ethics of human research. Thus, the argument is based on a reasonable and evidenced assumption that the limitation of acceptable risks in non-beneficial research involving persons able to give consent, will reduce the number of incidents of volunteers' deaths or disabling injuries, and—as a result—will support the existence, stability, and effectiveness of the research practice.

The second argument, namely the argument from the protection of the weaker party, stems from the recognition of an inherent inequity in power between researchers and research participants (in particular healthy volunteers). I distinguish three sources (aspects) of this inequity: asymmetry in allocation and control over information regarding research, asymmetry in allocation and control over research risks, and asymmetry in economic position (Różyńska 2015, 2019). I argue that—due to these asymmetries—informed consent does not provide competent participants with a reliable and sufficient safeguard against being exposed to excessive risks. Research participants often do not fully understand and appreciate research risks. This failure results from a combination of "natural" and social factors—the former include (but are not restricted to) various cognitive and emotional biases. The latter factors encompass poor education, lack of scientific knowledge, and the fact that the subject's understanding of the study design and risk–benefit profile depends on the quality of the information provided by the researcher, which is sometimes poor, or at least insufficient. Referring to Hélène Hermansson and Sven O. Hansson's model of three risk-related roles (2007), I further argue that individuals participating in non-beneficial research are the risk-exposed, but they are not the risk beneficiaries or the risk decision-makers (Różyńska 2015, 2018, 2019). They consent to passively enduring risky and burdensome procedures performed by others (investigators) for the benefit of others (science and society). Since subjects are mainly passive, they have very limited control over crucial risk-affecting aspects of the research, such as selecting enrollment criteria, performing research procedures, and implementing risk minimization and management strategies. Moreover, due to their socioeconomic position, "they have no power to negotiate terms and conditions of their participation in research; and are very often susceptible to acceptance of the risk they would ordinarily view as unacceptable, due to their poor financial situation or dependence on research income" (Różyńska 2015: 600). Therefore, I claim that research participants are always the weaker party of the research practice, and they deserve additional protection, even against their wishes. The imposition of an upper limit on risk in non-beneficial research should be one such protective measure. Setting such a risk ceiling is fully consistent with a risk-limiting function of the principle of the primacy of the human subject adopted by almost all international guidelines and legal regulations for biomedical research involving human subjects (World Medical Association 1975–2013; World Health Organization 1995; Council of Europe 1997, 2005a; United Nations Educational, Scientific and Cultural Organization 2005; International Conference on Harmonisation of Technical Requirements for Registration of Pharmaceuticals for Human Use 2016; cf. Różyńska 2021).

14.3 Regulatory Approaches to Setting a Maximal Risk Threshold in Research

The argument from the protection of the research enterprise and the argument from the weaker party's protection provide a sound justification for setting an upper limit on acceptable risk in non-beneficial biomedical research involving consenting individuals. However, they do not solve a crucial conceptual and practical problem of how to determine such a maximal risk threshold. As already mentioned, this question has been rarely addressed by international ethical and legal standards governing biomedical research. Except for the Nuremberg Code (1949), the Additional Protocol to the Convention on Human Rights and Biomedicine concerning Biomedical Research (Council of Europe 2005a), and the International Ethical Guidelines for Health-related Research Involving Humans developed by the Council for International

Organizations of Medical Sciences in collaboration with the World Health Organization (CIOMS 2016), no other international instrument imposes an absolute limit on acceptable research risks.

This section examines two strategies for setting a risk ceiling endorsed by the abovementioned regulatory instruments.

14.3.1 No Catastrophic Risk/Harm Approach

The first approach was suggested by the drafters of the so-called Nuremberg Code. It is a set of ten principles for medical research involving human subjects formulated in August 1947, in Nuremberg (Germany) by the US Military Tribunal I sitting in judgment of Nazi physicians accused of planning and performing experiments without the subjects' consent, and of committing murders, cruelties, tortures, and other inhuman acts in the course of these experiments (Annas and Godin 2008). Originally, the Code was a part of the verdict issued by the Military Tribunal. However, it soon began to be treated as the first international ethical standard for research involving humans, or even the "primary foundational document informing all ethical codes on research with humans" (Annas and Godin 2008: 136). Today its value is mainly historical and symbolic.

Paragraph 5 of the Nuremberg Code reads: "No experiment should be conducted where there is an a priori reason to believe that death or disabling injury will occur, except, perhaps, in those experiments where the experimental physicians also serve as subjects." The passage explicitly indicates the kinds of catastrophic harms against which research subjects must be protected. However, despite its apparent clarity, it can hardly serve as a guide for setting an upper-risk threshold for at least two reasons (cf. Miller 2003; Miller and Joffe 2009; Resnik 2012; Rid 2012).

First, it is unclear what is the meaning of the phrase "there is an a priori reason to believe that death or disabling injury will occur." Does the phrase refer to research projects that involve procedures inevitably leading to disabling diseases, mutilations, injuries, or death of the participants? Nazi physicians conducted such horrendous medical experiments during World War II. Thus, this interpretation of paragraph 5 of the Code may be historically justified, but it is far from providing a genuinely protective risk ceiling. Under this reading, only projects exposing participants to certain and catastrophic harms (disability or death) would be forbidden. But such studies are already universally regarded as unethical and illegal, even if competent individuals would be willing to give fully informed consent for participation.

One may argue that the contested phrase should be interpreted less stringently. Researchers have "an a priori reason to believe that death or disabling injury will occur" if they have evidence-based reasons to expect that the study's participation will involve some risk of death or disabling injury. At first glance, this reading seems attractive. Unfortunately, it would provide no help for RECs struggling with setting the limits of research risk. The phrase analyzed does not specify the degree of probability of death or disabling injury of subjects that makes the research impermissible.

Thus, it would not be against the literal wording of the provision to claim that a study should not be conducted if there is even the slightest chance of catastrophic harm to the subjects. However, if this reading were correct, the Code would prohibit all research involving invasive procedures, including taking a blood sample, as they all pose at least a minimal probability of very serious harm.

The second problem with paragraph 5 of the Nuremberg Code regards the provision's final clause: "except, perhaps, in those experiments where the experimental physicians also serve as subjects." Both the normative strength ("perhaps") and the normative justification of the exception are unclear. The Code does not explain why the research risk (harm) ceiling should not apply to auto-experimentation. Some commentators suggest that the drafters of the Code added this clause to provide a post-factum justification for famous high-risk research on a yellow fever vector led by Walter Reed and his team at the beginning of the XXth century in Cuba (Miller 2003; Miller and Joffe 2009; Resnik 2012). The study confirmed a hypothesis that yellow fever is transmitted by mosquitos rather than by direct contact. However, the cost of this breakthrough discovery was high. To test the hypothesis, two of Reed's collaborators allowed infected mosquitoes to feed on them. They both developed a severe attack of yellow fever. One of them died (Lederer 2008).

14.3.2   Pure Procedural Approach

The Additional Protocol to the Convention on Human Rights and Biomedicine concerning Biomedical Research is the second international regulatory instrument that directly addresses the problem of a risk ceiling in non-beneficial research on competent subjects (Council of Europe 2005a). The Protocol was prepared under the auspices of the Council of Europe as an addendum to the Convention for the Protection of Human Rights and Dignity of the Human Being with regard to the Application of Biology and Medicine: Convention on Human Rights and Biomedicine (Council of Europe 1997). It is a legally binding international treaty for eleven (out of 47) member states.

Article 6, paragraph 2 of the Additional Protocol stipulates that "where the research does not have the potential to produce results of direct benefit to the health of the research participant, such research may only be undertaken if the research entails no more than acceptable risk and acceptable burden for the research participant." Paragraph 27 of the Explanatory Report to the Additional Protocol adds that "whether or not the risk and burden are acceptable will be considered carefully by the ethics committee and competent body that approves the research project" (Council of Europe 2005b). Thus, instead of setting the maximal risk (harm) threshold, as the Nuremberg Code did, the Additional Protocol adopts a pure procedural approach that leaves the judgment of risk acceptability to REC discretion (Rid 2014; Różyńska 2015). It is assumed that such a judgment will be the result of a careful, multidisciplinary and deliberative process of ethical evaluation of a research project, drawing on "an appropriate range of expertise and experience adequately reflecting professional and lay views" (Article 9.3. of the Additional Protocol). Unfortunately, neither the Additional Protocol nor the Explanatory Report set more detailed procedural requirements or substantive criteria for determining the acceptability of research risk. They contain only one additional piece of information. Risks involved in non-beneficial research involving competent subjects may be higher than a minimal risk since the protective minimal risk threshold is prescribed for research on individuals unable to give free and informed consent (see: Article 6.2, 15.2.ii, and 17 of the Additional Protocol).

The newest 2016 version of the CIOMS International Ethical Guidelines for Health-related Research Involving Humans appears to favor the pure procedural approach similar to that of the Council of Europe's Additional Protocol. The concept of an absolute upper limit of risks in research involving volunteers is not, however, expressed in any guideline, but only mentioned in a commentary to guideline 4. The relevant passage states that researchers, sponsors, and RECs must ensure that research risks are justified in relation to the social and scientific value of the research, and that they do not exceed an upper limit. The commentary does not set any specific criteria for defining the threshold. It only gives two examples of research involving unacceptably high risks: "a study that involves deliberately infecting healthy individuals with anthrax or Ebola—both of which pose a very high mortality risk due to the absence of effective treatments—would not be acceptable even if it could result in developing an effective vaccine against these diseases" (CIOMS 2016: 10). Thus, the final evaluation of acceptability of the risk is left to the REC's judgment.

The pure procedural approach has many advantages. It is context-sensitive and flexible. It takes into consideration specific features of a research project under review as well as the unique deliberative potential and dynamic of a reviewing body. As Sigmund Simonsen rightly notes, it allows for an overall risk–benefit assessment "where both quantitative and qualitative factors, objective and subjective circumstances, and expert and lay-views count" (2012: 153). However, the approach also has significant disadvantages. Since it does not provide any procedural or substantive guidelines for RECs on evaluating research risks, it may result in great variations of RECs' opinions regarding the acceptability of risks involved in very similar or even the same research protocols. Of course, there will always be—and should be—a certain degree of variation and difference in the ethical judgments made by RECs (Edwards et al. 2007; Trace and Kolstoe 2017; Friesen et al. 2019). Several empirical studies have shown that institutional and local contexts in which RECs operate (in particular local knowledge, culture, laws, and trust relations) play an important role in the ethical review process (Edwards et al. 2007; Hedgecoe 2012; Jaspers et al. 2013; Trace and Kolstoe 2017). Thus, the situated character of the ethical review must be acknowledged. It is unavoidable, and—to a certain extent—desirable as it allows all relevant ethical issues and contextual factors to be taken into account (Friesen et al. 2019). Nevertheless, it does not mean that no efforts should be taken to achieve greater consistency and accountability, especially in RECs' opinions regarding the acceptability of the risk–benefit profile of the research. On the contrary, arbitrariness and unwarranted variations should be reduced for at least five reasons.

Firstly, to guarantee equal protection of the dignity, rights, and well-being of all research participants, which is the main role of RECs. Secondly, to facilitate the realization of ethically sound, scientifically significant, and socially valuable research. Empirical studies suggest that the lack of adequate guidelines on risk assessment creates a danger that more "risk-tolerant" RECs will accept biomedical research exposing subjects to excessive risks, while more "risk-aversive" RECs will block the implementation of valuable projects (Lenk et al. 2004; Shah et al. 2004; Van Luijn et al. 2006). Thirdly, greater consistency in RECs' opinions is crucial for the conduct of multi-center trials that often need multiple local ethical approvals. Fourthly, unwarranted differences in RECs' risk judgments may undermine trust in the mandate, objectivity, and competence of RECs. It has been observed that without proper guidelines, REC members' views on risks are largely based on intuitions, emotional and cognitive biases, and other non-evidence-based factors (Shah et al. 2004; Rid 2014; Resnik 2017). Fifthly, greater predictability and consistency in RECs' judgments reduce the need for "robust decision procedures" in terms of appeals or revision processes that are time and resource-consuming as they create additional administrative and financial burdens (Rid 2014).

Clearly, to address the abovementioned problems of the pure procedural approach, its "purity" must be abandoned. In other words, RECs need a method for determining a maximal risk threshold in non-beneficial research involving competent individuals. Such a method should meet three criteria:

1. It should be conceptually clear and easy to use in practice.
2. It should offer an ethically sound explanation for a proposed upper limit of risk, i.e., free from logical and normative fallacies, or unsupported factual claims, and consistent with existing ethical and legal requirements for an acceptable risk–benefit ratio in research, in particular the proportionality requirement.
3. To build and maintain trust among researchers, participants, other social stakeholders, and the general public, the method should balance subjects' interests against societal interests at stake in a socially acceptable way.

14.4   Conceptual Approaches for Limiting Research Risks

This section considers two main strategies for setting an upper limit of risk developed in the bioethics literature: the numerical approach and the comparative approach (or the analogical strategy, as it has been called by Kimmelman 2009: 43). Each of the strategies is presented and critically evaluated against the three features of a robust method for determining a maximal risk threshold set above.

14.4.1 The Numerical Approach

The numerical strategy aims to quantify a fixed maximal level of risk permissible in non-beneficial research on consenting subjects. David Resnik—the most renowned proponent of the approach—argues for setting the risk threshold at a 1% chance (1 in 100) of serious harm, such as death, permanent disability, or severe illness or injury (2012). Sigmund Simonsen cautiously suggests that "the highest risk of harm that may be legally accepted in normal nontherapeutic research on ordinary healthy volunteers is probably somewhere between 1 in 10,000 and 1 in 1000 for fatal harm, provided the potential benefits to others are great" (Simonsen 2012: 161). The two proposals diverge significantly. Alas, it is impossible to explain the numerical difference between them and evaluate their relative merit since neither of the authors provides any normative or empirical justification for his suggestion.

Resnik claims that the 1% proposal adequately weighs the subjects' rights and interests with the social interest in conducting valuable research, on the one hand, and the need to protect researchers and research practice (as a whole) from the loss of public trust and support, on the other. Thus, it represents "a fair compromise between over-protectiveness and under-protectiveness" (Resnik 2012: 144). However, in the absence of any arguments supporting the claim, it is hard to resist the impression that the 1% proposal is a pragmatic limit, rather than a normative solution that carefully balances and weighs the different interests at stake. Indeed, Resnik admits that the 1% limit may seem arbitrary, but he stresses that if the limit were higher, it would allow for research exposing research subjects to excessive risk of serious harm. If it were lower (say, set at 0.1%), it would prohibit a great deal of valuable clinical research, e.g., environmental health research on the consequences of air pollution involving a transbronchial biopsy (which poses a probability of death lower than 1%). He also adds that the 1% limit is not an absolute rule, but rather flexible guidance for RECs. It can be slightly exceeded in the case of studies addressing compelling public health or social problems.

Simonsen supports his proposal by making references to Albert Jonsen's discussion on the ethics of exposing healthy volunteers to liver biopsy for research purposes (1989); to a scale of research procedures divided into risk groups developed by a Swedish insurance company in the late 1970s; and to anecdotal opinions of researchers and members of the local RECs on the acceptability of risks involved in selected medical procedures. Clearly, none of these sources explain why the maximal limit of risk in biomedical research should be set between a 0.0001 and 0.001% chance of the subject's death. Perhaps the proposal's arbitrariness stems from the fact that, deep in his heart, Simonsen is very skeptical about the possibility of setting an absolute risk threshold that would apply to all kinds of biomedical research. He believes that his numerical suggestion can provide at best a valuable guide to a starting point for further deliberation by RECs.

In conclusion, none of the numerical strategy proponents seems to be genuinely attached to the idea of setting a fixed upper limit on risk. This should come as no surprise if one considers all the conceptual, practical, and ethical challenges the strategy poses. Although the numerical approach seems clear and easy to apply, it is in fact very difficult to implement and is incomplete. The approach would only be easy to use if relevant data about severe risks associated with particular research interventions were readily available. Unfortunately, there is, and always will be, a problem with adequate scientific evidence of risks that volunteers face in research. This is due to the inherent uncertainties associated with studies typically involving healthy subjects, namely phase I clinical trials, which are the very first stage of a long process of gathering and evaluating scientific data on novel interventions. Moreover, since the numerical strategy allows for exceptions, which are not well-defined, the approach does not provide RECs with comprehensive and clear guidelines for setting risk limits in all high-risk research.

The numerical approach—at least in a form advocated by Resnik or Simonsen—also does not meet the second and third criteria for a robust method of setting a maximal risk threshold in research as described in the preceding section. None of the authors provides a sound ethical or empirical rationale behind the suggested risk limit. Thus, the limits are arbitrary, which begs the question of why any of them should be widely socially accepted. Additionally, by indicating an absolute upper limit of risk, without making any reference to the potential scientific and social value of research, the approach seems to ignore the basic ethical requirements for an acceptable risk–benefit ratio.

14.4.2   The Comparative Approach

The comparative strategy has been developed by Alex J. London as a part of his "Integrative Approach" to risk–benefit assessment in biomedical research (2006, 2007). The Integrative Approach is based on a commitment to equal regard for the basic interests of study participants and the members of the larger community who are potential beneficiaries of the research. The term "basic interests" stands for every person's interests in cultivating and exercising fundamental capacities necessary for developing a conception of the good, and for formulating and pursuing particular life goals, aspirations, and projects that are an expression of this conception. Based on this normative foundation, London argues that the cumulative risks to the basic interests of research participants, which are not compensated by the expected direct benefits, must not be greater than the risks that are permitted in the context of "other socially sanctioned activities that are similar in structure to the research enterprise" (London 2006: 2881).

At first glance, London's proposal is very appealing and seems well-grounded. To safeguard equality between research participants and non-participants, it allows the former to assume risks that are no greater than risks experienced by the latter when involved in other socially sanctioned activities that are similar to research. However, as critical scrutiny of the comparative strategy will show, the strategy raises more questions than it answers.

The first challenge to the comparative approach lies in identifying an appropriate social activity comparable to non-beneficial biomedical research. London admits that it is not easy to define reasonable requirements for structural similarity necessary to identify appropriate comparators. Nevertheless, he suggests four such criteria.

First, since the primary goal of biomedical research is to benefit future patients, the primary objective of (potential) comparator activities should also be to benefit others than those directly involved in them. Second, risks to participants of comparator activities should be viewed as necessary evils that must be minimized and avoided as far as possible, rather than something that increases the attractiveness of the practices. This feature eliminates from the scope of London's analysis dangerous sports and hobbies, undertaken by people who share a "no risk, no fun" view. The third criterion refers to a fiduciary "principal-agent relationship" between research subjects and investigators. Although London does not discuss this feature of research practice in detail, he explains that "participants put their interests in the hands of identifiable parties who possess a particular expertise that creates limited but important responsibilities on the part of those experts to protect and to safeguard the interests of those participants" (London 2006: 2882). London believes that such a "principal-agent relationship" should be present also between parties of a relevant comparator activity. Finally, the fourth criterion of the structural similarity proposed by London is the presence of effective institutional or/and regulatory oversight. As in the case of biomedical research, an appropriate comparator activity should be subjected to "active public oversight so that the risk-profile associated with the activity can be seen, at least prima facie, as representing a level of risk that is deemed socially acceptable after due reflection" (London 2006: 2882).

Having these four criteria in mind, London suggests firefighting and paramedic services as relevant comparators. However, he acknowledges that although both these activities are intended to benefit others and are the subject of active public oversight, none of them involves the required "principal-agent relationship". Firefighters and paramedics are professionals equipped with knowledge, skills, and tools to actively and effectively minimize, manage, and face the risks involved in their work. They are not passively exposed to risky interventions performed by others, to whom they must entrust their health or life. In other words, they are the risk-exposed, but also the risk-decision makers responsible for their safety (Hermansson and Hansson 2007). Therefore, risks associated with firefighting and paramedic services should not be used to determine an upper limit of risk in non-beneficial biomedical research.

Following London's approach, Miller and Joffe (2009) explore whether live-organ donation by unrelated donors can serve as an adequate comparator activity. This practice has important similarities to participation in non-beneficial studies. Donors take short- and long-term health risks for the benefit of unrelated recipients. These risks and burdens are substantial and are treated as necessary evils. The fiduciary relationship between donors and physicians is analogous to the one between participants and investigators. Finally, the practice of organ transplantation intervivos, including live-organ donation between unrelated individuals, is the subject of well-developed regulatory and institutional oversight.

Although live-organ donation by unrelated donors meets all of London's structural similarity criteria, Miller and Joffe are very reluctant to accept the idea that the risks experienced by organ donors should serve as a benchmark for an upper limit of risk in research. Their reluctance is due to a significant and ethically relevant difference between the values of social benefit each of these practices normally delivers. Live-organ donation involves a high and quantifiable probability of substantive health benefits for an identifiable recipient. These benefits justify the level of risks and burdens experienced by a donor. In order words, the risks and burdens associated with organ donation are ethically acceptable because they are necessary and proportional to the expected health benefits for a recipient. In contrast, in "research there is an inherent and unquantifiable uncertainty that any given study will produce social benefit" (Miller and Joffe 2009: 447–448). Each study incrementally contributes to the development of generalizable medical knowledge that in turn, may have (or may not have) a potential for being used to improve future medical or public health interventions. Thus, it is extremely difficult, if not impossible, to assess a priori the probability and magnitude of social benefit associated with a particular research project. It is even harder to evaluate the incremental contribution to that benefit made by an individual research participant. Undoubtedly, the value of an individual subject's input is often uncertain and very modest compared to the highly probable and substantial health benefit of each organ donation and transplantation. Therefore, reference to the value of this input is usually insufficient to justify exposing research participants to risks similar to the risk borne by organ donors.

Miller and Joffe disqualify live-organ donation by unrelated donors from being a relevant comparator activity by pointing to the most severe deficiency of London's approach, namely its lack of sound ethical justification.

The comparative strategy is based on two question-begging assumptions: (a) risks involved in a socially sanctioned activity are ethically acceptable; (b) the structural similarity between a given socially sanctioned activity and non-beneficial biomedical research makes the risks involved in the comparator activity ethically acceptable in the research context.

The first premise is unwarranted because being socially accepted is not the same as being ethically sound (Miller and Joffe 2009; Resnik 2012; Rid and Wendler 2010; Rid 2012). For centuries slavery, human trafficking, or child labor were socially accepted practices. Today, no decent person would consider the risks (and evident harms) associated with these practices to be ethically justified. Additionally, one must remember that not all risks actually involved in socially and ethically accepted practices (e.g. firefighting, paramedic services, or organ donation) are justified. As London rightly notes, the level of risk actually associated with many activities is often higher than what is considered to be ethically and even socially acceptable. Many workers and professionals face risks that have not been minimized to a possible and desirable extent. This is due to many factors, such as lack of resources, poor risk management, or ineffective risk reduction policies. Quoting Kimmelman, the actual risks involved in a comparator activity may reflect "an imperfect political process rather than careful moral examination" (2009: 44). Therefore, instead of making groundless assumptions, the comparative strategy advocates must demonstrate that the risks involved in a given comparator activity are indeed ethically and socially acceptable. (Perhaps this is what London had in mind when he suggested that social acceptability of risks involved in comparator activities should be preceded by "due reflection"; 2006: 2882).

The second premise of the comparative strategy boils down to the claim that the structural similarity between two activities, defined by London's criteria, is sufficient to justify exposing participants of one activity (i.e., biomedical research) to a certain level of high risk simply because it is ethically acceptable to expose participants of the other activity to the same (or similar) level of risk. But this claim is false, as Miller and Joffe have rightly (though implicitly) noted. The comparative strategy fails to recognize that the risks associated with social activities aimed at promoting common goods (e.g., safety, health, or knowledge) are justified by the value of their potential social benefits. In general, the greater the expected benefits are, the higher the risks involved can be. London seems to forget that the logic of proportionality provides evaluation criteria for the *risk–benefit profile* of an activity, rather than for the "*risk-profile* associated with the activity*" (London 2006: 2882). He argues that if volunteer firefighting and paramedic services met all the similarity criteria, it would be justified to expose consenting participants of non-beneficial research to the level of risk that firefighters or paramedics face on a routine basis. However, this conclusion would be correct only if the expected social benefits of a given non-beneficial research project were equally probable and great as the benefits of firefighting or paramedic services. If the expected research benefits were uncertain or smaller (as is usually the case), the comparative approach would yield results exposing research subjects to excessive risks.

Finally, it is worth noting that—like Resnik and Simonsen—also Miller and Joffe seem to find the idea of a fixed limit of research risk ill-founded. At the end of their essay, they claim that in studies addressing acute public-health emergencies, such as dire epidemics, it could be justified to expose consenting subjects to considerably higher risk than those involved in routine clinical research. "The urgent importance of knowledge relating to the ability to prevent or minimize the public-health threat, for rapid translation into substantial health benefits for the large populations, must be weighed in the balance" (Miller and Joffe 2009: 448). However, they also suggest—unfortunately without any justification—that in such cases, it would be reasonable to set a limit of acceptable risk by appealing to the risks faced by firefighters or rescue workers involved in a large-scale disaster, or the risks faced by battlefield medical personnel. Thus, despite their objections to the concept of an absolute risk limit and their recognition of the inherent limitations of the comparative strategy, they consider reasoning by analogy to be somehow helpful in thinking about acceptable limits to research risk (although, as they note, "appeal to the comparator is, at best, suggestive"; Miller and Joffe 2009: 447).

In conclusion, the comparative approach may seem attractive, but it fails to meet the above-stated requirements for a robust method of setting a maximal risk threshold in research. This is due to the conceptual problems associated with defining proper criteria of similarity between research practice and other socially sanctioned activities, and the resulting difficulties with identification of an adequate comparator, but mainly due to the lack of a strong normative justification. London commits a *petitio principii* fallacy by assuming what should be proved, i.e. (a) that a chosen comparator activity involves risks that are both socially acceptable and ethically justified; and (b) that structural similarity between two higher-risk activities, as identified by the criteria developed by London, justifies the exposition of participants of each of them to the same (or similar) level of risk. As a result, the comparative approach ignores the requirement of proportionality between risks and benefits of the research activity and creates a danger of exposing non-beneficial research participants to unjustified and excessive risks. For obvious reasons, there could be no ethical and social acceptance for such an outcome.

14.5 Towards a More Standardized Procedural Approach

I have argued that neither the numerical approach nor the comparative approach has provided adequate guidelines for RECs' decisions regarding an upper limit of acceptable risk in non-beneficial research involving consenting participants. Although, at first sight, each strategy seems to offer a clear maximal risk threshold that is easy to apply, both of them have significant conceptual and normative drawbacks that undermine their value as guiding frameworks for RECs.

This section outlines an alternative strategy for helping RECs in identifying an upper limit of risk in research. Instead of determining a fixed and quantified risk ceiling (either directly or indirectly by an appeal to the risk of a comparator activity), the strategy offers procedural recommendations that will help RECs assess whether the high risks involved in particular non-beneficial research on volunteers are ethically and socially acceptable. Thus, the proposed strategy may seem close to Annette Rid's "constrained pure procedural" approach (2014). However, Rid addresses her proposal to the drafters of research regulations and guidelines. She aims to provide them with general guidelines on how to specify both minimal and maximal thresholds of research risks (defined by likelihood and magnitude of research harms) by using analogy to relevant comparator activities, and numeric information about risks involved in these activities. Contrary to Rid's proposal, the approach presented is targeted at individual RECs and has mainly a procedural character. Since the approach encompasses three interdependent evaluative steps: ethical appraisal, legal appraisal, and social appraisal of the acceptability of high risk involved in a study, it is further referred to as the "ELS procedural approach."

14.5.1 Background Assumptions

The ELS procedural approach is based on an assumption shared by most authors quoted in this essay that placing an upper limit on risks in non-beneficial research involving consenting subjects is necessary to protect research participants from excessive risks and exploitation, as well as to maintain public confidence in the research enterprise. It is also in line with the majority opinion among bioethicists that it would be unjustified or imprudent to set a fixed maximal limit of risk and apply it to all human studies. Assessment of research risk must be connected to considerations regarding the social and scientific value of research. This is because the expected benefits of a study provide the necessary justification for the associated risks. Moreover, evaluating risk acceptability always requires a good understanding of the broader context in which a study is to be conducted. Thus, it must always be context-sensitive and—to a certain degree—flexible.

Additionally, the ELS procedural approach takes the CIOMS Guidelines on community engagement seriously. It assumes that, when feasible, representatives of relevant communities should be engaged in a meaningful participatory process of identifying an acceptable maximum risk threshold in high-risk non-beneficial research. Their engagement is crucial for at least two reasons. First, because "a community's values and preferences are relevant in determining what constitutes benefits and acceptable risks" (CIOMS 2016: 10, Commentary to Guideline 4). Second, because "an open and active process of community engagement is critical for building and maintaining trust among researchers, participants, and other members of the local community" (Council for International Organizations of Medical Sciences 2016: 10, Commentary to Guideline 7). Following the CIOMS Guidelines 4 and 7, the ELS procedural approach adopts a very broad understanding of "the community". The term refers not only to people living in the geographic area(s) where research is to be carried out, but also to populations from which research participants will be recruited, populations of potential research beneficiaries (patients), and other stake-holders, i.e., various individuals, groups, organizations, and institutions that have a stake in the proposed research (cf. CIOMS 2016: 25, Commentary to Guideline 7).

14.5.2   Outlines of the ELS Procedural Approach

The ELS procedural approach suggests that to identify an ethically and socially acceptable upper limit of risks in high-risk non-beneficial research involving competent subjects, an REC should evaluate the risks involved from three perspectives: ethical, legal, and social. These perspectives are complementary, interdependent, and often overlapping. However, for analytical clarity, they are presented as three steps of the procedure.

Step 1—Ethical appraisal: The REC should comprehensively describe and carefully assess a risk–benefit profile of a study reviewed. This involves: (a) identifying and quantifying all risks and expected benefits of the project (both their magnitude and likelihood); (b) determining whether the risks are necessary to obtain the potential benefits, and whether they are adequately minimized; (c) evaluating whether the benefits are maximized; and (d) assessing whether the risks involved are proportional to the expected benefits. The REC should also carefully scrutinize the project's recruitment procedures, including informed consent documents, to assure that potential subjects will be selected fairly and will be provided with comprehensive, accurate, and understandable information about the project, particularly about its benefits and high risks involved. RECs routinely perform all these judgments as they apply to all evaluated projects, no matter how risky they are. Therefore, it is well known that they are usually complex and challenging (Martin et al. 1995; Van Luijn et al. 2002; Grinnell et al. 2017; Resnik 2017). They require expertise in research methodology as well as scientific and clinical expertise with respect to the targeted problem.

Step 2—Legal appraisal: The REC should establish whether in a relevant legal system the high risks involved in a reviewed project fall within the scope of the *volenti non fit injuria* doctrine, i.e., whether in a given legal system a competent person may effectively consent for being exposed to such a level of risk. In all jurisdictions there are some limits to consent as a defense against criminal and civil liability for harm-doing or for exposing a consenting individual to the risk of serious harm. The limits of consent may be viewed as reflecting a basic social consensus on what constitutes morally impermissible treatment of a human being. If the risks involved in the project are so high that they fall outside the *volenti non fit injuria* defense, the project is illegal and cannot be realized, regardless of the potential participants' consent. However, it must be stressed that since respect for personal autonomy is one of the most important values in all modern liberal-democratic societies, consent is usually deprived of its excusing power only in cases of grave harms such as serious bodily injury, disability, or death. Yet, there might be significant differences between particular legal systems regarding what constitutes serious harm beyond the afflicted person's autonomous choice (Baker 2009). This step is not a part of a standard RECs evaluative practice. It requires expert knowledge of a relevant legal system viewed as a whole. Therefore, it may require employment of external legal advice.

Step 3—Social appraisal: If the high risks involved in a reviewed project are considered ethically and legally permitted, the REC should establish whether they are socially acceptable. This step responds to the concerns regarding public trust in research and the protection of research participants from exploitation and excessive risks that justifies imposing an upper limit on research risk. Its main goal is to ensure that representatives of relevant communities to be involved in the project—especially populations from which research participants are to be recruited and populations of potential research beneficiaries (often represented by relevant patient organizations)—consider the high risk involved in the study: (a) reasonable in relation to its potential benefits, (b) not too high per se, and (c) reasonable to take. Risks are considered "reasonable to take" if a significant number of representatives of relevant social groups, adequately informed about a project (particularly its goals, design, risks, benefits, recruitment strategies, and financial or other non-clinical gratifications for the participants), believe that the voluntary exposition to such risks can be a reasonable choice, rather than a decision made under undue inducement or other exploitative influences.

The social appraisal step requires the REC to have access to the perspectives of the representatives of the relevant communities. Thus, it presupposes that the communities will be engaged in a meaningful, collaborative, and transparent process of evaluating the risks and potential benefits of a proposed project before a study is initiated or even before the project is submitted for review. How the relevant communities will be defined, by whom they will be represented, and how they will be engaged in the process will depend on the nature of a particular research project and the context in which it is to be carried out. In general, community engagement in research may take various forms, e.g., community consultations, meetings with community advisory boards, focus group interviews, and public opinion surveys (Dickert and Sugerman 2005).

RECs should request researchers to provide information about the relevant communities' views on the acceptability of research risks acquired from such consultations, interviews, or surveys. The information should include a description of the community engagement strategies and activities undertaken or proposed (cf. CIOMS 2016; Commentary to Guideline 7). When investigators are unable to provide such information, RECs should solicit the communities' perspectives by ensuring active participation of a broad representation of the relevant social groups in the process of reviewing high-risk projects (see Kimmelman 2009; Shah et al. 2020). As noted above, the involvement of relevant communities in the process of risk–benefit assessment will not only allow RECs to better understand and evaluate the level of social acceptability for the risks involved in high-risk research, it will also help build public trust in the research 'enterprise' by raising social awareness about the practice, increasing transparency, conferring ethical legitimacy on high-risk projects, and even conferring on communities consulted some degree of moral responsibility for such studies (Dickert and Sugerman 2005).

Admittedly, the ELS procedural approach, as sketched above, needs further refinements. However, it seems a promising approach in comparison to the regulatory and conceptual alternatives discussed in the previous sections. It is conceptually clear, well-grounded, and applicable. It provides RECs with general and context-sensitive recommendations that encourage RECs to engage in systematic ethical-legal analysis of high research risks and a "dialog" with relevant communities about their perception and evaluation of these risks. Thus, the ELS procedural approach can help RECs to determine ethically, legally, and socially acceptable limits of research risks.

14.6 Conclusion

In this essay, I have claimed that there are good reasons for placing some limits on risks that consenting participants face in non-beneficial biomedical research. I have also argued that RECs should consider carefully whether or not risks involved in a particular research project exceed an upper limit of acceptable risk. However, to minimize arbitrariness and unwarranted variations between judgments by RECs, they should be provided with guidelines for limiting research risks. I have examined two main approaches to setting a maximal risk threshold developed in the bioethics literature—the numerical approach and the comparative approach—and found that both suffer from significant conceptual, normative, and practical shortcomings. Therefore, I have proposed an alternative framework for guiding RECs' deliberations and decisions regarding risk limits.

The proposed approach provides RECs with procedural recommendations for identifying an ethically, legally, and socially acceptable upper limit of risk in non-beneficial research on volunteers. It consists of three interdependent steps: ethical appraisal, legal appraisal, and social appraisal. For this reason, I have called it "the ELS procedural approach". The last step recommends engaging relevant communities in the process of assessing the social acceptability of high risk associated with a project. It is assumed that the involvement of the relevant social groups in the review process will enhance research participants' protection against excessive risks and exploitation and contribute to maintenance of public confidence in research.

Funding This analysis was supported by a grant of the National Science Centre, Poland, no. 2014/15/B/HS1/03829.

References

Annas, George, J., and Michael A. Gordin. 2008. The nuremberg code. In *The oxford textbook on clinical research ethics*, ed. Ezekiel J. Emanuel et al., 136–140. New York: Oxford University Press.

Baker, Dennis J. 2009. The moral limits of consent as a defense in the criminal law. *New Criminal Law Review* 12: 93–121.

Bergkamp, Lucas. 2004. Medical research involving human beings: Some reflections on the main principles of the international regulatory instruments. *European Journal of Health Law* 11: 61–69.

Binik, Ariella. 2014. On the minimal risk threshold in research with children. *The American Journal of Bioethics* 14: 3–12.

Binik, Ariella. 2020. What risks should be permissible in controlled human infection model studies? *Bioethics* 34: 420–430.

Council of Europe. 1997. *Convention for the protection of human rights and dignity of the human being with regard to the application of biology and medicine: Convention on human rights and biomedicine*. ETS No.164. Oviedo. https://rm.coe.int/168007cf98. Accessed 1 June 2020.

Council of Europe. 2005a. *Additional protocol to the convention on human rights and biomedicine concerning biomedical research*. ETS No.195. Strasburg. https://rm.coe.int/168008 371a. Accessed 1 June 2020.

Council of Europe. 2005b. *Explanatory report: Additional protocol to the convention on human rights and biomedicine concerning biomedical research*. https://rm.coe.int/16800d3810. Accessed 1 June 2020.

CIOMS, Council for International Organizations of Medical Sciences. 2016. *International Ethical Guidelines for Health-related Research Involving Humans*. Geneva. https://cioms.ch/wp-content/uploads/2017/01/WEB-CIOMS-EthicalGuidelines.pdf. Accessed 1 June 2020.

DeGrazia, David, Michelle Groman, and Lisa M. Lee. 2017. Defining the boundaries of a right to adequate protection: A new lens on pediatric research ethics. *Journal of Medicine and Philosophy* 42: 132–153.

Dickert, Neal, and Jeremy Sugarman. 2005. Ethical goals of community consultation in research. *American Journal of Public Health* 95: 1123–1127.

Edwards, Sarah JL., Simon Kirchin, and Richard Huxtable. 2004. Research ethics committees and paternalism. *Journal of Medical Ethics* 30: 88–91.

Edwards, Sarah JL., Tracey Stone, and Teresa Swift. 2007. Differences between research ethics committees. *International Journal of Technology Assessment in Health Care* 23: 17–23.

Emanuel, Ezekiel J., David Wendler, and Christine Grady. 2000. What makes clinical research ethical? *JAMA* 283 (20): 2701–2711.

Emanuel, Ezekiel J., and Franklin G. Miller. 2007. Money and distorted ethical judgments about research: Ethical assessment of the TeGenero TGN1412 trial. *The American Journal of Bioethics* 7 (2): 76–81.

Freedman, Benjamin, Abraham Fuks, and Charles Weijer. 1993. In loco parentis minimal risk as an ethical threshold for research upon children. *The Hastings Center Report* 23: 13–19.

Friesen, Phoebe, Aimi Nadia Mohd. Yusof, and Mark Sheehan. 2019. Should the decisions of institutional review boards be consistent? *Ethics & Human Research* 41: 2–14.

Glass, Kathleen Cranley, and Ariella Binik. 2008. Rethinking risk in pediatric research. *The Journal of Law, Medicine & Ethics* 36: 567–576.

Grinnell, Frederick., John Z. Sadler, Victoria McNamara, Kristen Senetar, and Joan Reisch. 2017. Confidence of IRB/REC members in their assessments of human research risk: A study of IRB/REC decision making in action. *Journal of Empirical Research on Human Research Ethics* 12(3): 140–149.

Hedgecoe, Adam M. 2012. Trust and regulatory organisations: The role of local knowledge and facework in research ethics review. *Social Studies of Science* 42: 662–683.

Hermansson, Hélène., and Sven Ove Hansson. 2007. A three-party model tool for ethical risk analysis. *Risk Management* 9: 129–144.

Hope, Tony, and John McMillan. 2004. Challenge studies of human volunteers: Ethical issues. *Journal of Medical Ethics* 30: 110–116.

Jaspers, Patricia, Rob Houtepen, and Klasien Horstman. 2013. Ethical review: Standardizing procedures and local shaping of ethical review practices. *Social Science & Medicine* 98: 311–318.

Jonsen, Albert R. 1989. The ethics of using human volunteers for high-risk research. *The Journal of Infectious Diseases* 160: 205–208.

Kimmelman, Johnatan. 2009. *Gene Transfer and the Ethics of First-in-Human Experiments: Lost in Translation*. New York: Cambridge University Press.

Kopelman, Loretta M. 2004. Minimal risk as an international ethical standard in research. *The Journal of Medicine and Philosophy* 29: 351–378.

Lederer, Susan E. 2008. Walter Reed and the Yellow Fever Experiments. In *The Oxford Textbook on Clinical Research Ethics*, ed. Ezekiel J. Emanuel, et al., 9–17. New York: Oxford University Press.

Lenk, Christian, K. Katrin Radenbach, Morten Dahl, and Claudia Wiesemann. 2004. Non-therapeutic research with minors: How do chairpersons of German research ethics committees decide? *Journal of Medical Ethics* 30: 85–87.

London, Alex J. 2006. Reasonable risks in clinical research: A critique and a proposal for the integrative approach. *Statistics in Medicine* 25: 2869–2885.

London, Alex J. 2007. Two dogmas of research ethics and the integrative approach to human-subjects research. *The Journal of Medicine and Philosophy* 32 (2): 99–116.

London, Alex J. Jonathan Kimmelman, and Marina Elena Emborg. 2010. Beyond access v. protection in trials of innovative therapies. *Science* 328: 829–830.

Martin, Douglas, K., Eric M. Meslin, Nitsa Kohut, and Peter A. Singer. 1995. The incommensurability of research risks and benefits: practical help for research ethics committees. *IRB: Ethics & Human Research* 17 (2): 8–10.

Miller, Franklin G. 2003. Ethical issues in research with healthy volunteers: Risk-benefit assessment. *Clinical Pharmacology & Therapeutics* 74: 513–515.

Miller, Franklin G., and Alan Wertheimer. 2007. Facing up to paternalism in research ethics. *Hastings Center Report* 37: 24–34.

Miller, Franklin G., and Stephen Joffe. 2009. Limits to research risks. *Journal of Medical Ethics* 35: 445–449.

Nelson, Robert M., and Lainie Friedman Ross. 2005. In defense of a single standard of research risk for all children. *The Journal of Pediatrics* 147: 565–566.

Nuremberg Code. 1949. In *Trials of war criminals before the Nuremberg military tribunals*, vol. 2, 181–182. Washington DC: US Government Printing Office.

Paquette, Erin T., and Seema K. Shah. 2020. Towards identifying an upper limit of risk: A persistent area of controversy in research ethics. *Perspectives in Biology and Medicine* 63 (2): 327–345.

Rajczi, Alex. 2004. Making risk-benefit assessments of medical research protocols. *The Journal of Law, Medicine & Ethics* 32: 338–348.

Resnik, David B. 2005. Eliminating the daily life risks standard from the definition of minimal risk. *Journal of Medical Ethics* 31: 35–38.

Resnik, David B. 2012. Limits on risks for healthy volunteers in biomedical research. *Theoretical Medicine and Bioethics* 33: 137–149.

Resnik, David B. 2017. The role of intuition in risk/benefit decision-making in human subjects research. *Accountability in Research* 24(1): 1–29.

Rid, Annette, and David Wendler. 2010. Risk–benefit assessment in medical research—Critical review and open questions. *Law, Probability & Risk* 9: 151–177.

Rid, Annette, and David Wendler. 2011. A framework for risk-benefit evaluations in biomedical research. *Kennedy Institute of Ethics Journal* 21: 141–179.

Rid, Annette. 2012. Risk and risk-benefit evaluations in biomedical research. In *Handbook of risk theory. Epistemology, decision theory, ethics, and social implications of risk*, eds. Sabine Roeser, et al. 179–211. Springer Science + Business Media B.V.

Rid, Annette. 2014. Setting risk thresholds in biomedical research: Lessons from the debate about minimal risk. *Monash Bioethics Review* 32: 63–85.

Rossi, John, and Robert M. Nelson. 2017. Minimal risk in pediatric research: A philosophical review and reconsideration. *Accountability in Research* 24: 407–432.

Różyńska, Joanna. 2015. On the alleged right to participate in high-risk research. *Bioethics* 29: 451–461.

Różyńska, Joanna. 2018. What makes clinical labour different? The case of human guinea pigging. *Journal of Medical Ethics* 44: 638–642.

Różyńska, Joanna. 2019. Passivity, research risks, and worker-type protections for research subjects. *The American Journal of Bioethics* 19: 46–48.

Różyńska, Joanna. 2021. Taking the principle of the primacy of the human being seriously. *Medicine, Health Care and Philosophy* 24: 547–562. https://doi.org/10.1007/s11019-021-10043-2

Shah, Seema K., Franklin G. Miller, Thomas C. Darton, Devan Duenas, Claudia Emerson, Holly Fernandez Lynch, Euzebiusz Jamrozik, Nancy S. Jecker, Dorcas Kamuya, Melissa Kapulu, Jonathan Kimmelman, Douglas MacKay, Matthew J. Memoli, Sean C. Murphy, Ricardo Palacios, Thomas L. Richie, Meta Roestenberg, Abha Saxena, Katherine Saylor, Michael J. Selgelid, Vina Vaswani, and Annette Rid. 2020. Ethics of controlled human infection to study COVID-19. *Science* 368 (6493): 832–834.

Shah, Seema, Amy Whittle, Benjamin Wilfond, Gary Gensler, and David Wendler. 2004. How do institutional review boards apply the federal risk and benefit standards for pediatric research? *JAMA* 291: 476–482.

Shaw, David. 2014. The right to participate in high-risk research. *Lancet* 383: 1009–1011.

Simonsen, Sigmund. 2012. *Acceptable risk in biomedical research. European perspective.* Dordrecht Heidelberg London New York: Springer Science+Business Media BV.

Steel, Robert. 2020. Reconceptualising risk–benefit analyses: The case of HIV cure research. *Journal of Medical Ethics* 46: 212–219.

Steinbrook, Robert. 2002. Protecting research subjects—The crisis at Johns Hopkins. *New England Journal of Medicine* 346 (9): 716–720.

Steinbrook, Robert. 2008. The Gelsinger case. In *The oxford textbook on clinical research ethics*, ed. Ezekiel J. Emanuel, et al., 110–120. New York: Oxford University Press.

Trace, Samantha, and Simon E. Kolstoe. 2017. Measuring inconsistency in research ethics committee review. *BMC Medical Ethics* 18. https://doi.org/10.1186/s12910-017-0224-7.

United Nations Educational, Scientific and Cultural Organization (UNESCO). 2005. Universal declaration on bioethics and human rights. http://portal.unesco.org/en/ev.php-URL_ID=31058%26URL_DO=DO_TOPIC%26URL_SECTION=201.html. Accessed 1 June 2020.

Wendler, David. 2005. Protecting subjects who cannot give consent: Toward a better standard for "minimal" risks. *Hastings Center Report* 35: 37–43.

Wendler, David, and Leonard Glantz. 2007. A standard for assessing the risks of pediatric research: Pro and con. *The Journal of Pediatrics* 150: 579–582.

Wendler, David. 2018. The ethics of net-risk pediatric research: implications of valueless and harmful studies. *IRB: Ethics & Human Research* 40: 13–18.

World Health Organization (WHO). 1995. Guidelines for good clinical practice (GCP) for trials on pharmaceutical products. Geneva.

Van Luijn, Heleen E.M., Albert W. Musschenga, Ronald B. Keus, Walter M. Robinson, and Neil K. Aaronson. 2002. Assessment of the risk/benefit ratio of phase II cancer clinical trials by Institutional Review Board (IRB) members. *Annals of Oncology* 13: 1307–1313.

Van Luijn, Heleen E.M., Neil K. Aaronson, Ronald B. Keus, and Albert W. Musschenga. 2006. The evaluation of the risks and benefits of phase II cancer clinical trials by institutional review board (IRB) members: A case study. *Journal of Medical Ethics* 32: 170–176.

Joanna Różyńska is an Assistant Professor at the Faculty of Philosophy, University of Warsaw,and a researcher at the Center for Bioethics & Biolaw affiliated at the Faculty. She was a cocoordinator and senior faculty member of the Advanced Certificate Program in Research Ethics funded by the Fogarty International Center, NIH USA. Prof. Rozynska is the Chairperson of the Bioethics Committee of the Polish Academy of Sciences, and a member of the UNESCO International Bioethics Committee. Her research interests focus on ethics of biomedical research involving humans, in particular normative models of risk-benefit assessment, ethics of compassionate use and reproductive law and ethics. She is an author and/or editor of numerous publications, including four books.

Chapter 15
Ethical Issues in the Use of Risk Assessment

David Shapiro

15.1 Introduction

> I learned that murderers die for their crimes, even if we make a mistake some time. (Paxton 1962)

Tom Paxton was not a health care professional nor was he involved in doing risk assessments of the potential for future violence. Nevertheless, these two lines from a song he wrote in 1962 "*What did you learn in school today?*" highlights some of the concerns examined in this article concerning the ethical use of risk assessment by health care professionals. Paxton's reference in the song was to highlight what happens when someone mistakenly identifies someone as a rapist or murderer, now seen as being more common than once thought since the use of DNA. Scientific studies of risk assessment described below also demonstrate the difficulty of predicting future violence using only clinical interview methods. Yet the laws in the U.S. and Europe are asking health care professionals to do so, even though they may not pay attention to their limitations. Applying these new scientific techniques to the identification and prevention of future violence has limited research at this time. This research is even more difficult to carry out when the base rates of the behavior (in this case commission of a violent act) is very low. The prediction of someone's future dangerousness has never been particularly accurate even when using the scientific methods of psychology. Very often people displaying the same characteristics signifying potential for violence actually do not engage in violence. This chapter, then, will explore some of the difficulties in researching the bases for such acts given our limited ability to predict violence. Ethical and moral questions become important when demanding that health care professionals violate their long established ethical principles regarding confidentiality and only making statements when there

D. Shapiro (✉)
Nova Southeastern University, Fort Lauderdale, USA
e-mail: psyfor@aol.com

© The Author(s), under exclusive license to Springer Nature Switzerland AG 2023 287
T. Zima and D. N. Weisstub (eds.), *Medical Research Ethics: Challenges in the 21st Century*, Philosophy and Medicine 132,
https://doi.org/10.1007/978-3-031-12692-5_15

is sufficient data to back up the statements. Nevertheless, laws in the United States and in Europe are asking, and in fact encouraging health care professionals to do such predictions without paying adequate attention to the limitations. There is, in fact, very limited research on terrorist activities, and, consistent with other studies of violence, such activities have a very low base rate; further, few terrorists survive an attack, and those who do generally do not cooperate with authorities. There seems to be a compelling desire to find better ways of assessing the risk of terrorist activities in the future, but prediction of future violence has never been particularly accurate., and, in fact, very few people who display characteristics that are consistent with the potential for future violence, actually do act out violently. The research is made even more complicated by the fact that it involves parameters other than those tradition-ally used in studies of risk assessment. For instance, there are different dynamics involved in political terrorism as part of religious fervor, mass murder due to intense anger, and acting out of a delusional disorder. It is a very complex area with limited research and one size "does not fit all". There is a premature rush to judgment to find simple answers because of the ever increasing threats of violent terroristic acts. However these additional unresearched parameters are being added to an area in which our predictions are at present only slightly better than chance. Developing effective interventions when the state of knowledge is so limited, raises serios ethical issues.

The need for confidentiality protection unequivocally stands out, since the disclo-sure of sensitive information is only acceptable under extraordinary conditions, which must be verified and proven by the professionals involved. But it is apparent that public safety and security needs may outweigh the right to absolute confidentiality and privacy. However, if our methodology regarding prediction is flawed, how many individuals would fall into the category of "false positive" someone who has many characteristics in common with someone who will be violent, but does not in fact act violently? Here, the breach of confidentiality is even more compelling, since there cannot be any justification for breaking the confidence. There is a frequent argument heard that public safety and security may outweigh the need for privacy, but that is based on having made an accurate assessment in the first place. A well known case in the United States is called Tarasoff v. Regents of the University of Cali-fornia (1976), dealing with the need for a therapist to break confidentiality when a patient threatens harm to an identifiable third party. In its final decision, the California Supreme Court stated "The protective privilege ends where the public peril begins". However, there is no discussion of how to determine public peril; that is left up to the clinical judgment of mental health professionals. Research has demonstrated repeatedly that such clinical judgment of the potential for future violence is more often inaccurate than it is accurate. When people demand that therapists develop and devise effective prevention and treatment standards, we are faced with the moral and ethical dilemma that a mental health professional whose goal is to help alleviate the suffering of their patient, becomes unwittingly an arm of the law. These professionals are asked to provide law enforcement with information that will prove useful in the fight against terrorism, but the nature of this information is rarely specified, leading to a "slippery slope" in which a therapist may inadvertently provide the authorities with confidential information.

15.2 Criminal Laws in the U.S. and Risk Assessment

There are many areas of the law in the United States on which predictions of the potential for violence play a significant role. These include criminal cases, workplace and school violence, and domestic violence. One of the most common areas is in criminal law cases, especially during sentencing options that depend upon the person's propensity for continued violence. Aggravating as well as mitigating factors are frequently the subject of mental health testimony in death penalty cases and have led to various United States Supreme Court (U.S.S.C.) decisions. Testimony about the potential for violence with regard to decisions made during bail hearings has also been allowed in *United States v. Salerno* (1987). Further, decisions regarding probation/parole conditions often entail a concern for whether the person will pose a danger to others. Since the landmark case of *Lessard v. Schmidt* (United States U.S.S.C. 1972), the process of involuntary civil commitment has been based on the concept of "danger to self or others" by reason of mental illness.

For example, of the 29 states in the U.S. in which the death penalty is currently permitted, 26 of them allow information to be presented to the jury about an individual's potential for future violence. This information (often regarded as an aggravating circumstance) is provided during the sentencing phase, in which the jury is tasked with determining whether the appropriate sentence for the crime is death or life in prison. Although the research in this area is limited in the ability to predict violent behavior, the question of the defendant's future dangerousness is frequently asked of mental health professionals who serve as expert witnesses in capital cases. Despite the paucity of available research to back up the validity of such predictions, many experts have no compunction about rendering their opinions (Shapiro and Noe 2018).

15.3 History of Risk Assessment for Prediction of Future Violence

The range of the areas mentioned above, in which the prediction of violent behavior carried great weight, assumed that clinicians were able to make such assessments with a high degree of accuracy. In fact, the early terminology involved the 'prediction of dangerousness', a dichotomous opinion. In other words, the mental health professional decided if the person was dangerous or not dangerous; there was no 'middle-ground'. Indeed, prior to critical analyses of the research and additional studies highlighting the flaws in these judgments, an untested assumption existed that mental health professionals had special clinical abilities that could ferret out 'dangerousness' during an extended clinical interview. There was, however, little attention directed to what clinicians assessed during their evaluations, or the methods utilized. Further, such assessments were highly idiosyncratic in that most clinicians did not follow any sort of standardized protocol for their evaluations, which made it difficult for researchers to evaluate the accuracy of the prediction methodology used.

During the 1960s, the case of Baxstrom v. Herold (1966), in the state of New York, revealed some groundbreaking news about the low risk of violent behavior posed by psychiatric patients. In this case, a class action suit was filed on behalf of inpatients at a psychiatric hospital in New York State. The court ordered 966 of these patients released or transferred to less secure facilities. There were dire predictions made by a number of people that there would be a great deal of violent acting out based on previous clinical assessment of these individuals. Yet, as it turned out, very few of the cohort from the Baxstrom study actually committed a violent act after their release. Steadman and Cocozza (1974) found that only 20% had been reconvicted, predominantly for nonviolent offenses. This served as the beginning of a new era, in which the unquestioned assumption that mental health clinicians could predict future violence was challenged. Based on their research, Steadman and Cocozza (1976) concluded that the assumption that clinicians had the ability to predict future violent behavior using their clinical skills had little scientific basis.

During the late 1970s and into the early 1980s, psychologist John Monahan reviewed the research in this area. Interestingly, as of the late 1970s, there were only five research studies on the accuracy of clinicians predicting violent behavior. Monahan (1981) reported an unsettling discovery: In predictions of future violent behavior, clinicians were essentially wrong two out of three times. That is, based on clinical interviews alone, psychiatrists and psychologists were accurate in no more than one out of every three predictions made concerning violent behavior. Obviously, this is even less than chance. These findings were based on results obtained over several years with institutional populations that had both committed violence in the past (and, thus, had high base rates) and who were diagnosed as mentally ill (Monahan 1981, p. 47). This was a rather dramatic finding that challenged the foundation upon which much of the clinical work on risk assessment was based, up until that time.

Monahan further noted in his work that the best predictors of violent behavior among the mentally ill were, in fact, the same predictive factors found in populations that were not mentally ill. Of these factors, he concluded that the strongest was a history of past violence. The poorest predictors were those valued most by clinicians, which included diagnosis and personality structure. He was also critical of mental health professionals for ignoring base rates in their research. A base rate is the percentage that identifies the occurrence of a particular behavior in a certain group. For instance, the base rate of violent behavior in the general population is about 2%. That is, 2 out of 100 individuals—on average—will act in a violent manner. Ultimately, the assessment of the prevalence of violent behavior among mentally ill patients is only meaningful when compared to the general population base rate.

Although questions were being raised regarding the accuracy of these predictions of future violent behavior, the judicial system moved on as if these issues were not troubling. Whether this was due to the fact that researchers did not want to raise the issues in court or they were raised and the courts ignored them is a matter of some speculation that is beyond the scope of this discussion Suffice it to say, several major court decisions in the U.S. ignored the concerns that psychological research was raising regarding the accuracy of the prediction of violent behavior.

In *Jurek v. Texas* (1976), for instance, the U.S.S.C. reviewed the three statutory aggravating factors for capital cases in Texas. The first involved the individual's probability of committing criminal acts of violence that would constitute a continuing threat to society. The other two factors were whether the conduct was deliberate with a reasonable expectation that death would result and that the conduct was an unreasonable response to any provocation. If the State proved, beyond a reasonable doubt, that all three were present, the recommended sentence was death. If any of the three were not applicable, then the recommended sentence would be life imprisonment. The appellants in *Jurek* argued that the aggravating factor about future violence was 'unconstitutionally vague' and, therefore, lacked merit. The U.S.S.C. rejected this argument, stating that future dangerousness is commonly addressed and answered throughout the American Criminal Justice System and, further, that it is no different from any other prediction of future behavior. They went on to cite the relevance of this factor in many other legal decisions, such as bail, sentencing and parole as mentioned above. The Court noted that such prediction may be difficult, but not impossible; the Court was essentially saying, "Do not bother me with the research; this is the way we have always done it."

The issue about prediction of future violence was again raised in a subsequent case, *Estelle v. Smith* (1981). However, the issue was addressed in a peripheral way, dealing with the inadequacy of the examining psychiatrist's methodology of predicting future violent behavior based on a diagnosis. Based on a brief clinical interview, the psychiatrist in question, Dr. James Grigson, concluded that Smith showed no remorse and, therefore, he was a sociopath. He further elaborated that since sociopaths, by definition, commit criminal acts, Smith would be violent in the future. As distressing as this inadequate examination and conclusion is, it should be kept in mind that is not the nature of the concerns that may be argued in front of the U.S.S.C. Rather, the U.S.S.C. addresses matters in which there may be a violation of the Constitution; therefore, the Court would only address issues about inadequate assessment if it were directly linked to a constitutional violation.

In Smith's case, the court was focussed on possible violations of the Fifth and Sixth Amendments to the U.S. Constitution. The Fifth Amendment deals with the fact that one need not incriminate oneself in their statements and the Sixth Amendment deals with having legal representation at a 'critical stage of the proceedings'. Smith was not told that Dr. Grigson would be testifying for the State in his sentencing hearing and was led to believe that he was being interviewed as part of a competency evaluation. Smith's statements, later taken out of context and used by Grigson to establish the fact that he was a sociopath (and, therefore, dangerous), essentially amounted to self-incrimination. Generally, a competency evaluation was not seen as a critical stage of the proceedings; however, an evaluation to determine whether he would be violent and, consequently, eligible for the death penalty would be considered a critical stage of the proceedings and would require the availability and assistance of counsel. Ultimately, the Court found that Mr. Smith being led to believe that the interview was only a competency evaluation was a violation of both his Fifth and Sixth Amendment rights. Smith's case was reversed and remanded for a new trial, with Smith eventually sentenced to life in prison rather than to death.

Shortly afterwards the Court again dismissed the concerns raised by mental health professionals in *Barefoot v. Estelle* (1983). In this case, the same Dr. Grigson, along with a Dr. Holbrook, testified for the State, never having examined the defendant. Not surprisingly, their response to hypothetical questions reflected their opinion that Barefoot would probably commit further acts of violence and, thus, would represent a continuing threat to society.

Psychiatrists and psychologists both recognize that the rendering of an opinion about an individual without having examined him or her is usually considered professionally unethical. Nevertheless, Dr. Grigson testified that he was 100% certain and absolute that Barefoot would kill again, even though he had never even examined him. Barefoot's attorneys argued that the use of psychiatric testimony at a sentencing hearing about future violent behavior was unconstitutional, as such professionals are not competent to predict future violent behavior. Given that their predictions were likely to result in erroneous judgments, it was argued that their usage constituted violations of his Constitutional rights under the Eighth Amendment (cruel and unusual punishment) and Fourteenth Amendment (equal protection under the law). Ultimately, the U.S.S.C. rejected all of the petitioner's arguments and affirmed the conviction and the sentence.

The Court made many observations in response to the arguments in the case of *Barefoot v. Estelle* (1983). For example, it was noted that not letting psychiatrists testify about future dangerousness would be equivalent to 'disinventing the wheel.' Once again, the research was ignored and the flawed testimony admitted based on the fact that it had always been done in this manner. The Justices in their Opinion cited the case of *Jurek v. Texas* as affirming the fact that potential for violence was a legitimate issue to be considered in capital cases and that any methodological deficiencies could be 'cured by rigorous cross-examination'. The Court also chided the American Psychiatric Association (which, in an amicus brief, argued for the elimination of such testimony), by stating that there are many psychiatrists who disagreed with the position outlined in the Brief. The certainty with which Dr. Grigson made his predictions was also addressed. More specifically, the Court stated that such a declaration of certainty is a matter of weight to be given to the testimony, not to its admissibility. Finally, addressing the issue of hypothetical questions, the Court observed that hypotheticals and responses to them are an acceptable part of the judicial process. In essence, once again, the Court rejected the concerns of mental health professionals about the true ability (or lack thereof) to predict future violent behavior.

15.4   The MacArthur Foundation Research on Risk Assessment

As previously mentioned, Monahan conducted research to determine the reasons for the poor predictive validity of violence from clinical work. He suggested that the use of actuarial assessments would be a better methodology than the way clinical interviews were being used as an assessment. The clinical assessments he studied were typically conducted individually, using a diagnostic formulation approach. Actuarial assessments, on the other hand, provided results based on a strict algorithmic formula (or "equation"). These assessments look for the presence of a specific combination of static, fixed and largely demographic risk factors that had been 'weighted', meaning the number of points assigned to each item was based on the purported strength of its relationship with violent behavior. For example, on one specific actuarial measure, the Violence Risk Appraisal Guide (VRAG), the endorsement of past violent behavior contributes more points (+6) to the final numerical score than, say, marital status (+1), a personality disorder diagnosis (+3) or history of alcohol problems (+2) (Harris et al. 2015).

The specific risk factors on actuarial measures require the evaluator to focus on variables that do not change over time and, perhaps more importantly, are able to be studied far more precisely than clinical judgment; however, this approach deliberately ignores individual variations that do not fit the pattern. Using the VRAG as an example, the item, "lived with both biological parents until age 16" ("Yes" is assigned a weight of −2 and No is assigned a + 3) generalizes the statistics from a specific group of subjects in such a way that the person who did not grow up with both biological parents until age 16 is assumed to be more likely to engage in violent behavior than someone who did live with both parents until age 16. Fundamentally, these instruments ignore individual variations, such as the fact that there are different reasons a person might have lived in a home without both parents such as a divorce due to abuse or parental separation due to military service. These variations simply do not matter when looking at violence from a purely actuarial point of view based on group statistics and therefore may skew the results into false positives or negatives.

The MacArthur Risk Assessment Studies (Monahan et al., 2005) grew into a comprehensive research-based assessment model and began to incorporate many clinical variables into the actuarial data. This model became one of the bases for an approach called 'Structured Professional Judgment' (SPJ). This approach combined the positive parts of the clinical interview technique together with specific questions that were seen as significant in predicting violence from the actuarial data. The MacArthur research primarily involved monitoring psychiatric populations, as well as incorporating several assumptions: All relevant information was available, 'prediction' applied to a specific behavior within a limited time period and training in the method could yield high inter-rater reliability.

In time, Monahan discovered three factors that seemed to be limiting the original violence prediction research: (1) A constricted range of risk factors—That is, clinicians would try to predict future violent behavior after looking at only a few demographic variables or an impression from a clinical interview or the results from only one psychological test. This meant predictions were stated in rather broad statements based on too little data; (2) Weak criterion measures—Monahan found that most of the studies were using 're-arrest' as the outcome variable (violent behavior) for inclusion in the study. This is a requirement that is far too narrow and restricts the number of individuals who can be included. Most violent acts do not lead to an arrest, especially in cases of domestic violence, and therefore underestimate the repeated violence. Another variable that could also have been used would be re-hospitalization, which was more common with the patient population Monahan's studies were monitoring (i.e., males with a history of previous violence). So, once again, only a narrow segment of that specific population was being considered for participation. Finally, the research was based on findings from a single site, meaning generalizations were made that ignored the possible effects of differences in socio-economic status or cultural background.

To rectify these limitations, the MacArthur research made a conscious decision to improve the study's reliability and validity by: (1) Studying a large and diverse array of risk factors; (2) Expanding the criterion variable of re-arrest to also include patient's self-report and input from a large variety of collateral sources, such as family interviews and hospital records; (3) Including both male and female participants; and (4) Collecting data from multiple sites. Greater focus needed be placed on risk factors that were relatively robust predictors. Among these were: Psychopathy, difficulties regulating anger, the presence of delusions and/or hallucinations, diagnosis, gender, the endorsement of violent thoughts, abuse as a child, a history of previous violence and, lastly, a variety of contextual variables. It is important to add that these efforts ultimately led to the replacement of the commonly used terminology 'prediction of dangerousness' with the more accurate term 'risk assessment'. Using the continuum of assessing the risk of violence occurring was more accurate that simply stating whether it would or would not occur. Further, assessment of risk could be stated in probabilities of violence occurring rather than yes or no terms.

The investigators noticed that a main effects regression model (which is part of the development of the actuarial 'equation' described earlier, including the weighting assigned to specific items) did not improve predictive accuracy. Consequently, they opted for a 'classification tree model' in place of a linear regression model approach. The classification tree model is similar to the regression model, except it involves identifying discrete categories rather than numerical scores. The specific questions asked at any point during the assessment were based on the respondent's answer to the question that preceded it rather than a fixed list of questions to ask all of the participants (which is typically the case for regression model approaches). Basically, it was hypothesized that the specific, empirically-based risk factors that would be important to explore during one individual's risk assessment may, in fact, not be relevant to the risk assessment of another.

Overall, the MacArthur research identified several different parameters that should be used to structure any examination of risk factors. This new perspective included: (1) Demographic features, such as age and gender; (2) Historical factors, including

prior mental hospitalization and/or history of violent behavior; (3) Contextual variables, such as the degree of social support an individual has, the extent of disinhibiting
influences, the degree of inhibiting influences; (4) Clinical (psychological) traits,
including a broad array of factors having to do with diagnosis and specific symptoms and (5) Biological variables, especially regarding the sequelae of head trauma.
This model resulted in the identification of well over one hundred risk factors which
could be combined in different ways using the decision trees. As noted earlier, such an
approach would use a broader range of information sources, including not only arrest
records, but also hospitalizations and input from self-report and collateral sources.
In addition, violence would be more clearly specified in terms of types of violence,
targets of violence, location of the violence and timing of the violence.

Essentially, violent behavior was conceptualized as the product of a complex array
of interactions among a number of factors (i.e., demographic, psychological, sociological, biological and contextual). Subsequent to this realization, a research protocol
was developed to identify the specific features and interactions the researchers should
explore further. While this was helpful for determining which factors were, in fact,
risk factors, it also proved to be useful for ruling out the risk factors that were simply
assumptions. An ideal example of this can be drawn from the demographic category:
Race was originally identified as a demographic factor for consideration; however,
when socio-economic variables were controlled for, the racial factor proved to be
unremarkable. The researchers highlighted an elevated risk in males between the
ages of 18 and 24, especially those who were members of a lower socio-economic
group and poorly educated. It should be clarified that these are purely static variables
and, for the most part, do not change very much over time.

The researchers also explored a wide variety of psychological variables in an
effort to identify measurable risk factors for future violent behavior. Included were an
examination of the various diagnostic groupings, the impact of certain kinds of mental
illnesses (or, perhaps more importantly, the influence of specific symptoms) and the
contributions of substance abuse. The diagnostic groupings and mental disorders
evaluated were not restricted to the 'serious mental illnesses' (e.g., major depressive
disorder, bipolar disorder, schizophrenia) but, also, included the potential influence
of personality disorders on violent behavior. Specific focus was given to the traits
characteristically seen in narcissistic, histrionic, borderline and antisocial personality
disorders, with quite a bit of focus dedicated to the subset of antisocial individuals
known as psychopaths (Monahan and Steadman 1996). Indeed, it was around this
time that a significant amount of research began to emerge dedicated to exploring
the relationship between Hare's (1980, 1985, 1991, 2003) construct of psychopathy
and violent behavior. According to Hare, the construct of psychopathy could be
broken down into two factors: Factor 1 pertained to the interpersonal/affective deficits
characteristic of the prototypical psychopath; Factor 2, on the other hand, involved
a general anti- social behavior pattern and chronically unstable lifestyle. It should
be noted that others are developing different ways of assessing psychopathy using
statistical techniques based on research. These are not yet being used in the clinical
population.

A link between delusions/hallucinations and inevitable violence is an assumption endorsed by many, especially within the context of the media. For example, it is often presumed that a mentally ill person will eventually act out in response to delusions of persecution or at the behest of auditory hallucinations. The research substantiating this belief is, at best, mixed. In this specific study, (Macarthur research) delusions (in general) were not found to be highly correlated with violent behavior; however, a rather powerful association did appear when the experience involved a specific type of delusion; what the investigators called 'thought control override'." This occurred when there were delusions in which the patient felt that another was controlling his or her mind and, as a result, felt powerless to resist the influence of the delusions and/or hallucinations. Additionally, they reported that patients also acted out in response to delusions in which they perceived others as trying to harm them.

Another particularly interesting finding from this study involved the influential power of hallucinations. The original belief was that auditory command hallucinations were very rarely followed. Subsequent research suggested that they were followed more frequently. This discrepancy was puzzling until the researchers observed that the mere presence of auditory command hallucinations was not enough to prompt violent behavior. Rather, only voices characterized as familiar and benevolent were found to be influential. For example, if the patient heard the familiar voice of a loving family member instructing them to take revenge on someone, the patient would be more likely to act out; however, if the voice was unfamiliar or associated with a hostile individual, such compliance was far less likely.

The role of anger in violent behavior was also studied during the MacArthur research. The researchers chose to use the *Novacco Anger Scale and Provocation Inventory* (NAS-PI; 1994) to assess for this trait, which is a measure that conceptualizes anger as the interaction of three components: cognitive, arousal and behavioral indicators. Indeed, they reported that those with high scores on the Novacco Anger Scales were twice as likely to act violently after discharge than those who did not obtain high scores.

The sociological point of view entailed the exploration of a number of environmental variables, such as the family's approach to problem solving. One important area was whether the family encouraged or inhibited expressions of violent behavior as a legitimate means of solving an issue. Similarly, another important area was whether the individual's peer group typically supported or tried to prevent him or her from engaging in violent behavior. Another psychologist, Meloy (2000) further highlighted the role sociological factors or 'popular cultural influences', played in violent behavior and especially urged clinicians to examine an individual's propensity to use weapons (primarily firearms) carefully. Meloy built on his recommendation by developing the Weapons History Assessment Model (WHAM), which assessed for the person's attraction to firearms, his or her skill in using the weapon, the selection of reading materials preferred by the individual and the nature of his or her affective response to firearms. Loss of employment and general employment instability were also identified as being correlated with the potential for future violence.

Recently, Meloy (2018) published a new assessment device which he calls the TRAP (Terrorist Radicalization Assessment Protocol) which he describes as a structured professional judgment instrument for threat assessment of the individual terrorist. Meloy derives, based on a theoretical model, eight proximal warning behaviors, and ten distal characteristics. It is of note that Meloy clearly recognizes this instrument as post-dictive, rather than predictive, studying and classifying the characteristics of individuals whose attacks have been thwarted and those that have been successful. There is clear recognition that the instrument should not be used in its current form as a way of predicting future violence. Nevertheless, there is little doubt that some will try to use this instrument in a manner inconsistent with the research on, and proper use of it, a clear ethical violation, according to the Code of ethics in Psychology.

The biological point of view was also seen as quite important in the original MacArthur research, especially given the advancements in neurological imaging today. This biological perspective is especially relevant when making distinctions between affective (or reactive) violence and predatory violence. People who have a history of traumatic brain injury (TBI) may become quite irritable, over-react to minor irritations and respond to such irritations in an explosive and potentially violent manner. Today, some neuropsychologists, such as Dr. Adrian Raine (1993), have argued that there is a neurological component underlying even more violent behavior than seen in the MacArthur studies. According to Raine (2014), there may even be abnormalities in the brain structure that facilitate or promote predatory violence. For example, the size of the amygdala has been found to be smaller in individuals categorized as psychopaths compared to that seen in non-psychopaths (Yang et al. 2009). This region of the brain is linked with feelings of fear, suggesting that psychopaths experience less fear than do others (including fear of consequences and fear of shame/scrutiny). Additionally, individuals with deficits in pre-frontal lobe functioning, such as psychopaths (Yang et al. 2009), often present with compromised judgment and poor ability to appreciate the long-term effects of their decisions. Predatory behavior, therefore, may be endorsed more often in individuals who do not fear the consequences of their actions, are not concerned with the opinions of others and are primarily interested in instant gratification.

Finally, the MacArthur studies placed a great deal of emphasis on contextual variables. In other words, they found that violent behavior frequently occurred in a specific setting or under specific circumstances, such as workplace settings, family settings or after the ingestion of drugs and/or alcohol. Random acts of violence, on the other hand, were found to occur far less often. Looking at an individual's particular pattern can assist with decisions about interventions.

In summary, the structure suggested by the MacArthur research involves probabilities concerning specific types of violence occurring within a specific context in contrast with the earlier "predictions of dangerousness". This provides a far more precise picture of an individual's potential for future violence, in that it qualifies the circumstances under which such behavior is more likely to occur. Again, this represents a major departure from the antiquated, "Yes, he or she will be dangerous" or "No, he or she will not be dangerous" that was once so heavily relied upon.

15.5   Violence Predictor Variables

15.5.1   The Role of Mental Illness and Violence

Although already outlined in the section covering the MacArthur research, stereotypes concerning violence among the mentally ill is pervasive enough to warrant additional discussion.

There is a widespread public perception that the mentally ill are inherently dangerous. Whenever there is an act of seemingly senseless violence, such as the ever-increasing number of active shooter situations emerging in the United States, the role of mental illness is quickly brought into the picture as an explanatory factor.

The assumption that a violent person must be mentally ill is often grossly inaccurate. For example, Adam Lanza, the 20-year-old active shooter who murdered his mother and then attacked the Sandy Hook Elementary School in Newtown, Connecticut, was reported to have been diagnosed with Asperger syndrome. Immediately after this diagnosis was made public, hotlines began being flooded with calls from anxious parents who were concerned that their children (who were also diagnosed with Asperger syndrome) would soon become violent, too. However, the paucity of data that actually support this hypothetical link between Asperger syndrome and violence suggests that it is, yet again, another assumption meant to explain "unexplainable" behavior.

In a similar manner, a well- known media psychologist was interviewing the parents of an adolescent boy with certain behavioral problems. The psychologist observed that the boy had nine of the characteristics of potential serial killers, while Jeffrey Dahmer (who had killed and cannibalized a number of young men), only had six. This psychologist was instilling much fear in the parents, and failing to consider the fact that no such research (about the characteristics of potential serial killers) exists. These statements were ethically questionable in that they were not based on any research and could well cause emotional harm to the parents. The parallels to individuals who might be inaccurately identified as potential terrorists is relevant here as well.

Of course, there are some cases in which symptoms of mental illness do contribute to the violent behavior. It should be clarified, though, that behavior does not occur in a vacuum and requires outside influences before it can emerge. Although such influences may vary, an overwhelming amount of stress and/or a recent significant loss is frequently present. The presence of child abuse and domestic violence in the history of violent actors may also contribute to the development of certain forms of mental illness.

Up until the late 1980s, the general consensus within the mental health community was that there was no statistically significant correlation between mental illness and violent behavior. Basically, violence among the mentally ill had about the same base rate as violence in the general population when demographic and other historical factors were controlled (around 2%). However, in an influential article published in the *American Psychologist*, Monahan (1992) described a conclusion he said he did not want to reach. After looking at data from several different perspectives, he discovered a relationship between certain mental illnesses and violence, suggesting the possibility that mental illness in some cases may be a predictor of violent behavior. When mental illness was defined as psychosis or a major mood disorder, 13% of the sample reported acting violently. This, of course, is a notable increase from the 2% noted above. Yet, Monahan made it a point to emphasize that severe mental illness is rare and any relationship between it and violence is moderate, at best.

While it is true, for example, that in general mentally ill patients rarely act out violently, there are some select groups for which there is, in fact, a higher likelihood of such behavior, such as paranoid delusional disorders and paranoid schizophrenia, that being the case, it is important to consider other moderating variables that could also influence the outcome, such as culture, ethnicity, socio-economic status, gender and substance abuse. It is not uncommon for some of these variables to cancel out the effects of others. Admittedly, it is a very complex undertaking and any attempt to establish clear correlations will fail because there is no way to separate out such discrete factors.

15.5.2 Predicting Child Abuse and Domestic Violence

The prediction of child abuse and domestic violence after commission of an act that has come to the attention of the legal authorities has had a history similar to that of criminal and sexual violence. Interestingly, since the violent acts were mostly within families, there was less attention paid to assessment of future violence. Nor was there sufficient attention paid to continuing violence especially when coercive techniques were psychological in nature rather than causing only physical harm (Walker 2018). Child abuse, especially child sexual abuse, by definition consists of physical, sexual, and psychological harm yet assessment of risk within the family is rarely done by courts in most countries around the world. Although the research indicates that child abuse also occurs somewhere between 40 and 60% of homes where domestic violence (interpersonal partner violence or IPV) occurs (Walker et al. 2013), rarely are children protected from further abuse when parents separate or divorce.

The importance of the use of risk assessment procedures when violence and abuse are reported in the family cannot be overstated here. The research on criminal violent offenders indicates a high percentage of those incarcerated were abused as children. (Of some interest is that while a history of child abuse is one of the parameters explored by means of structured professional judgment, it is notably absent from actuarial assessment of future violent behavior.) Many also lived in families where domestic violence occurred including high rates of IPV. Studies of children and youth seduced into sex trafficking also indicate a high rate of child abuse and IPV in their homes (Walker et al. 2019).

Child Abuse and Risk of Violence

Research has shown that there is a significant correlation between abuse history and delinquent behavior, especially for boys (Chesney-Lind 2013). The more severe or chronic the child abuse, the poorer the psychological outcomes for the youth. They are less well liked by peers, have difficulty making friends, and have a lower self esteem (Jacob 2007). In fact, their attachments to other people and their community leave them more vulnerable to develop pro-social or anti-social behaviors depending on the characteristics and values of those with whom they associate (Browne et al. 2005; Catelono and Hawkins 1996).

The literature on juveniles who are at risk to develop delinquent behaviors is similar to that of the adults found in the Monahan study described. Simões and Batista-Foguet (2008) found that biology, genetics, physiology, cognitive and behavioral determinants of behavior all play a role in risk of future violence. It has long been known that children with a difficult temperament characterized by negative moods and difficulty in controlling emotions early in life have these markers for later anti-social behavior (Guerin et al. 1997). Add to these an individual's positive attitude towards violence, deficient self-control, delayed maturation, depression and withdrawn behavior and the risk for violence is significantly raised (Loeber et al. 2003).

It is important to remember that the adolescent brain is not fully developed until most youth are in their early to mid twenties. The brain undergoes both further myelination and pruning of neurons before youth have full control of many of the functions needed to make good decisions especially in prosocial areas in groups and to control their impulses and plan ahead. The U.S.S.C. case, *Graham v. Florida* (2010) has now prohibited anyone under 18 who commits a serious violent crime to be eligible for the death penalty and *Miller v. Alabama*, (2012) does the same for life without parole. These legal decisions are primarily based on the fact that the brain of a youthful offender can be expected to mature further although whether or not that youth will then engage in pro-social behavior is still not known.

15.6   Risk Assessment Instruments

There are large amounts of data regarding risk assessments from the point of view of demographic, psychological, sociological, biological and contextual bases. There also are a variety of different techniques to be utilized by the mental health professional all the way from the unaided standard clinical interview method, the purely actuarial or adjusted actuarial model and the structured professional judgment (SPJ) approach. It is necessary to make judgment calls in fields as diverse as interpersonal violence in general or specific populations including domestic, workplace and school violence. The various approaches to risk assessment other than the unaided clinical interview are essentially equivalent in their ability to assess violent behavior despite the many arguments found in the literature. After an extensive and exhaustive review of the different risk assessment approaches, Heilbrun et al. (2009) concluded that carefully used, a combination of clinical interviewing and the statistically based actuarial approach may yield the most accurate results. Actuarial instruments alone have limitations both in using only past history that is static and that it cannot be changed. Actuarials are developed using complicated statistical methods that could be manipulated both in its use and its interpretation. In addition, the research regarding the accuracy of many of the assessment instruments is limited. On the other hand, clinical interviews without the careful addition of important factual data may have been found to yield the poorest results. Considering the wide-reaching consequences, there are troubling ethical issues raised here.

Basically, when we obtain a predictive value for someone in a group, depending on certain other variables, the actual risk for that person may be even higher or lower than the group mean. If, for instance, an individual scores in the moderate range on a risk assessment measure but has a mental illness, presents with features of psychopathy and paranoia, has a history of self-medicating with illicit substances, refuses to take his or her prescribed medication, is unemployed and has a history of traumatic brain injury, the risk of future violent behavior would more likely be high rather than moderate. But, someone who may have been raised in a home with only one parent and had a child out of wedlock, which are high risk demographic factors, but has never been in trouble with the law nor committed any violent act, would be low rather than moderate despite the group category within which their scores placed them. Consequently, the importance of modifications to the overall likelihood of future violence should considered as many variables as possible when calculating risk assessment.

A final issue about assessment methodology worth discussing is that there may not be sufficient data available to make an assessment in any given case. For example, we may not have a complete or factual mental health history, an accurate assessment of the presence of a personality disorder, incomplete records relating to previous violence (or the individual not telling the truth about it) and no records relating to the person's employment status. In the event that there is such a scarcity of information, we need to be able to say that there is insufficient data to make a complete assessment. That would be the only statement a truly conscientious practitioner could make, and yet courts tend to demand more definitive answers and often find mental health professionals willing to do so based on the inadequate data.

The assessment instruments that deal with prediction of violence in general and domestic violence appear to be the most relevant to the topics under consideration in this article even though there are others such as those used to assess sexual violence recidivism that may not be particularly relevant here.

The range of assessment techniques is very extensive and will not be described in this chapter. In general however, there are actuarial assessments, which, as noted earlier attempt to predict future violent behavior based on the presence (or absence) of certain demographic, static variables which are correlated with violent behavior. However, this results in group, not individual data. That is, the prediction is that this individual belongs to a group that has a certain percentage likelihood of acting out violently. It does not say that this particular individual is at a certain percentage of acting out violently.

Some practitioners attempt to deal with this by doing an "adjusted actuarial "assessment in which they use the actuarials to obtain an "anchoring point" and then get at the individual factors through the investigation of variables specific to that individual; however, this suffers from the same difficulties as the unstructured clinical interview, since there is no standardization of the inquiries.

The third approach is the one most frequently seen these days called "Structured Clinical Judgment". Here, the research literature is used to identify areas in need of inquiry, but how the inquiry is done is left to the clinical discretion of the examiner; it is, therefore, a blend of clinical and actuarial approaches.

Risk assessment instruments can generally be described as falling into one of these three approaches, though the specific topics to be covered may vary, such as violence in general, sexual violence, or domestic violence.

Observations and Limitations
Risk assessment measures, whether actuarial in nature or relying on structured professional judgment, typically have one major element in common: They produce results that are only slightly better than chance. This level seems to hold regardless of whether we are talking about sexual recidivism or violent recidivism It is not uncommon for studies concerning the predictive validity of risk assessment tools to produce results that conflict with the findings of other studies. Clearly, this is an exceedingly complex area and within this brief overview we cannot discuss these in any great detail. However, a few general comments are in order.

Sampling errors can occur when there are discrepancies between the characteristics of a sample chosen to represent a population as well as when findings are generalized despite it being inappropriate to do so. Sampling errors are known to occur when different groups of violent or sexual offenders are assumed to represent all such offenders. For example, a study that utilizes a sample composed of rapists to determine the predictive validity of a risk assessment instrument for sex offenders is unlikely to produce the same results as would be obtained from a sample of child molesters due to the differences in the known risk factors that predict recidivism in both groups despite some similarities (Firestone et al. 2000; Hanson and Morton-Bourgon 2006).

Even the kinds of predictions made may vary. Becker (2014), in a presentation at the APA-ABA conference on Violence and the Family, noted that certain risk factors may predict violence in general, while others may predict an initial sexual offense, while still others may predict sexual recidivism. Becker notes that no one etiological theory is widely accepted in the field. Some theories include, but are not limited to, brain structure, hormonal abnormalities, cognitive distortions, poor social learning skills and a variety of personality variables. Those who commit violence are a heterogeneous group despite some similarities so Becker suggests that future research follow a model that widens the range of both predictor and criterion variables being studied. Thus, specificity of the person's past behavior may be critical to factor into risk assessment.

Another area that limits some of the actuarial instruments has to do with those who commit violence having a history of some form of psychological intervention or psychotherapy. While there is no known treatment effective with all violent offenders, some may benefit from insight- oriented therapy while others may respond to a more cognitive or behavioral approach. In most of the actuarial measures discussed, there is little room for improvement from treatment.

An offender's race, culture and ethnicity has also been found to influence the outcome when examining an instrument's predictive validity. For instance, Varela et al. (2013) found that the Static-99 (and Static-99R; Hanson and Thornton 1999, 2012) were far less effective in predicting sexual re-offending with Latino offenders than they were with African Americans and Caucasians. It may well be premature to assume that these tools are equally predictive across the different diversity classifications. A frequent occurrence, is the tendency of forensic examiners to either be oblivious to these findings or just dismissive of them (Shapiro et al. 2019). The country in which a research study took place has also been found to influence the reported predictive ability of an instrument. In a meta-analysis examining the accuracy of assessment measures for identifying an offender's risk of sexual re-offending, Hanson and Morton-Bourgon (2009) determined that the effect sizes produced by such instruments were strongest in the United Kingdom and rather weak in the United States. Since the majority of the assessments were derived in Canada and the United Kingdom, this is not particularly surprising since the United States has a different diverse ethnic make-up. Additional research is needed to determine whether these differences are or are not exclusively attributable to ethnic or differences.

Research validity describes how valid a measure is when applied by researchers who have received training in the administration and coding of the assessment. Field validity has to do with the predictive validity of the same measure when administered and scored by clinicians. Often, the research validity surpasses the field validity because the researchers have greater familiarity with the operational definitions and coding criteria used during a tool's development. Murrie et al. (2012), for instance, found this to be the case in some of their evaluators scoring the PCL-R. Similarly, in one recent study (Singh et al. 2013), it was found that studies authored by an instrument's designers reported predictive validity rates almost twice as high as those found by independent researchers. They identified this as a result of "authorship bias."

Another area in need of more research is the interactive effect of various risk factors. Dvoskin (2014) in the APA/ABA conference on Violence and the Family, noted the misperception propagated by the media that mental illness and violence are highly correlated. In fact, only 5–10% of violent crime can be attributed to severe mental illness alone. While most mentally ill individuals do not act violently, they often have other risk factors, such as child abuse, victimization, unemployment and substance abuse. Co-occurrence with substance abuse, for instance, is known to multiply the base rate by a factor of 3. Dvoskin also notes a high correlation between those few mentally ill individuals who do act violently and suicidal ideation/behavior on their parts. He recommends doing a suicide risk assessment as part of any violence risk assessment. Clearly, a more complex analysis, such as the MacArthur studies, needs to be utilized on a widespread basis.

15.7 Duty to Protect or Duty to Warn Laws

The issues of disclosure and privacy have become a problem in the U.S. as the laws in some states are changing from requiring disclosure of personal health care information upon the health care professional's reasonable belief of the person's potential for harm to him or herself or others to a duty to report such suspicions to the authorities. Previously, the health care professional had a duty to protect the patient from harm to self or others by choosing from a variety of other options including report. These could include involuntary or voluntary hospitalization, warning an intended victim, or discussion with family members among other options. In some states the health professional's judgment is no longer important; the potential for violence shall be reported under the penalty of criminal arrest as well as loss of the license to practice as a health care professional. In fact, under certain circumstances, the clinician is not even responsible for doing a formal risk assessment. Rather, a reasonable belief that the patient is a danger to himself or others would be sufficient for the breach of confidentiality despite Federal laws like HIPAA that allegedly protect the patient's privacy.

When deciding whether the public interest in disclosing information outweighs the patient's and the public interest in keeping the information confidential, the clinician must consider:

a. the potential harm or distress to the patient arising from the disclosure—for example, in terms of their future engagement with treatment and their overall health
b. the potential harm to trust in doctors generally—for example, if it is widely perceived that doctors will readily disclose information about patients without consent
c. the potential harm to others (whether to a specific person or people, or to the public more broadly) if the information is not disclosed
d. the potential benefits to an individual or to society arising from the release of the information
e. the nature of the information to be disclosed, and any views expressed by the patient
f. whether the harms can be avoided or benefits gained without breaching the patient's privacy or, if not, what is the least intrusive method?

15.8 Do Health Care Providers Have a Duty to Prevent Violent Behavior?

Recent trends in the United States appear to be relying more on mandatory reporting and taking professional judgment away from doctors and therapists. These new laws seem to be a response to try to prevent violence. For example, in Florida, the legislature passed a new law that makes reports mandatory after a particularly devastating school shooting killing 17 students and teachers. The shooter, a known mentally ill and violent individual had been expelled from school, and had been in treatment at a local clinic, but the authorities did not follow up on his violence potential. This type of report to authorities remains controversial amongst health care professionals but the general public cries out each time a shooter's prior mental health history is revealed in the media and will blame the therapist, believing that disclosure will reduce the likelihood of violent acting out.

There is limited research on the effectiveness of duty to warn or protect laws. Beck (1982) presented some evidence based on case studies that mandatory warning of third parties had little or no apparent effect on preventing violence. He observed that of 19 cases studied, mandatory warning had a positive effect (preventing violence) in only 2, a negative effect in 4 (violence occurred despite the warning) and no effect in 13 cases. Givelber et al. (1984) noted in their research, that notifying an intended victim by a therapist served to reduce violence by only one percentage point. Despite these limitations, there continues to be a popular misperception that if therapists reported more frequently, then violence would be reduced. Some more recent research is also consistent with Givelber's and Beck's earlier findings. Edwards (2010) found that mandatory duty to warn laws actually increased the rate of homicide by 8.9%, because patients would not share their violent fantasies and therapists would not pursue inquiry into them. Givelber hasd also alluded to this, in a somewhat different manner, that patients who had problems with violent behavior would not even seek psychotherapy for fear of disclosure, and that if they did reveal their violent fantasies to therapists there would be an increased number of unnecessary involuntary commitments. In a subsequent paper, Edwards (2013) also observed an increase in the rate of adolescent suicide of approximately the same size when there was a mandatory duty to warn. Once again, these findings lead us to some troubling ethical issues regarding the actions we take based on inadequate data, including violation of confidentiality, especially when the prediction is a false positive, improper use of assessment techniques, having sufficient data on which to base an assessment, and failure to take reasonable steps to protect the people with whom we work.

The parallel to a therapist's having a mandatory duty to report a threat of potential terrorist activities is worthy of note. Just like in the previous examples dealing with mandatory duty to warn laws, we have the concern about therapists, based on inadequate research data, making predictions that can result in serious harm to individuals, especially in the case of false positives. We simply do not have the available research yet to distinguish those who may profess allegiance to violent terrorism and those who will actually carry it out. In fact, since the feeling of being betrayed by society appears to be one of the common themes in those who do perpetrate violence, the same or increased sense of betrayal may occur when a therapist makes an inaccurate prediction of future terrorist activities; it could, in fact be argued that this heightened sense of betrayal by a therapist could incite a desire to retaliate (Meloy 2018).

What remedies, if any could be effective in stopping this racing train, that demands disclosure of possible terrorist activities, based on inadequate data? First, of course, is to educate the profession and to educate the legislators. As noted earlier in the discussion of some states in the United States that have moved from a standard involving professional discretion to one of mandatory notification of law enforcement, it is unlikely that any such laws had been informed by psychological research. Perlin (2016) has discussed the heuristic which he calls 'Ordinary common sense" which is what laypeople believe the truth to be, regardless of research to the contrary. Legislators will often pass laws that seem "reasonable," but are actually contrary to any established scientific research; Some examples are laws restricting the insanity defense based on the "ordinary common sense" that the defense is misused very frequently; research has demonstrated the misperception that successful insanity defenses occur between 45 and 75% of the time, when in fact they occur about 0.0025% of the time. In a similar manner, many legislatures have passed legislation dealing with "sexually violent predators" based on the belief that there is some" mental abnormality" that predisposes an individual to predatory acts of sexual violence when in fact there is no such disorder. Why is there so little influence on legislation of scientific research? Sadly, it appears to rest in the fact that mental health professionals do not involve themselves adequately in communicating the results of research to lawmakers; certainly the professionals involved in the Therapeutic Jurisprudence movement try to do so, but even there, the impact on legislation is not as great as we would hope. The mental health community needs to devise more effective ways of making themselves alert to pending legislation and make interventions as necessary to inform legislators of the science underlying areas of law influencing mental health.

15.9   Conclusions

This article has attempted to answer the questions (1) whether it is possible to make an accurate prediction of future use of violence; (2) if so, what are the potential outcomes of such disclosures; and (3) what are the ethical concerns for the health care profession in general as well as the individual professional? In reviewing the history of the field of prediction of violent behavior the research shows that it is not a yes or no answer but rather one of how accurate is the risk assessment of future violence based on the particular methodology utilized. There are some variables that are known about offenders committing various types of violence, but it needs more research to determine levels of accuracy. As might be expected, politics makes the obtaining of such data difficult if not impossible. Nor has there been sufficient research on the outcomes of disclosure of previously considered private information. The little that has been cited does not suggest that reporting potential for future violence has had much impact on either prevention or recidivism of violent acts by offenders.

Ethical concerns include the methodology used to assess the risk of violence, the need to augment the typical clinical interview with the research, the inability to predict from group data, and the role of health care professionals in treatment approaches. We have noted that even with the most sophisticated methodology our ability to predict future violence is at best only slightly better than chance. This of course is of concern because of our ethical standard that allows us to make conclusions, diagnoses, and recommendations only when we have data sufficient to justify the conclusion. Does data that is only slightly better than chance meet this criterion?

It is also of concern that the current research regarding the accuracy of violence risk assessment is based on extensive use of and integration of collateral data with clinical judgment (review of past records, interviews with family and friends, etc.). Rarely, if ever, is such extensive research done in settings where the professional contact is one psychotherapy.

Some of the benefits of the use of actuarial instruments as well as the difficulties were presented. Some of these difficulties include lack of precise definitions of the terms being measured resulting in poor validity as well as reliability of predictions of recidivism, laws that create standards that are at odds with the mental health profession, unchallenged myths such as the lack of cause and effect relationship between violence and diagnosis of serious mental illness, ethical issues with professionals use of inadequate data to come to decisions of risk, and need for more research across populations of diversity.

References

Aarten, P.G., E. Mulder, and A. Pemberton. 2017. The narrative of victimization and deradicalization: An expert view. *Studies in Conflict & Terrorism* 41 (7): 557–572. https://doi.org/10.1080/1057610x.2017.1311111.

Acheson, S. K., J.W. Payne, and D.J. Olmi. 2005. Review of the psychopathy checklist-revised (2nd ed.). In *Sixteenth mental measurements yearbook* ed. B.S. Plake, J.C. Impara, and R.A. Spies. Lincoln: Buros Institute of Mental Health.

American Medical Association. 2016. *Confidentiality: Code of medical ethics opinion 3.2.1.* Retrieved from https://www.ama-assn.org/delivering-care/confidentiality.

American Psychiatric Association. 2013. *Diagnostic and statistical manual of mental disorders*, 5th ed. Arlington: American Psychiatric Association.

American Psychological Association. 2002, 2010, 2016. Ethical principles of psychologists and code of conduct. *American Psychologist* 57 (12), 1060–1073. https://doi.org/10.1037/0003-066X.47.12.1597.

Anti-radicalisation strategy: Confidentiality and doctors' responsibilities | The BMA. 2018, December 06. Retrieved from https://www.bma.org.uk/advice/employment/ethics/confidentiality-and-health-records/anti-radicalisation-strategy

Arbisi, P.A. 2003. Review of the HCR-20: Assessing risk for violence. In *The fifteenth mental measurements yearbook* ed. B.S. Plake, J.C. Impara, and R.A. Spies.

Barefoot v. Estelle, 463 U.S. 880 (1983).

Baxstrom v. Herold, 383 U.S. 107 (1966).

Beck, A.T., and R.A. Steer. 1990. *Manual for the beck anxiety inventory*. San Antonio: Psychological Corporation.

Becker, J. 2014. *Recidivism in sex offenders*. Presentation at APA/ABA Conference on Family Violence.

Beck, A.T., R.A. Steer, and G.K. Brown. 1996. *Manual for the beck depression inventory-II*. San Antonio, TX: Psychological Corporation.

Behnke et al. v. Hoffman et al. . 2017. CA 005989 B (DC Superior Court, filed August 28, 2017).

Borum, R. 2007. Psychology of fighting. *Mental Health Law & Policy Faculty Publications., 570.*

Borum, R. 2011. *Psychology of terrorism*. Lexington: University of South Florida.

Bowen, E. 2011. An overview of partner violence risk assessment and the potential role of female victim risk appraisals. *Aggression and Violent Behavior* 16 (3): 214–226. https://doi.org/10.1016/j.avb.2011.02.007.

Boyle, D.J., K.D. O'Leary, A. Rosenbaum, and C. Hassett-Walker. 2008. Differentiating between generally and partner-only violent subgroups: Lifetime antisocial behavior, family of origin violence, and impulsivity. *Journal of Family Violence* 23 (1): 47–55. https://doi.org/10.1007/s10896-007-9133-8.

Briere, J.N. 1996. *Trauma symptom checklist for children: Professional manual*. Odessa: Psychological Assessment Resources (PAR).

Briere, J., D.M. Elliot, K. Harris, and A. Cotman. 1995. Trauma Symptom inventory: Psychometrics and association with childhood and adult trauma in clinical samples. *Journal of Interpersonal Violence* 10: 387–401.

British Association of Social Workers. 2014. *The code of ethics for social work: Statement of principles*. Retrieved from www.basw.co.uk/codeofethics.

Brown, C., et al. 2005. Mediator effects in the social developmental model: An examination of constituent theories. *Criminal Behavior and Mental Health* 15: 221–235.

Canada Medical Association. 2018. *CMA code of ethics and professionalism*. Retrieved from https://www.cma.ca/cma-code-ethics-and-professionalism

Canadian Association of Social workers. 2005. *Canadian association of social workers code of ethics.*

Canter, D., S. Sarangi, and D. Youngs. 2012. Terrorists personal constructs and their roles: A comparison of the three Islamic terrorists. *Legal and Criminological Psychology* 19 (1): 160–178. https://doi.org/10.1111/j.2044-8333.2012.02067.x.

Catalano, R.F., and J.D. Hawkins. 1996. The social development model: A theory of antisocial behavior. In *Cambridge criminology series. Delinquency and crime: Current theories* ed. J.D. Hawkins, 149–197. New York: Cambridge University Press.

Cawood, J.S., and M.H. Corcoran. 2009. *Violence assessment and intervention: The practitioner's handbook-2nd edition*. Boca Raton, FL: CRC Press.

Chesney-Lind, M., & Pasko, L. (2013). *The female offender: Girls, women, and crime*. Los Angeles: SAGE.

Cleckley, H.M. 1964. *The mask of sanity: An attempt to reinterpret the so-called psychopathic personality*. St. Louis: Mosby.

Conroy, M.A., and D.C. Murrie. 2007. *Forensic assessment of violence risk a guide for risk assessment and risk management*. Hoboken: Wiley.

Demant, F., and B.D. Graaf. 2010. How to counter radical narratives: Dutch deradicalization policy in the case of Moluccan and Islamic radicals. *Studies in Conflict & Terrorism* 33 (5): 408–428. https://doi.org/10.1080/10576101003691549.

Department of Education. 2011, July 01. *Teachers' standards*. Retrieved from https://www.gov.uk/government/publications/teachers-standards.

Douglas, K.S., C. Shaffer, A.J.E. Blanchard, L.S. Guy, K. Reeves, and J. Weir. 2014. *HCR- 20 violence risk assessment scheme: Overview and annotated bibliography (HCR-20 Violence Risk*

Assessment White Paper Series, No. 1). Burnaby, British Columbia, Canada: Mental Health, Law, and Policy Institute, Simon Fraser University.

Dvoskin, J. 2014. *Clinical predictors of violent behavior.* Presentation at APA/ABA Conference on Family Violence.

Edwards, G.S. 2010. *Data base of state tarasoff laws.* https://ssrn.com/abstract=1551505, 2/11/2010.

Edwards, G.S. 2013. Tarasoff, duty to warn laws, and suicide. *International Review of Law and Economics* 34.

Estelle v. Smith, 451 U.S. 454 (1981).

Ethics Committee of the British Psychological Society. 2018. *BPS code of ethics.* Leicester: The British Psychological Society, St. Andrews House. European Federation of Psychologists' Association. (2015). Model-Code. Retrieved from http://ethics.efpa.eu/metaand-model-code/model-code/.

Firestone, P., J.M. Bradford, D.M. Greenberg, and G.A. Serran. 2000. The relationship of deviant sexual arousal and psychopathy in incest offenders, extrafamilial child molesters, and rapists. *Journal of the American Academy of Psychiatry and the Law* 28 (3), 303–308.

General Medical Council. 2017. *Confidentiality: Good practice in handling patient information,* §§ 64, 65 e 67. Retrieved from https://www.gmcuk.org/Confidentiality_good_practice_in_hand ling_patient_information___English_417.pdf_70080105.pdf).

Givelber, D.J., W.J. Bowers, and C.L. Blitch. 1984. Tarasoff, myth and reality: An empirical study of private law in action. *Wisconsin l. Rev.* 2: 443–497.

Graham v. Florida, 560 U.S. 48.

Guerin, D.W., A.W. Gottfried, and C.W. Thomas. 1997. Difficult temperament and behaviour problems: A longitudinal study from 1.5 to 12 years. *International Journal of Behavioral Development* 21 (1): 71–90.http://dx.doi.org/https://doi.org/10.1080/016502597384992.

Hanson, R.K., and D. Thornton. 1999, 2012. *Static 99: Improving actuarial risk assessments for sex offenders.* Ottaway, ONT.: Canada Communication Group.

Hanson, R.K., and K.E. Morton-Bourgon. 2006. The characteristics of persistent sexual offenders: A meta-analysis of recidivism studies. *Clinical Forensic Psychology and Law* 67–76. https://doi.org/10.4324/9781351161565-5.

Hanson, R.K., and K.E. Morton-Bourgon. 2009. The accuracy of recidivism risk assessments for sexual offenders: A meta-analysis of 118 prediction studies. *Psychological Assessment* 21 (1): 1–21. https://doi.org/10.1037/a0014421.

Hare, R.D. 1980. A research scale for the assessment of psychopathy in criminal populations. *Personality and Individual Differences* 1 (2): 111–119. https://doi.org/10.1016/0191-8869(80)900 28-8.

Hare, R.D. 1985. Comparison of procedures for the assessment of psychopathy. *Journal of Consulting and Clinical Psychology* 53 (1): 7–16. https://doi.org/10.1037//0022-006x.53.1.7.

Hare, R.D. 1991, 2003. *The Hare psychopathy checklist-revised: Manual.* Multi-Health Systems, Incorporated.

Hare, R.D. 1996. Psychopathy: A clinical construct whose time has come. *Criminal Justice and Behavior* 23 (1): 25–54. https://doi.org/10.1177/0093854896023001004.

Hare, R.D., S.D. Hart, and T.J. Harpur. 1991. Psychopathy and the DSM-IV criteria for antisocial personality disorder. *Journal of Abnormal Psychology* 100 (3): 391–398. https://doi.org/10.1037/0021-843x.100.3.391.

Harpur, T.J., R.D. Hare, and A.R. Hakstian. 1989. Two-factor conceptualization of psychopathy: Construct validity and assessment implications. *Psychological Assessment: A Journal of Consulting and Clinical Psychology* 1 (1): 6–17. https://doi.org/10.1037/1040-3590.1.1.6.

Harris, G.T., Rice, M.E., and Quinsey, V.L. 1993. Violent recidivism of mentally disordered offenders: The development of a statistical prediction instrument. *Criminal Justice and Behavior* 20 (4), 315–335. http://dx.doi.org/https://doi.org/10.1177/0093854893020004001.

Harris, G.T., M.E. Rice, V.L. Quinsey, and C.A. Cormier. 2015. *Violent offenders: Appraising and managing risk.* Washington: American Psychological Association.

Hart, S.D., P.R. Kropp, and R.D. Hare. 1988. Performance of male psychopaths following conditional release from prison. *Journal of Consulting and Clinical Psychology* 56 (2): 227–232. https://doi.org/10.1037/0022-006x.56.2.227.

Hawkins, J.D., T.L. Herrenkohl, D.P. Farrington, D. Brewer, R.F. Catalano, and T.W. Harachi. 1998. A review of predictors of youth violence. In *Serious and violent juvenile offenders: Risk factors and successful interventions,* ed. R. Loeber, and D.P. Farrington, 106–146. Thousand Oaks, CA: Sage Publications.

Helmus, L., and G. Bourgon. 2011. Taking stock of 15 years of research on the spousal assault risk assessment guide (SARA): A critical review. *International Journal of Forensic Mental Health* 10 (1): 64–75. https://doi.org/10.1080/14999013.2010.551709.

Heilbrun, K., T. Grisso, and A.M. Goldstein. 2009. *Foundations of forensic mental health assessment.* Oxford: Oxfird University Press.

Hilton, N.Z., G.T. Harris, and M.E. Rice. 2001. Predicting violence by serious wife assaulters. *Journal of Interpersonal Violence* 16 (5): 408–423. https://doi.org/10.1177/088626001016 005002.

Hilton, N.Z., G.T. Harris, and M.E. Rice. 2007. The effect of arrest on wife assault recidivism, controlling for pre-arrest risk. *Criminal Justice and Behavior* 34 (1334): 1344. https://doi.org/10.1177/0093854807300757.

Hilton, N.Z., G.T. Harris, M.E. Rice, R.E. Houghton, and A.W. Eke. 2008. An indepth actuarial assessment for wife assault recidivism: The Domestic violence risk appraisal guide. *Law and Human Behavior* 32 (2): 150–163. https://doi.org/10.1007/s10979-007-9088-6.

Hilton, N.Z., G.T. Harris, and M.E. Rice. 2010. *Risk assessment for domestically violent men: Tools for criminal justice, offender intervention, and victim services.* Washington: American Psychological Association.

International Federation of Social Workers. 2012. *Statement of ethical principles.* Retrieved from http://ifsw.org/policies/statement-of-ethical-principles/.

Jacob, B. 2007. Nature and nurture predispositions to violent behavior: Serotonergic genes and adverse childhood experience. *Neuropsychopharmacology.*

Jurek v. Texas, 428 U.S. 262 (1976).

Kalmus, D. 1984. The intergenerational transmission of violence in the family. *Journal of Marriage and the Family* 46: 11–19.

Katz, L. 2003. Prison conditions, capital punishment, and deterrence. *American Law and Economics Association* 5 (2): 318–343. https://doi.org/10.1093/aler/ahg014.

Kleinman, P.K. 2016. *Diagnostic imaging of child abuse.* Cambridge: Cambridge University Press.

Kropp, P.R., S.D. Hart, C.D. Webster, and D. Eaves. 1995. *Manual for the spousal assault risk assessment guide,* 2nd ed. Vancouver, British Columbia: British Columbia Institute on Family Violence.

Kropp, P.R., S.D. Hart, C.D. Webster, and D. Eaves. 1999. *Manual for the spousal assault risk assessment guide (Version 3).* Vancouver: British Columbia Institute Against Family Violence.

Kropp, P.R., and S.D. Hart. 2000. The spousal assault risk assessment (SARA) guide: Reliability and validity in adult male offenders. *Law and Human Behavior* 24 (1): 101–118. https://doi.org/10.1023/a:1005430904495.

Lessard v. Schmidt, 349 F. Supp. 1078 (E.D. Wis. 1972).

Loeber, R., D.P. Farrington, and D. Petechuk. 2003. Child delinquency: Early intervention and prevention. *PsycEXTRA Dataset.* https://doi.org/10.1037/e510482006-001.

Los Angeles County Department of Children and Family Services. (n.d.). Retrieved from http://dcfs.co.la.ca.us/.

Marczyk, G., D. DeMatteo, and D. Festinger. 2005. *Essentials of behavioral science series. Essentials of research design and methodology.* Hoboken: John Wiley & Sons Inc.

Marcus, D.K., S.L. John, and J.F. Edens. 2004. A Taxometric Analysis of psychopathic personality. *Journal of Abnormal Psychology* 113 (4): 626–635. https://doi.org/10.1037/0021-843x. 113.4.626.

McCauley, C., and S. Moskalenko. 2008. Mechanisms of political radicalization: Pathways toward terrorism. *Terrorism and Political Violence* 20 (3): 415–433.

McNiel, D.E., A.L. Gregory, J.N. Lam, R.L. Binder, and G.R. Sullivan. 2003. Utility of decision support tools for assessing acute risk of violence. *Journal of Consulting and Clinical Psychology* 71 (5): 945–953. https://doi.org/10.1037/0022-006X.71.5.945.

Meloy, J.R. 2000. *Violence risk and threat assessment: A practical guide for mental health and criminal justice professionals*. San Diego: Specialized Training Services.

Meloy, J.R. 2018. The operational development and empirical testing of the terrorist radicalization assessment protocol (TRAP). *Journal of Personality Assessment*.

Miller v. Alabama, 567 U.S. 460.

Milner, J.S., R.G. Gold, and R.C. Wimberley. 1986. Prediction and explanation of child abuse: Cross-validation of the child abuse potential inventory. *Journal of Consulting and Clinical Psychology* 54 (6): 865–866. https://doi.org/10.1037//0022-006x.54.6.865.

Monahan, J. 1981. The clinical prediction of violent behavior. *Crime & Delinquency Issues: A Monograph Series, ADM* 81–921: 134.

Monahan, J. 1992. Mental disorder and violent behavior: Perceptions and evidence. *American Psychologist* 47 (4): 511–521. https://doi.org/10.1037//0003-066x.47.4.511.

Monahan, J., and H.J. Steadman. 1996. *Violence and mental disorder: Developments in risk assessment*. Chicago: Univ. Chicago P.

Monahan, J., H.J. Steadman, P.C. Robbins, P. Appelbaum, S. Banks, T. Grisso, K. Heilbrun, E. Mulvey, L. Roth, and E. Silver. 2005. An actuarial model of violence risk assessment for persons with mental disorders. *Psychiatric Services* 56 (7): 810–815. https://doi.org/10.1176/appi.ps.56.7.810.

Murrie, D.C., M.T. Boccaccini, J. Caperton, and K. Rufino. 2012. Field validity of the psychopathy checklist-revised in sex offender risk assessment. *Psychological Assessment* 24 (2): 524–529. https://doi.org/10.1037/a0026015.

National Association of Social Workers. 2017. *Code of Ethics art*. 1.07. Retrieved from https://www.socialworkers.org/About/Ethics/Code-of-Ethics/Code-of-Ethics-English.

National Association of State Directors of Teacher Education and Certification. 2015. Model Code of Ethics for Educators. Retrieved from https://rm.coe.int/vol-4-codes-of-conduct-for-teachers-in-europe-a-background-study/168074cc72).

National Education Association. 1975. *Code of ethics*. Retrieved from http://www.nea.org/home/30442.htm.

Novaco, R.W. 1994, 2003. *The Novaco anger scale and provocation inventory: NAS-PI*. Los Angeles, CA: Western Psychological Services.

114th German Medical Assembly. 2011. (Model) *Professional code for physicians in Germany*. Retrieved from http://www.bundesaerztekammer.de/fileadmin/user_upload/downloads/MBOen2012.pdf.

Paxton, T. 1962. What did you learn in school today? (Recorded by T. Paxton).

Perlin, M., and A. Lynch. 2016. The four factors: Sanism, pretextuality, heuristics, and ordinary common sense. In *Sexual disability and the law Amazon*.

Pope, K.S. 2018. A human rights and ethics crisis facing the world's largest organization of psychologists. *European Psychologist*: 1–15. https://doi.org/10.1027/1016-9040/a000341.

Quinsey, V.L., G.T. Harris, M.E. Rice, and C.A. Cormier. 1998. *The law and public policy: Psychology and the social sciences series. Violent offenders: Appraising and managing risk*. Washington, DC, US: American Psychological Association. http://dx.doi.org/https://doi.org/10.1037/10304-000.

Quinsey, V.L., G.T. Harris, M.E. Rice, and C.A. Cormier. 2006. *The law and publicpolicy. Violent offenders: Appraising and managing risk*, 2nd ed. Washington, DC, US: American Psychological Association. http://dx.doi.org/https://doi.org/10.1037/11367-000.

Raine, A. 1993. *The psychopathology of crime*. Academic Press.

Raine, A. 2014. *The anatomy of violence*. Pantheon books.

Rogers, R., and D. Graves-Oliver. 2003. Psychological and psychiatric measures in forensic practice. *Principles and Practice of Forensic Psychiatry, 2Ed,* 621–630. https://doi.org/10.1201/b13 499-71.

Royal College of Psychiatrists. 2010. *Good psychiatric practice: Confidentiality and information sharing.* Retrieved from http://www.rcpsych.ac.uk/files/pdfversion/CR160.pdf.

Royal College of Psychiatrists. 2017. *Ethical considerations arising from the government's counterterrorism strategy.* Retrieved from http://www.rcpsych.ac.uk/pdf/PS04_16S.pdf.

Schuurman, B., and J.G. Horgan. 2016. Rationales for terrorist violence in homegrown jihadist groups: A case study from the Netherlands. *Aggression and Violent Behavior* 27: 55–63. https://doi.org/10.1016/j.avb.2016.02.005.

Shapiro, D.L., and A.M. Noe. 2015. *Risk assessment: Origins, evolution and implications for practice.* New York: Springer.

Shapiro, D.L., L. Mixon, M. Jackson, and J. Shook. 2016. Ethical issues in forensic psychology and psychiatry. *Ethics, Medicine and Public Health* 2 (1): 45–58. https://doi.org/10.1016/j.jemep.2016.01.015.

Shapiro, D., and A. Noe. 2018. Risk assessment: Law, theory, and implementation. In *Handbook of behavioral criminology* ed. V. Van Hasselt, and M. Bourke. New York, Springer.

Shapiro, D., and L. Walker. 2019. *Forensic practice for the mental health clinician.* The Practice Institute Press.

Simões, C., J.M. Batista-Foguet, M.G. Matos, and L. Calmeiro. 2008. Alcohol use and abuse in adolescence: Proposal of an alternative analysis. *Child: Care, Health and Development* 34 (3): 291–301. https://doi.org/10.1111/j.1365-2214.2007.00808.x.

Singh, J.P., M. Grann, and S. Fazel. 2013. Authorship bias in violence risk assessment? A systematic review and meta-analysis. *PLoS ONE* 8 (9). https://doi.org/10.1371/journal.pone.0072484.

Smith, I., and T. Hamilton. 2016. *ETINED council of Europe platform on ethics, transparency and integrity in education.* Strasbourg: Council of Europe.

Spielberger, C.D., G. Jacobs, S.F. Russell, and R.S. Crane. 1983. Assessment of anger: The state-trait anger scale. In *Advances in personality assessment*, vol. 2, ed. J.N. Butcher and C.D. Spielberger, 159–187. Hillsdale: Lawrence Erlbaum.

Steadman, H.J., and J.J. Cocozza. 1974. *Careers of the criminally insane: Excessive social control of deviance.* Lexington: D C Heath.

Steadman, H.J., and J.J. Cocozza. 1976. Failure of psychiatric predictions of dangerousness—Clear and convincing evidence. *Rutgers Law Review* 29: 1084–1101.

Steadman, H.J., and J. Cocozza. (1978). Psychiatry, dangerousness and the repetitively violent offender. *The Journal of Criminal Law and Criminology (1973-)* 69 (2): 226. https://doi.org/10.2307/1142396.

Straus, M.A., and R.J. Gelles. 1990. Researching family violence. *Physical Violence in American Families*: 1–92. https://doi.org/10.4324/9781315126401-1.

Tarasoff v. Regents of University of California, 1976. 17 Cal. 3d 425, 551 P. 2d 334, 131 Cal. Rptr. 14.

The Policy, Ethics and Human Rights Committee. 2014. *The code of ethics for social workers.* Retrieved from https://www.basw.co.uk/about-basw/code-ethics.

United States v. Salerno. 1987. 481 U.S. 739.

Varela, J.G., M.T. Boccaccini, D.C. Murrie, J.D. Caperton, and E. Gonzalez Jr. 2013. Do the Static-99 and Static-99R perform similarly for White, Black, and Latino sexual offenders? *The International Journal of Forensic Mental Health* 12 (4): 231–243. https://doi.org/10.1080/14999013.2013.846950.

Walker, L.E.A. 2018. *The battered woman syndrome*, 4th ed. New York: Springer.

Walker, L.E.A. 2012. Seven deadly sins in family court. In *Our broken family court system* ed. L.E.A. Walker, D. Cummings, and N. Cummings. Dryden, New York: Ithaca Press.

Walker, L.E.A., D.M. Cummings, and N.A. Cummings. 2013. *Our broken family court system.* Dryden: Ithaca Press.

Walker, L.E.A., G. Gaviria, and K. Gopal. 2019. *Handbook of sex trafficking.* New York: Springer.

Webster, C.D., K.S. Douglas, D. Eaves, and S.D. Hart. 1997. Assessing risk of violence to others. In *Impulsivity: Theory, assessment, and treatment*, ed. C.D. Webster and M.A. Jackson, 251–277. New York: The Guilford Press.

Weine, S., D.P. Eisenman, L.T. Jackson, J. Kinsler, and C. Polutnik. 2017. Utilizing mental health professionals to help prevent the next attacks. *International Review of Psychiatry* 29 (4): 334–340. https://doi.org/10.1080/09540261.2017.1343533.

Williams, M.J., and S.M. Kleinman. 2013. A utilization-focused guide for conducting terrorism risk reduction program evaluations. *Behavioral Sciences of Terrorism and Political Aggression* 6 (2): 102–146. https://doi.org/10.1080/19434472.2013.860183.

World Medical Association. 2006. *International Code of medical ethics*. Retrieved from http://www.wma.net/policies-post/wma-international-code-of-medical-ethics/.

Yang, et al. 2009. Prefrontal structure and functional brain imaging. In *Antisocial, violent, and psychopathic individuals: A meta-analysis, psychiatry research*, vol. 174, no. 2, 81–88.

David Shapiro is a pioneer in the use and production of forensic psychological assessments in legal settings. Previously he was a Professor in forensic psychology at John Jay College of Criminal Justice in New York and the University of Maryland. He is the author of numerous books in the field. A recipient of the Distinguished Contribution to Forensic Psychology Award from the American Academy of Forensic Psychology, he is known for his specialization in the ethical regulation of expertise in the context of various disciplines.

Chapter 16
PTSD and Biomedical Research: Ethical Conundrums

Lenore Walker

16.1 Introduction

The large number of traumatic events experienced by most humans today bring with them some of the best examples of how physical and mental illnesses and treatment are interrelated and influenced by intersectionality in social conditions. Much of our knowledge has been learned through the lens of the biomedical model that emphasizes the impact of biological determinants of disease amenable to intervention in the health care system. This chapter will address if it is ethical to use a biomedical model when studying PTSD or if there are better ways to understand the psychological distress from trauma.

The focus with a biomedical lens is on diagnosis so that a cure may be found. Social determinants that might influence the impact on individuals are often considered secondary or not relevant. While this model may be helpful under some medical treatment conditions, when studying PTSD and other related trauma treatment it raises ethical conundrums, especially since there may be other research models that may better account for the interactions of various social conditions that modify the individual's trauma experiences. One example is the biopsychosocial approach where biology, psychology, and sociological influences carry equal weight in understanding etiology and treatment of diseases that have a strong mental health component as does PTSD. Another is a modified epidemiological approach where environmental causes in populations are often understood from the point of these intersectionalities impacting biology, (psychology), and sociology as it looks at impact on groups, too (Krieger 1994). In many cases, such as when studying the impact from traumatic stressors, the intervention must involve more than the individual. If the community in which the trauma occurs is not understood, including how it affects others,

L. Walker (✉)
Nova Southeastern University, Fort Lauderdale, FL, USA
e-mail: lewalker@nova.edu

© The Author(s), under exclusive license to Springer Nature Switzerland AG 2023
T. Zima and D. N. Weisstub (eds.), *Medical Research Ethics: Challenges in the 21st Century*, Philosophy and Medicine 132,
https://doi.org/10.1007/978-3-031-12692-5_16

understanding the impact on the individual may not be adequately understood. If that occurs, then any intervention will likely not be successful. The conundrum, of course, is that intersectionalities themselves may bring about bidirectional influences with interaction effects that then modify the individual, so the impact may then be different than expected and difficult if not impossible to understand. It is not that these intersectionalities cannot be studied; rather, the right questions are not being asked when using a biomedical approach that doesn't acknowledge the importance of these influences.

Understanding trauma and its impact on humans has called for the best scientific research that may include biological, psychological, sociological and epidemiological approaches. The more what is learned about body-brain interactions that intersect with a person's race, gender, socioeconomic status and environment, the more necessary it is to focus on ethical ways of knowing what traumatic experiences mean for different people at different times in different places. This may call for newer methods that incorporate what is being called an 'ecosocial' model in public health literature using all these theories together (Krieger 1994, 2003). This chapter will focus on what has been learned about the impact from normal and unusual stressors on the physical and mental health of populations as well as individuals and how mitigating and preventing its toll might be approached ethically. The fact that trauma, by definition, occurs in a social milieu, makes it different from traditional diseases studied and therefore requires a more ecological and social perspective to understand its impact on an individual or a population. One example made clearly visible during the current COVID-19 pandemic are the disparities of who gets access to health care, with minority and economically disadvantaged people being less likely to get care at all due to systemic discrimination and oppression.

It is well accepted that life today exposes most individuals to a variety of different stressors. A recent study using data from the National Study of Daily Experiences suggests that the number and types of daily stressors have been slowly increasing during the last fifty or more years with a particularly high spike from 1990 to 2010 (Almeida et al. 2002). These stressors may be viewed as traumatic for some individuals depending on a number of factors including genetics, neurological factors, physical health, job stress, living conditions among others. These data did not take into account the 2020 pandemic caused by the coronavirus which has intersected with all other stressors experienced by humans around the world. All of these stressors including systemic discrimination and oppression influencing living conditions must be understood in order to persuade people to follow behavior that might lessen the pandemic for the community as a whole. For example, many people live in crowded housing that makes the recommended social distancing impossible. Or, battered women urged to quarantine at home with a batterer may face two different dangers: dying from the virus or from their abuser. It should also be anticipated that teenagers may have a difficult time socially distancing simply because their brain development may not be sufficient for them to choose long term gratification over immediate reward. A public health perspective would provide more data than a biomedical approach in these situations and also encourage population health prevention strategies for high risk groups that are identified.

Traumatic events experienced by individuals independently, but also in groups such as mass shootings and terrorist explosions, can impact people throughout the rest of their lives. Other stressors may also rise to the level of traumatic events in some populations such as civil unrest due to racial injustice and political divisiveness in the U.S. and around the world. Such oppression may cause people to voluntarily leave homes in search of a better life or in some cases be responsible for the forced flight of refugees fearing death. Disturbing images that appear on the news as well as other social media may cause the reexperiencing of previously thought-to- be-healed traumatic memories. The old adage that focused on keeping traumatic memories buried with admonitions not to talk about the experience has been found to perpetuate trauma symptoms, not heal people. In the 21st Century those in the health professions are beginning to learn anew, how to understand and treat people who have been harmed by deadly and harmful events that occur; some accidental and others perpetrated by another person or persons.

The understanding that traumatic experiences can cause some people to have severe psychological distress long after they have occurred is not new. Such angst is portrayed in the ancient Greek tragedies as well as in other stories from the past. As psychology began to develop as a discipline, traumatic experiences from war began to be studied, opening up a new field for research. Van der Kolk (2015) describes how the research on soldiers in World War II who were originally labeled as having developed 'shell shock' back during World War I helped define the current category now labeled Post Traumatic Stress Disorder (PTSD). Within a relatively short time, further study helped expand the category to include other traumatic experiences including those uncovered by Freud and his followers on childhood sexual abuse, rape, and other intentional and accidental events causing death and destruction (c.f. Walker 2015). The most recent *APA Diagnostic and Statistical Manual of Mental Disorders (DSM-5)* (APA 2013) has moved PTSD and related trauma disorders from the anxiety category to its own category emphasizing the growth in both diagnostic tools and available treatment.

Benjet and 30 colleagues recently surveyed almost 70,000 respondents in 24 countries located in six continents to determine exposure to 29 named traumatic event types. Over 70% of those who responded reported having experienced a traumatic event. Statistical techniques were utilized to try to determine if events were clustered into interpretable factors. Five types experienced by over half of those who responded were found; witnessing death or serious injury, the unexpected death of a loved one, being mugged, experiencing a life-threatening accident, and experiencing a life-threatening illness or injury. Exposure varied by country, sociodemographic factors, and history of prior traumatic events. Exposure to interpersonal violence had the strongest association with experiencing multiple traumatic events (Benjet et al. 2016).

Research is being conducted around the world as people are dealing with the effects of the quarantine and economic collapse from the new coronavirus pandemic. In the U.S. the addition of trauma due to racial unrest makes the distress even more intense prompting the American Psychological Association Chief Executive Officer, Dr. Arthur Evans to testify before U.S. Congress calling it a 'syndemic' to emphasize the intersectionality of all these problems. At this time, it is difficult to separate out which factor causes what type of trauma using traditional research methods. Most biomedical scientists prefer the research method called 'random control trials' (RCT)[1] as first level evidence of an intervention's impact under ideal conditions. However, it may be unethical to depend upon RCT as the proper method for several reasons. First, knowing effect sizes between two groups is not useful in understanding risk factors that intersect with both the trauma experience and the intervention studied. Most researchers using this method simplify their study by removing 'outliers' from their sample who often exhibit the intersectionalities discussed earlier. There is also a second important ethical issue raised when choosing who gets into the experimental group and who only gets a placebo as treatment. RCTs are usually conducted with strict laboratory guidelines that often exclude people with multiple problems; sometimes they are the very group who need an intervention being studied.

Thirdly, the interventions available may be limited by focusing only on studies that meet a particular standard, especially if that standard is not similar to the entire population needing the intervention. This occurred when the American Psychological Association chose only research studies using RCTs when recommending clinical guidelines creating a controversy among many psychotherapists who found other interventions equally or more effective (see Kudler 2019; Silver and Levant 2019 for more information). Many trauma survivors have experienced multiple traumatic events that can best be understood using clinical, observational and qualitative research tools that are frequently used by those in the other sciences and social sciences. A fourth important ethical issue is the fact that trauma impacts individuals differently given their background and occurs in a context that also may interact with its strength. The statistical term for this variability is causal effect heterogeneity. When RCTs treat this heterogeneity as error variance, they oversimplify the complexity of psychological processes and assume every trauma in the study is experienced at the same intensity. While it is possible researchers are just not controlling their trials adequately, an ethical violation itself, it is more likely the RCT method has limited usefulness when studying some biopsychosocial problems, especially those disorders caused by trauma, not disease. Newer research methods that control for these multiple variables using within-subject repeated measures designs rather than between-subject designs (Bolger et al. 2019) may give us more accurate ways to assess what portion each intersecting variable plays in the results as well as avoiding the ethical issues noted above. Today, many trauma research projects move forward using some quantitative and some qualitative data using such sophisticated statistical factor analyses.

Consistent with what Benjet et al. (2016) found, the majority of clients or patients with whom psychologists work in professional practice—regardless of the setting and the presenting mental or physical health diagnoses—have experienced psychological trauma at some point in their lives (Kessler et al. 2017; McLaughlin et al. 2013) as well

---

[1] A typical random control trial (RCT) is one where subjects matched on a number of demographic and other variables are randomly placed in the experimental or control group. The experimental group gets the treatment being studied while the control group gets nothing (like remaining on a waitlist) or a placebo designed to make the person think he or she is getting a treatment.

as potentially traumatic victimization at some point in their childhoods (Finkelhor et al. 2009; Teicher et al. 2016). This may be true for the majority of patients who are seen in medical settings, too. Rates of various forms of gender violence such as domestic violence, sexual assault, harassment and exploitation, sex trafficking, rape and child sexual abuse range between 10 and 50% of the population depending on the study (Walker 2017). Research has also found that over half to three-quarters of those sentenced to prison have experienced trauma, often child abuse. There is some evidence that this may also be true for terrorists and those who commit mass shootings (Shapiro 2020).

Although many children and adults who experience psychological trauma recover without experiencing chronic behavioral or physical health problems, at least 15% (and as many as 50–75% who experience traumatic violence, abuse, or exploitation) develop PTSD and many more (33 + %) develop or have a significant increase in the severity of other psychiatric or behavioral and physical health disorders (Alisic et al. 2014; Bryant 2019; McElroy et al. 2016; Pietrzak et al. 2011a, b, c; Santiago et al. 2013; Shonkoff et al. 2009; Smith et al. 2019). Therefore, all health professionals must be prepared to recognize and effectively assess and treat the trauma-related problems that are causing or contributing to the mental, behavioral, and physical health conditions with which their patients are presenting – or to provide appropriate and timely referrals for specialized trauma-related services that are outside of the scope of their expertise and practice.

These data suggest that experiencing a traumatic event that could cause significant physical and mental distress that disrupts a person's ability to function is more commonplace than ever before expected. And both preventing and healing from most traumatic events will take more than physical medicine, calling upon the integration of the social sciences to make lasting change in our social systems. The clearest example is the need to overcome the disparities in health care access by changing social structures maintained by outmoded attitudes and values or else only a fraction of those who need services will be able to receive them; and those who do, may be retraumatized by those who are less privileged.

What is a Traumatic Experience?

Definitions

Psychological trauma often originates from "an event, series of events, or set of circumstances that is experienced by an individual as physically or emotionally harmful or life threatening and that has lasting adverse effects on the individual's functioning and mental, physical, social, emotional, or spiritual well-being" (Ford and Courtois 2020). Broadening this definition finds that besides placing a heavy burden on individuals, families and whole communities, research indicates that traumatic experiences are associated with both behavioral health and chronic physical health conditions, especially traumatic events that occur during childhood or interpersonal events that involve intentional violence or the infliction of other severe harm at any point in the lifespan. The Adverse Childhood Experiences study (ACEs) by Felitti (2001) had documented the lifelong impact of child abuse and other childhood traumas on the person's health and future behavior. Exposure to psychological trauma increases the risk and severity of substance use problems (e.g., smoking, excessive alcohol use, taking drugs), mental health conditions (e.g., depression, anxiety, post-traumatic stress disorder [PTSD], dissociative and psychosomatic disorders, disruptive behavior disorders), other risky behaviors (e.g., self-injury, suicidality, risky sexual encounters), and physical health problems (e.g., cardiovascular, gastrointestinal, metabolic, and immune disorders). These data are consistent with the view that trauma-related disorders inherently involve biopsychosocial aspects (see Trauma and Violence Program (SAMHSA), retrieved from https://www.samhsa.gov/trauma-violence.

Impact from trauma may take many different forms and their psychological impact must be measured as it interacts with other physical, mental health, and social conditions experienced by each individual. Some traumatic events only occur one time but have a long- lasting psychological impact on one person and a lesser impact on others depending on a host of factors including other health conditions, prior traumatic experiences, socioeconomic status, and culture. Some chronic traumatic events also may have a different impact on an individual, again depending on similar factors. Experiencing multiple traumas may affect one person differently than another given other intersecting conditions, such as disparities in racial and socioeconomic access to health care. The current world-wide COVID-19 pandemic has made these disparities very clear (see APA Policy Statement on COVID-19 2020; Webb Hooper et al. 2020). Some traumatic events have more of an impact on biological and neurological functions often depending on what stage of a person's development during which they occur. For example, children may be more likely to develop more complex PTSD or trauma-related disorders due to the impact on their developing brain and biochemical structures (Shonkoff et al. 2009).

Events causing trauma reactions can be classified in different ways. One way is to look at the categories of incidents themselves. Some of the events more commonly expected to cause a traumatic reaction in anyone who experiences them are:

- Disasters that are usually unpredictable such as hurricanes, earthquakes, airplane crashes, and car accidents (some call these 'shock traumas' as they are usually sudden, single-incidents).
- Terrorist actions based on religious or other ideology and mass shootings.
- Combat trauma such as those experienced during war or civil unrest.
- Family trauma such as child abuse, domestic violence, elderly abuse.
- Gender-related trauma such as rape, sexual assault, sexual harassment and sexual exploitation.

Another way to classify trauma is to look at areas where a traumatic reaction might not be expected from the event itself although it does happen. These categories might include:

- Medical trauma where permanent injuries unexpectedly occur.
- Trauma from an unknown virus like the recent COVID-19 pandemic.
- Trauma from immigration or refugee status.
- Trauma from racial or religious injustices.

16.1.1   Interventions for Healing

How traumatic events are defined and classified often determines the types of interventions that may be recommended to assist in their healing. If the classification is only by the type of event as listed above, and treatment does not address the interactions with other factors in a person's life, healing may not occur. Many of those interactions become complex when social conditions are understood. For example, many known terrorists and mass shooters have childhood abuse in their backgrounds (Shapiro 2020; Moldinari et al. 2019). In some cases, a person has experienced multiple traumatic events. Meichenbaum (2019) describes the loss of resiliency each time a different trauma may occur. His examples are with sexually trafficked victims who often have experienced child abuse and domestic violence in their childhood homes. Walker (2017) has found, in cases of car accidents, one person may develop a chronic disc problem in their back while another in the same car walks away with a slight headache. Those who may slip and fall on a broken sidewalk or unseen puddle of water may develop regional pain syndrome resulting in loss of a limb while others heal quickly from a simple bruise. Often prior trauma may partly complicate healing from these injuries.

The most common types of interventions that are known to assist traumatic injuries to heal are pharmacology, psychotherapy, biological devices, telehealth, restorative justice, and civil lawsuits.

16.1.2   Psychopharmacology

Medicine is often seen as the first line treatment for any injury whether it is physical or psychological. However, it is not necessarily the most effective for healing trauma memories although there is some research to suggest memory consolidation can occur if certain pharmaceutical agents such as propranolol are given within a short time after the injury occurs (Ford and Courtois 2020). Obviously if there is repeated or chronic trauma such as occurs in combat veterans or domestic violence survivors, consolidating memories will not be useful unless the medication is administered at the first event. Nor would it likely be helpful if a rape trauma victim also experienced child abuse or other ACEs growing up. There is research suggesting that some individual's genetics are more amenable to this type of treatment than others which may then block development of PTSD itself (Ford and Courtois 2020).

When the treatment focus is on enhancing trauma memory processing, pharmacological agents have shown promise. For example, N-Methyl-3,4-methylenedioxyamphetamine (N-MDMA) "enhances release of monoamines (serotonin, norepinephrine, dopamine), hormones (oxytocin, cortisol), and other downstream signaling molecules (BDNF) to dynamically modulate emotional memory circuits (Mithoefer et al. 2013). By reducing activation in brain regions implicated in the expression of fear- and anxiety-related behaviors, namely the amygdala and insula, and increasing connectivity between the amygdala and hippocampus, MDMA may allow for reprocessing of traumatic memories and emotional engagement with therapeutic processes). Repeated infusions of the N-MDMA receptor agonist ketamine, originally developed as an anesthetic, have been reported to result in remission in 80–98% of adults with treatment resistant PTSD and depression in a pilot clinical trial, although relapse occurred within 3 weeks for depression and 6 weeks for PTSD for 50% of the patients leading to concern about the durability of the benefits. Ketamine is now being used for treatment resistant PTSD and depression as are some psychedelics. These medications may be useful because they enhance relational connection and feelings of safety provided they are administered under proper supervision.

Relief for anxiety symptoms including panic disorders may respond well to anxiolytics and benzodiazepines that are often prescribed without much monitoring by primary care physicians. They work quickly to calm down some of the most annoying symptoms but can become addictive and usually require increasing dosage over time. They also may be dangerous when combined with use of alcoholic substances or opioids and death from intentional or accidental overdoses are reported. Other medications may be used 'off-label' to assist with traumatized individuals who dissociate and need more time to notice what is happening such as a low dose of Nalttrexone prior to a psychotherapy session usually used to reduce the effects of and cravings for alcohol (Bohus et al. 1999). Some reports have been using prazosin (generally used to treat high blood pressure) to improve sleep, including sleeping through nightmares although it has failed some drug trials (http://www.scientificamerican.com/article/a-drug-widely-used-to-treat-ptsd-symptoms-has-failed-a-rigorous-trial/).

Other types of psychotropic medications are frequently prescribed for PTSD, such as antidepressants in the Selective Serotonin Reuptake Inhibitors (SSRIs) classification or even atypical antipsychotics in cases of people with severe symptoms. However, their efficacy as reported in the research is very low despite the fact that doctors prescribe them and people want to take them (Ford and Courtois 2020). This raises an ethical question of whether and how efficacy is being measured if people feel better even though the science does not explain why that happens. We discuss issues about standards and criteria that can be used in research training at the end of this chapter.

16.1.3   Psychotherapy

When the pills, often prescribed by general practitioners who are not specialists in trauma, do not work by themselves, talk therapy is usually the next attempt at trying to heal from trauma. However, the research shows that it probably should be the first line treatment as trauma-focused psychotherapy is the most effective with or without medication. This multi-modal intervention is the usual protocol for treatment in centers specific for PTSD and other related trauma reactions. It is not unusual for some trauma survivors to talk about what happened to them to anyone who will listen, especially when the trauma is not attributed to something they did or did not do or has no shame attached to it. First responders in an earthquake or terrorist event are trained to just talk with the person. Religious clergy are similarly trained. Recently, as more has been learned about PTSD and other trauma reactions, various types of psychotherapy have been provided by psychologists and other mental health providers.

Choosing which type of psychotherapy is often difficult, as measuring effectiveness depends on the type of criteria used. For example, most of the psychotherapy research on effectiveness suggests that lasting relief usually requires a particular methodology that deals with the interpersonal relationship between the psychotherapist and the patient. These are referred to as 'common factors' and described by Norcross and Wampold (2019), Gelso (2014) and others. These conditions are usually found in some type of relationship therapy and often require regular sessions over time. They are discussed below. However, when looking specifically at the reduction of PTSD symptoms, trauma-focused cognitive behavioral therapy (CBT) might be recommended, which is more efficient at reducing the time to eliminate some of the more annoying problems (Briere and Scott 2007). Meichenbaum (2019) adds a component to traditional CBT that includes rebuilding a person's resiliency to treatment to reestablish a feeling of well-being and psychological growth from the experience. Research suggests that resiliency is a factor in preventing repeated victimization that may be seen in victims who do not completely heal from one trauma before experiencing another.

The VA/DOD Working Group (2017) working group has suggested clinical algorithms using a step-by-step decision tree approach to diagnosis and selection of a treatment plan. It includes an ordered sequence of care, recommended observations and examination, decisions to be considered, and a plan of actions to be taken (p. 41). As discussed earlier, in 2017 the APA produced a set of clinical guidelines that were biased towards RCT studies that were rejected by a large number of clinicians who cited other types of psychotherapy as equally or more effective in working with trauma survivors that were not always found in the limited scientific literature reviewed by that task force. It is not known whether the bias occurred because RCT research was favored by the Institute of Medicine where it was published or the task force itself. However, the decision-makers decided other research methods were not deemed appropriate for recommending what was evidence-based treatment for PTSD and other trauma reactions. A special issue of *Psychotherapy: Research and Practice* (September, 2019) suggests that clinical journals and entire literatures were excluded during their guideline searches because they were not seen as relevant for treatment. The APA Clinical Guidelines found cognitive behavioral training (CBT) that addressed symptom reduction was a useful type of psychotherapy basically ignoring other types of treatment being successfully used by psychotherapists including psychodynamic and feminist clinical approaches that did not appear in the literature that they reviewed. Another set of professional guidelines are in the process of being produced by APA that will deal more with the complex and intersectional issues of how and what clinicians treating clients with PTSD and other trauma reactions should be aware.

Most important is a tripartite model that considers the scientific literature, the common factors generated by the therapist-patient relationship, and the wishes, values and preferences of the person (American Psychological Association 2005). This tripartite model has been adopted as part of APA's policy on Evidence-Based Treatment but the guidelines for professional psychologists separated its three parts into clinical guidelines where the RCT literature was reviewed and forthcoming professional guidelines where the therapy relationship and patient preferences will be found. Ethical concerns may be raised about the limitations of manualized treatment that does not recognize the necessary elements for actual change; it just focuses on the elimination of symptoms. This singular biomedical focus continues to be debated as inadequate in psychology especially since the research on patient adherence suggests it is low. However, there are programs to assist psychotherapists that are evidence-based and include the intersection between symptoms, interpersonal relationships and skill building in treatment such as the Survivor Therapy Empowerment Program (STEP) (Jungersen et al. 2018).

There is a body of research on the efficacy of what is sometimes referred to as the 'contextual model' that measures the factors necessary to produce change in psychotherapy (Wampold and Imel 2015). These factors are part of the common factors model mentioned above and include an established alliance between the psychotherapist and the patient, empathy, mutual expectations that the treatment will work, adaptation of the intervention to the patient's culture and values, and specific treatment techniques that include assessment of adherence to competent methods (Wampold and Imel 2015). Researchers involved in measuring success in this way comment on it being superior to what they label as measuring tiny effect size differences between manualized psychotherapy treatments in RCT research. Wampold and colleagues have had impact on the development of APA policy on Evidence-based Practice in Psychology (APA, EBPP 2006). However, it is not known to what extent this research has taken into account the impact of contextual psychotherapy models on healing from PTSD and other trauma reactions, specifically.

Ford and Courtois (2020) suggest yet another model for the treatment of PTSD and related disorders that is similar to what the contextual model requires. They base it on the principles of humanism, professionalism, and science, with four key philosophical foundations for the treatment including (1) respect for the individual, empowering patients based on their personal strengths, (2) fostering and sustaining a therapeutic alliance, (3) providing the tripartite evidence-based treatment within the framework of trauma-informed care, and (4) ensuring that treatment is provided with expertise based upon professional training, qualifications, and ongoing supervision and consultation. They add a caveat that treatment of PTSD is never one- size-fits-all: rather, each patient is assessed, and treatment is planned differentially according to the specific needs and preferences of the individual which can vary considerably for different persons and over the course of treatment, as the patient's internal experience and external contexts evolve.

16.1.4 Biological Devices

As mentioned earlier, trauma often impacts the body as well as the mind so typical symptoms may include physiological activation, mental and emotional signs of hypervigilance, and an inability or slowness to process information or to engage in movement. In many cases people develop a continued fear of the reminders of the event, called triggers, that interrupts their normal life functions. Several different methods of relaxation training help calm this activation. More specialized methods such as Eye Movement Desensitization Reprocessing (EMDR) (Shapiro 1995), that directly process traumatizing experiences are available to change how memories are stored in the brain. While the exact mechanism is unclear, there is ample evidence to suggest her method uses biopsychosocial mechanisms to reduce or eliminate the distressing emotions attached to the traumatic memories.

Biofeedback and neurofeedback using very diverse methods may assist in gaining better control over persistent anxiety by reducing emotional reactivity. There are also other more specialized services available such as use of a hyperbaric chamber that can send more oxygen to the body and brain to induce neuroplasticity that repairs damaged areas. Some of these interventions are based on the theory that traumatic memories are stored in the mid-brain area, particularly in the hippocampus and can be reached through altering brain waves through use of these mechanisms. This implies that remembering and processing abuse is a part of the healing process. It should also be noted that there are also other types of specialized services being developed as well (Baldwin 2013).

In more complex PTSD reactions, there may be more affect dysregulation, lowered arousal and dissociation that directly involves the parasympathetic branch of the autonomic nervous system (ANS). The ANS responds to constant threats even when the actual danger has passed (Porges 2011). Often memories causing re-experiencing parts, or all of the traumatic events are part of these threats. Complex PTSD is more likely to stem from repeated or chronic instances of inescapable trauma or abuse, often occurring during childhood, that require immobility rather than active defenses. The signs of severe immobility defenses, such as being unable to regulate emotions, lowered arousal and dissociation that are common in complex PTSD, often involve parasympathetic autonomic states as well as physical and somatoform symptoms that are commonly comorbid with trauma (irritable bowel syndrome (IBS), fibromyalgia, chronic fatigue, varied sorts of chronic pain, and other unexplained medical symptoms). More specialized training may be necessary to treat these patients as they move through different defense states especially since it is important to understand the interaction with any earlier traumas that the person has experienced.

Complex trauma usually implies some form of dissolution or dissociation, as cortical areas of the brain shut down during the severe immobility defenses during both the initial trauma experience and perhaps again, as it is relived during the reexperiences. These primitive reactions often appear impulsive or dysregulated and may include features of the transiently dominant brain areas, networks or hemispheres. The dissolution and dissociation permit more primitive brain areas to respond autonomously to threats, real or imagined, without the cortical regulation of normal states of ventral vagal safety. Complex traumatizing experiences (i.e., that cannot be escaped or overcome with active defenses) may disrupt immunological, lymph, connective tissues, and other physical processes. Cases of complex PTSD may require specialized relational or somatic treatments including consultation with medically trained providers. For example, sleep may be disturbed; this might be addressed by processing traumatizing experiences in therapy; or via sleep hygiene, exercise, neurofeedback training, acupuncture or craniosacral therapy or a trial administration of prazosin.

16.1.5 Telehealth Delivery and Apps

Treatment of PTSD has not always been available worldwide especially during periods of civil unrest, wars, and mass terrorist and shooting events. People must feel safe for specific trauma treatment as described here as opposed to social support or psychoeducation. If they are in a war zone or mass shooting, the best treatment would be to get them out to someplace where they can begin to feel safe. Further, health crises such as the recent COVID-19 pandemic, have also placed more demand on services for trauma-related symptoms around the world. The American Psychological Association has been meeting with leaders of other international psychological associations and disseminating information in the various languages. Although the delivery of services via the internet has been controversial, given the need to provide psychotherapy to those who were quarantined during this pandemic, psychologists have learned to provide counseling and health related services via computer.

Obviously, not all families have access to a computer or iPad, so once again, disparity exists in some of the same people who have no access to other medical care. In the U.S. many communities purchased IPads or laptops for school children so they could continue their lessons at home. Many of the parents 'borrowed' their child's devices so they could learn more about healing from the stressors caused by the pandemic. For some, the choice was to stay-at-home and be safe from the virus but be traumatized by intimate partner abuse or child abuse while trapped with the perpetrator. (Bettinger-Loopez and Faro 2020).

One of the ways to try to expand the number of people exposed to interventions has been the development of digital applications for IPhones or other Smart phones. The U.S. Veterans Affairs National Center for PTSD has been a leader in this area with several apps for self-care as well as companions to those also engaged in psychotherapy. A recent app has been to help victims of sexual harassment (the Me-too group) identify and heal, especially those women who were exposed during their service in the military. Other apps to help emotional regulation, especially anxiety, have been produced commercially. This is an emerging area with little research on its use or efficacy either alone or in conjunction with psychotherapy at this time.

16.1.6   Legal Remedies

While not exactly health-care treatment, legal actions sometimes have an important role to play in helping trauma victims become survivors. Civil lawsuits have provided the opportunity for compensatory damages or even awards for punitive damages if large corporations knowingly cause harm, often from toxic substances causing health problems in entire communities. The money collected from large class action lawsuits such as those against tobacco companies have been used to give victims compensation and engage in prevention campaigns such as no-smoking advertisements to targeted high risk segments of the population. In some communities, money collected from those applying for marriage licenses has been used to fund domestic violence treatment programs; a scheme that both called awareness to domestic violence in that selected population while also finding a new funding stream for a necessary community program. Restorative justice schemes are another way some communities apply principles of social justice to try to make things feel more just or right. Some of these methods come from practices in indigenous communities where one or more persons with certain needed characteristics are selected as healers.

16.1.7   Biomedical versus Ethical Biopsychosocial and Ecosocial Models

Let's look at some differences between using a strictly biomedical model and a model that allows for flexibility in analysis using biopsychosocial or ecosocial structures seen in population health strategies. This would permit cross-analysis between and among variables that prevent and cause what we call PTSD or other trauma reactions. Two examples that come to mind call for analysis either way; child abuse and terrorism. Those who study child abuse from a biomedical model focus on the individual or sometimes the interaction between the parent and the child. It does not focus on relational or subjective aspects. Nor does it include general preventive approaches such as educating an entire population as to the short- and -long term effects of child abuse which has been done quite effectively by others including Felitti et al. (2001). A selective prevention approach might target a subpopulation where there are high levels of child abuse reported to the various authorities or even emergency rooms. And a more specific indicated prevention approach might target the next generation in a family where there is known child abuse. Programs for each of these groups have existed with prevention as a primary goal (Walker 2017). A broader approach might include targeting some who do not have indicators of child abuse in their family background, but perhaps have witnessed their father abusing their mother. Research looking at more than one variable has found a high relationship or correlation between child abuse and domestic violence (Walker 2017). It does not yield information about cause and effect but rather an association that is important if we widen the target group we wish to prevent experiencing trauma.

If we widen the net even further, we might include those who experienced incest from a father or father-like person. Focus on this subgroup of abused children are known to develop complex trauma symptoms, many of which impact on personality development. Here we may run into difficulty especially if the child is young and has difficulty in making reports or if the examiner is untrained to recognize signs of child sexual abuse. One research study in the U.S. indicated approximately 58,000 children are allowed to live with a parent after reports are made to authorities each year (Leadership Council on Child Abuse and Interpersonal Violence 2008). We do not really know the extent of the problem that goes unrecognized or undiagnosed because special methods are not readily taught in health care or legal settings.

A recent study of over a thousand custody decisions by many different judges found that mothers who reported child abuse were likely to lose custody in anywhere between one-quarter to one-half the time (Meier et al. 2019). If the father cross-complained charging parental alienation (a controversial designation with little empirical support) the mother was even more likely to lose access to the child. Some might believe that all these mothers were coaching the child or somehow fabricating the abuse, but the data show that false reports occur in less than 10%, about the same rate as most criminal acts. The gender bias is shown by the difference between these data when men make the complaint that the child is abused; they are still more likely to win custody. And, perhaps even more problematic, is that when the supposedly trained mental health custody evaluators were involved, the decision to disbelieve the woman was even higher (Meier et al. 2019). Obviously, any research about the rate of child abuse would have to take into account several of these social variables including bias against mother's reports and towards father's complaints of being discriminated against.

Another relationship with child abuse and domestic violence that has little empirical data on cause and effect is found in studies of children who are lured into sex trafficking (Barron and Frost 2019; Gill and Gaviria 2019; Walker et al. 2019). The association appears to have something to do with the difficulty some abused children have in attaching to others in their families. Cassidy et al. (2013) have researched how the ability to form meaningful loving relationships or secure attachment styles are based on the ability to have your needs consistently met during childhood. Many who perpetrate abuse on others have what is called 'inconsistent or disorganized' attachment. They may be highly anxious, fearing their needs will not be met or avoidant, having stopped trying. Thus, studying at risk youth who have insecure attachment styles might be a subgroup that can yield results in being able to better protect youth from the lure of the sex trafficker. But, an ecosocial or population health approach, might also suggest looking at a subgroup within that population who also have experienced trauma from racial injustice or experience living in poverty with little access to necessary life-sustaining materials, as these are also conditions found when examining both girls and boys vulnerable to the lure of the sex trafficker (Gill and Gaviria 2019) as well as those who identify as being discriminated against as LGBTQ (Barron and Frost 2019). Just addressing one of those different factors would most likely be inadequate in an intervention for a teenager with a sexually transmitted disease; they would need special treatment addressing all their particular social conditions or they will probably be back in due time.

The association with child abuse and terrorism is another area of interest here. When analyzing the information gained from known terrorists, both those who shoot and bomb in multiple settings like in the U.S. schools and those who ideologically blow-up bridges and people in European cities, we find information about child abuse and domestic violence in their backgrounds. Of course, this doesn't mean all male child abuse survivors become terrorists. Rather, there are other known factors that intervene to create the willingness to kill others as we see from training soldiers for combat including a passion for their particular cause. But, there are data to suggest that the feelings of betrayal and lack of trust, often untreated components from child abuse, are present in those who have been examined (Montanari et al. 2019). Poland (2014) has similar findings as well as data about insecure attachment as infants and young children in the histories of school shooters and suicide.

We also find most terrorists, like in domestic violence and sex trafficking, are males. We also know that more male children are reported as child abuse victims and that males are more likely to inflict deadly blows to children. Females are more likely to become victims of other forms of intrafamily abuse such as domestic violence and elder abuse although the latter can change from victim to perpetrator if the formerly abused woman becomes that man's caregiver. Females who kill their babies have histories often reveal mental illness, especially postpartum depression while those who kill their abusive partners often do so in self-defense (Walker 2017). Obviously, gender is an important variable in understanding the impact of trauma on both men and women.

For example, in 2017, the U.S. FBI found that of the 9,576 arrests for murder, 1193 were women. While only 50% of their dataset permitted analysis of the relationship between the woman and who she killed, for the half whose relationship was known indicated it was usually an intimate partner or someone in their family. Nothing is known about prior histories or other pertinent facts to compare them with men except that five times more men killed women intimate partners during that same time period. International cases suggest similarities in other countries with more women being killed by abusive men. Nonetheless, the media, including movies, inaccurately portray women killers as angry psychopathic vengeful murderers, often missing their trauma backgrounds.

Looking at people incarcerated in the U.S. prisons, for example, also show a relationship with child abuse and domestic violence that is probably common in other countries as well (Walker 2017). Close to 80% of women prisoners and over half of the men surveyed mention that they experienced one or both of those forms of trauma. Many of those who become substance abusers also mention their earlier traumatic family abuse histories. Teenage runaways, gang members, and those who join terrorist cells often have been traumatically abused without receiving any treatment for its psychological effects.

So, what does all this mean. Certainly, this discussion raises more questions than it answers. What is the role of gender in both perpetrating and being victimized by violence? What about the newer ways of understanding gender; rather than looking at it from a binary lens, the expectation is that there is a continuum of what is typically considered femaleness and maleness. Do we use such a non-binary gender analysis in our research and if so, to what end? Much of the social science research into non-binary roles continues to argue about how much is due to biology and how much is due to social learning conditions (Krieger 2020). This chapter will not attempt to resolve that issue. But it does have implications in our understanding of how trauma is viewed by different groups of people. And how untreated or inadequately treated trauma, such as the effects of child abuse can influence the later response to new traumatic experiences.

16.2 Future Considerations

One of the important issues raised by the discussion here is how to change the focus of future research from a strictly biomedical model, often represented in this chapter by using RCTs to assess cause and effect of independent variables, to a more epidemiological population or ecosocial research approach that looks at relationships among intersectional variables. The RCT approach is easier to understand when one independent variable has a particular effect size. But, as described here, that may be meaningless as outside the laboratory most people have more than one variable interacting with others, especially when reacting to a traumatic experience. Instead, it will become necessary to train health care researchers across disciplines using the biopsychosocial model together with health care methodologists and statisticians. It may be important to include such methodologists on research grant submissions as one way to encourage non-RCT studies. Journal editors and other traditional publication outlets will need encouragement to find appropriate reviewers to disseminate and publish deserving research.

Medical and allied health schools must commit to an interdisciplinary model with professors from the social sciences and psychology alongside the biomedical faculty. Methodologists must encourage biopsychosocial and ecosocial models where biomedical models have become outdated, unethical where they promote health care inequities. Changing from a biomedical model with emphasis on physical and neurological health to one that equally values all intersectional variables including social conditions such as poverty, racial injustice, access to health care and behavioral problems will not be easy. Intersectionality can be messy with many variables, some easier to identify and measure than others. However, as can be seen with one disorder, PTSD, it can and must be done.

16.3 Conclusion

In conclusion, this chapter has described why the biomedical model is inadequate and even unethical while proposing different models of research for PTSD and trauma including biopsychosocial and epidemiological models from what is often called an ecosocial lens emphasizing public health approaches. This model permits the inclusion of intersecting social factors into the medical diagnoses as an integral part of the traumatic impact. It calls for new training of health care workers to be integrated with social factors that may underlie or even cause violence, although much of that research still needs to be done. Rather, acceptance of sophisticated correlational statistics will need to be considered as acceptable methodology along with carefully analyzed qualitative studies and move away from the standard RCT models measuring effect-sizes that are ill-suited to clarify the intersectional conditions especially in describing effective treatment. This does not mean all our knowledge gathered to understand cause and effect should be discarded. Rather, the approach to understanding trauma will need both a broadened approach in some areas and narrowly defined targeted subgroups in other areas based on our data. The arguments presented here demonstrated the need for this new research approach by illustrating how the intersectionalities were missed using a biomedical model in discussing the impact from two traumatic events that involve PTSD; child abuse and terrorism. Without effective treatment, child abuse can and has given rise to other intrafamily trauma and community disruption through terrorism. It would be unethical to do less.

References

American Psychiatric Association. 2013. *Diagnostic and statistical manual of mental disorders 5th Ed* (DSM-5). Washington, DC: APA.
American Psychological Association. 2005. *Report of the task force on evidence based psychology.* Washington, DC: APA.
American Psychological Association. 2016. Ethical principles of psychologists and code of conduct. http://www.apa.org/ethics/code/
American Psychological Association. 2020. *Policy on COVID-19*. Washington, DC: APA.
American Psychological Association. 2022. *Policy on population health*. Washington, DC: APA.
Almeida, D.M., E. Wethington, and R.C. Kessler. 2002. The daily inventory of stressful experiences (DISE): An interview-based approach for measuring daily stressors. *Assessment* 9: 41–55. https://doi.org/10.1177/1073191102091006.
Alisic, E., A.K. Zalta, F. van Wesel, S.E. Larsen, G.S. Hafstad, K. Hassanpour, and G.E. Smid. 2014. Rates of post-traumatic stress disorder in trauma-exposed children and adolescents: Meta-analysis. *British Journal of Psychiatry* 204: 335–340. https://doi.org/10.1192/bjp.bp.113.131227.
APA 2006. Report of the presidential task force on Evidence-based practice in psychology (EBPP). American Psychologist, p.271-283.
Atwoli, L., D.J. Stein, K.C. Koenen, and K.A. McLaughlin. 2015. Epidemiology of posttraumatic stress disorder: Prevalence, correlates and consequences. *Current Opinion in Psychiatry* 28 (4): 307–311. https://doi.org/10.1097/YCO.0000000000000167.
Baldwin, D.V. 2013. Primitive mechanisms of trauma response: An evolutionary perspective on trauma-related disorders. *Neuroscience & Biobehavioral Reviews* 37 (8): 1549–1566. https://doi.org/10.1016/j.neubiorev.2013.06.004.
Bandura, A. 1999. Self-efficacy: Towards a unifying theory of behavioral change. In Men, boys, and LGBTQ: Invisible victims of human trafficking, ed. R.F. Barron, C. Frost, C. In *Handbook of sex trafficking*, ed., L.E. Walker, G. Gaviria, and K. Gopal, 73–84. New York: Springer Nature.

Benjet, C., E. Bromet, E.G. Karam, R.C., Kessler, K.A. McLaughlin, A.M.V. Ruscio, V. Shahly, D.J. Stein, M. Petukhova, E. Hill, J. Alonso, and K.C. Koenen. 2016. The epidemiology of traumatic event exposure worldwide: Results from the World Mental Health Survey Consortium. *Psychological Medicine* 46 (2): 327–343. https://doi.org/10.1017/S0033291715001981

Bolger, N., K.S. Zee, M. Rossignac-Milon, and R.R. Hassin. 2019. Causal processes in psychology are heterogeneous. *Journal of Experimental Psychology: General* 148 (4): 601–618. https://doi.org/10.1037/xge0000558.

Baumeister (ed.). *The self in social psychology*, 285–298. NY: Psychology Press.

Briere, J., and Scott. 2007. *Principles of trauma therapy: A guide to symptoms, evaluation and treatment.* Thousand Oaks, CA: Sage.

Bryant, R.A. 2019. Post-traumatic stress disorder: A state-of-the-art review of evidence and challenges. *World Psychiatry* 18 (3): 259–269. https://doi.org/10.1002/wps.20656.

Courtois, C.A., J. Sonis, L.S., Brown, J. Cook, J. Fairbank, A.A.M. Friedman, and P. Schulz. 2017. Clinical practice guideline for the treatment of posttraumatic stress disorder (PTSD) in adults. Washington, DC: American Psychological Association. Retrieved from http://www.apa.org/monitor/2017/11/ptsd-guideline.aspx

Felitti, V.J. 2001. Reverse alchemy in childhood: Turning gold into lead. *Health Alert* 8: 1–4.

Finkelhor, D., R.K. Ormrod, and H.A. Turner. 2009. The developmental epidemiology of childhood victimization. *Journal of Interpersonal Violence* 24 (5): 711–731. https://doi.org/10.1177/0886260508317185.

Ford, J., and C. Courtois. 2020. *Treating complex traumatic stress disorders in adults*, Rev ed. New York: Guilford.

Gelso, C. 2014. Tripartite model of the therapeutic relationship: Theory, research & practice. *Psychotherapy Research* 24 (2): 117–131.

Gill, K.A., and Gaviria, G. 2019. Vulnerability factors when women and girls are trafficked. In *Handbook of sex trafficking*, ed. L.E.A. Walker, G. Gaviria, and K. Gopal, 63–72. New York: Springer Nature

Jungersen, T., L.E. Walker, T.P. Kennedy, R. Black, and C. Groth. 2018. Trauma treatment for intimate partner violence in incarcerated populations. *Practice Innovations.* https://doi.org/10.1037/pri0000083.

Kessler, R. C., S. Aguilar-Gaxiola, J. Alonso, C. Benjet, E.J. Bromet, G. Cardoso, L. Degenhardt, G. de Girolamo, R.V. Dinolova, F. Ferry, S. Florescu, and K.C. Koenen. 2017. Trauma and PTSD in the WHO World Mental Health Surveys. *European Journal of Psychotraumatology, 8*(sup5), 1353383. https://doi.org/10.1080/20008198.2017.1353383

Krieger, N. 1994. Epidemiology and the web of causation: Has anyone seen the spider? *Social Science Medicine* 39 (7): 887–903.

Krieger, N. 2003. Does racism harm health? Did child abuse exist before 1962? On explicit questions, critical science, and current controversies: An ecosocial perspective. *American Journal of Public Health* 93: 194–199.

Krieger, N. 2020. Measures of racism, sexism, heterosexism, and gender binarism for health equity: From structural injustice to embodied harm—An ecosocial analysis. *Annual Review of Public Health* 41: 37–62.

Kudler, H.S. 2019. Clinical practice guidelines for posttraumatic stress disorder: Are they still clinical? *Psychotherapy* 56 (3): 383–390. https://doi.org/10.1037/pst0000236.

McElroy, E., M. Shevlin, A. Elklit, P. Hyland, S. Murphy, and J. Murphy. 2016. Prevalence and predictors of Axis I disorders in a large sample of treatment-seeking victims of sexual abuse and incest. *European Journal of Psychotraumatology* 7: 30686. https://doi.org/10.3402/ejpt.v7.30686.

McLaughlin, K.A., K.C. Koenen, E.D. Hill, M. Petukhova, N.A. Sampson, A.M. Zaslavsky, and R.C. Kessler. 2013. Trauma exposure and posttraumatic stress disorder in a national sample of adolescents. *Journal of the American Academy of Child and Adolescent Psychiatry* 52 (8): 815–830 e814. https://doi.org/10.1016/j.jaac.2013.05.011

Meichenbaum, D. 2019. Resilience: A constructive narrative perspective. In *Handbook of sex trafficking*, ed. L.E.A. Walker, G. Gaviria, and K. Gopal, 213–216. New York: Springer Nature.

Mithoefer, M.C., M.T. Wagner, A.T. Mithoefer, L. Jerome, S.F. Martin, B. Yazar-Klosinski, et al. 2013. Durability of improvement in post-traumatic stress disorder symptoms and absence of harmful effects of drug dependency after 3.4-methylenediooxymethamphetamine-assisted psychotherapy: A prospective long-term follow-up study. *Journal of Psychopharmacology* 27: 28–39.

Montanari, D.L. Shapiro, L.E. Walker, V. Mastronardi, M. Caalderara, C.I.S. Santonico, R. Bracadenti, E. Marinelli, and S. Zaami. 2019. Health care providers ethical use of risk assessment to identify and prevent terrorism. *Journal of Ethics, Medicine, & Public Health* 12. https://doi.org/10.1016/j.jemep.2019.100436

Norcross, J.C., & M.J. Wampold. 2019. Evidence-based psychotherapy responsiveness: The third task force chapter in Norcross & Wampold. *Psychotherapy relationships that work: Vol. 2. Evidence-based therapy responsiveness*, 3rd ed. NY: Oxford. https://doi.org/10.1093/med-psych/9780190843960.003.0001

Pietrzak, R.H., Goldstein, R.B., Southwick, S.M., & Grant, B.F. (2011a). Personality disorders associated with full and partial posttraumatic stress disorder in the U.S. population: Results from Wave 2 of the National Epidemiologic Survey on Alcohol and Related Conditions. *Journal of Psychiatric Research* 45 (5), 678–686. https://doi.org/10.1016/j.jpsychires.2010.09.013

Pietrzak, R.H., R.B. Goldstein, S.M. Southwick, and B.F. Grant. 2011b. Medical comorbidity of full and partial posttraumatic stress disorder in US adults: Results from Wave 2 of the National Epidemiologic Survey on Alcohol and Related Conditions. *Psychosomatic Medicine* 73 (8): 697–707. https://doi.org/10.1097/PSY.0b013e3182303775.

Pietrzak, R.H., R.B. Goldstein, S.M. Southwick, and B.F. Grant. 2011c. Prevalence and Axis I comorbidity of full and partial posttraumatic stress disorder in the United States: Results from Wave 2 of the National Epidemiologic Survey on Alcohol and Related Conditions. *Journal of Anxiety Disorders* 25 (3): 456–465. https://doi.org/10.1016/j.janxdis.2010.11.010.

Poland, S. 2014. *Suicide Intervention in the schools.*

Porges, S.W. 2011. *The Polyvagal theory: Neurophysiological foundations of emotions, attachment, communication, and self-regulation.* New York, NY: W W Norton & Co.

Santiago, P.N., R.J. Ursano, C.L. Gray, R.S. Pynoos, D. Spiegel, R. Lewis-Fernandez, M.J. Friedman, and C.S. Fullerton. 2013. A systematic review of PTSD prevalence and trajectories in DSM-5 defined trauma exposed populations: intentional and non-intentional traumatic events. *PLoS One,* 8(4), e59236. https://doi.org/10.1371/journal.pone.0059236

Shapiro, F. 1995. *Eye movement desensitization and reprocessing: Basic Principles.* New York: Basic Books.

Shapiro, D.L. 2020. *Psychotherapy, risk assessment, and terrorism: Ethical Issues.* In this volume.

Shonkoff, J.P., W.T. Boyce, and B.S. McEwen. 2009. Neuroscience, molecular biology, and the childhood roots of health disparities. *JAMA* 301 (21): 2252–2259. https://doi.org/10.1001/jama.2009.754.

Silver, K.E., and R.F. Levant. 2019. An appraisal of the American Psychological Association's clinical practice guideline for the treatment of posttraumatic stress disorder. *Psychotherapy* 56 (3): 347–358. https://doi.org/10.1037/pst0000230.

Smith, P., T. Dalgleish, and R. Meiser-Stedman. 2019. Practitioner review: Posttraumatic stress disorder and its treatment in children and adolescents. *Journal of Child Psychology and Psychiatry* 60 (5): 500–515. https://doi.org/10.1111/jcpp.12983.

Teicher, M.H., J.A. Samson, C.M. Anderson, and K. Ohashi. 2016. The effects of childhood maltreatment on brain structure, function and connectivity. *Nature Reviews Neuroscience* 17 (10): 652–666. https://doi.org/10.1038/nrn.2016.111.

VA/DOD Working Group. 2017.

Van der Kolk, B. 2015. *The body keeps score: Brain, mind and body in the healing of trauma.* New York: Penguin.

Walker, L.E.A. 2017. *Battered woman syndrome*, 4th ed. New York: Springer.

Walker, L.E., G. Gaviria, and K. Gopal, eds. 2019. *Handbook of sex trafficking*. New York: Springer Nature.

Walker, L.E.A. 2015. Looking back and looking forward: Psychological and legal interventions for domestic violence. *Ethics, Medicine & Public Health: A Multidisciplinary Journal*. Inaugural Issue. https://doi.org/10.1016/j.jemep.2015.02.002

Wampold, B.E., and Z.E. Imel. 2015. *The great psychotherapy debate: The research evidence for what works in psychotherapy*, 2nd ed. NY: Routledge.

Webb-Hooper, M., A.M. Napoles, and E.J. Perez-Stable. 2020. COVID-19 and racial/ethnic disparities. *JAMA* E1–E2. https://doi.org/10.1001/jama.2020.8598

Lenore Walker was formerly chair of the forensic psychology program at Nova Southeastern University. She has been a pioneer in the field of interpersonal partner violence having been the creator of the *Battered Woman Syndrome*. She recently has served as the chairperson of the American Psychological Association Task Force on professional guidelines on PTSD and other trauma related conditions. She is the author of many books and articles and is frequently called upon internationally to serve as an expert witness.

Part VIII
Security and Pandemic Threats

Chapter 17
State Interventions During the COVID-19 Pandemic: The Case for Mask Mandates Under Human Rights Law

Jonathan Hafetz

17.1 Introduction

The SARS-CoV-2 (COVID-19) global pandemic has surfaced important questions regarding the human rights obligations of states when confronting a public health emergency. Countries have imposed various measures to curb COVID-19's spread, including travel restrictions; stay-at-home orders; the closure of schools, government offices, and private businesses; and requirements to wear face coverings (masks) in public. A state's multiple human rights obligations, however, may be in tension with one another in this context, given the state's duty not only to protect human life but also to protect fundamental freedoms. While a state, for example, may be compelled to adopt measures to protect life, including those that substantially restrict commercial activity and freedom of movement, it must also not unduly restrict liberty or act in a manner that is arbitrary or discriminatory. Further, states must calibrate the effects—both intended and unintended—of potentially life-saving measures that could undermine socio-economic rights, including the right to education, housing, and an adequate standard of living. Scientific research on the efficacy of these measures not only underpins decisions by public officials, but is central to their legality as well.

This chapter focuses on one particular public health measure: mask mandates. Compared with other state interventions to curb the spread of COVID-19, mask mandates involve few, if any, human rights trade-offs. They should be non-controversial, if not openly welcomed, at least during periods where the risk of transmission is high. Masks have tremendous potential to curb the virus's spread and save lives, while imposing relatively limited infringements on personal liberty. Indeed, masks can help reduce the necessity of other, far more drastic and costly

J. Hafetz (✉)
Seton Hall Law School, Newark, NJ, USA
e-mail: jonathan.hafetz@shu.edu

interventions by still allowing a substantial degree of movement and daily activity to continue. Yet mask mandates have proven controversial, particularly in countries that prioritize individual liberty, such as the United States.

The chapter argues that human rights law warrants and potentially obligates states to impose mask mandates and that scientific research plays an important role in framing, justifying, and implementing them.[1] First, the chapter examines the human rights duties of states regarding public health measures during a pandemic. It also explains how states frequently face competing demands in safeguarding human rights, a tension heightened by a devastating global pandemic like COVID-19. The chapter next argues that scientific research not only supports the imposition of mask mandates to protect the right to life, but also suggests how such mandates might lessen reliance on other, more burdensome public health measures that potentially infringe on other human rights. The chapter then discusses the role of scientific research in both framing and justifying mask mandates. It examines, for example, the importance of transparency and clear messaging by public health officials, the absence of which hindered the effective early use of masks in response to COVID-19 in numerous countries and engendered confusion that impeded later attempts at course correction. Finally, the chapter discusses some implications for future scientific research, including on how mask mandates can best be formulated, applied, and enforced to help maximize compliance, overcome resistance, and reinforce public education. While the marginal utility of mask wearing to address COVID-19 will continue to decrease with the increased availability of safe and effective vaccines, masks remain a critical element of any preparedness strategy for future pandemics, which an independent panel of experts has described as an existential threat to humanity that should be elevated to the highest level.[2]

17.2   A State's Human Rights Obligations During a Global Pandemic

A global pandemic like COVID-19 threatens multiple human rights guarantees. Above all, it jeopardizes the right to life.[3] The right to life is a fundamental principle of international human rights law. The Universal Declaration of Human Rights provides that "everyone has the right to life, liberty, and security of person.[4] The right

---

[1] The chapter uses the umbrella term "scientific research" to capture the different types of research discussed herein. Scientific research is intended to include, for example, both epidemiological research (which informs public health recommendations on mask wearing) and social science research (which includes behavioral research and informs policy recommendations and implementations on mask wearing).

[2] The Independent Panel for Pandemic Preparedness & Response (2021).

[3] Bennouhe (2020, 667–668).

[4] Universal Declaration of Human Rights (1948).

to life is recognized by leading international[5] and regional human rights treaties.[6] The U.N. Human Rights Committee (HRC) characterizes this guarantee as the "supreme right," one that has "crucial importance both for individuals and for society as a whole" and that "constitutes a fundamental right, the effective protection of which is the pre-requisite for the enjoyment of other rights."[7] Although the right to life is typically framed in negative terms—i.e., prohibiting "arbitrary deprivations of life" by acts and omissions—it also imposes affirmative duties on states to take "appropriate measures to address the general conditions in society that may give rise to direct threats to life," including "the prevalence of life-threatening diseases"[8] like COVID-19. Protecting the right to life includes, where necessary, measures intended to guarantee access without delay to health care.[9] The European Court of Human Rights, for example, has recognized that a state's duty to protect the right to life not only includes ensuring access to health care to safeguard the lives of those within its jurisdiction,[10] but also potentially encompasses "preventive operational measures" that may be required in "dangerous situations of specific threat to life which arise from risks posed by... man-made or natural hazards."[11]

A state's obligation to protect life during a pandemic is augmented by its obligation to ensure that individuals have the highest attainable standard of physical as well as mental health care, which similarly imposes duties on states to adopt measures necessary to prevent, control, and treat epidemic diseases.[12] This obligation, moreover, informs a state's obligation to prevent the transmission of COVID-19 not only to protect life but also to prevent the harmful consequences for the significant number of people who survive the virus but who nonetheless suffer harmful long-term effects of the disease.[13]

The obligation to protect human rights is also baked into international public health law. For example, the International Health Regulations (IHR) of the WHO, adopted in 1969 and revised in 2005 to improve the global capacity to prevent the spread of infectious diseases,[14] states that public health measures during a pandemic must be implemented with "full respect for the dignity, human rights, and fundamental freedoms of persons."[15]

---

[5] International Covenant on Civil and Political Rights, art. 6, Dec. 16, 1966, 999 U.N.T.S. 171 (1967).

[6] American Declaration on the Rights and Duties of Man (1948); European Convention on Human Rights for the Protection of Fundamental Freedoms (1981).

[7] UN Human Rights Comm. (2019).

[8] *Id.* para. 26.

[9] *Id.*

[10] *Cyprus v. Turkey,* App. No. 25781/94, para. 219 (Eur. Ct. H.R. May 10, 2001).

[11] *Stoyanovi v. Bulgaria*, App. No. 42980/04, para. 61, (Nov. 10, 2010).

[12] International Covenant on Economic, Social, and Cultural Rights, art. 12(2)(c), Dec. 16, 1966, 993 U.N.T.S. 3; Bennouhe, "'Lest We Sleep,'" 668.

[13] Higgins et al. (2020).

[14] Taylor and Habibi (2020).

[15] World Health Org. (2005).

The human right to life, moreover, is non-derogable. It therefore cannot be suspended in time of emergency, in contrast with most other human rights which can be suspended or relaxed if such derogation is done in conformance with applicable legal rules, such as necessity, proportionality, and non-discrimination.[16] A state's obligation to protect public health provides a basis for temporarily limiting the exercise of other fundamental rights, as long as the measures are strictly necessary for the promotion of welfare in a democratic society.[17]

COVID-19, and the measures taken to address it, have also impacted other human rights. These rights include: freedom of movement (both internally and across borders); freedom of assembly and association; freedom of expression; freedom of religious exercise[18]; and a panoply of socio-economic rights, including the right to education and to an adequate standard of living, which encompasses the right to adequate food and nutrition, water, and housing.[19] These rights remain subject to derogation and limitations, in accordance with the restrictions described above.

17.2.1 Potentially Competing Human Rights Obligations

COVID-19 has often placed a state's human rights obligations in tension, if not direct conflict. To protect life, government authorities have adopted various measures, including restricting travel and movement, imposing stay-at-home orders, closing government offices, and restricting or prohibiting certain businesses from operating—measures frequently grouped under the general rubric of "lockdowns."[20] These measures have generally enjoyed the broad support of public health bodies and medical experts.[21]

These measures can help fulfill a state's obligations to protect life and to attain the highest available health care. Requiring individuals to stay at home significantly reduces the transmission of COVID-19, as does limiting opportunities for people to congregate, thus substantially reducing fatalities and severe illness.[22] The measures also minimize the strain on overburdened hospitals and medical personnel and help conserve public health resources and ensure treatment is provided to those who most need it. But the measures also can have significant adverse economic consequences,

---

[16] Human Rights Comm. (2020).

[17] UN Committee on Economic, Social and Cultural Rights, "CESCR General Comment No. 14: The Right to the Highest Attainable Standard of Health," Aug. 11 m 2000, UN Doc E/C.12/2000/4, paras. 28–29.

[18] Spadaro (2020).

[19] Human Rights Watch (2020a).

[20] Morris (2020).

[21] Flaxman et al. (2020).

[22] Lurie (2020), Fowler et al. (2020).

especially in industries most directly impacted by the pandemic, such as the restaurant industry, tourism, and retail.[23] While the spread of the virus and the general waning of consumer confidence are partly responsible for these consequences,[24] restrictions also play a role, especially the longer they remain in effect. Further, emergency measures imposed without sufficient notice or coordination can exacerbate the economic harm, as in India, where millions of migrants were stranded and confronted a risk of famine.[25] Additionally, pandemic restrictions can impact other rights, including freedom of association, freedom of assembly, and religious exercise.[26]

Trade-offs, both perceived and real, help inform decisions in confronting a global pandemic like COVID-19. While these decisions are often driven by a cost-benefit analysis,[27] they may also be understood as state responses to competing human rights obligations. In the United States and Europe, for example, governments have sought to minimize the collateral impact of stay-at-home orders, business closures, and other restrictions, by exempting essential businesses and workers, thus enabling infrastructure, health care, food services, energy, and other critical functions continue to operate.[28] These exemptions, however, have also increased the risk for essential workers in these areas who have been required by employers to continue laboring during the pandemic, often without adequate protective equipment and precautions.[29] This increased risk further implicates the human rights guarantee of non-discrimination because these workers are disproportionately from marginalized racial, national, and ethnic groups.[30]

Some government officials, to be sure, have ignored scientific research and cynically manipulated COVID-19 restrictions for political gain, whether by imposing draconian restrictions to increase surveillance and quash dissent[31] or removing restrictions to curry favor with a pandemic-weary public, against the advice of public health experts.[32] But despite the tendency of many Western human rights responses to focus on the opportunity the pandemic has presented for state political suppression,[33] and despite the need to guard against such authoritarian manipulation, the overriding human rights challenge presented by COVID-19 has been to navigate among what are often difficult choices about how to best protect human rights, a process in which science plays a critical role.

---

[23] National Restaurant Association (2021). Pietsch (2021).

[24] Gibson and Sun (2020). Available at SSRN: https://ssrn.com/abstract=3601108

[25] Bennouhe (2020), 669–670.

[26] Congressional Research Service (2020), The Conversation (2020).

[27] Rowthorn and Maciejowski (2020).

[28] *See,* e.g., Xu (2021); Executive Order No. 202.6 (2020).

[29] Editorial (2020b).

[30] *Id.*; U.N. Hum. Rts. Off. of High Comm'r (2020); Williams et al. (2020) (noting the high percentage of people of color in essential, poorly protected positions).

[31] United Nations Secretary-General (2021).

[32] Lethang (2021).

[33] Bennouhe (2020) (discussing Western human rights responses to COVID-19).

Thus, even if "following the science" merely informs and does not determine state responses,[34] scientific research remains critical to how public officials, courts, and other actors assess and frame COVID-19 mitigation measures. It likewise informs how courts evaluate legal challenges to those restrictions, from attacks on infringements on religious freedom and other individual liberties, to requests for relief on behalf of particularly vulnerable populations, such as incarcerated persons.[35] As explained above, these measures not only involve traditional political trade-offs, but also competing human rights considerations. The next section explores the science around masks and the human rights obligations that flow from it as well as some implications for future scientific research.

17.3   Scientific Research and Mask Mandates Under Human Rights Law

Masks are one of the most important effective tools for limiting the transmission and spread of COVID-19. Scientific research has repeatedly confirmed their efficacy.[36] Public mask wearing is most effective, moreover, when compliance is high.[37] While the benefits of masks are enhanced when they are coupled with other measures, such as social distancing, hygiene, and contact tracing, masks remain the most single effective tool for reducing transmission.[38] One study estimates, for example, that universal mask use could have saved over 130,000 lives in the United States during the first year of the pandemic.[39] Masks also provide a significant degree of protection when social distancing is not possible or is inadequate, such as in enclosed spaces, given airborne transmission of the virus.[40] Further, although certain types of masks (N95 respirator masks and surgical masks) offer greater protection, even simple cloth coverings offer significant protection if worn properly, and thus provide a pragmatic alternative when higher quality masks are unavailable.[41]

---

[34] Maani (2021).

[35] Garrett and Kovarsky (2021) (contrasting how courts assessed the rights-versus-safety trade-off in the context of restrictions on religious worship and protection of incarcerated individuals).

[36] Bai (2020). McCabe (2020).

[37] Howard et al. (2021).

[38] *Id.*

[39] Boodman (2021).

[40] Centers for Disease Control and Prevention (2020).

[41] Howard et al. (2021).

17.3.1 Mask Mandates

Given their effectiveness in reducing the transmission of COVID-19, saving lives, and reducing stress on overburdened health care systems, states are not only justified in mandating masks under human rights law as a means of ensuring compliance but arguably have an obligation to do so to satisfy their paramount duty to protect life. A state, moreover, might not fully discharge this obligation merely by announcing a mask mandate; states also have a duty to ensure mask requirements are effectively implemented, whether through sanctions, public education, or other means.

While an obligation to impose mask mandates to protect the right to life under human rights law has not yet been established,[42] the case for such an obligation is strong, at least absent a showing that substantially the same degree of compliance could be achieved without such a requirement. A mask mandate could be triggered by what the UN Human Rights Committee has described as the "due diligence obligation" for States to "undertake reasonable positive measures... in response to reasonably foreseeable threats to life originating from private persons or entities"—an obligation that, as Alessandra Spadaro has observed, could extend to "protecting individuals from threats to life posed by others carrying an infectious and deadly disease, such as COVID-19."[43]

Despite criticisms of masks mandates as infringing personal freedom, scientific research demonstrates that the benefits masks provide greatly outweigh the minor inconvenience or mild discomfort they may cause, easily satisfying a proportionality requirement. Further, as long as the mandate applies to all persons and is even-handedly enforced it would satisfy requirements of non-discrimination.

17.3.2 Use of Masks to Lessen Reliance on Other, More Restrictive State Interventions

Masks not only protect lives; they also help mitigate COVID-19's impact on other human rights by reducing the need for governments to rely on more drastic and costly measures, such as stay-at-home orders, travel restrictions, and business closures. Masks do not merely allow for but make possible a significant degree of continued daily activity and thus mitigate the economic effects of deadly diseases like COVID-19, including the impact on the right to an adequate standard of living. They also help make it possible to more safely engage in political protest, religious activity, and the exercise of other fundamental freedoms. The efficacy of masks is enhanced when coupled, where possible, with restrictions like social distancing, that are less burdensome than other state interventions like stay-at-home orders. Epidemiologists underscore that countries that have been hard-hit by COVID-19, such as Brazil, could

---

[42] Hathaway et al. (2020).

[43] Spadaro (2020).

have avoided additional lockdowns if the government had promoted the use of masks and social distancing.[44]

While states have an absolute duty to protect life, they must also protect other fundamental rights, and the measures states take in protecting life must be proportionate. As explained above, mask mandates and other measures to ensure universal mask wearing do not merely satisfy a state's obligation to protect the lives of individuals within its territory. They enable states to further the paramount end of protecting life while reducing the impact on other fundamental rights. Too often, however, mask mandates are framed as rights-restricting, rather than rights-enabling. Such a framing not only misconstrues the nature of the trade-offs surrounding pandemic restrictions, but also obscures the need to view human rights responses to a global pandemic like COVID-19 holistically. Measures that might seem like balancing rights (i.e., safety versus individual liberty) may in fact be protecting both rights. Mask mandates underscore how when a state acts to protect life it may be protecting other fundamental rights as well.

Several additional policy recommendations flow from this more holistic framing of mask requirements and a state's human rights obligations. Mask requirements, even if rigorously implemented and observed, may do less to blunt the impact of COVID-19 in certain sectors of the economy, where mask wearing is either impracticable due to the nature of the activity (e.g., dining in restaurants) or insufficient due to heightened risk (e.g., tourism). This suggests that states should formulate mask mandates and other measures to increase mask-wearing in conjunction with their obligation to protect other rights, such as the right to an adequate standard of living, by targeting financial assistance and other support to industries, small business owners, and workers in sectors that are still likely to suffer significant impacts even with widespread mask-wearing.[45] Similarly, states should utilize masking measures to minimize the burden imposed by capacity limitations and other restrictions that affect protected activities such as freedom of association and religious exercise.[46]

Some progress in this direction may be observed. In the United States, as states gradually reopened following the issuance of stay-at-home orders, they began to implement mask requirements, with 37 states requiring mask wearing in certain public settings, places of business, and places of employment.[47] While some states prematurely lifted mask mandates as part of a broader loosening of COVID-19 restrictions prior to the widespread availability of safe and effective vaccines (under the banner of "reopening"),[48] others maintained them, even as they gradually modified

---

[44] Londoño and Casado (2021).

[45] For a discussion of the need to target short-term assistance based on the particular vulnerabilities of certain sectors and local economies, see Chávez (2020).

[46] For example, in *Roman Catholic Diocese of Brooklyn v. Cuomo*, 141 S. Ct. 63 (2020), the U.S. Supreme Court addressed capacity restrictions on attendance at religious services in New York. The Court divided over whether restrictions unduly infringed on constitutionally protected religious freedom, which a majority enjoined. The debate over the capacity requirement would have been informed by demonstrated compliance by houses of worship with a mask requirement.

[47] Cloud et al. (2020).

[48] CBS News (2021).

other restrictions to lessen the economic impact of COVID-19 and in response to public pressure to resume normal daily activities.[49] Further, even in states that abandoned mask mandates, many retailers (both large and small) maintained them, filling in where states abandoned their human rights obligation to protect life.[50]

17.3.3   Additional Duty to Make Masks Widely Available

States have more than an obligation to enforce mask-wearing. As a group of U.N. human rights experts stated, "Everyone has the right to life-saving interventions."[51] Given masks' enormous life-saving potential and relatively low cost, states also have a duty to make masks available to everyone.[52] This duty is particularly pressing with respect to vulnerable populations.[53] The obligation to ensure the availability of masks, moreover, is central to a mask mandate, as it ensures that individuals do not face sanctions due to lack of access to masks.

Human rights law further informs plans for mask distribution. Governments should prioritize groups such as first responders and emergency health care workers employed in high-risk settings as well as particularly vulnerable sub-groups, such as the elderly, and those detained by the state along with those who interact with them.[54]

One obstacle that governments faced in ensuring that masks were widely available, especially during the initial stages of the COVID-19 pandemic, was a shortage of supply. States, however, either knew or should have been known beforehand of the significant risk of a pandemic like COVID-19 given the prior threats posed by SARS and MERS, along with the mounting risk of coronaviruses spilling over from animals into humans that scientists had documented for decades.[55] Certainly now, after the devastating impact of COVID-19, it is untenable for states not to prepare for the outbreak of another deadly virus, the occurance of which is a question not of whether but when. To fulfill their obligation to protect life, states should implement the best strategies *ex ante* to avoid a repeat of the flawed response to COVID-19. This includes not only ensuring an adequate stockpile of masks,[56] but also designing

---

[49] Gold (2021).

[50] Halkias (2021).

[51] U.S. Human Rights, Office of the High Commissioner (2020).

[52] U.S. Senator Bernie Sanders (D-VT), for example, introduced legislation that would have directed the Trump administration to send three, high-quality reusable masks to every person in the United States and to provide $5 billion for mask production. Editorial (2020a).

[53] U.S. Human Rights (2020).

[54] Hathaway et al., "COVID-19 and International Law Series." States likewise have an obligation to ensure other forms of Personal Protective Equipment (PPE) are made available to such individuals.

[55] "The Kind of Outbreak Scientists Knew Would Happen," *Columbia Mailman School of Public Health* (Mar. 13, 2020), https://www.publichealth.columbia.edu/public-health-now/news/kind-out break-our-scientists-knew-would-happen; Henig (2020).

[56] Kirubarajan et al. (2020).

legal frameworks and policies to ensure universal mask wearing through an effective combination of carrots and sticks.

17.4   Implications for Scientific Research and Human Rights

17.4.1   Centrality of Scientific Research to Mask Mandates and Related Measures

Scientific research on the efficacy of masks is central to their evaluation under human rights law. The research on the life-saving potential of masks supports a state's implementation of a mask mandate, use of sanctions, public education, and other measures to ensure widespread compliance, and adoption of measures to ensure masks are widely available within its territory. Scientific research further suggests how masks can help minimize reliance on additional measures, ranging from capacity requirements to the closure of businesses and stay-at-home orders, that burden fundamental rights. Additionally, research is critical to evaluating—and conesting—legal challenges to mask requirements, as opponents will continue to argue that masks are either unnecessary or a disproportionate response to the threat.

17.4.2   Importance of Transparency and Clear Messaging by Public Health Officials

Mask requirements have sparked considerable resistance in some countries. Early missteps by both the World Health Organization (WHO) and leading public health experts inadvertently contributed to this resistance. In the United States, for example, the surgeon general told the public that masks were not necessary for protecting the public while also insisting that health care workers needed the dwindling supply due to lack of emergency preparedness by the U.S. government,[57] a message echoed by Dr. Anthony Fauci, director of the U.S. National Institute of Allergy and Infectious Diseases.[58] Experts compounded the problem by seeking to bolster this initial message and cautioning the public that masks, especially medical-grade respirator masks, needed to be properly fitted and that ordinary people would therefore not

---

[57] Tufekci (2020), Gostin et al. (2020). For a discussion of the lack of preparedness in the United States, including the shortage of surgical grade masks, see, for example, Gostin et al. (2020).

[58] Jankowicz (2020). "Fact check: Outdated video of Fauci saying 'there's no reason to be walking around with a mask,'" *Reuters* (Oct. 8, 2020). https://www.reuters.com/article/uk-factcheck-fauci-outdated-video-masks/fact-checkoutdated-video-of-fauci-saying-theres-no-reason-to-be-walking-around-with-a-mask-idUSKBN26T2TR.

benefit from them.[59] Additionally, the WHO and Centers for Disease Control and Prevention (CDC) told the public to wear masks when they were sick, despite early evidence of asymptomatic transmission of COVID-19.[60] Indeed, the WHO did not endorse the use of face masks by the public to reduce the transmission of the virus until June 2020, and, even then, signaled its ambivalence by noting several disadvantages of masks, such as "difficulty communicating clearly" and "potential discomfort."[61] To be sure, some of the early, flawed advice from public health regarding mask wearing may have been the result of genuine uncertainty about the efficacy of masks in confronting COVID-19[62]—at least outside some countries in Asia where mask wearing was already normalized, partly as a protection against polluted air and partly as a result of the prior SARS and MERS outbreaks.[63] But even so, the precautionary principle should have led those officials to be more aggressive during any such initial stage of uncertainty, especially given that masks are "simple, cheap, and potentially effective" in curbing the spread of respiratory viruses generally.[64]

Public health officials did later clarify their guidance on the efficacy of masks, including by advocating the use of cloth masks when surgical grade masks were unavailable. But these early missteps sowed confusion among the broader public, particularly in the United States. They also complicated efforts for government officials who sought to persuade people to wear masks and created opportunities for other officials bent on exploiting opportunities to downplay the virus for political gain.[65] The impact of these inconsistent messages was particularly significant in places, such as the United States and Europe, that lacked the tradition and wide acceptance of mask-wearing that existed in many Asian nations due to past experience in managing other public health threats.[66] Countries like Hong Kong and Taiwan that moved quickly to implement universal mask requirements had far greater success in gaining control over the pandemic.[67]

COVID-19 underscores the need for transparency and clear messaging on mask wearing and other prophylactic measures. Health officials and politicians in the United States, for example, would have better served the public—and more effectively advanced their human rights obligation to protect life—if they had been more candid with the public up front.[68] Likewise, clear messaging on masks can help build support for advanced preparedness so that nations do not experience the same type of shortages they experienced during COVID-19 if and when another pandemic arrives. The importance of transparency and effective communication is particularly

---

[59] Tufekci (2020).

[60] *Id.*

[61] Greer et al. (2021).

[62] Sunjaya and Jenkins (2020).

[63] Leung (2020).

[64] Greenhalgh (2020).

[65] Kehane (2021), North (2020).

[66] Syal (2020), Beswick (2020).

[67] Tufekci (2020).

[68] *Id.*

critical given the challenges posed by disinformation in the digital age, which has affected the response of countries to the COVID-19 pandemic by undermining trust, magnifying fears, and sometimes leading to harmful behaviors.[69]

17.4.3 Implication for Scientific Research

A state's human rights obligation to ensure universal mask-wearing during the COVID-19 pandemic has several implications for scientific research. As explained above, human rights law not only justifies the imposition of mask mandates but also requires them to the extent that their efficacy is clearly demonstrated and that any mandate is accompanied by measures to ensure masks are universally available. In the United States, for example, research demonstrates that states with mask mandates witnessed sizeable increases in mask-wearing.[70] Further, while public education is important, education alone has proven insufficient.[71] Scientific research should continue to examine the efficacy of masks in preventing transmission, both in conjunction with other measures, such as capacity requirements, and without such measures, in order for states to better calibrate restrictions and minimize the impact on other human rights.

Research should additionally focus on measures for achieving universal masking-wearing. This includes research into the following: how best to formulate and frame mask requirements to increase compliance and public acceptance (with past public safety campaigns such as seatbelts mandates serving as an example);[72] whether sanctions are necessary (in addition to consistent public messaging) and, if so, what types of sanctions should be used to optimize compliance (for example, by evaluating the relative efficacy of fines, closures of businesses for non-compliance, court-ordered education classes or educational programs, and other penalties);[73] how to ensure any system of sanctions is coherent and applied consistently (as opposed, for example, to the isolated use of sanctions only against flagrant violators, such as individuals who hold large parties, or the disproportionate use of sanctions against marginalized groups); how to utilize community engagement and the inclusion of multiple stakeholders to increase mask wearing[74]; how to decrease the politicization surrounding mask wearing;[75] and how legal requirements can provide social-behavioral cover (for example, by allowing businesses to require masks without having to defend

---

[69] OECD Policy Responses to Coronavirus (COVID-19) (2020).

[70] Kehane, "Politicizing the Mask.".

[71] Gee and Gupta (2020).

[72] Gostin et al. (2020).

[73] See Anderson and Scott, "Is the Law Working? A Brief Look at the Legal Epidemiology of COVID-19, in *Addressing Legal Responses to COVID-19*," at 20–21.

[74] See, e.g., World Health Organization (2017).

[75] Anderson & Burris, "Is the Law Working?" at 21.

them on philosophical or health grounds).[76] It further includes research on how to increase mask use through public education, such as examining what factors most motivate people to wear masks, which remains a significantly under-studied area,[77] despite evidence that resistance to mask-wearing can be deeply rooted in some countries[78] and can track existing internal political divisions.[79] Additionally, research should explore emerging evidence that the cautious lifting of mask mandates can be used to encourage vaccinations and overcome vaccine hesitancy in countries where safe and effective vaccines have become broadly available[80] as well as more targeted use of mask mandates while vaccines are gradually made available (such as continued mandates in schools where younger students may not yet be eligible for vaccination).[81]

Finally, the foregoing analysis supports a related but broader claim: that states not only have human rights obligations related to masking itself, but also have an obligation under human rights law to support and conduct research on masking (and other public health interventions), as such research is necessary to fully protect the right to life. The WHO has argued that there is an ethical obligation to conduct research in the context of other public health emergencies. It has stated, for example, that during an infectious disease outbreak, states have "a moral obligation to learn as much as possible as quickly as possible, in order to inform the ongoing public health response, and to allow for scientific evaluations of new interventions being tested."[82] Moreover, the WHO maintains that learning this information "also improve[s] preparedness for similar future outbreaks."[83] While an analysis of a state's obligation under human rights law to conduct such research is beyond the scope of this chapter, the enormous utility of masking in reducing the spread of a deadly and destructive disease like COVID-19, coupled with the likelihood of future outbreaks of other highly transmissible diseases—a threat experts warned against long before the COVID-19[84]—suggests that states may be obliged to conduct additional research on masking as part of their broader duty to protect human life under international law.

17.5 Conclusion

Measures taken by governments to prevent the transmission and spread of COVID-19 can help fulfill their obligation under human rights law to protect life. These measures,

---

[76] Ibid.

[77] Shelus (2020).

[78] Navvaro (2020).

[79] Haddow et al. (2020), at 73.

[80] *See* Leonhardt (2021).

[81] *See* Zimmerman and Benjamin (2021).

[82] World Health Organization (2016).

[83] *Id.*

[84] *See* Bennett and Terry (2014).

however, can also negatively impact the exercise of other fundamental freedoms as well as socio-economic rights. Mask mandates and other policies designed to achieve universal mask-wearing have the unique ability to protect life while imposing relatively minimal burdens on the exercise of other rights, especially compared with other, more heavy-handed measures such as stay-at-home orders and business closures. The role of masks in protecting human rights during the current pandemic suggests the importance of scientific research to determine how best to design and implement laws and policies to achieve universal masking. In addition, employing effective, uniform educational and enforcement measures can increase acceptance of mask requirements among the population. Finally, while prior experiences with COVID-19 should inform scientific research on masking, this research should be undertaken not only to combat the current pandemic, but to better prepare for future pandemics as well.

References

African Charter on Human and Peoples' Rights, OAU Doc. CAB/LEG/67/3/ rev.5 (1981).

American Declaration on the Rights and Duties of Man, OEA/ser./L./V./II.23, doc. 21 rev. 6 (1948).

Bai, N. 2020, June 26. Still Confused about Masks? Here's the Science Behind How Face Masks Prevent Coronavirus. University of California San Francisco. https://www.ucsf.edu/news/2020/06/417906/still-confused-about-masks-heres-science-behind-how-face-masks-prevent.

Bennett, B. & Carney, T. 2014, July 8. Planning for pandemics: Lessons from the past decade. *Journal of Bioethical Inquiry* 12 (3). https://www.ncbi.nlm.nih.gov/pmc/articles/PMC7089178/.

Bennouhe, K. 2020. 'Lest We Sleep': COVID-19 and human rights. *American Journal of International Law* 666.

Beswick, E. 2020, July 15. Coronavirus: How the wearing of face masks has exposed a divided Europe. *Euronews*. https://www.euronews.com/2020/07/14/coronavirus-how-the-wearing-of-face-masks-has-exposed-a-divided-europe.

Boodman, E. *Stat.* 2021, October 20. Universal mask use could save 130,000 U.S. lives by the end of February, new study estimates. https://www.statnews.com/2020/10/23/universal-mask-use-could-save-130000-lives-by-the-end-of-february-new-modeling-study-says/.

CBS News. 2021. Workers worry for their safety as more states lift mask mandates, March 8. https://www.cbsnews.com/news/mask-mandates-workers-health-safety-concerns/.

Centers for Disease Control and Prevention. 2020, October 5. Science Brief: SARS-Cov-2 and Potential Airborne Transmission. https://www.cdc.gov/coronavirus/2019-ncov/more/scientific-brief-sars-cov-2.html.

Chávez, B.V. 2020, April. Planning resilient local communities. *Fulbright Split Screen.* https://fulbrightsplitscreen.com/articles/basilio-verduzco-chavez/.

Cloud, L.K., et al. (Public Health Law Watch, August 2020). "A Chronological Overview of the Federal, State, and Local Response to COVID-19," in *Assessing Legal Responses to COVID-19* https://static1.squarespace.com/static/5956e16e6b8f5b8c45f1c216/t/5f4d65782257 05285562d0f0/1598908033901/COVID19PolicyPlaybook_Aug2020+Full.pdf

Columbia Mailman School of Public Health. 2020, March 13. The kind of outbreak scientists knew would happen. https://www.publichealth.columbia.edu/public-health-now/news/kind-outbreak-our-scientists-knew-would-happen

Congressional Research Service. 2020a, April 16. Freedom of association in the wake of coronavirus. https://crsreports.congress.gov/product/pdf/LSB/LSB10451.

Cyprus v. Turkey, App. No. 25781/94 (Eur. Ct. H.R. May 10, 2001).

Editorial. 2020a, July 29. Bernie Sanders is asking for masks for all. *N.Y. Times.* https://www.nyt imes.com/2020/07/29/opinion/us-coronavirus-masks.html.

Editorial. 2020b. The plight of essential workers during the pandemic. *The Lancet* 36 (10237). https://www.thelancet.com/journals/lancet/article/PIIS0140-6736(20)31200-9/fulltext.

European Convention on Human Rights for the Protection of Fundamental Freedoms, *opened for signature* Nov. 4, 1950, 213 U.N.T.S. 221 (1981).

Executive Order No. 202.6, "Continuing Temporary Suspension and Modification of Laws Relating to the Disaster Emergency" (2020, May 23). https://www.governor.ny.gov/sites/default/files/atoms/files/EO202.6.pdf

Fact check: Outdated video of Fauci saying 'there's no reason to be walking around with a mask,' *Reuters.* https://www.reuters.com/article/uk-factcheck-fauci-outdated-video-masks/fact-checko utdated-video-of-fauci-saying-theres-no-reason-to-be-walking-around-with-a-mask-idUSKB N26T2TR.

Flaxman, S., et al. 2020, June 8. "Estimating the effects of non-pharmaceutical interventions on COVID-19 in Europe." https://www.nature.com/articles/s41586-020-2405-7?fbclid=IwAR37 gxOwxZ1CbmN1guU-9sn6EYg65_4RxIFeoP8sqSYMaa7A0fYqvTwHw2I.

Fowler, J.H., et al. 2020, May 12. The effect of stay-at-home orders on COVID-19 cases and fatalities in the United States. MedRXiv. https://www.medrxiv.org/content/https://doi.org/10.1101/2020. 04.13.20063628v3.full-text.

Garrett, B.L. & Kovarsky L. 2021, February 22. Viral injustice. *California Law Review*, Forthcoming. https://ssrn.com/abstract=3790859 or https://doi.org/10.2139/ssrn.3790859.

Gee, R.E., & V. Gupta. 2020, October 5. Mask mandates: A public health framework for enforcement. *Health Affairs.* https://www.healthaffairs.org/do/https://doi.org/10.1377/hblog20201002. 655610/full/.

Gibson, J., & X. Sun. Understanding the economic impact of COVID-19 stay-at-home orders: A synthetic control analysis. https://doi.org/10.2139/ssrn.3601108.

Gold, M. 2021, March 4. Connecticut lifts capacity on many businesses but keeps mask mandate. *N.Y. Times.* https://www.nytimes.com/2021/03/04/nyregion/connecticut-coronavirus-restrictions.html

Gostin, L.O. et al. 2020. Universal Masking in the United States: The Role of Mandates, the Health Education, and the CDC. *JAMA Network* vol. 324 (9). https://jamanetwork.com/journals/jama/fullarticle/2769440.

Gostin, L.O., et al. 2020, March 26. Responding to COVID-19: How to navigate a public health emergency legally and ethically. *The Hastings Center Report.* https://doi.org/10.1002/hast.1090.

Greenhalgh, T., et al. 2020, April 9. Face masks for the public during the covid-19 crisis. *BMJ.* https://doi.org/10.1136/bmj.m1435.

Greer, S., et al. 2021. Coronavirus politics: The comparative politics and policy of COVID-19. *University of Michigan Press.* https://doi.org/10.3998/mpub.11927713.

Haddow, K., et al. 2020. "Preemption, public health, and equity in the time of COVID-19," in *Assessing Legal Responses to COVID-19.*

Halkias, M. 2021, March 3. Kroger, Costco and other big and small retailers stick with required masks. *Dallas News.* https://www.dallasnews.com/business/retail/2021/03/03/kroger-and-other-big-and-small-retailers-stick-with-required-masks/.

Hathaway, O. et al. 2020, November 18. COVID-19 and international law series—Human rights law: Right to live. *Just Security.* https://www.justsecurity.org/73426/covid-19-and-international-law-series-human-rights-law-right-to-life/.

Henig, R.M. 2020, April 8. 2020. Experts warned of a pandemic decades ago. Why weren't we ready? *National Geographic.* https://www.nationalgeographic.com/science/article/experts-warned-pandemic-decades-ago-why-not-ready-for-coronavirus.

Higgins, V. et al. 2020, December 21. *Covid-19:* "From an acute to chronic disease? Long-term health consequences. *Taylor and Francis* 58 (5). https://www.tandfonline.com/doi/full/https://doi.org/10.1080/10408363.2020.18608950.

Howard, J., et al. 2021, January 26. An evidence review of face masks against COVID-19. *PNAS* 118, no. 4. https://www.pnas.org/content/118/4/e2014564118.

Human Rights Watch. 2020a, June 29. Protecting economic and social rights during and post-COVID-19. https://www.hrw.org/news/2020/06/29/protecting-economic-and-social-rights-during-and-post-covid-19.

Human Rights Comm. 2020b, April 24. Statement on Derogations from the Covenant in Connection with the COVID-19 Pandemic, UN Doc. CCPR/C/128/2.

International Covenant on Civil and Political Rights. (1966), 999 U.N.T.S. 171.

Jankowicz, M. 2020, June 15. Fauci said US government held off promoting face masks because it knew shortages were so bad that even doctors couldn't get enough. *Business Insider*. https://www.businessinsider.com/fauci-mask-advice-was-because-doctors-shortages-from-the-start-2020-6.

Kehane, L.H. 2021, January 5. Politicizing the mask: Political, economic and democratic factors affecting mask wearing behavior in the USA. *Eastern Economic Journal*. https://link.springer.com/article/10.1057%2Fs41302-020-00186-0.

Kirubarajan, A., et al. 2020. Mask shortage during epidemics and pandemics: a scoping review of intentions to overcome limited supply. *BMJ Open Access*. https://bmjopen.bmj.com/content/bmj open/10/11/e040547.full.pdf..

Leonhardt, D. 2021, June 2. Covid hope over fear. *N.Y. Times*. https://www.nytimes.com/2021/06/02/briefing/covid-19-guidance-vaccinations.html.

Lethang, M. 2021, March 4. Which states have dropped mask requirements and why. *ABC News*. https://abcnews.go.com/Health/states-dropped-mask-mandates/story?id=76249857.

Leung, H. 2020, March 12. Why wearing a face mask is encouraged in Asia but Shunned in the U.S. *Time*. https://time.com/5799964/coronavirus-face-mask-asia-us/.

Londoño, E, and L. Casado. 2021, March 28. A collapse foretold: How Brazil's Covid-19 outbreak overwhelmed hospitals. *N.Y. Times*. https://www.nytimes.com/2021/03/27/world/americas/virus-brazil-bolsonaro.html.

Lurie, M.N. 2020, November. Coronavirus disease 2019 epidemic doubling time in the United States before and during stay-at-home restrictions. *Journal of Infectious* Diseases 222 (10) https://doi.org/10.1093/infdis/jiaa491.

Maani, N. 2021, February 2. What science can and cannot do in a time of pandemic. *Scientific American*. https://www.scientificamerican.com/article/what-science-can-and-cannot-do-in-a-time-of-pandemic/.

McCabe, C. 2020, August 13. Face masks really do matter. The scientific evidence is growing. *The Wall Street Journal*. https://perma.cc/2S89-4R5T.

Morris, S. 2020, March 31. 31 states now have stay at home orders amid the coronavirus outbreak. *Newsweek*. https://perma.cc/D8YH-26X9.

Navvaro, A.J. 2020, July 13. Mask resistance during a pandemic isn't new—in 1918 many Americans were 'Slackers,' *The Conversation*. https://theconversation.com/mask-resistance-during-a-pandemic-isnt-new-in-1918-many-americans-were-slackers-141687.

National Restaurant Association, "State of the Restaurant Industry Report". 2021, January 26. https://restaurant.org/news/pressroom/press-releases/2021-state-of-the-restaurant-industry-report.

North A. 2020, July 22. Why masks are (still) politicized in America. *Vox*. https://www.vox.com/2020/7/21/21331310/mask-masks-trump-covid19-rule-georgia-alabama.

OECD Policy Responses to Coronavirus (COVID-19). 2020, July 3. Transparency, communication, and trust: The role of communication in responding to the first wave of disinformation about the new Coronavirus. https://www.oecd.org/coronavirus/policy-responses/transparency-communication-and-trust-bef7ad6e/.

Pietsch, B. 2021. 20.5 million people lost their jobs in April. Here are the 10 job types that were hardest hit. *Business Insider*. https://www.businessinsider.com/jobs-industries-careers-hit-hardest-by-coronavirus-unemployment-data-2020-5.

Roman Catholic Diocese of Brooklyn v. Cuomo, 141 S. Ct. 63 (2020).

Rowthorn, R., and J. Maciejowski, 2020, August 29. A cost-benefit analysis of the COVID-19 disease. *Oxford Review of Economic Policy* 36. https://academic.oup.com/oxrep/article/36/Sup plement_1/S38/5899017?login=true/.

Shelus, V.S. 2020. Motivations and barriers for the use of face coverings during the COVID-19 pandemic: Messaging insights from focus groups. *Int'l J. Environ. Research & Pub. Health*, vol. 17(24). https://www.mdpi.com/1660-4601/17/24/9298/htm.

Spadaro, A. 2020, June. COVID-19: Testing the limits of human rights. *European Journal of Risk Regulation* 11 (2). https://doi.org/10.1017/err.2020.27.

Stoyanovi v. Bulgaria, App. No. 42980/04 (Nov. 10, 2010).

Syal, A. 2020, July 1. Wearing a mask has become politicized. Science says it shouldn't be. *NBC News*. https://www.nbcnews.com/health/health-news/wearing-mask-has-become-politicized-sci ence-says-it-shouldn-t-n1232604.

Sunjaya, A.P., and C. Jenkins, 2020, April 30. Rationale for universal face masks in public against COVID-19. *Respirology*. https://www.ncbi.nlm.nih.gov/pmc/articles/PMC7267357/.

Taylor, A.L., and R. Habibi. 2020, June 5. The collapse of global health cooperation under the WHO international health regulations at the outset of COVID-19: Sculpting the future of global health governance. *American Society of International Insights 24* (15). https://www.asil.org/insights/ volume/24/issue/15/collapse-global-cooperation-under-who-international-health-regulations.

The Conversation. 2020b, June 2. Do you have a right to protest? The coronavirus's impact on freedom of assembly. *The Conversation*. https://theconversation.com/do-you-have-a-right-to-pro test-the-coronaviruss-impact-on-freedom-of-assembly-139363.

The Independent Panel for Pandemic Preparedness & Response. 2021. COVID-19: Make it the last pandemic. https://theindependentpanel.org/wp-content/uploads/2021/05/COVID-19-Make-it-the-Last-Pandemic_final.pdf.

Tufekci, Z. 2020, March 17. Op-Ed, "Why Telling People They Don't Need Masks Backfired". *N.Y. Times*. https://www.nytimes.com/2020/03/17/opinion/coronavirus-face-masks.html.

UN Committee on Economic, Social and Cultural Rights. 2000, August 11. CESCR General Comment No. 14: The Right to the Highest Attainable Standard of Health, UN Doc E/C.12/2000/4.

U.N. Hum. Rts. Off. of High Comm'r. 2020, June 22. Racial discrimination in the context of the COVID-19 crisis. https://www.ohchr.org/Documents/Issues/Racism/COVID-19_and_Rac ial_Discrimination.pdf.

UN Human Rights Comm. 2019, September 3. General Comment No. 36, para 2, UN Doc. CCPR/C/GC36.

U.S. Human Rights, Office of the High Commissioner.2020, March 26. No exceptions with COVID-19: 'Everyone has the right to life-saving interventions. U.N. experts say. https://www. ohchr.org/EN/NewsEvents/Pages/DisplayNews.aspx?NewsID=25746&LangID=E&fbclid= IwAR1vFZDTrmlWSQXRq5BMJF3144OrKv9HRQevSO_SH1mqOEJXcXDM4KqUHM4.

United Nations Secretary-General. 2021, February 22. Secretary-General's Message to the Opening of the 46th Regular Session of the Human Rights Council.

Universal Declaration of Human Rights, G.A. Res. 217A, U.N. GAOR, 3d Sess., 1st plen. mtg., U.N. Doc. A/810 (1948, December 12).

Williams, C.J., et al. 2020, October 1. Reopening the United States: Black and hispanic workers are essential and expendable again. *American Journal of Public Health* 110 (10). https://doi.org/ 10.2105/AJPH.2020.305879.

World Health Org. 2005, May 23. Revision of the International Health Regulations, WHA58.3. http://www.who.int/gb/ebwha/pdf_files/WHA58/WHA58_3-en.pdf.

World Health Organization. 2016. Guidance for managing ethical issues in infectious disease outbreaks. https://www.who.int/publications/i/item/guidance-for-managing-ethical-issues-in-inf ectious-disease-outbreaks.

World Health Organization. 2017. WHO community engagement framework for quality, people-centered and resilient health services. https://apps.who.int/iris/bitstream/handle/10665/259280/ WHO-HIS-SDS-2017.15-eng.pdf.

Xu, D. 2021, January. Physical mobility under stay-at-home orders: A comparative analysis of movement restrictions between the United States and Europe. *Economics and Biology* 40. https://doi.org/10.1016/j.ehb.2020.100936.

Zimmerman, Kanecia and Benjamin, Danny Jr. 2021, August 10. Op-Ed, "We Studied One Million Students. This Is What We Learned About Masking." *N.Y. Times.* https://www.nytimes.com/2021/08/10/opinion/covid-schools-masks.html.

Jonathan Hafetz is an expert on constitutional law, national security, and international justice issues. He joined Seton Hall Law School in 2010. Professor Hafetz is the author of multiple books and academic articles. Prior to joining Seton Hall, he was a senior attorney at the American Civil Liberties Union, a litigation director at New York University's Brennan Center for Justice, and a John J. Gibbons Fellow in Public Interest and Constitutional Law at Gibbons, P.C. From 2014–15, Professor Hafetz was a Visiting Research Scholar in the Program in Law and Public Affairs at Princeton University. He is the recipient of Fulbright fellowships from the U.S. Government for study in Mexico and Japan.

Chapter 18
Research Ethics in Exceptional Times: What Lessons Should We Learn from Covid19?

Søren Holm

18.1 Introduction

The COVID-19 pandemic has led to an unprecedented acceleration of research. Thousands of papers have been published in a very short time across a wide range of academic disciplines. This has already led to instances of research misconduct and articles have been withdrawn from prominent journals very soon after being published. The web-site Retraction Watch lists 145 COVID-19 related publications that have been retracted at the time of writing of this chapter in October 2021, including papers in the New England Journal of Medicine and The Lancet.[1] The pandemic has also led to calls for the relaxation of generally accepted research ethics rules and rules concerning the protection of personal data during what is claimed to be an exceptional time of crisis where research need to be speeded up and not hampered by rules more suitable to normal times. This chapter will provide an analysis of whether research ethics processes and rules should have been relaxed during COVID-19, and consider the implications of that answer for future health crises. It will proceed in three steps. It will first analyse to what extent the COVID-19 pandemic can be claimed to be truly exceptional, and will argue that although the pandemic is a very significant health crisis, it is neither unique nor truly exceptional. But, perhaps a crisis does not need to be exceptional. Perhaps a significant, global health crisis is sufficient justification for modifications of research ethics processes and principles. The second part of the chapter provides an analysis of this question both in relation to processes and principles, and further explores the general issue of when a state of exception truly pertains and the effects of declaring such a state of exception. The final section briefly draws out the lessons for how we should handle

---

[1] https://retractionwatch.com/retracted-coronavirus-covid-19-papers/ (accessed October 15, 2021).

S. Holm (✉)
University of Manchester, Manchester, UK
e-mail: soren.holm@manchester.ac.uk

© The Author(s), under exclusive license to Springer Nature Switzerland AG 2023   355
T. Zima and D. N. Weisstub (eds.), *Medical Research Ethics: Challenges in the 21st Century*, Philosophy and Medicine 132,
https://doi.org/10.1007/978-3-031-12692-5_18

future health crises from a research ethics perspective. Throughout the analysis it is important to keep in mind that one of the only things we know for certain about the next global health crisis is that it will not be identical to the COVID-19 crisis in all its details. It has been claimed by no less an authority than Winston Churchill that "Generals are always prepared to fight the last war" (although that attribution may be apocryphal) and early on during COVID-19 we saw this happening in the public health response where assumptions going into the epidemiological modelling was based on earlier experiences with Influenza pandemics and to a lesser extent the SARS 2002–3 outbreak (see for instance House of Commons 2021). It is important not to make the same mistake in preparing research ethics for the next big crisis.

18.2   Is the COVID-19 Pandemic Exceptional?

One major trope in the justification of emergency curtailments of ordinary freedoms and liberties during the COVID-19 pandemic has been that the pandemic is exceptional, special or unique, either in itself or in its impacts on society. This is often stated as though it was an obvious or self-evident truth, and it might be backed by reference to the rapid spread of the disease, the prevalence, the morbidity and mortality, the impact on the health care system, or the societal and economic impacts. It is, however worth considering whether and to what extent this is true.

It is undoubtedly true that the pandemic has had a major impact on life in many countries around the globe, partly as a result of the travel restrictions imposed by many countries, partly due to the physical distancing measures introduced and their impact on business activities and daily life, and partly due to the impact on the health and social care systems of large numbers of very ill patients needing care. These impacts, especially the economic impacts will continue to grow for some time to come.[2]

Recent estimates of the global excess deaths due to COVID-19 puts this figure at more than 6 million, whereas the excess mortality of the 1957–59 Asian influenza pandemic is estimated at 3.1 million when population is scaled to 2020 size, and the estimate for the 1918–19 influenza pandemic is approximately 75 million when scaled to current population numbers (Simonsen and Viboud 2021). Historical plague pandemics were much more serious, causing absolute death tolls far exceeding COVID-19 within a much smaller global population (Piret and Boivin 2020).

However, what has made the pandemic seem truly exceptional is not only the size of its effects, but also that it has happened to 'us', i.e. it has affected almost all of those who live in affluent societies in some way. It is probably the first global outbreak of infectious disease to have this pervasive effect since the Spanish flu pandemic of 1918–19, and almost all of us are too young to remember the impact of that pandemic.

---

[2] From a parochial UK point of view it is perhaps allowable to point out the long term economic effects of Brexit are much more significant for the UK than the effects of COVID-19 (Latorre et al. 2020).

But, there are several other infectious diseases that are globally epidemic or endemic and continue to cause large number of infections, morbidity, mortality, and negative economic and social impact in those societies where they occur, especially if the society does not have resources to provide effective prevention and treatment. This includes age old killers like Tuberculosis and Malaria, newer ones such as HIV/AIDS and Ebola, and somewhat overlooked conditions such as Hepatitis B. Combined, all of these regionally or globally epidemic or endemic infectious diseases add up to a very large burden of disease and a very large unmet health care and research need. In 2019, for instance 1.74 billion people were reported by the WHO to require mass or individual treatment and care for neglected tropical diseases.[3] And, this is even before we come to the big killers, lower respiratory infections and diarrheal diseases that kill millions, primarily children, every year in low income countries.

So, the COVID-19 pandemic is a significant, global health crisis. But, it is not unique in its impact, and not even likely to be seen as particularly exceptional when considered retrospectively in a hundred years' time. Our current impending fear of COVID-19 doom may be more a perception than a reality (Ferguson 2021).

18.3 Should Research Ethics Rules Be Exceptionally Relaxed During COVID-19?

As mentioned in the introduction there has been calls for many different relaxations of research ethics procedures and rules during COVID-19. These can be divided into several categories according to what relaxation or changes that is being proposed:

1. Purely procedural changes and relaxation of safeguards unlike to lead to significant harm or rights violations
2. Relaxation of safeguards directly related to possibilities of significant harm or rights violations.

The distinctions between categories are not always sharp, but the categories are never the less useful for organising the analysis.

Some of the changes that have been asked for and that have been implemented are at least in principle purely procedural. Many research ethics systems have, for instance introduced a fast track for COVID-19 related applications, thereby ensuring that they can be processed and approved much faster than was the average for applications before COVID-19 and much faster than non-COVID related applications. If this change is purely procedural and does not affect the quality of the scrutiny of the applications and any required revisions before they are approved, this does not raise any research ethics issues in relation to the protection of potential research

---

[3] SDG Target 3.3|Communicable diseases: By 2030, end the epidemics of AIDS, tuberculosis, malaria and neglected tropical diseases and combat hepatitis, water-borne diseases and other communicable diseases. https://www.who.int/data/gho/data/themes/topics/indicator-groups/indicator-group-details/GHO/sdg-target-3.3-communicable-diseases.

participants and the upholding of their rights. If the procedural change also involves a lighter touch approach to the scrutiny of the applications, or a lowered bar for approval in relation to ethically significant issues we are moving beyond purely procedural changes and into changes to the system that may have substantive implications for research participants.

Let us briefly look at an example. Because of the physical/social distancing requirements imposed in many societies there has, for instance been a rapid shift to research interviews being conducted online via virtual platforms like Skype, Zoom, Teams etc. and this shift has often been approved by research ethics committees.[4] This raises data protection issues because the servers used by these platforms may be located anywhere in the world, but it also raises potential research ethics issues that may initially seem to be relatively inconsequential. In a normal face to face interview situation the interviewer has a considerable degree of control over the interview situation, even when interviewing a respondent in the respondent's own home. The interviewer can ensure that no one apart from the respondent is present during the interview, in *extremis* by refusing to start the interview or by terminating it if any other person demand to be present. Using virtual means, the control of the interview situation is decreased. Even if conducting a video interview the interviewer has no control over who else is present out of picture. This probably does not matter much if the interview is about mundane and non-sensitive issues, but if it is about sensitive or controversial issues, or issues about which there might be conflict in households, it does create potential research ethical issues. These issues cannot be fully solved simply by providing instructions to respondents that no one else should be present, and/or by asking them at the start of the interview to confirm that no one else is present. What seems like a simple change in research procedures necessitated by the (exceptional?) inability to conduct face to face interviews thus opens up a space of considerable ethical complexity.

There is, a residual broader research ethics issue about purely procedural changes if the focus on COVID-19 related research and the priority given to COVID-19 related applications directly or indirectly lead to a situation where other important applications are slowed down in the system, or not handled at all. The problem here is that the mere fact that a project is COVID-19 related does not tell us how important it is, comparatively. There are many trivial COVID-19 related research projects that are not particularly important *sub specie aeternitatis*. Using system resources on such trivial projects may lead to a sub-optimal outcome overall because more important research will be delayed. It is therefore important not to lose sight of the comparative importance of other research and not to focus exclusively on COVID-19 relevance as a proxy for overall importance or value.

The most problematic relaxations are those that may lead to significant harm to research participants or that entail that important rights of research participants are violated. Let us look at the issue of rights violations first. In one sense it has very rarely been the case that the rights of research participants have been violated in a legal

---

[4] Including the one the author of this paper Chair at the University of Manchester.

scnsc. What has more often happened is that legal rights that existed before COVID-19 have been abolished or suspended by the state with reference to COVID-19 causing a state of exception. However, insofar as such legal rights are legal instantiations of moral rights linked to the protection of important personal interests or choices of research participants, it still makes sense to talk about rights violations, even though the individuals whose rights are violated have no way to vindicate these rights in the legal system because the legal instantiation of the rights have been abolished or suspended. The most common rights that have been abolished or suspended are rights in relation to the collection and research use of personal information, including health information. In order to speed up research, many countries have suspended the normal regulation of data collection and data access and allowed researchers and epidemiologist easy access to enable data driven science (Stokel-Walker 2021). Types of data that were previously categorised as confidential, and which could only be accessed with the consent of the data subject or with special permission, were reclassified so that those safeguards no longer applied. This for instance happened in relation to clinical data that is seen in many countries as strictly confidential because it is divulged or produced within a patient—health care professional relationship that is based on confidentiality. If, for instance I tell my doctor about some risky behaviour that has contributed to my disease, this will usually be confidential information. And, now that the pandemic is slowly receding, we see calls for the very large health information databases that have been generated while normal rules have been suspended to be kept open for research use indefinitely, because they are valuable for future COVID and non-COVID related research (Stokel-Walker 2021).

It is perfectly understandable that governments who control health information data bases decide to suspend the normal rules for the use of these databases in research, but it still needs an ethical justification which cannot just be that the situation is exceptional or critical. The same types of information are valuable for non-COVID related research, but we had previously accepted that the right of individuals to control their confidential health information was more important than the benefits to others that can flow from the research use of such data. If there is a justification to change that position it cannot flow simply from the number of lives saved, since for instance the number of lives saved by the research efforts flowing from the 'War on Cancer' declared by Richard Nixon in 1971 far outweigh the number of lives likely to be lost to COVID-19 (DeVita 2004). And the War on Cancer has not in general been seen as a justification for breaching ordinary expectations of confidentiality. So, maybe the justification is that the need for access to the information is urgent in the sense that if researchers can get access to it now, they can prevent something very bad happening. At the time of making the decision to allow access this will be a prediction of future benefit, but it raises the interesting question of how big a benefit and how urgent a need is sufficient to create a full justification for suspending rights. This question does probably not have a definite answer. It is, however important to note that if the real justification is an urgent, very important need to provide access to otherwise confidential information, then that justification disappears when the urgency recedes and does not carry over to continued use of the datasets that have been created.

A further issue about information and data concerns the use of data that have been generated by compulsion. Many countries have at various points during the pandemic had COVID apps that at the same time collected data about their users, principally for contact tracing purposes, and was the way in which users could document various aspects of their COVID status. In a few countries the use of such apps have been strictly compulsory/legally mandated, whereas in most the use have been in principle voluntary, but with an element of *de facto* compulsion since access to important activities of daily life have been contingent on using the app. Many countries have also had more traditional paper forms that have had to be filled in by travellers, school children etc. If data has been generated by compulsion then there is no consent for the initial collection and definitely no consent of any kind for the use for research. If, as has often been the case the use of the apps or the collection of the data has been resented by some of those having to use apps or provide the data, there is not even a basis for assuming consent for further use.

This creates a significant research ethics issue that is further exacerbated when the exceptional situation that justified the data collection is resolved. Even those who think that COVID-19 is a truly exceptional crisis, and who argue that this justifies truly exceptional means of compulsory data collection or suspensions of confidentiality in relation to data, have a justificatory problem about the use of the data when the crisis is over. The exceptionality of the crisis cannot in itself justify the future use of the data for unconnected purposes post crisis. The argument for further use may then shift to be based on claims about the knowledge and value that can be gained by further analysis and mining of the data, but putting the value of science above the rights and interests of individuals is normally and for good reasons not accepted in research ethics.

There have also been instances where it has been argued that we should allow interventional research that we would not normally have allowed, or allow research designs that are otherwise seen as problematic because the COVID-19 pandemic is exceptional and requires us to allow science and development to proceed as the fastest rate possible. There are many examples of this, but here the analysis will focus on COVID-19 challenge studies for vaccine development. There is already an extensive literature on the ethical issues inherent in such trials and a lively debate about whether they are acceptable or not (Chappell and Singer 2020; Eyal et al. 2020; Holm 2020; Solbakk et al. 2021), so here the focus will be on the particular argument that they should be accepted because we are in an exceptional crisis situation. In a human challenge study (HCS) of a vaccine a group of healthy individuals are randomised to be either vaccinated or given a placebo vaccine and the whole group is then later inoculated with a sufficient does of the infectious agent to ensure a 100% infection rate in unvaccinated individuals. Human challenge studies have been used in vaccine development in relation to diseases such as flu, malaria, and dengue fever. Before the COVID-19 pandemic it was generally argued and held that a HCS could be ethically acceptable if the risk of serious disease or death is minimal. This can be the case if the disease itself creates a minimal risk (e.g. flu in healthy, young volunteers with access to supportive treatment), if a weakened strain of the infectious agent is used (e.g. dengue fever), or if there is an effective rescue treatment (e.g. malaria).

COVID-19 infection did at the time of the proposals to allow HCSs not fulfil the generally accepted requirements for an ethical HCS. The disease is serious and potentially lethal, there is no known weakened SARS-CoV-2 strain, and there was no rescue treatment. It was nevertheless proposed in the literature by several groups that COVID-19 HCSs should be allowed, and a very vocal advocacy group called '1 Day Sooner' was set up on the internet to promote such studies and recruit participants for them.[5] The underlying argument put forward is (1) that vaccine development is so important that even a few days delay in the development process is problematic, (2) that risks can be made low enough by careful participant selection, and (3) that people should be allowed to consent to these risks. The different contributors to this argument emphasise different aspects of this argumentative structure, the web-group 1 Day Sooner, for instance writing about 'high-risk' studies being allowable. The argument is essentially consequentialist in claiming that the current research regulations that would prevent such a HCS from being approved are a hindrance to the best aggregate outcome and should be set aside. There are many problems with this argument, including in relation to the prediction that great benefit will flow from any particular COVID-19 HCS or by HCSs in aggregate (Holm 2020; Solbakk et al. 2021), but the focus here will be on the underlying implicit and sometimes explicit assumption that the need for research and the need for speeding up research in a crisis situation justifies the suspension of research ethics rules designed for protecting research participants against harm.

Let us begin the analysis by remembering Standard 5 of the Nuremberg Code:

5. No experiment should be conducted where there is an a priori reason to believe that death or disabling injury will occur; except, perhaps, in those experiments where the experimental physicians also serve as subjects.

On the face of it, Standard 5 prohibits HCS with a potentially life-threatening disease with no rescue treatment, except if the researchers themselves are also research subjects. And, insofar as I know, it has not been proposed by any HCS researchers that they would themselves be in the subject pool. What Standard 5 expresses is the idea that we should never plan, approved or conduct research where it is likely that research participants will die or be permanently and significantly harmed as a direct result of participating in the research. It is an absolute, universal prohibition. Why would we hold such a view? First, perhaps because history teaches us that if we don't have absolute prohibitions against causing serious harm to research participants, there will be circumstances where we have strong incentives to put research participants at risk of harm, and we will find it difficult to draw any other line between the acceptable and the unacceptable causing of harm. Secondly, because we would often be trading a quantifiable risk of significant harm, against a predicted and much less precisely quantifiable chance of benefit. Given that the benefit calculations in such circumstances are highly malleable, there is a great risk of bias affecting the balancing in crisis situations. Third, there is a significant risk that the goal posts will shift. COVID-19 challenge studies were approved in England, but effective vaccines were developed before the studies took place. Did this mean that the studies were

---

[5] https://www.1daysooner.org/.

stopped? Of course not! Instead of being claimed to be necessary for the development of vaccines *per se*, new purposes were put forward.[6] These new purposes are scientifically important, but none of them is likely to create anything like the societal benefit that the development, and perhaps more importantly widespread roll out of the first effective vaccine is/was likely to create. And, it was that very large benefit which was the original justification for allowing COVID-19 HCSs, despite the risks to research participants. So, the justification for allowing COVID-19 HCS has weakened, but they still continue, once the principle prohibiting them has been breached.

To put it in other words, Standard 5 and its progeny in later normative documents puts in place an absolute prohibition exactly in order to prevent the temptation to exploit a less absolute standard. The willingness of ethicists, lawyers and politicians to find a way around the prohibition shows how strong that temptation can become when the crisis mood gets close to panic.

It is, again completely understandable that there is a strong desire to develop vaccines as quickly as possible and completely explainable why political decision-makers may decide to suspend or circumvent the research ethics that are seen as obstacles to this. But, understandability and explainability is not justification.

This argument can perhaps be further illuminated by considering an issue outside of research ethics. A prominent feature of the public health response to the pandemic has been vaccine nationalism. States and regions have often taken a 'beggar thy neighbour' approach to vaccine supplies (and supplies of other essential products), and have only allowed the export of significant amounts of vaccines when either (1) they have already vaccinated their own population, or (2) they have decided not to use a particular vaccine they have bought. Vaccine sharing has therefore primarily been of vaccines that have been surplus to requirements in the richer parts of the world. This has led to immense disparities in the speed of vaccine roll out and in vaccine coverage between resource rich and resource poor countries, and has created a deep issue of global (in)justice which perfectly illustrates the difference between a particular set of decisions being explainable or understandable, and the set of decisions being justifiable. It is not difficult to explain or understand why governments have prioritised their own populations. One of the primary purposes of

---

[6] https://hvivo.com/the-human-challenge-programme/ (accessed 15 October 2021).
"Are these studies still needed with approved vaccines available?

- First generation vaccines are not the final answer to eradicating the virus. Uneven vaccine delivery and breakthrough infections mean that outbreaks will continue to happen for months and years to come.
- New variants of the virus mean vaccines will likely have to be adjusted to ensure effectiveness but our capacity to test them in field studies alone will be difficult. Human challenge will be the fastest way to compare old and new vaccines.
- One of the benefits of a human challenge study is that it enables the rapid study of volunteers' natural immune response to Covid-19 infection which helps doctors better understand how the virus affects people, how to detect symptoms early, and also how each vaccine works.
- By testing vaccines side by side in a challenge study, it is possible to quickly identify which is the most effective."

the state is to protect its citizens, and this has been recognised in political philosophy since the inception of that discipline; and it is clearly also politically and electorally difficult to provide vaccines to other countries before 'your own' people have been fully vaccinated. But this does not mean that such government actions are fully justifiable. There is an obligation on governments to provide aid elsewhere in the world during a pandemic, and also a more general obligation to contribute to global justice. This is also recognised by many of the governments in resource rich countries who have been happy to pledge support to international vaccine distribution schemes, even though their pledges have not resulted in much concrete action. The fact that vaccine nationalism is both explainable and understandable as a government response to the pandemic does not make it justifiable. And, it does not become more justifiable just by claiming that the COVID-19 pandemic is exceptional and calls for exceptional measures. This does not affect the resulting global injustice or the evaluation of that injustice. It may be true that the pandemic calls for exceptional measures, but those measures are not a relaxation of our standards of justice, but a strengthening of the courage of our political leaders to act with integrity and conviction to discharge the existing obligations to global justice instead of perpetuating and deepening injustice.

A particular argument in relation to research ethics has been that the way we think about research ethics, and the regulations and systems that are the result of that thinking are for 'normal times' and that we should therefore expect them not to be fully applicable during a crisis. Some have even claimed that "The COVID-19 pandemic has upset ordinary moral assumptions" and then ramped up the rhetoric by stating that "When so much is at stake, complacency and moral inertia cost lives"(Chappell and Singer 2020, p. 2). The in-principle issue of exceptionality of COVID-19 has been analysed above, but there is another strand to the 'normal times' argument and that is the claim that our thinking about research ethics primarily developed in normal times and not during crisis and that this genealogy has influenced the regulations and systems. The problem with this second strand of the normal times argument is that it is simply false. Giving a full account of the complicated history of research ethics is far beyond the scope of this chapter, but a few pertinent facts should suffice. The first international document which is still referenced in current substantive discussions is the Nuremberg Code, enunciating *inter alia* the requirements of consent and of a right to withdrawal. The Nuremberg Code did not emanate from thinking about normal times, but of thinking about the research ethics that has to be established and adhered to during periods of profound crisis. A world war creates a series of military and civilian health crises, and also, for those countries who perceive their very existence as threatened, a strong temptation to think of this as a strong justification for instituting a state of exception. During World War II this was then overlaid by the racist ideologies of Nazi Germany and Imperial Japan, leading to the research atrocities committed by these two Axis powers. There were also research ethics problems on the Allied side, though much less serious (Rasmussen 2020), and later research ethics problems on all sides during the Cold War (see for instance Advisory Committee on Human Radiation Experiments 1996). The next set of important foundational documents in research ethics were also a response to crisis, and to something exceptional happening. Just like in other areas of research and technological development, the post-war years witnessed an amazing growth in biomedical research fuelled by rapid underlying developments in biological knowledge and technical developments. But, in the 1960 and early 1970s it became clear that this rapidly growing and very successful field of research had created an ethics crisis. Researchers were ignoring the interests and rights of research participants, despite the principles enunciated in the Nuremberg Code, or the rights established in the Universal Declaration of Human Rights. This was evidenced in the UK by Maurice Pappworth in his 1967 book 'Human Guinea Pigs', following on from a 1962 paper of the same title, and by Henry K. Beecher for the US in his 1966 article 'Ethics and Clinical Research' in the New England Journal of Medicine (Beecher 1966; Pappworth 1962, 1967). Together with the exposé of the Tuskegee syphilis study in 1972, this led to the introduction of Institutional Review Boards in the USA, and Research Ethics Committees in the 1975 revision of the World Medical Association Declaration of Helsinki which also introduced numerous other changes to the Declaration. So, both the shape of current research ethics system and the underlying principles have come out of the recognition of and reflection on crises, and not primarily out of considerations of research activities in normal times.

18.4   What Should We Learn from the COVID-19 Pandemic?

There are many things we can learn from the COVID-19 pandemic when it comes to research ethics. The first is that our research ethics system is perhaps more fragile than we thought it was. Research ethics has been institutionalised for a considerable number of years, and there is a core of generally recognised declarations and conventions[7] that spell out the core, basic principles of research ethics as applied to biomedicine. But, the pandemic has shown that in a crisis it is worryingly easy to find reasons to abrogate those principles.

The second lesson to learn therefore follows from the fragility of the system that has been uncovered during the pandemic. Because the system is fragile and potentially undermined by claims that exceptional circumstances justify exceptional exemptions, it needs to be protected.

The third lesson is that although a specific problem is very important, and research to understand, solve or ameliorate it is also therefore very important, this does not mean (a) that every research project with these aims is important, and even less (b) that every research project with these aims is lexically more important than all research projects aimed at understanding, solving or ameliorating other problems. COVID-19

---

[7] Including the Nuremberg Code, the World Medical Association Declaration of Helsinki, the CIOMS Guidelines, the Council of Europe Convention for the Protection of Human Rights and Dignity of the Human Being with regard to the Application of Biology and Medicine: Convention on Human Rights and Biomedicine and its protocols, and the UNESCO Universal Declaration on Bioethics and Human Rights.

research is important, but so is research into many other infectious and non-infectious diseases. We should never give absolute priority to 'crisis research' in our research ethics processes. If we do, we will automatically neglect, under-prioritise, and delay other more important research.

In summary the main lesson we ought to learn from COVID-19 in relation to research ethics is to 'Keep calm and carry on'. Our research ethics system was borne out of crises in the past, and one major purpose of the set of rules we have developed over many years is to provide stability whenever there is pressure on the system.

References

Advisory Committee on Human Radiation Experiments (Ruth Faden, Chair). 1996. The Human radiation experiments: Final report of the President's Advisory Committee on Human Radiation Experiments. New York: Oxford University Press.

Beecher, H.K. 1966. Ethics and clinical research. *The New England Journal of Medicine* 274 (24): 1354–1360.

Chappell, R.Y., and P. Singer. 2020. Pandemic ethics: The case for risky research. *Research Ethics* 16 (3–4): 1–8.

DeVita, V.T. 2004. The 'War on Cancer' and its impact. *Nature Clinical Practice Oncology* 1 (2): 55–55.

Eyal, N., M. Lipsitch, and P.G. Smith. 2020. Human challenge studies to accelerate coronavirus vaccine licensure. *The Journal of Infectious Diseases* 221 (11): 1752–1756.

Ferguson, N. 2021. *Doom—The Politics of Catastrophe*. London: Penguin Books.

Holm, S. 2020. Controlled human infection with SARS-CoV-2 to study COVID-19 vaccines and treatments: Bioethics in Utopia. *Journal of Medical Ethics* 46 (9): 569–573.

House of Commons—Health and Social Care, and Science and Technology Committees 2021 Coronavirus: lessons learned to date. London: House of Commons. https://committees.parlia ment.uk/publications/7496/documents/78687/default/

Latorre, M.C., Z. Olekseyuk, H. Yonezawa, and S. Robinson. 2020. Making sense of Brexit losses: An in-depth review of macroeconomic studies. *Economic Modelling* 89: 72–87.

Pappworth, M.H. 1962. Human guinea pigs: A warning. *Twentieth Century* 171: 67–75.

Pappworth, M.H. 1967. *Human Guinea Pigs*. London: Routledge & Kegan Paul.

Piret, J., and G. Boivin. 2020. Pandemics throughout History. *Frontiers in Microbiology* 11: 631736.

Rasmussen, L.M., ed. 2020. *Human guinea pigs, by Kenneth Mellanby: A reprint with commentaries*. Cham: Springer.

Simonsen, L., and C. Viboud. 2021. Mortality: A comprehensive look at the COVID-19 pandemic death toll. *eLife* 10: e71974.

Solbakk, J.H., H.B. Bentzen, S. Holm, A.K.T. Heggestad, B. Hofmann, A. Robertsen, et al. 2021. Back to WHAT? The role of research ethics in pandemic times. *Medicine, Health Care and Philosophy* 24 (1): 3–20.

Stokel-Walker, C. 2021. How health data have been used during covid-19, and whether the changes are here to stay. *BMJ* 372: n681. https://doi.org/10.1136/bmj.n681

Søren Holm is Professor of Bioethics at the University of Manchester and Professor of Medical Ethics (part-time) at the University of Oslo. He is a former President of the European Society for the Philosophy of Medicine and Health Care and of the International Association of Bioethics. He was Editor in Chief of the Journal of Medical Ethics from 2004 to 2021. He currently serves as editor of the Journal Clinical Ethics. He has been researching and writing about issues in research ethics since the early 1990s. During the Covid crisis he advised the Norwegian and Danish health authorities on priorities in relation to the initial vaccine roll out.

Chapter 19
Ethics of Expanded Access During the COVID-19 Pandemic

Eline Bunnik and Marleen Eijkholt

19.1 Introduction

After the novel coronavirus disease (SARS-CoV-2, hereafter: COVID-19) was first identified in Wuhan in China in December 2019, it rapidly spread across the globe. On 11 March 2020, the World Health Organization (WHO) characterized the outbreak of the coronavirus as a pandemic. It had dramatic effects on populations on all continents. While most patients experienced mild symptoms such as fever, cough, shortness of breath and muscle ache, others were dying (Huang et al. 2020). As the virus was unknown, no treatments were available that were proven to be safe and effective for patients suffering from infection with COVID-19. Limited supplies and shortages of hospital and intensive-care beds, ventilators and personal protective equipment were further putting pressure on frontline health care workers (Ranney et al. 2020). An urgent need was felt to develop new drugs or to repurpose existing drugs to treat COVID-19 infections. Around the world, physicians were testing candidate therapeutic compounds in clinical trials. Some of these compounds had been approved for the treatment of other conditions, such as (hydroxy)chloroquine and sofosbuvir for malaria and hepatitis C (Ahn et al. 2020), respectively. Other drugs, such as Remdesivir, were investigational, and had not received marketing authorization at all.

In the pandemic setting, ethical questions arose in relation to physicians' duties to inform patients about—and provide—investigational treatments and in relation to access to such treatments, which may have been variable across categories of patients. Some severely ill COVID-19 patients could be included in clinical trials and thus

E. Bunnik (✉)
Erasmus MC, Rotterdam, The Netherlands
e-mail: e.bunnik@erasmusmc.nl

M. Eijkholt
Leiden University Medical Centre, Leiden, The Netherlands

© The Author(s), under exclusive license to Springer Nature Switzerland AG 2023
T. Zima and D. N. Weisstub (eds.), *Medical Research Ethics: Challenges in the 21st Century*, Philosophy and Medicine 132,
https://doi.org/10.1007/978-3-031-12692-5_19

receive investigational compounds. Other patients, however, presented at hospitals at which no clinical trials were under way, or failed to meet the in- and exclusion criteria of ongoing trials. Some groups of patients, including ethnic minorities, children, pregnant women (Einav et al. 2020), and patients with co-morbidity, were less likely than other groups to be enrolled in clinical trials. When patients cannot participate in clinical trials, their treating physicians may apply for so-called expanded access: access to an unapproved, investigational drug outside the context of a clinical trial. Expanded access is associated with burdens and risks, as is trial participation, because evidence of the safety and efficacy of the investigational treatment is lacking. In contrast to clinical trials, however, expanded access is uncontrolled, ethical oversight is less robust, and as patients may be more severely ill or have co-morbidities, they may be exposed to greater risks. Expanded access differs from clinical research also in its aim, which is therapeutic, not scientific.

Expanded access programmes are traditionally open exclusively to patients who are suffering from serious or life-threatening disease, have exhausted standard treatment options, and are not eligible for trial participation. Patients qualify for expanded access only when the investigational treatment is their sole remaining chance. Applying for expanded access requires time, effort and paperwork, and not all treating physicians may be willing or able to pursue it for their patients, especially in the context of a pandemic. Moreover, investigational drugs are often in limited supply. In many countries, including our country, the Netherlands, practices of expanded access seemed variable and inequitable during the 'first wave' of the pandemic (Eijkholt et al. 2020). Unequal distribution of access to potentially beneficial reasons is undesirable for moral reasons, and it may also lead to concerns among patients and families, difficulties for the physician–patient relationship, and societal turbulence. Some of this variability may be explained by current uncertainty regarding physicians' moral obligations: What are treating physicians expected to do (or not to do) in helping patients obtain access to investigational treatments—outside clinical trial settings?

Below, we describe existing legal, ethical and professional frameworks that determine physicians' duties regarding the provision of investigational treatments, including those for COVID-19, and their limitations. Also, we analyse the experiences in the Netherlands, during the first wave of the COVID-19 pandemic, with expanded access to two investigational treatments: Remdesivir and hydroxychloroquine. We identify important gaps in current frameworks, and analyse their ethical implications. We discuss one issue more in-depth, namely physicians' responsibilities for information provision about expanded access, and we illustrate how current policies and practices do not promote fairness. We conclude by arguing that in spite of prevailing uncertainty and potential practical barriers to access, physicians should inform patients or their proxies about relevant opportunities for expanded access.

19.2 Frameworks for Expanded Access in the Netherlands, and Their Limitations

Expanded access has always been associated with a range of ethical issues, including safety issues, informed consent, equitable access, and consequences for the clinical development of investigational drugs (Bunnik et al. 2018). For instance, as there is usually little evidence available on the efficacy and safety of investigational drugs, there are concerns that patients might be confronted with undesirable side effects of the drug. At the same time, lack of evidence renders it difficult for physicians to obtain informed consent from patients, who may be in desperate need of treatment, and may fare on (false) hope. Expanded access is not at all part of standard care, but an option that may be difficult to achieve in practice, and pursued only in exceptional circumstances. Consequently, in most countries, access to unapproved interventions is offered only to few patients. As patients who are health literate, well-connected, wealthy or otherwise privileged may know better how to navigate the healthcare system, they are often more likely to obtain expanded access than patients of lower socioeconomic status—which results in inequality. As a final example, more widespread use of investigational drugs through expanded access programmes might jeopardize clinical trials: patients may become less willing to enrol in studies in which they might be randomized to placebo or (unsatisfactory) standard of care, when they are given the opportunity to access investigational drugs directly, in a therapeutic setting.

In a pandemic context, some of these issues may be exacerbated. While legal, ethical and professional frameworks have been developed to provide some degree of guidance for physicians, pharmaceutical companies, payers (i.e. health insurers) and regulatory authorities in dealing with these issues, these frameworks do not fully clarify these actors' duties and responsibilities.

19.2.1 Legal Frameworks

Legal frameworks in the Netherlands distinguish between expanded access and off-label access. Physicians can prescribe drugs that have been approved for marketing for a specific indication (e.g. malaria) and use it off-label for a different indication (e.g. COVID-19). Expanded access, however, refers to the use of investigational drugs that have not been approved for marketing at all in a given country. Both legal frameworks enable some prescription liberties. Off-label prescription is deemed justified when it is in line with existing protocols or clinical guidelines issued by professional or scientific physicians' associations (Wigersma and Babovic 2010). If such protocols or guidelines are not (yet) available, the treating physician must consult with the pharmacist. If off-label drugs are considered to be best available care, treating physicians may be expected to prescribe them.

In the Netherlands, expanded access can be offered to individual patients through a regulatory route called 'levering op artsenverklaring' (literally: delivery on doctor's statement), which is comparable to so-called named-patient programmes in other countries. This route can be taken when the treating physician wants to prescribe an unapproved drug for an individual patient, and is convinced that the potential benefits of treatment outweigh potential risks. He or she must apply at the Health Inspectorate beforehand, demonstrate in a written statement that there are no registered alternatives, support this claim with evidence, acknowledge to bear full responsibility for the consequences of treatment, acknowledge that the treatment has not been tested for efficacy, safety or quality, and obtain the informed consent of the patient (IGJ 2020). In line with a disciplinary court case (RTG 2019), the Health Inspectorate adapted its guidance for informed consent for expanded access. The new guidance requires physicians to stress explicitly that the treatment has not been tested for efficacy, safety or quality as intended in the Dutch Medicines Act (IGJ 2020). Physicians must ensure that patients understand the 'experimental nature' of the treatment. Also, the pharmacist must supply product information about the drug, and the manufacturer must have a Good Manufacturing Practices (GMP) certificate and record the name of the physician, the number of patients treated and doses delivered, and any side effects or adverse events (IGJ 2020).

Another route to expanded access is through compassionate use programmes that are open to groups of patients. Manufacturers usually take the initiative to set up compassionate use programmes for cohorts of patients in the light of expected demand and medical need. Compassionate use programmes require approval by the Dutch Medicines Evaluation Board (CBG 2020). Programmes are usually initiated to bridge the gap between successful phase III clinical trials and marketing authorisation (EURORDIS 2020), to allow patient to use the drug while commercial availability of the drug is pending. Drugs are supplied free of charge.

For both legal routes, three conditions must be met that apply equally to expanded access programmes in most other countries: (1) the patient must be suffering from a serious or life-threatening condition; (2) there is no approved alternative; (3) the patient is not eligible for participation in a clinical trial. The latter condition helps to ensure that drug development and clinical research are not hindered by expanded access, and that expanded access does not run counter to the interests of other patients and future generations of patients.

Although existing legal frameworks enable expanded access, they fail to assign responsibilities to key stakeholders (Bunnik et al. 2018). First, they do not mandate manufacturers to set up compassionate use programmes. Second, they do not order health insurers or other third-party payers to reimburse expanded access. Third, they do not specify the roles and responsibilities of treating physicians in relation to expanded access. Hence several questions remain unanswered: Under what conditions are doctors expected to pursue expanded access for their patients? How should clinical decisions about expanded access be made in the absence of protocols and guidelines? What if protocols and guidelines are in conflict, while uncertainties, time pressures and resource constraints prevail? How should conflicts between treating physicians and pharmacists about the appropriateness of off-label use be resolved?

When must patients be informed about opportunities for expanded access? At present, no practical guidance helps to clarify these ambiguities. Consequently, variation in practices of expanded access are inevitable.

19.2.2 Ethical Frameworks

During the COVID-19 pandemic, ethical frameworks have been further developed for the use of investigational drugs in public health emergency settings. These frameworks, too, do not solve all ambiguities. In a scientific brief on the off-label use of medicines for COVID-19, for instance, the WHO stresses the importance of conducting controlled clinical trials, but acknowledges that "it can be ethically appropriate to offer individual patients experimental interventions on an emergency basis outside clinical trials", provided that clinical outcomes are monitored, recorded, and "shared in a timely manner with the wider medical and scientific community" (WHO 2020). This statement built on the ethical framework for 'monitored emergency use of unregistered and experimental interventions' (MEURI) developed in 2014–2016 for infectious disease outbreaks (WHO 2016). The MEURI framework stresses the importance of ethical oversight, effective resource allocation, minimising risk, collection and sharing of meaningful data, informed consent, community engagement, and fair distribution in the face of scarcity (WHO 2016). It sets a series of requirements. First, approvals of emergency use of investigational drugs by a national health authority and a qualified ethics committee should be in place. Emergency use of the investigational drug should not hinder or delay the clinical research that is required to gather robust safety and efficacy data, or counteract effective public health measures. Also, known risks should be minimized and patients should be monitored, as is customary in clinical trials. In order not to let a research opportunity to go waste, physicians have a "moral obligation to collect all scientifically relevant data on the safety and efficacy of the intervention" (p. 36–37). As part of the informed consent process, patients "should be made aware that the intervention might not benefit them and might even harm them" (p. 37). Established conditions for informed consent apply, such as voluntariness, and the provision of information in an intelligible and culturally sensitive fashion in order to ensure understanding. In order for the provision of investigational treatments in emergency settings to be respectful of local norms and practices, early community engagement is recommended. Finally, healthcare systems should develop frameworks for fair allocation of emergency use treatments, as scarcity will likely preclude the treatment of all patients who might benefit.

While most of these requirements are broadly supported, the requirement of data collection raises questions. Data collection has traditionally *not* been seen as a primary aim of expanded access; rather, the aim of expanded access has always been *therapeutic*, not research (Bunnik et al. 2018). There is a growing consensus, however, that opportunities to collect robust real-world data within expanded access programmes should not be left untouched (Walker et al. 2014; Webb et al. 2020).

In practice, however, it may be difficult to ensure good-quality data collection and meet the methodological challenges of COVID-19 research (Wolkewitz and Puljak 2020). For doctors who are caring for patients with COVID-19 in time- and resource-constrained settings, it may not always be feasible to record outcome data in a standardised manner, so that the data can be re-used for scientific analyses. Hospitals may not have the infrastructure in place to make clinical data available for sharing with national or international research groups (Eijkholt et al. 2020). Uncertainty exists as to the question what physicians are expected to do when they cannot guarantee the collection of useful data when prescribing an investigational treatment: Should they then refrain from pursuing it?

19.2.3 Professional Frameworks

In the Netherlands, professional frameworks were developed by various professional bodies to guide the provision of investigational treatments for COVID-19. Physicians' associations, such as the Federation of Medical Specialists (FMS) and the Dutch Intensive Care Association (NVIC) issued guidelines, as did other health organizations, such as the Centre for Infectious Disease Control (LCI) of the National Institute for Public Health and the Environment (RIVM), and local teams of physicians in university hospitals (for an overview of Dutch protocols and guidelines, see FMS 2021). However, these professional frameworks failed to take away uncertainties regarding the appropriate use of investigational treatments.

Guidelines issued by professional associations and groups were contradictory at times and/or changed over time. For instance, in March 2020, the Dutch Working Party on Antibiotic Policy (SWAB) and NVIC recommended the use of (hydroxy)chloroquine in hospitalised patients requiring mechanical ventilation (NVIC 2020). (Hydroxy)chloroquine, an antimalarial agent which is indicated also for patients with systemic lupus erythematosus and rheumatoid arthritis, has been approved for marketing for decades, and can thus be prescribed off-label to COVID-19 infected patients. There were several reasons for its recommended off-label use: (a) preliminary data from China; (b) the urgent need for a potentially effective antiviral intervention; (c) the drug was cheap, relatively safe, and available; and (d) there were no alternatives (Coumou and de Vries 2020). However, not all hospitals were willing to provide it (Erasmus MC 2020). Many patients or their family members insisted on trying (hydroxy)chloroquine, and felt wronged when they could not gain access. It seemed incomprehensible to some of them why the local intensivist was refusing to offer (hydroxy)chloroquine to their loved one who was undergoing mechanical ventilation, while it was routinely given to patients in other hospitals, or made available by certain physicians, even general practitioners, outside hospital contexts (Kleijne 2020).

A similar dynamic was seen with Remdesivir, an investigational compound, which was considered "by far the most promising drug" (Tu et al. 2020) for the treatment of COVID-19. Remdesivir was developed by the American drug manufacturer Gilead

Sciences, and tested initially during the outbreak of Ebola in the Democratic Republic of Congo in 2014–2016, against which it was deemed insufficiently effective. In February 2020, phase III clinical trials of Remdesivir were under way in the United States of America (US), Europe and Asia. Some academic hospitals took part in Gilead's trials and routinely offered Remdesivir to COVID-19 patients, while in other hospitals, Remdesivir was not available. Early results suggested that COVID-19 patients with oxygen saturation of 94% or less experienced improved clinical outcomes and fewer side effects when treated with Remdesivir (Antinori et al. 2020; Grein et al. 2020). Patients admitted to community hospitals that did not take part in Gilead's trials, however, were unlikely to be able to access the drug. These disparities were met with frustration among patients and families.

After the Food and Drug Administration (FDA) of the US issued an Emergency Use Authorization for Remdesivir on May 1, 2020, the Minnesota State Department of Health employed an especially developed ethical framework for a fair and equitable allocation of Remdesivir (Lim et al. 2020). In our country, however, variability in practices of expanded access to (hydroxy)chloroquine and Remdesivir in COVID-19 infections existed, due to scientific uncertainty, disagreement among experts, changing professional guidelines and wavering supply. Together, this led to societal turmoil. The legal, ethical and professional frameworks for expanded access to investigational drugs leave many issues unaddressed. Especially the issues around informational duties merit attention from a medical-ethical and research-ethical perspective. These issues lie at the heart of many others.

19.3 Informational Duties in Relation to Expanded Access

In the Netherlands, as in many other countries, treating physicians are required by law to inform patients about relevant treatment options. Article 448 of the Dutch Medical Treatment Act (WGBO) obliges the healthcare professional to inform the patient clearly—and, if so desired, in writing—about everything "the patient reasonably needs to know" about (a) the nature and purpose of the proposed treatment; (b) the expected consequences and risks thereof for the health of the patient; (c) alternative treatments, including those offered by other providers; (d) the patient's health status and prospects (7:448 BW), and; (e) the timing and duration of the treatment.

The Royal Dutch Medial Association (KNMG) explains in relation to requirement (c) that doctors should inform patients about any alternative treatment options that are within the 'domain of competence' of the treating physician (if they are not, the physician may need to refer to a colleague) and are *relevant* to the patient's medical problem, even when the physician may not him- or herself favour or support the alternative (Witmer and de Roode 2004) (p. 37). Changes made in this provision (effectuated in January 2020) require doctors to inform patients about the scientific foundations of alternative treatment options (Legemaate 2018). Still, it remains uncertain whether options for expanded access fall under requirement (c)—whether expanded access is (or should be) thought of and treated as an alternative treatment—and whether such treatment falls under the treating physician's informational duties. In the Netherlands, this has not been subject to court assessment (yet).

19.3.1 Proposing a Prima Facie Moral Duty to Inform Patients About Expanded Access

We argue that while it is unclear whether opportunities for expanded access are covered by the legal informational obligations of treating physicians in the Netherlands, there is a prima facie *moral* duty to provide such information. A moral duty to inform patients about expanded access is in line with the four classic principles of medical ethics: respect for autonomy, beneficence, non-maleficence, and justice (Beauchamp and Childress 2008). Firstly, patients should be informed about relevant treatment options because they need this information to decide autonomously about their medical treatment and to provide informed consent (the principle of respect for autonomy). Secondly, patients have an interest in receiving information about relevant treatment options because these options may offer medical benefit (the principle of beneficence). We understand an option for expanded access to be relevant when the potential benefits are believed to outweigh the risks. This aligns with one of the criteria of the FDA: expanded access may be appropriate when the "potential patient benefit justifies the potential risks of treatment" (FDA 2020). Third, the moral obligation to inform patients about expanded access options we propose should be seen as *prima facie*—at first sight—because it may be overridden by other, more important obligations. For instance, patients need not be informed about expanded access options that conflict with the obligation of physicians to refrain from interventions that cause harm (the principle of non-maleficence). Treatments that are associated with serious risks of harming patients (that are not outweighed by potential benefits) should not be considered relevant and are consequently beyond the scope of a duty to inform.

Fourth, a duty for treating physicians to inform patients about relevant options for expanded access might help to resolve some of the concerns related to equal access to expanded access (the principle of justice). In the context of a new pandemic, differences in policies and practices across hospitals in any given country will be inevitable, as healthcare providers are grappling with uncertainty and strained resources. Nevertheless, some degree of consistency in the treatment of patients is widely believed to be desirable. The principle of justice requires doctors to treat equal patients equally. Today, as said, not all patients are being informed equally about existing opportunities for expanded access to unapproved drugs. This discretion in informational duties may disproportionately affect patients of lower socio-economic status or health literacy, who are less likely to search and find information about existing options for expanded access themselves. At an online conference on expanded access in October 2020, Bettina Ryll, founder of the Melanoma Patient Network Europe, said: "Expanded access should not be a game of whom you know," implying that, thus far, it has been. Indeed, we agree, expanded access should not be reserved for patients who are friends or spouses of doctors working in university hospitals. A duty to inform would give *all* patients a greater voice in clinical decision-making about expanded access—serving both the principle of justice and the principle of respect for autonomy.

It should be noted that when providing information, doctors should acknowledge any supply issues and other practical and financial hurdles, and offer a truthful assessment of the feasibility of opportunities for expanded access. Naturally, when the drugs are not available or not relevant at all (for instance, in a particular country, or due to a particular patient's underlying illness or co-morbidities), they should not be (falsely) presented as such, and had better not be mentioned. Also, the proposed duty to inform does not extend beyond the classic criteria for expanded access, to reiterate: patients suffering from serious or life-threatening disease for whom standard treatment options are not available and who cannot be enrolled in clinical trials. This latter criterion is important, as expanded access should not interfere with drug development, clinical trials, and marketing approval, as it should not come at the cost of future generations of patients.

19.3.2 Three Reasons Why Doctors May not Inform Patients About Expanded Access

We have argued that there is a moral duty for physicians to inform patients about unapproved treatments they are aware of and that are being investigated in clinical trials and for which some level of scientific evidence on safety and efficacy is available. This implies that doctors should—at minimum—make an effort to inform patients about relevant options for expanded access and involve them in the clinical decision-making process. While only scarce empirical data are available on physicians' experiences and *modi operandi* regarding expanded access (Moerdler et al. 2019; Bunnik and Aarts 2021; Vermeulen et al. 2021) and no data specifically in relation to COVID-19, it seems that information on expanded access has not been and is not being routinely provided. Below, we describe some of the experiences in the Netherlands during the first wave of the COVID-19 pandemic, and propose three prominent reasons why in practice, doctors may not inform patients about expanded access (or pursue these options): scientific uncertainty, practical and logistical difficulties, and value judgments.

19.3.2.1 Scientific Uncertainty

First, drugs that are not approved for marketing (for the indication concerned) have not been proven to be effective, beneficial or safe. Therefore, it may be difficult for treating physicians to decide whether or not a drug may benefit the patient. This applies especially in a pandemic context, when doctors are confronted with a new hitherto unknown virus, and this uncertainty is exacerbated when expert opinion may be contradictory and change over time. For instance, while (hydroxy)chloroquine was routinely used in most hospitals in the Netherlands, in the beginning of the pandemic, in accordance with clinical guidelines, some hospitals and individual physicians refused to prescribe it (Erasmus MC 2020; Nieuwenhuis 2020), citing lack of data. Later, experts began to worry about safety and warn against the general side effects of hydroxychloroquine and a potentially increased risk of adverse cardiac events in COVID-19 patients (Gevers et al. 2020; Van den Broek 2020). In May 2020, a paper in The Lancet suggested increased risk of in-hospital mortality among COVID-19 patients treated with (hydroxy)chloroquine and lack of benefit (Mehra et al. 2020). After that, SWAB guidelines changed and off-label use of (hydroxy)chloroquine was no longer recommended in Dutch hospitals (SWAB 2020), except in research settings. The Lancet paper was later retracted.

Likewise, the state of knowledge of Remdesivir has been changing. On 3 July 2020, based on positive results from phase III clinical trials in COVID-19 patients, Remdesivir was granted conditional marketing authorisation by the European Medicines Agency (EMA), which made way for its provision in the Netherlands for COVID-19 patients aged 12 or older who require supplemental oxygen (Lamb 2020), pending further studies. In a double-blind, randomised, placebo-controlled trial, published on 8 October 2020, the drug was demonstrated to be effective in shortening time to recovery in adult patients who were hospitalised with COVID-19 and had evidence of lower respiratory tract infection (Beigel et al. 2020). Nevertheless, interim results from the Solidarity Therapeutics Trial, the world's largest randomised-controlled trial, posted on 15 October 2020, indicated inter alia that Remdesivir has no effect on 28-day mortality or the in-hospital course of COVID-19 among hospitalised adult patients (WHO 2020a). On 20 November 2020, the WHO issued recommendations against the use of Remdesivir in COVID-19 patients (WHO 2020b). In the face of such uncertainty, unfolding over months, it may not have been clear to doctors whether they should bring up investigational treatments with patients or their proxies.

19.3.2.2 Practical and Logistical Difficulties

Secondly, access to unapproved treatments can be difficult to arrange, practically and logistically. This may be especially so during a pandemic, which puts exceptional strain on manpower and capacity, and causes illness also among healthcare workers. Requests for expanded access may not always be granted, as each stakeholder has an ability to block the process; the manufacturer may turn out unable or unwilling to supply the drug, the health insurer or the hospital may refuse to fund the costs of treatment, or the regulatory authorities may not approve of its intended use. As the outcomes of attempts at access are uncertain, treating physicians may hesitate to consider (or offer information about) expanded access options (Vermeulen et al. 2021). Remaining healthcare providers face increased burdens in the provision of (regular) care and safety precautions, including donning and doffing of personal protective equipment, and may simply lack the time to even attempt requesting expanded access. Moreover, some expanded access options require monitoring (e.g. ECG monitoring in COVID-19 patients treated with hydroxychloroquine) (Van den Broek 2020) and specialized medical-technical skills for their administration and application, which may not be available in strained settings. Similarly, expanded access interventions might require additional administrative tasks, e.g. to ensure reimbursement. Manufacturers may require outcome data, and doctors may anticipate not being able to manage to collect and report such data. Taken together, these practical obstacles might easily dissuade providers to consider or inform their patients about expanded access options.

Furthermore, the unapproved compound itself may be in short supply. Remdesivir, for instance, was less controversial than (hydroxy)chloroquine, but scarce. Following recommendations by the EMA (EMA 2020), Gilead, the manufacturer, set up compassionate use programmes in many countries, including the Netherlands, in parallel with its clinical trials, to offer patients who could not be enrolled access to the compound. Initially, the programmes were open to any COVID-19 infected patients, then, in response to diminishing stocks, exclusively to pregnant women and children under 18 years of age undergoing mechanical ventilation, and then again for somewhat wider groups of patients. But even after Remdesivir was granted conditional marketing authorisation by the EMA, the European stock of Remdesivir remained limited. Since then, physicians have had to apply for each individual patient at the National Institute for Public Health and the Environment (RIVM 2020). As there was not enough Remdesivir available for all COVID-19 patients in the Netherlands, not all requests could be granted.

Also, reimbursement issues created moral dilemmas. For instance, in March, Dutch health insurers stated that they would collectively reimburse off-label use of drugs, including (hydroxy)chloroquine in COVID-19 patients, on the condition that this was in accordance with guidelines from the SWAB and NVIC (ZN 2020). As these guidelines changed every few weeks, reimbursement policies did, too, leading to uncertainty regarding availability of and funding for investigational drugs among doctors and patients. Doctors may not want to embark on a journey that is uncertain or unlikely lead to access (Vermeulen et al. 2021), and may not consider the (theoretical) option of expanded access realistic enough to discuss with their patients.

19.3.2.3 Value Judgments

Doctors may choose to forego expanded access or withhold information about expanded access from their patients on the basis of moral values or dispositions. Some Dutch doctors have principled moral objections to expanded access (Bunnik and Aarts 2021). Those who are risk-averse may steer clear from trying unproven medical treatments, and, in clinical decision-making, focus on safety issues and potential side effects (Veatch 2020), taking a slightly paternalistic stance. Of the four classic principles of biomedical ethics (beneficence, non-maleficence, respect for autonomy, and justice) (Beauchamp and Childress 2008), these doctors prioritize non-maleficence. Those who are less risk-averse, on the other hand, and more prone to trying, may prioritize beneficence in clinical decision-making, focusing on the potential benefits of treatment. And while both types of physicians may be working towards the same aims—namely, to improve the patient's health while avoiding unnecessary harms—they may weigh (existing evidence on) benefits and risks differently.

Also, doctors may have differing interpretations of the requirements of the medical-ethical principle of justice. While there is no consensus on what this ethical principle means for fair allocation in resource-constrained settings, four interpretations are broadly supported: maximising the benefits produced by scarce resources, treating people equally, promoting and rewarding instrumental value, and giving priority to the worst off (Emanuel et al. 2020). It is not always possible to reconcile these interpretations, and reasonable people may differ in their judgments of which interpretation should take priority in specific cases. Some doctors, for instance, may believe that it is unfair to pursue expanded access for individual patients during a pandemic, as this requires time, effort and manpower, which is scarce and had better be allocated to medical treatments (for other patients) that have been proven safe and effective. They may cite the importance of maximising benefits or saving the most lives. Expanded access would be at odds with these principles if it were to displace regular care. Experts noted that "panic prescribing" of hydroxychloroquine, for instance, led to drug hoarding (Caplan and Upshur 2020), especially in the US, and created distressing and potentially dangerous shortages for patients who needed the drug for approved indications (Peschken 2020). Others may favour tending to the needs of the worst off, and may accordingly justify allocating relatively high quantities of resources to provide expanded access to severely ill patients who are running out of options. Also, doctors must negotiate between professional responsibilities towards individual patients under their care and concerns about other patients and society at large. Even if expanded access may benefit an individual patient, it should not lead to displacement of resources such that other patients, who may have equally significant medical needs, may be harmed. Doctors may also wish to prioritize clinical trials. They could (rightfully) support the generation of scientifically valid and useful evidence on the drug's efficacy and safety, to the benefit of future patients. Value judgments may thus lead to categorical rejections of opportunities for expanded access. Some believe that expanded access, in principle, is flawed, and have argued, for instance, that it is "cause for deep concern" and "at odds with the rational use of medicines" (Paumgartten and Oliveira 2020).

Thus, doctors may be hampered in pursuing expanded access for eligible patients because of lack of evidence on safety and efficacy, limited supplies, lack of funding, and other practical obstacles, or personal values. There may be other reasons, as well. Healthcare providers may have (very) little knowledge of and experience or familiarity with expanded access, or they may fail to identify or be aware of potentially relevant treatments (Moerdler et al. 2019; Bunnik and Aarts 2021). Doctors are not expected to have up-to-date knowledge about innovative treatments or treatments offered outside of their hospital (Witmer and de Roode 2004), and may not (always) be blamed for failing to know about ongoing clinical trials or expanded access programmes running elsewhere. In practice, expanded access might not always be considered or recognised a relevant alternative treatment option, and doctors might not always inform their patients about these treatments.

19.3.2.4   The Case for a Prima Facie Duty to Inform

If doctors are aware of relevant and available options for expanded access—that is, of investigational treatments of which the potential benefits may outweigh potential risks and to which access could be arranged—we argue that they should not withhold this information based on scientific uncertainty, financial or feasibility concerns, or principled moral objections. First, as said, in the absence of scientific evidence, it may be difficult to determine whether the balance of benefits and risks is favourable. We feel, however, that this is reason to give patients a *larger role* in clinical decision-making about expanded access rather than a smaller role. As long as patients have the capacity to consent, they have a right to be adequately and comprehensively informed, so that they can decide autonomously about medical treatment, in line with their personal goals and values (Beauchamp and Childress 2008). In fact, the weighing of potential benefits and risks of an investigational treatment is—and should be—within the purview of the patient, not (exclusively) the physician. *Patients* must have a say in whether they are willing to take risks in exchange for a chance at benefit, however uncertain. Second, patients have an interest in being informed about relevant options, even when there is no certainty that they will be able to obtain them because of limited supply or funding issues. Patients may prove resourceful; if the investigational agent is not reimbursed by the health insurer, for instance, the patient might try to find other ways to fund the treatment. As said, if arranging expanded access is plainly impossible, a duty to inform does not apply. But if there is a chance, even if that chance is slim, patients may need to be informed about it. Third, we contend that categorical moral objections to expanded access are not tenable. Although it is perfectly reasonable for a healthcare professional *not* to be convinced that the potential benefits of a particular investigational drug will outweigh the risks for a particular patient in a particular context, it is not reasonable to hold that expanded access is always—or in principle—objectionable. Healthcare systems have set up routes and regulations for expanded access with clear eligibility criteria. When patients meet these criteria, their doctors can lawfully prescribe drugs that might benefit them. In many countries, this has been standing practice for decades. More importantly, however, doctors should not withhold information because the principle of respect for patient autonomy implies that doctors should not impose their personal values or disposition towards risk on patients, who may not share these values or valuations.

In sum, we argue that expanded access may be a relevant alternative to patients, and that three main reasons why doctors might refrain from informing patients about relevant opportunities for expanded access, may not always be valid, creating a prima facie duty to inform about expanded access options.

19.4   Conclusion

This chapter proposes a *prima facie* duty for treating physicians who are caring for severely ill patients for whom standard treatments are not (or no longer) available and who cannot be enrolled in clinical trials: a duty to inform these patients about existing opportunities for expanded access to (unapproved) investigational drugs. Our proposal fills some of the void within or between current legal, ethical and professional frameworks for expanded access, which have been discussed in this chapter. These voids has given rise to ethical issues during the first wave of the COVID-19 epidemic, and have led to variability in access to (information about) investigational drugs, potential loss of opportunity for patients to benefit from investigational drugs, and societal unrest.

Legal frameworks place the responsibility for the use of investigational treatments on individual doctors, but leave room for variability in practices. Physicians must decide whether or not to pursue expanded access based on their clinical judgment, balancing potential benefits and risks. In the face of uncertainty, individual physicians and patients may weigh risks and benefits differently, but they may also differ in their assessments of the feasibility of (theoretical) options for expanded access, and of whether or not these may be worth pursuing. Patients need to be given a stronger voice in decision-making about expanded access. A *prima facie* right to be informed may help to counter variations of practices and inconsistencies and (seeming) arbitrariness in the expanded access landscape. When treating physicians inform their patients more actively about expanded access, it may become less reserved for patients with greater health literacy or socio-economic status, and less conditional upon the personal values of treating physicians. A *prima facie* obligation to engage and inform patients may help counter existing inequality of opportunity to obtain expanded access across patient groups. While our proposal reiterates traditional ethical requirements for informed consent and expanded access, it is meant to promote transparency and consistency and contribute to equal opportunity, even in the context of the COVID-19 pandemic.

Acknowledgements   This chapter builds on previous work undertaken by the authors for an essay which was published in a Dutch-language collection of essays on the ethical implications of the corona crisis by the Centre for Ethics and Health (CEG 2010) of the Dutch Health Council, in collaboration with Prof. Dr. Martine de Vries and Dr. Marie-Astrid Hoogerwerf.

References

Ahn, D.-G., H.-J. Shin, M.-H. Kim, S. Lee, H.-S. Kim, J. Myoung, B.-T. Kim, and S.-J. Kim. 2020. Current status of epidemiology, diagnosis, therapeutics, and vaccines for novel coronavirus disease 2019 (COVID-19). *Journal of Microbiology and Biotechnology* 30 (3): 313–324. https://doi.org/10.4014/jmb.2003.03011.

Antinori, S., M.V. Cossu, A.L. Ridolfo, R. Rech, C. Bonazzetti, G. Pagani, G. Gubertini, M. Coen, C. Magni, A. Castelli, B. Borghi, R. Colombo, R. Giorgi, E. Angeli, D. Mileto, L. Milazzo, S. Vimercati, M. Pelliciotta, M. Corbellino, … and M. Galli. 2020. Compassionate Remdesivir treatment of severe Covid-19 pneumonia in intensive care unit (ICU) and Non-ICU patients: Clinical outcome and differences in post-treatment hospitalisation status. *Pharmacological Research* 158: 104899. https://doi.org/10.1016/j.phrs.2020.104899

Beauchamp, T.L., and J.F. Childress. 2008. *Principles of Biomedical Ethics (Principles of Biomedical Ethics*, 6th ed. USA: Oxford University Press.

Beigel, J.H., K.M. Tomashek, L.E. Dodd, A.K. Mehta, B.S. Zingman, A.C. Kalil, E. Hohmann, H.Y. Chu, A. Luetkemeyer, S. Kline, D. Lopez de Castilla, R.W.Finberg, K. Dierberg, V. Tapson, L. Hsieh, T.F. Patterson, R. Paredes, D.A. Sweeney, W.R. Short, … and H.C. Lane. 2020. Remdesivir for the treatment of Covid-19—final report. *New England Journal of Medicine* 0(0):null. https://doi.org/10.1056/NEJMoa2007764

Bunnik, E., and N. Aarts. 2021. The Role of physicians in expanded access to investigational drugs: A mixed-methods study of physicians' views and experiences in the Netherlands. *Journal of Bioethical Inquiry.* 2021 February 15. https://doi.org/10.1007/s11673-021-10090-7. Online ahead of print.

Bunnik, E.M., N. Aarts, and S. van de Vathorst. 2018. Little to lose and no other options: Ethical issues in efforts to facilitate expanded access to investigational drugs. *Health Policy (amsterdam, Netherlands)* 122 (9): 977–983. https://doi.org/10.1016/j.healthpol.2018.06.005.

Caplan, A.L., and R. Upshur. 2020. Panic prescribing has become omnipresent during the COVID-19 pandemic. *The Journal of Clinical Investigation* 130 (6): 2752–2753. https://doi.org/10.1172/JCI139562.

College ter Beoordeling van Geneesmiddelen (CBG). 2020. *Compassionate use programma.* https://www.cbg-meb.nl/onderwerpen/hv-compassionate-use-programma [in Dutch]. Accessed 9 Mar 2021.

Coumou, P., and P. de Vries. 2020. Chloroquine als mogelijk behandeling van COVID-19. *Nederlands Tijdschrift Voor Geneeskunde* 164: D4936.

Eijkholt, M., E. Bunnik, M.A. Hoogerwerf, and de Vries M. Ethische knelpunten rondom de inzet van experimentele geneesmiddelen buiten onderzoeksverband in een pandemie: Een plicht tot informatievoorziening? In: *Ethiek in Tijden van Corona.* Den Haag: Centrum voor Ethiek en Gezondheid (CEG) 2020: 67–75. [in Dutch] Available at: https://www.ceg.nl/documenten/pub licaties/2020/12/15/ethiek-in-tijden-van-corona. Accessed 9 Mar 2021.

Einav, S., M. Ippolito, and A. Cortegiani. 2020. Inclusion of pregnant women in clinical trials of COVID-19 therapies: What have we learned? *BJA: British Journal of Anaesthesia* 125(3): e326–e328. https://doi.org/10.1016/j.bja.2020.05.020

Emanuel, E.J., G. Persad, R. Upshur, B. Thome, M. Parker, A. Glickman, C. Zhang, C. Boyle, M. Smith, and J.P. Phillips. 2020. Fair allocation of scarce medical resources in the time of Covid-19. *New England Journal of Medicine* 382 (21): 2049–2055. https://doi.org/10.1056/NEJMsb200 5114.

Erasmus MC. 2020. *Behandeladvies COVID-19, beleid Erasmus MC.* https://www.vvzg.nl/images/20200316_Erasmus_behandeladvies_COVID-19.pdf [in Dutch]. Accessed 9 Mar 2021.

European Medicines Agency (EMA). 2020. *EMA provides recommendations on compassionate use of Remdesivir for COVID-19.* EMA. https://www.ema.europa.eu/en/news/ema-provides-recomm endations-compassionate-use-remdesivir-covid-19. Accessed 9 Mar 2021.

EURORDIS (EURORDIS-Rare Diseases Europe). 2020. *Main characteristics of CUPs in different EU Member States.* https://www.eurordis.org/content/main-characteristics-cups-differ ent-eu-member-states. Accessed 9 Mar 2021.

Federatie Medisch Specialisten (FMS). 2021 Overzichtspagina COVID-19: Richtlijnen, handreikingen, leidraden. https://www.demedischspecialist.nl/onderwerp/details/richtlijnen-han dreikingen-leidraden [in Dutch]. Accessed 9 Mar 2021.

Food and Drug Administration (FDA). 2020. *Expanded Access.* https://www.fda.gov/news-events/ public-health-focus/expanded-access. Accessed 9 Mar 2021.

Gevers, S., M. Kwa, E. Wijnans, and K. van Nieuwkoop. 2020. Hebben covid-19-patiënten iets aan chloroquine en hydroxychloroquine? *Medisch Contact.* https://www.medischcontact.nl/ nieuws/laatste-nieuws/artikel/hebben-covid-19-patienten-iets-aan-chloroquine-en-hydroxychlor oquine-.htm [in Dutch]. Accessed 9 Mar 2021.

Grein, J., N. Ohmagari, D. Shin, G. Diaz, E. Asperges, A. Castagna, T. Feldt, G. Green, M.L. Green, F.X. Lescure, E. Nicastri, R. Oda, K. Yo, E. Quiros-Roldan, A. Studemeister, J. Redinski, S. Ahmed, J. Bernett, D. Chelliah, and T. Flanigan. 2020. Compassionate use of Remdesivir for patients with severe Covid-19. *New England Journal of Medicine.* https://doi.org/10.1056/NEJ Moa2007016

Huang, C., Y. Wang, X. Li, L. Ren, J. Zhao, Y. Hu, L. Zhang, G. Fan, J. Xu, X. Gu, Z. Cheng, T. Yu, J. Xia, Y. Wei, W. Wu, X. Xie, Y. Yin, H. Li, M. Liu, and B. Cao (2020). Clinical features of patients infected with 2019 novel coronavirus in Wuhan, China. *Lancet (London, England)* 395(10223): 497–506. https://doi.org/10.1016/S0140-6736(20)30183-5

Inspectie Gezondheidszorg en Jeugd (IGJ). 2020. *Leveren op artsenverklaring.* https://www.igj.nl/ zorgsectoren/geneesmiddelen/geneesmiddelen-zonder-handelsvergunning/leveren-op-artsenver klaring [in Dutch] [Accessed 9 March 2021].

Kleijne, I. 2020. Actiegroep wil dat patiënt huisarts vraagt om andere behandeling covid-19. *Medisch Contact* 16 juli 2020.

Lamb, Y.N. 2020. Remdesivir: First Approval. *Drugs* 80 (13): 1355–1363. https://doi.org/10.1007/ s40265-020-01378-w.

Legemaate, J. 2018. Aanpassingen van de WGBO. *Tijdschrift Voor Gezondheidsrecht* 6: 556.

Lim, S., D.A. DeBruin, J.P. Leider, N. Sederstrom, R. Lynfield, J.V. Baker, S. Kline, S. Kesler, S. Rizza, J. Wu, R.R. Sharp, and S.M. Wolf. 2020. Developing an ethics framework for allocating Remdesivir in the COVID-19 pandemic. *Mayo Clinic Proceedings* 95 (9): 1946–1954. https:// doi.org/10.1016/j.mayocp.2020.06.016.

Mehra, M.R., S.S. Desai, F. Ruschitzka, and A.N Patel. 2020. RETRACTED: Hydroxychloroquine or chloroquine with or without a macrolide for treatment of COVID-19: a multinational registry analysis. *The Lancet* 0(0). https://doi.org/10.1016/S0140-6736(20)31180-6

Moerdler, S., L. Zhang, E. Gerasimov, C. Zhu, T. Wolinsky, M. Roth, N. Goodman, and D.A. Weiser. 2019. Physician perspectives on compassionate use in pediatric oncology. *Pediatric Blood & Cancer* 66 (3): e27545. https://doi.org/10.1002/pbc.27545.

Nederlandse Vereniging voor Intensive Care (NVIC). 2020. *Handreiking COVID-19 op de intensive care.* https://ctgnetwerk.com/wp-content/uploads/2020/05/Handreiking-COVID-19-op-de-intensive-care.pdf [in Dutch]. Accessed 9 Mar 2021.

Nieuwenhuis, M. 2020, March 22. *Longartsen met handen in haar: 'Aanbevolen medicijnen hebben niet veel effect, maar wel bijwerkingen.'* AD.nl. https://www.ad.nl/binnenland/longar tsen-met-handen-in-haar-aanbevolen-medicijnen-hebben-niet-veel-effect-maar-wel-bijwerkin gen~a21b9548/ [in Dutch]. Accessed 9 Mar 2021.

Paumgartten, F.J.R., and A.C.A.X. de Oliveira. 2020. Off label, compassionate and irrational use of medicines in Covid-19 pandemic, health consequences and ethical issues. *Ciencia & Saude Coletiva* 25 (9): 3413–3419. https://doi.org/10.1590/1413-81232020259.16792020.

Peschken, C.A. 2020. Possible consequences of a shortage of hydroxychloroquine for patients with systemic lupus erythematosus amid the COVID-19 pandemic. *The Journal of Rheumatology* 47 (6): 787–790. https://doi.org/10.3899/jrheum.200395.

Ranney, M.L., V. Griffeth, and A.K. Jha. 2020. Critical supply shortages—The need for ventilators and personal protective equipment during the Covid-19 pandemic. *New England Journal of Medicine* 382 (18): e41. https://doi.org/10.1056/NEJMp2006141.

Rijksinstituut voor Volksgezondheid en Milieu (RIVM). 2020. *Beschikbaarheid en bestelprocedure van remdesivir*. RIVM. https://lci.rivm.nl/remdesivir [in Dutch]. Accessed 9 Mar 2021.

Regionaal Tuchtcollege voor de Gezondheidszorg (RTG) Amsterdam 2019/353. ECLI:NL:TGZRAMS:2020:97. 24 September 2020. Available at: https://tuchtrecht.overheid.nl/ECLI_NL_TGZRAMS_2020_97 [in Dutch]. Accessed 9 Mar 2021.

Saleh, M., J. Gabriels, D. Chang, B. Soo Kim, A. Mansoor, E. Mahmood, P. Makker, H. Ismail, B. Goldner, J. Willner, S. Beldner, R. Mitra, R. John, J. Chinitz, N. Skipitaris, S. Mountantonakis, and L.M. Epstein. 2020. Effect of chloroquine, hydroxychloroquine, and azithromycin on the corrected QT interval in patients with SARS-CoV-2 infection. *Circulation. Arrhythmia and Electrophysiology* 13(6):e008662. https://doi.org/10.1161/CIRCEP.120.008662

Stichting Werkgroep Antibioticabeleid (SWAB). 2020. *Medicamenteuze behandeling voor patiënten met COVID-19 (infectie met SARS–CoV-2)*. SWAB. https://swab.nl/nl/covid-19 [in Dutch]. Accessed 9 Mar 2021.

Tu, Y.-F., C.-S. Chien, A.A. Yarmishyn, Y.-Y. Lin, Y.-H. Luo, Y.-T. Lin, W.-Y. Lai, D.-M. Yang, S.-J. Chou, Y.-P. Yang, M.-L. Wang, and S.-H. Chiou. 2020. A Review of SARS-CoV-2 and the Ongoing Clinical Trials. *International Journal of Molecular Sciences* 21(7). https://doi.org/10.3390/ijms21072657

Van den Broek, M.P.H., J.E. Möhlmann, B.G.S. Abeln, M. Liebregts, V.F. van Dijk, and E.M.W. van de Garde. 2020. Chloroquine-induced QTc prolongation in COVID-19 patients. *Netherlands Heart Journal*: 1–4. https://doi.org/10.1007/s12471-020-01429-7

Veatch, R.M. 2020, April 29. Clinical trials vs. right to try: Ethical use of chloroquine for Covid-19. *The Hastings Center*. https://www.thehastingscenter.org/clinical-trials-vs-right-to-try-ethical-use-of-chloroquine-for-covid-19/. Accessed 9 Mar 2021.

Vermeulen, S.F., M. Hordijk, N. Aarts, and E.M. Bunnik. 2021. Factors of feasibility: An interview study of physicians' experiences of expanded access to investigational drugs in three countries. *Humanities and Social Sciences Communications* 8: 275.

Walker, M.J., W.A. Rogers, and V. Entwistle. 2014. Ethical justifications for access to unapproved medical interventions: An argument for (limited) patient obligations. *The American Journal of Bioethics: AJOB* 14 (11): 3–15. https://doi.org/10.1080/15265161.2014.957416.

Webb, J., L.D. Shah, and H.F. Lynch. 2020. Ethically allocating COVID-19 drugs via pre-approval access and emergency use authorization. *The American Journal of Bioethics* 20 (9): 4–17. https://doi.org/10.1080/15265161.2020.1795529.

World Health Organization (WHO). 2020. *Solidarity Therapeutics Trial produces conclusive evidence on the effectiveness of repurposed drugs for COVID-19 in record time*. WHO. https://www.who.int/news/item/15-10-2020-solidarity-therapeutics-trial-produces-conclusive-evidence-on-the-effectiveness-of-repurposed-drugs-for-covid-19-in-record-time#:~:text=In%20just%20six%20months%2C%20the,the%20treatment%20of%20COVID%2D19. Accessed 9 Mar 2021.

Wigersma, L., and M. Babovic. 2010, March 25. Off-label voorschrijven. *Medisch Contact*.

Witmer, J.M., and R.P. de Roode. (eds.). 2004. *Van wet naar praktijk. Implementatie van de WGBO. Deel 2 Informatie en toestemming*. Utrecht: KNMG 2004. [in Dutch]. Available at: https://www.knmg.nl/advies-richtlijnen/knmgpublicaties/wgbo-1.htm. Accessed 15 Sep 2022.

Wolkewitz, M., and L. Puljak. 2020. Methodological challenges of analysing COVID-19 data during the pandemic. *BMC Medical Research Methodology* 20 (1): 81. https://doi.org/10.1186/s12874-020-00972-6.

World Health Organization (WHO). 2016. Emergency use of unproven interventions outside of research. In *Guidance for Managing Ethical Issues in Infectious Disease Outbreaks* (Chap. 9) https://apps.who.int/iris/handle/10665/250580. Accessed 9 Mar 2021.

World Health Organization (WHO). 2020a. *Off-label use of medicines for COVID-19.* https://www.who.int/news-room/commentaries/detail/off-label-use-of-medicines-for-covid-19. Accessed 9 Mar 2021.

World Health Organization (WHO). 2020b. *WHO recommends against the use of Remdesivir in COVID-19 patients.* https://www.who.int/news-room/feature-stories/detail/who-recommends-against-the-use-of-remdesivir-in-covid-19-patients. Accessed 9 Mar 2021.

Zorgverzekeraars Nederland (ZN). 2020. *Vergoeding van off-label toepassing van add-on geneesmiddelen bij de behandeling van Covid-19 patiënten.* https://www.znformulieren.nl/nieuwsbericht?newsitemid=4741365760 [in Dutch]. Accessed 9 Mar 2021.

Eline Bunnik is an Associate Professor at the Department of Medical Ethics, Philosophy and History of Medicine at Erasmus MC in Rotterdam. She has received research grants from The Dutch Research Council, The Netherlands Organisation for Health Research and Development and The Dutch Cancer Society. She is the author of over 60 articles on a wide spectrum of current topics in bioethics. Her recent work focuses on ethical aspects of access to unapproved and newly approved investigational drugs.

Marleen Eijkholt is a senior lecturer in the Department of Medical Ethics and Health Law at Leiden University Medical Center in the Netherlands. A recipient of grants from the Stem Cell Network and the Canadian Research Council, she has worked for policy makers, clinical bodies, and universities in the US, Canada, France and England. She has an expansive range of interests in bioethics and has numerous publications including stem cells, deep brain stimulation and ethical reproducibility.

Chapter 20
Bioethics and Its Relation to Medical Research in Japan: Historical Influences and Contemporary Pressures

Darryl R. J. Macer

20.1 Introduction

How can we relate the unique ethos of a country to the policy and laws that are enacted? This chapter will address the way that the spirit of Japan has shaped the ways that influences of international bioethics, civil rights and legal reforms have shifted the ethical norms associated with medical research in Japan through phases dominated by Confucian benevolence, structured paternalism, impunity from accountability, and fear of legal punishment. Japan has developed some of its own medical ethics, merging Buddhist and Confucian rules into a Shinto background with a recent importation of Western values (Kimura 1995; Macer 1999, 2003). Japanese ethics could be said to be now rather pragmatic and centered on the authorities. What was the historical background that allowed ethical abuses to be committed by medical researchers in the World War II (WWII) era, including in Unit 731 and in medical schools in Japan? Why do contemporary research agendas and policies embrace bioethical guidelines but sometimes struggle to apply them?

20.2 Values and Medical Research in Pre-Western Contact Japan

There are a number of influences on the type of medicines that were used in ancient Japan. There is a long history of traditional medicines used by the indigenous people, the Ainu and the Ryūkyūans of Okinawa, and also of many medicines and healthy lifestyle habits used by the general Japanese people (Fujikawa 1911). There were

D. R. J. Macer (✉)
American University of Sovereign Nations, Sacaton, USA
e-mail: provost@ausn.info

© The Author(s), under exclusive license to Springer Nature Switzerland AG 2023   387
T. Zima and D. N. Weisstub (eds.), *Medical Research Ethics: Challenges in the 21st Century*, Philosophy and Medicine 132,
https://doi.org/10.1007/978-3-031-12692-5_20

periods where the grounds for treatment were based on scientific understandings of disease and gathering medical evidence, as well as periods where the roots of illness were seen primarily as spiritual. Tradition records experiments upon monkeys to determine the action of certain vegetable substances possessing supposed remedial virtues, of which 37 were tested and employed in treatment of sickness. These consisted chiefly of roots and the barks of trees and were represented in the Japanese materia-medica of that period, about two to three thousand years ago (Berry 1912). The use of experimental monkeys for medical research at least two thousand years ago is a significant indicator of inquiry-based reasoning, and they tended to use a four element model of disease similar to in Greece. We may recollect that around a similar period there were some human vivisection experiments in Egypt, although generally Greek Physicians of that period used animals. From around two thousand years ago there are influences of Chinese medicine in the mainland of Japan adding to the pharmacological substances and also to the theoretical understanding of disease.

There are written records from the invasion of Korea in 201 A.D. showing that Korean and Chinese medical knowledge was brought to Japan. From 608 A.D., records show young Japanese physicians were sent to China for extended periods of study, which included research and practice on the use of Chinese medicine based especially on the theory of disease causation as the balance of yin and yang. Around the same time Buddhism was also brought to Japan, and Buddhist teaching, increasingly emphasized the older theory that all human suffering arose from the discord of the spirits of the four elements. This meant that the treatment of disease became increasingly a religious rite, and the priests were religious healers.

In 669 A.D. a school of learning was established by the Emperor Tenshi and by 690 A.D. a medical department was added which taught the Chinese system of internal medicine, materia-medica, cultivation and curing of medicinal plants, acupuncture, massage, diseases of the skin, and bone complaints. There was a 12 year program of study for medical students (Berry 1912), and their clinical skills were based on accumulating experience from the treatment of various patients, and from the accounts of new developments in the art of medicine over time. At least that amount of research was conducted, and new treatments and substances were added over time.

Since the fifth and sixth centuries A.D., the medical profession in Japan has been generally restricted to the privileged classes (Kimura 1995). With the centralization of government in the seventh and eighth centuries, a bureau of medicine was established, with the Yoro penal and civil codes creating an official physician class. After the Heian period (800–1200 A.D.), the government-sponsored health service was replaced by professional physicians.

The earliest written record of the hospital in Japan appears in 724 A.D. This included the establishment of Seyaku-in (a free dispensary) and Hidden-in (an infirmary for the poor and orphaned) within Kohfukuji (Sakai 2010). Subsequently, free dispensaries were established during the Heian Period (794–1185). At this stage, for example, cold water was early employed in the treatment of fevers but the use of this remedy was later abandoned for the period from the twelfth to the nineteenth century A.D.

During the Kamakura Period regent Houjou Tokimune (1251–1284) established a free medical care facility at the Kuwagayatsu. Records show that Ryokanbo-Ninsho, Chief Priest of the Gokurakuji Temple, continued these efforts by establishing a medical care facility within the temple (Sakai 2010). Logic would suggest that the development of medical treatment in institutions involving both medical education and treatment would continue to involve research on what medicinal compounds were effective, and the dosages that worked.

20.3 Samurai Tradition

Although we can trace the origins of bioethics in Japan back through millennia, with the Ainu indigenous practices and relationships to nature, the sophisticated hierarchical relationships expressed in the seventh century A.D. book the *Kojiki*, the conversion of forest to rice paddies, emergence of an integrated religious system including Shintoism and Buddhism, and the sophisticated class system supported by linguistics and social relationships, one of the most interesting influences to understand bioethics in Japan is the Samurai tradition (Macer 2003).

In the sixteenth century, a code of practice called the "Seventeen Rules of Enjuin" was drawn up that is very similar to the Hippocratic code (Kimura 1995). This code, developed by practitioners of the Ri-shu school, also emphasized a priestly role for the physician. The physicians *"should always be kind to people…. [they] should always be devoted to loving people."* There is a very strong paternalistic attitude by doctors even today (Hamano 2003). The code also has a directive to keep the art secret and to be concerned about quacks. No abortives are allowed, or poisons. The code contains a number of rules for virtue, such as, *"You should rescue even such patients as you dislike or hate"* and *"You should be delighted if, after treating a patient without success, the patient receives medicine from another physician, and is cured."*

The ancient and revered Japanese warrior code of *Bushido* emphasized the nobility of the warrior, and the necessity to treat the enemy with courtesy and honor (Harris 2003).

Bushido is very relevant to the bloodstain on Japanese medical ethics, the human experimentation during WWII. There are no accounts of mistreatment of prisoners of war until 1937 with the invasion of China, and then for the 8 years until 1945, the end of WWII, there are significant reports of maltreatment of prisoners which represents a rejection of Bushido. Actually, the reason for the change to unethical behaviour is the subject of speculation, but it may have been linked to nationalistic racism and militarism (Harris, 2003). There was moral decay in the 1920s in the whole of society, not only in the military, and a growing culture of impunity where military officers increasingly killed Japanese senior politicians and officers who were seen to be not promoting Japanese nationalism. This is actually ironic, that the samurai tradition would have guided a more honorable military ethics than the one that emerged. We will come back later to that time.

Before I depart from discussion of the Samurai tradition, it is important to understand that this tradition values both autonomy and informed choice, while also allowing the legacy of tolerance of the elite classes to make decisions that may sacrifice some persons for the greater good. One of the interesting questions that face is all cultures is how to ease the pain and suffering in the terminal stages of life. We see some traditions in the yoga and yogis in Indian medicine where meditation techniques are used to control the heartbeat and respiration rates. In the case of a samurai one of the most advanced techniques that was available to end life was the use of meditation to control respiration rate and eventually to end one's life (Macer 2003). Only the most advanced samurai could exercise such mental control, so a more bloody method is more familiar. Although there must have also been some type of experimentation into the most effective methods for suicide by samurai, using first a self-imposed dagger wound, followed by a fellow samurai who may assist later with a sword was commonly known as an honorable method.

20.4 Arrival of the Southern Barbarians and Western Influences

There are records from the sixteenth century that when the early Europeans (so-called "southern barbarians") came to Japan, in addition to their Christian missionary activities they established facilities for the sick in Kyoto and in Oita Prefecture on Kyushu. In 1555, Dr. Luis de Almeida is recorded as introducing Western medical practice to complement the Japanese and Chinese medical theories active at that time (Joshi and Kumar 2002). However, these facilities were later eradicated during the prohibition on Christianity (Sakai 2010), except for Dutch who were permitted in the trading island of Dejima in the city of Nagasaki. Among the records from Dejima, it appears that there was significant surgical innovation from the arrival of a German surgeon Dr. Casper Schamberger. Various other fields of medical knowledge were spread from Dejima under what was called "Dutch studies" (rangaku), through Japanese medical doctors who spread the knowledge (Joshi and Kumar 2002).

There was also reverse medical information flow with Dr. Engelbert Kaempfer who published books back in Holland to circulate some of the Japanese medical knowledge and history (Joshi and Kumar 2002). In 1570 a 15-volume medical work had been published in Japan by Dr. Menase Dōsan, and these provide written evidence of extensive medical knowledge. Dr. Nagata Tokuhun, published *I-no-ben* (1585) and the *Baika mujinzo* (1611), and wrote that the chief aim of the medical art was to support the natural force and, consequently, that it was useless to persist with stereotyped methods of treatment unless the physician had the cooperation of the patient. This implies that consent and participation in the healing were ethical values at the time, and thus a part of medical ethics at the time.

The exchange of medical information also saw the creation of many new Chinese words and phrases which provides evidence of the exchange of knowledge of science, research and practice. There was significant research and capacity in Japan itself as well, with the maintenance of large urban populations from the sixteenth century with water systems and hygiene that better avoided some of the large outbreaks of infectious diseases seen in European cities even in the nineteenth century. There is a record of an imperial permission for dissection of an executed criminal for anatomical research in 1771 (Berry 1912), which suggests that the deceased human bodies were also treated with respect. Japan was admitted to the Geneva Convention of the Red Cross Society in 1886.

Kampo is the Japanese medical tradition of using herbs, and also continued to be the subject of research prior to the Meiji era. Ahn et al. (2020) found a number of early research studies on the use of cannabis. Cannabis was prescribed in Meiji-era Japan to alleviate pain, and there were regulations also to limit its recreational use.

20.5 Meiji-Era to the Beginning of the Sino-Japan War

Japan opened up to Western influence and trade more widely after 1853, under threat of United States expansionism. In 1857 a group of Dutch-trained Japanese physicians founded a medical school in Edo (Tokyo) to begin the medical faculty of the Imperial University of Tokyo. Another group founded a medical school attached to a hospital in Nagasaki. Selected medical students were also sent to Holland and Germany for postgraduate medical training as well. Since then and throughout the twentieth century, Japan has been a centre of research and innovation. By the time of the Meiji restoration many physicians had knowledge of European Medicine through the scholars of the Dutch school discussed above.

The Meiji era (1868–1912 A.D.) is named after Emperor Meiji, and is the time when many western ideas were imported into Japan. It's also associated with the introduction of western style universities and it stimulated the development of the newly established western style medical schools. Japanese doctors and scientists attended international conferences to discuss their research, and there were some notable successes. Since the Meiji restoration in the nineteenth century, the doors of Japan have been opened to all countries. Recently, traditional ideas have undergone rapid change with globalization, itself driven by the communications devices that Japanese industry has exported around the world. Modern Western medicine took hold in Asia in the nineteenth century.

By the end of this period it is clear that Japanese medical scientists were conducting research using internationally accepted methodology and that one would also expect international standards of ethics. The successful development of the vaccine against yellow fever by Dr. Nguyen Noguchi is clearly based on having conducted substantial medical research on the use of vaccines, and that this research was also conducted in international field research. Other important medical firsts by the Japanese include the discovery of the plague bacillus in 1894, the discovery of a dysentery bacillus in 1897, the isolation of adrenaline (epinephrine) in crystalline form in 1901, and the first experimental production of a tar-induced cancer in 1918. By 1912 there were 50 medical magazines regularly published in Japan, including many Western articles. Dr. Nguyen Noguchi was credited with the 1911 discovery of Syphilis bacteria as the cause of progressive paralytic disease, and led a research team at the Rockefeller Institute in New York.

In the nineteenth century, some philosophers, such as Nakae Chomin, introduced concepts of human rights into Japan (Macer, 1999). He reinterpreted Confucianism by injecting concepts of popular sovereignty and democratic equality, and provided an internal tradition of human rights. Macer (1999) would place the origin of informed choice with the older samurai tradition, as discussed above.

In addition, the concept of informed consent is seen in the writings of Hanaoka Seishu on breast cancer from the nineteenth century. The records of Hanaoka Seishu who documented the surgical removal of tumors in breast cancer using anesthetic are some of the earliest examples of both the use of anesthetics and of the documentation of something approaching informed consent (Macer 2003). These are earlier than the records we have in the twentieth century use of informed consent in the United States of America (Annas and Miller 1994). It is also documents the use of anesthetics which must have involved experimentation and research in order to show their effectiveness as an alternative to pain control such as the use of acupuncture that was also used in Chinese and Japanese medicine. Interestingly in the twentieth century Japanese acupuncture and Japanese kampo were more readily adopted internationally than Chinese acupuncture and Traditional Chinese Medicine, because of the greater extent of communication between Japan and the West. Politically both the West and Japan were also united in their fight against communism, although Germany, Japan and Italy, joined together on the losing side of WWII. Japan had close ties in both medicine and constitutional legal systems with Germany, and both were subject to sanctions on resources that contributed to the emergence of strong nationalist movements.

During the years leading up to WWII and throughout the war, Japanese military and civilian medical personnel conducted experiments on human subjects without their consent (Harris, 2003). Their crimes, which are estimated to have resulted in the deaths of thousands of individuals, fell under the rubric of official Japanese government policy covering biomedical research with human subjects, beginning as early as 1930 and lasting until 1945. The concerns of the researchers were to develop viable chemical and biological warfare weapons to be employed in the current and future wars. The various chemical and biological programs alone ultimately involved many thousands of technically trained people, both civilian and military. Hundreds of other physicians and scientists participated in the freelance actions.

Medical schools, dental schools, and veterinary schools supplied their best students for the biological warfare (BW) and chemical warfare (CW) programs. Directors of these laboratories recruited students at top Japan's medical schools, e.g. Tokyo Imperial University and Kyoto Imperial University, by holding public lectures and by showing motion pictures and photographs of human experiments (Harris 2003).

Naito Ryoichi, founder of the Green Cross Company, a major pharmaceutical company, said, "Most microbiologists in Japan were connected in some way or another" to the human experimentation programs. In the case of support staff, many joined in the work because "the pay was good. At eighteen or nineteen years of age, we were getting higher salaries than the teachers who had educated us a long time ago, back in school." (Harris 2003).

20.6 Unit 731 and WWII Medical Research

By 1935 the Japanese medical research establishment was leading medical research in Asia. Because of the occupation of Manchuria and other parts of China as the war to gather mineral resources and oil took the Japanese nation on an expansionist journey. By the end of the 1930s the most well equipped and well-funded medical research laboratory in Asia was Unit 731 in Harbin China, with three thousand staff and around 150 buildings. For a number of years until 1945 when Japan lost the war, this was the leading medical research facility in Asia, and it is a measure of its success that the lead scientists were granted immunity by the American occupation forces in exchange for exclusive secret sharing of their medical findings, that included the results of live vivisection experiments on what is estimated to be three thousand persons (Harris 1994; Tsuchiya 2000, 2003; Nie et al. 2003). In addition, similar human experiments and vivisections were done at four branches of Unit 731, four other "Boeki Kyusui Bu" (Anti-Epidemic Water Supply and Purification Bureaus), "Gunju Boeki Sho" (Anti-Epizootic Protection Units) including Unit 100, the Manchuria Medical School, and army hospitals (Tsuchiya 2000; Harris 1994, 2003).

It is interesting to reflect on a story of medical morals that occurred a century earlier during the war of restoration. Dr. William Willis, an English naval surgeon, accompanied the government forces. At the close of the first battle he was informed of the wish of the officers of the army to have the wounded government forces treated first, and the wounded enemy attended to later. Willis was reported emphatical that he would not allow his instruments to be unpacked unless all the wounded could be treated alike (Berry 1912). His ethical principles were accepted, and thus those training with him in the medical profession were aware of medical ethical principles for wartime treatment of all equally. Thus, Japanese physicians were aware that prisoners of war should be treated medically. After the war, he became Professor at Surgery at a medical school in Tokyo and continued to be a significant influence in the Meiji restoration. This only complemented the *Bushido* tradition of samurai that would also have served against the inhumane treatment of prisoners.

The activities at Unit 731 included vivisection practice for newly qualified army surgeons, intentional infection, trials of non-standardized treatments, and tests to discover the tolerances of the human body (Tsuchiya 2000). These unethical research practices included:

(1) Vivisections for training newly employed army surgeons

At army hospitals in China, army surgeons carried out vivisections on Chinese prisoners. These doctors performed appendectomies and tracheostomies on the prisoners, shot them and took bullets from their bodies, cut their arms and legs and sewed up the skin around the wounds, and finally killed them. This surgical practice was part of the training program of newly employed army surgeons to teach them how to treat wounded soldiers at the front lines.

(2) Intentional infection of diseases

At the research faculties of the "Boeki Kyusui Bu," including Unit 731, researchers infected prisoners with many kinds of diseases, for example, plague, cholera, epidemic hemorrhagic fever, tuberculosis, typhoid, tetanus, anthrax, glanders, typhus, and dysentery. The purpose of this intentional infection was to seek the pathogen of the disease, to measure the infectiousness of the pathogen, to select more infectious strains, and to investigate the effect of bacteriological weapons. The subjects were dissected after their death or vivisected to death.

(3) Trials of non-standardized treatments

Many prisoners were killed during trials of unestablished and unusual "treatments." Many kinds of vaccines in the development stage were tried directly on prisoners, with no prior trials on animals. There were also experiments on recovery from frostbite, and using horse blood for blood transfusion.

(4) Learning tolerance of the human body

"There were deadly experiments with airtight chambers at Unit 731, the same ones as those conducted at the Nazi concentration camps. Some prisoners were forced to breathe poison gas. Others were killed by lowering the air pressure. In addition, there were doctors who only wanted to know how much air could be injected intravenously, how much bleeding brought prisoners to death, how many days prisoners could live with no food or water or only water without food, or how high electric current or voltage human beings could bear. There were also many trials of newly developed weapons with human subjects." (Tsuchiya 2000).

All of the prisoners that were killed in these experiments would have been executed by the Kwantung Army Military Police (the Japanese Army Manchuria division), so the logic was that it was better to gain some medical knowledge before their execution. The medical doctors who refused to participate would be blamed as "Hikokumin" (traitors) if they refused. Most accepted their fate without trying to resist it, even when they knew what they would be assigned to do in China. They were also ordered by their academic superiors to go to China. In Japanese medial schools, even now, head professors exercise supreme power over their staff (Tsuchiya 2000). In return, Universities and professors were willing to be cooperative with the army and Lt. Surgeon General Ishii (Director, Unit 731) because they would be provided research funds and facilities. In addition, the pay for those who went was high.

Tsuchiya (2000) concluded that, *Japanese and East-Asian values, such as respect for authority and harmony, in the Japanese medical profession that not only made possible the massacre by human experimentation in China during the period of 1933–1945 but also prevented a public investigation after the war.*

Another lesson is that in times when atrocities are committed normally the subjects of the atrocity are called not human, and different words are used the action. For example, the Japanese in China called the subjects maruta, "logs of wood" (Macer 2001). Not all the ethical standards were lost however, as the research scientists still conducted traditional Japanese annual memorial services for the experimental mice and other laboratory animals—but not for the human subjects.

The preoccupation with technical as opposed to moral and spiritual issues of medicine is consistent with the mentality of medical experimentation that led to such extremes. The fact that the medical records were destroyed as it became clear that the Unit would be taken over with the pending defeat of the Japanese military, suggests that the staff knew what they were doing was unethical. Staff were also made to promise not to talk about what they did after the war, although prior to WWII, recruitment videos and pictures had included human vivisection experiments. Having said that, the ally of Japan was Nazi Germany, where medical scientists were also conducting human vivisection. Unit 731, was a military research establishment, so it was also natural that the research would be secret. There were also very advanced medical research facilities in the universities in Japan, and some of these also noted to have used prisoners of war for vivisection experiments, and these were also hidden from the public. It was not until the mid-1990s that some members of these Units started to confess and apologize for their actions as they reached old age.

20.7   The Tokyo and Moscow War Crime Tribunals and Immunity and the Nuremberg Code

In 1945 at the end of WWII there were two main war crimes tribunals held to discuss alleged war crimes by Japan, who lost the war. For some staff from Unit 731 who were captured by the Russian army, they were taken to face war crime tribunals in Moscow, if they survived labor camps in Siberia. However, in Japan, immunity from prosecution was provided by the Americans to those medical scientists who conducted research in the Japanese army research facilities, and the item was removed from the agenda of the Tokyo War Crimes Tribunal. These decisions were made at very senior level, and for utilitarian reasons that there was alleged military utility of some of the Japanese medical research related to biological warfare. By excluding these items from the war crimes tribunal, the details did not need to shared publicly nor put on the public record.

Although the Allied Nazi War Crimes Tribunal issued the Nuremberg Code, it has been found subsequently that even American researchers and the government were breaking that code. Some Nazi rocket scientists were provided American passports to establish NASA, so science was valued more highly than criminal punishment for war crimes.

A large number of scientists were conducting research in the Japanese army funded medical research during World War II, not necessarily because they were sympathetic to military uses of research, but they may have been forced to conduct research. Some researchers were also attracted by the unlimited budget to conduct research, especially in the late stages of the war when research would have been very difficult at traditional universities that were not connected to the military. Following WWII, universities became the center for medical research again, and the trained staff who had been serving the military research machine were now back in the civilian world.

There were at least two reasons why they may not want to talk about the military research they had conducted during the war. One was a promise to keep that research secret in order to avoid possible damage to their own reputation, and to the reputation of the country. Another was that part of the deal for the immunity was that the research only be provided to the US military and not to anybody else. This was a type of technology transfer to the military. Over the following decades there had been rumors of some of the atrocities that had been conducted, and of course they had been details on trial in Moscow, and in China, that had been published. However, because of extreme propaganda that had been presented by every side during the Cold War, the facts were questionable. The real revelations came with the opening of documents as the U.S. statute of limitations expired on keeping them secret, and the publication of the books from the 1980s.

After WWII, and after the Nuremberg Code, there were still unethical research experiments continued in Japan (Tsuchiya 2003). Perhaps this is not surprising given that human experiments on unconsenting people continued in the United States, and in many other countries. The fact that it occurred everywhere does not forgive the unethical acts, but it raises questions of how bioethics can assist in the reduction of the occurrences of unethical experiments, and unethical medical practices such as the forced sterilization of women that was practiced almost everywhere.

Among those discussed in Japan include a March 1951 paper titled "Studies on the Carbohydrate Metabolism in Brain Tissue of Schizophrenic Patients" (Utena and Ezoe (1951). In this study a professor of psychiatry of Tokyo University, Dr. Hiroshi Utena and colleagues used a lobotomy take a small piece of brain tissue from 70 inmates of a psychiatric hospital, with some deaths. Over twenty years later, the "Committee Investigating Mr. Ishikawa's Criticism of Mr. Utena" of the Japanese Society for Psychiatry and Neurology (Seishin Iryo Henshu Iinkai, JSPN) (1973) issued "the Proposed Principles for Human Experimentation" in 1973 in its final report. Although JSPN's General Assembly did not adopt them, these principles were important in terms of including not only general principles but also principles for experiments on patients with mental disabilities, cited from the Nuremberg Code and the Declaration of Helsinki (Tsuchiya 2003).

Inside the Asian bioethics community there are repeated calls for open recognition of the unethical practices of the twentieth century as a means to better educate future researchers (Macer 2001; Nie 2001; Tsuchiya 2003). The culture not to question senior professors, and a pro-technology attitudes still exist in current Japan, and the evolution of bioethics has been shaped by that.

20.8   Current Medical Law and Guidance on Human Experimentation

Currently, professional responsibility is outlined by laws and guidelines. In Japan, there are several basic laws, including the Doctor's Act (Macer 2003). The Japan Medical Association (JMA) approved the concept of informed consent in 1991, which superseded the Physician's Code of Ethics of 1951, which was more paternalistic (Kimura 1995). There are professional guidelines issued for members of academic societies to follow, but they can still practice medicine and research outside of the professional society. Consensus is often more important than passing a law (Bai et al. 1987; Shinagawa 2000). Research that is independent of government or academic associations has less strict guidelines.

Physicians are required to obtain consent for medical treatment according to the 1971 Medical Practitioner's Act, Article 23. The obligation for treatment is based on assessing what can reasonably be expected in view of the knowledge and experience that characterizes the average physician. The obligation for consent means that the patient's will is to be respected when medical opinions are divided as to the necessity of the treatment. These also apply to research. Arbitrary administration of medical care violates Articles 204 and 205 of the criminal code. Article 202 of the criminal code forbids a person to help in another person's suicide.

The Council of Medical Ethics, established under the provisions of Article 25 of the Medical Practitioner's Act, is an advisory body supervised by the Minister of Public Welfare and consists of the presidents of the Japan Medical Association, the Japanese Dental Association, and scholars and staffs from related administrative departments. It functions to take administrative measures to eliminate physicians and dentists who commit malpractice or act unethically. Article 211 of the penal code states that if a physician injures a patient and the injuries cause death by mistreatment, the physician is liable for up to 5 years' imprisonment and/ or up to a 500,000-yen fine (US$4500). According to Article 7–2 of the Physicians Law, if a doctor is sentenced to imprisonment or a fine, the Ministry of Health, Labor and Welfare (MHLW) can remove his license or suspend his practice for a certain period of time. This action follows the decision of the Medical Practice Council according to Article 7–4. However, in practice, many Japanese are still reluctant to seek damages for malpractice (Feldman 1985; Macer 2003). There are separate laws outlining the activities of health professionals, including physicians, dentists, nurses, acupuncturists, masseurs, and other health care professionals.

The Medical Service Law and the Health Center Law are two important laws in a series that control the operation of medical facilities, which is a common location of research facilities. For research laboratories in university medical schools, they are subject to the ethical guidelines for the Ministry of Education, Culture, Sports, Science and Technology (MEXT), as well as MHLW.

In 1998, the government altered the Medical Service Law to allow physicians some compensation for taking the extra time that is needed to seek informed consent (Macer 2003). The obligation for consent means that the patient's will is to be respected when medical opinions are divided as to the necessity of the treatment (Tokyo District Court 1971). The patient must be competent; a person over 15 years is considered competent in most cases. For infants and the mentally ill, the consent of the person exercising parental authority is required. A Tokyo District Court in 1992 upheld a case brought against Tokyo University Medical School involving informed consent. The operation was a medical success, but the patient was not informed of the chances of failure and brought a case against the hospital. However, in other cases since then, the courts did not uphold informed consent (Tanida 1991). In practice, the concept of fully informed consent is now the norm in both Japanese medical practice and research.

20.9   Introduction of Bioethics Committees

Informed consent is now widely accepted, and bioethics is part of a transition which is transforming Japanese society from a paternalistic society to an individualistic one (Macer 1999).

Since the 1970s, people have become more conscious of their right to informed consent, which could be attributed to the importation of civil rights debates that occurred in the United States and Japan in the 1960s (Kimura 1995). Article 13 of the Japanese constitution guarantees each adult individual's right to self-determination. This has been interpreted by the courts to include the right to refuse life-sustaining care and medical intervention, including refusal of blood transfusion by adults. There are however, still issues faced by self-management of chronic diseases including balancing self-determination with what the National health Insurance may provide (Enzo et al. 2016).

Concurrent with the broader acceptance of informed consent has been the rise of ethics committees (Shirai 2003). This was also stimulated by the Human Genome Project (Macer 1992), and the Japanese government's adoption of the UNESCO Declaration on the Human Genome and Human Rights (UNESCO 1997). In March 2001, "Ethics Guidelines for Human Genome/Gene Analysis Research" was issued as jointly prepared ethics guidelines by three Ministries (the Ministry of Education, Culture, Sports, Science and Technology (MEXT), the Ministry of Health, Labor and Welfare (MHLW), and the Ministry of Economy, Trade and Industry). This was followed, in June 2002, by the issue of "Ethics Guidelines for Epidemiological Studies" jointly presented by the MEXT and the MHLW. According to these ethics guidelines, every research institute and hospital which intends to conduct genomic research involving human subjects is required to establish an ethics review committee for reviewing and monitoring the appropriateness of conducting a research protocol from ethical and scientific viewpoints. When these ethics guidelines were issued, it was implicitly premised on condition that research institutes and hospitals had set up their own ethics review committees and these committees could fulfill their roles properly. This was a significant stimulus to the establishment of ethics review committees at hundreds of institutions (Shirai 2003), which were then present to govern all forms of human research. Even social science research has gradually been included under their mandate, similar to the broadening of the scope of ethics review seen globally. Still there are challenges in the full implementation of ethical review systems (Kanayama 2015).

Although some cases of unethical research may provide the impression that the value of life is low in Japan, a more common challenge is the overtreatment at the end of life, which is considered by some as an afront to human dignity (Akabayashi 2002). One of the arguments used to prolong active treatment is giving a high value to life. There are a number of research trials of unproven drugs and treatments that consenting patients agree to, and usually pay from their own pocket, as attempts to prolong their life (Fukuyama et al. 2017). This is consistent with self-determination, and informed choice, but may also promote medical experimentation and the medical industry. Some of these trials are approved by ethics committees, but small clinics may still not have ethics committees. This dilemma is not unique to Japan, but the balance may be to argue that the government has no right to prevent someone's last chance at prolonging their life. Conversely it is argued that "death with dignity" is almost impossible in Japan (Asai et al. 2012).

Since the early 2000s the ethical review of large population research studies has also included methods for community engagement, which provide an additional level of ethical reflection to ensure that research studies have broader community consent, not simply the individual consent of participants who provide samples to the research (Suda et al. 2009).

20.10 Restrictive Policies on Access to Advanced Technologies

Although as discussed above, self-determination is paramount, perhaps as an over-reaction against advanced technology the policies in Japan towards prenatal diagnosis of genetic disease, surrogacy, and brain death and organ donation are among the restrictive side globally. Certain areas of advanced technology find themselves included in particular guidelines. Surveys of the members of the Japan Association of Bioethics in 1995 (Macer et al. 1996) found that they were significantly less supportive of prenatal diagnosis and assisted reproductive technology than the general public. This may be a consequence of the negative reactions to the pro-science and technology policy of the Japanese government. During the past three decades the implementation of bioethics guidelines to regulate science and technology in Japan has tended to develop more restrictive policies because of the influences of lawyers and philosophers in bioethics circles, who want to put some brakes on the rapid advancement of science and technology.

In terms of the use of stem cell therapies, however, Japan lies in between very positive countries and the restrictive countries, allowing the use of embryonic stem cell lines. On the overhand the Medical Association has successfully resisted the

voices of some bioethicists who called for laws to regulate the use of assisted reproductive technology and associated research. Such practice is regulated by academic
associations, such as the Japan Society of Obstetrics and Gynecology (JSOG). JSOG
announced on February 2002, "ethics committee opinion regarding surrogacy" that
at the moment surrogate motherhood is not acceptable for the following reasons: (1)
The welfare of the child to be born should be the primary consideration; (2) Surrogacy involves physical risk and psychological burden; (3) Family relationships will
become complex; (4) Surrogacy contracts are against social morals. Still, there are
people who wish to have their own child, but cannot bear them except using surrogate
mothers. These people either use surrogacy services in countries where surrogacy is
openly accepted, or use the services of some Japanese physicians who decide not to
belong to the JSOG (Kodama 2014).

Medical research on brain death is actually developed more than in some countries,
because there has been a widespread discussion and a number of arguments made
against acceptance of brain-dead donors for organ donation. Japan has a low rate of
organ transplants from brain dead donors (Yasuoka 2015) although there is an Organ
Transplantation Law permitting it. As a response to this shortage of donor organs,
there has been significant research on the use of live donors, and some techniques
such as living liver and kidney donation are well developed because the alternative
source of cadavers is difficult.

20.11   The Future

In Japan we can see a country with advanced biomedical research capacity in private
and public spheres, the generally well accepted regulation through ethics committees,
and a balance between a generally positive attitude to science and technology and
low levels of trust in the government. Japan has moved significantly from being
a country reluctant to implement bioethics guidelines (Macer 1992) to a country
with widespread consciousness of informed consent and ethics committee guidance.
The independent press and strongly held principles of self-determination ensure that
scientists are subject to public scrutiny.

A key turning point issue for bioethics in Japan was the unprecedented social
debate over the law to allow organ transplants from brain-dead donors. It is rare
to see a debate between the public and the policy makers over any issue, but this
issue led to the introduction of informed consent and the need for medical policy
to be more sensitive to public concerns. Patients' rights have been increasing since
the 1990s (Macer 1992; Annas and Miller 1994; Morikawa 1994). Japan is still
in a transition from paternalism to informed consent to informed choice (Macer
2003). Research by medical anthropologists (Lock 2002; Yasuoka 2015) follows
international ethical guidelines. The bioethics debate served as a catalyst to transform
Japan from a "paternalistic democracy" into one in which the individual has a greater
voice in health care decisions. People of any country may resist the rapid change
and globalization of their ethics, ideals, and paradigms because ethnic and national

identities may be changed or lost, especially in countries with a long history of culture. Part of the bioethical development in Japan includes rediscovering, importing and developing ethical approaches that can be debated, but a more important part is involving the public in discussion and development of a bioethics policy that takes into account the country's diverse ethical traditions.

We can see a compliant population with public health measures such as a high degree of mask wearing in the current COVID-19 pandemic, based on principles of responsibility to others. We can also see some hesitancy in the approval of vaccines against COVID-19, with approvals some months after other major industrialized countries. For the past three decades there have been about one thousand members of the Japan Association of Bioethics, and thus significant research on ethical issues of research is available.

References

Ahn, B.-S., S. Kang, K.H. Lee, S.Y. Seoyoon Kim, J.S. Park, and H.-S. Seo. 2020. A literature analysis on medicinal use and research of cannabis in the Meiji Era of Japan. *Journal of Pharmacopuncture* 23(3): 142–157 https://doi.org/10.3831/KPI.2020.23.3.142

Akabayashi, A. 2002. Euthanasia, assisted suicide, and cessation of life support: Japan's policy, law, and an analysis of whistle blowing in two recent mercy killing cases. *Social Science and Medicine* 55: 517–527.

Annas, G.J., and F.H. Miller. 1994. The empire of death: how culture and economics affect informed consent in the U.S., the UK and Japan. *American Journal of Law and Medicine* 20: 357–394.

Asai, A., K. Aizawa, Y. Kadooka, and N. Tanida. 2012. Death with dignity is impossible in contemporary Japan: Considering patient peace of mind in end-of-life care. *Eubios Journal of Asian and International Bioethics (EJAIB)* 22: 49–53.

Bai, K., Y. Shirai, and M. Ishii. 1987. In Japan, consensus has limits. *Hastings Center Report* (June) (special suppl.): 18–20.

Berry, J.C. 1912. Medicine in Japan: Its development and present status. *The Journal of Race Development* 2(4) (April): 455–479. https://www.jstor.org/stable/29737930

Enzo, A., T. Okita, and A. Asai. 2016. Japanese bioethical challenges concerning self management support for patients with chronic conditions: an analysis of quality of life & autonomy. *Eubios Journal of Asian and International Bioethics (EJAIB)* 26: 175–180.

Feldman, E. 1985. Medical ethics the Japanese way. *Hastings Center Report* (October): 21–24.

Fukuyama, F., A. Asai, T. Hanada, K. Sakai, and Y. Kadooka. 2017. Factors influencing the decision-making of elderly acute leukemia patients in Japan regarding their treatment. *Eubios Journal of Asian and International Bioethics (EJAIB)* 27: 106–113.

Fujikawa, Y. 1911. *Geschichte der Medizin in Japan: kurzgefasste Darstellung der Entwicklung der japanischen Medizin mit besonderer Beruecksichtigung der Einfuehrung der europaeischen Heilkunde in Japan*. Tokyo: Kaiserlich-Japanisches Unterrichtsministerium.

Hamano, K. 2003. Should euthanasia be legalised in Japan? The importance of attitude towards life. In *Bioethics in Asia in the 21st Century*, ed. S.Y. Song, Y.M. Koo, and D.R.J. Macer, 110–117. Christchurch, N.Z.: Eubios Ethics Institute.

Harris, S.H. 1994. *Factories of Death: Japanese Biological Warfare, 1932–45, and the American Cover-Up*. New York: Routledge.

Harris, S.H. 2003. Japanese biomedical experimentation during the World-War-II Era. In *Military Medical Ethics*, vol. 2, ed. T.E. Beam, L.R. Sparacino, E.D. Pellegrino, A.E. Hartle, and

E.G. Howe, 463–506. Washington, DC: TMM Publications, Borden Institute, Walter Reed Army Medical Center. http://hdl.handle.net/10822/1004389

Joshi, J.S., R. Kumar. 2002. The dutch physicians at Dejima or Deshima and the rise of western medicine in Japan. In *Proceedings of the Indian History Congress*, 63, 1062–1072. https://www.jstor.org/stable/44158176

Kanayama, A. 2015. Deficiencies in Japan's medical ethics review system. *Eubios Journal of Asian and International Bioethics (EJAIB)* 25(2): 52–7.

Kimura, R. 1995. History of medical ethics: Contemporary Japan. In *Encyclopedia of Bioethics*, ed. W.T. Reich, 1496–1505. New York: Simon and Schuster Macmillan.

Kodama, M. 2014. The current state of surrogate conception in Japan and the ethical assessment of Dr. Yahiro Netsu: An ethical investigation of Japanese reproductive medicine (Surrogacy). *Eubios Journal of Asian and International Bioethics (EJAIB)* 24(1): 12–18.

Lock, M. 2002. *Twice Dead: Organ Transplants and the Reinvention of Death*. Berkeley: University of California Press.

Macer, D. 1992. The 'Far East' of biological ethics. *Nature* 359: 770.

Macer, D. 1999. Bioethics in and from Asia. *Journal of Medical Ethics* 25: 293–295.

Macer, D.R.J. 2001. What is our bioethics? *Eubios Journal of Asian and International Bioethics (EJAIB)* 11: 1–2.

Macer, D.R.J. (2003). Regional perspectives in bioethics: Japan. In *Annals of Bioethics: Foundational Volume on Regional Perspectives*, ed. J. Peppin, 321–337. Lisse: Swets & Zeitlinger.

Macer, D., Y. Niimura, T. Umeno, and K. Wakai. 1996. Bioethical attitudes of Japanese university doctors, and members of Japan association of bioethics. *Eubios Journal of Asian and International Bioethics (EJAIB)* 6: 33–48.

Morikawa, I. 1994. Patient's rights in Japan: Progress and resistance. *Kennedy Institute of Ethics Journal* 4: 337–343.

Nie, J.-B. 2001. Challenges of Japanese doctor's human experimentation in China for East Asian and Chinese bioethics. *Eubios Journal of Asian and International Bioethics (EJAIB)* 11 (1): 3–7.

Nie, J.B., T. Tsuchiya, H.M. Sass, and K. Tsuneishi. 2003. A call for further studies on the ethical lessons of Japanese doctors' experimentation in wartime China for Asian and international bioethics today. *Eubios Journal of Asian and International Bioethics (EJAIB)* 13: 106–107.

Sakai, S. 2010. History of medical care at inpatient facilities in Japan. *Journal of the Japan Medical Association* 139 (7): 1453–1458.

Seishin Iryo Henshu Iinkai (Editorial Committee of Psychiatry). 1973 *Seishin Iryo (Psychiatry)* 3(1) [Summer 1973]. Iwasaki Gakujutsu Shuppan Sha.

Shinagawa, S. 2000. Tradition, ethics and medicine in Japan. Berliner Medizinethische Schriften. Dortmund: Humanitas Publishing House, numbers 40/41.

Shirai, Y. 2003. The status of ethics committees in Japan. *Eubios Journal of Asian and International Bioethics* 13: 130–134.

Suda, E., D. Macer, and I. Matsuda. 2009. Challenges to public engagement in science and technology in Japan: Experiences in the HapMap Project. *Genomics, Society and Policy* 5 (1): 40–59.

Tokyo District Court. 1971. Decision 5.19.

Tanida, N. 1991. Patient's rights in Japan. *Lancet* 337: 894.

Tsuchiya T. 2000. Why Japanese doctors performed human experiments in China 1933–1945. *Eubios Journal of Asian and International Bioethics (EJAIB)* 10: 179–180. http://www.eubios.info/EJ106/EJ106C.htm

Tsuchiya, T. 2003. In the shadow of the past atrocities: Research ethics with human subjects in contemporary Japan. *Eubios Journal of Asian and International Bioethics (EJAIB)* 13: 100–102.

UNESCO. 1997. *Universal Declaration on the Human Genome and Human Rights*.

Utena, H., and T. Ezoe. 1951. Studies on the carbohydrate metabolism in brain tissue of schizophrenic patients. Report I and II. The aerobic Metabolism of Glucose. *Psychiatria et Neurologica Japonica* 52(6) (March 1951):204–232.

Yasuoka, M.K. 2015. *Organ Donation in Japan: A Medical Anthropological Study*. Lanham: Lexington Books.

Darryl R. J. Macer is President of the American University of Sovereign Nations and widely known as Director of the Eubios Ethics Institute. He has a Ph.D. in molecular biology from the MRC Laboratory of Molecular Biology, University of Cambridge. He is also Director of the International Peace and Development Ethics Centre at Kaeng Krachan. He has been a visiting Professor and Research Fellow at various universities throughout the world, and is the former UNESCO Regional Adviser for Asia and the Pacific. Dr. Macer is coordinator of the IAB Genetics and Bioethics Network. During various periods he has been a member of many international committees including the UNESCO International Bioethics Committee, the Board of International Association of Bioethics and the HUGO Ethics Committee. The author of over 30 books, he also the editor and author of some hundreds of books, chapters and articles. In 2009 he was awarded an Honorary Doctorate from Kumamoto University in Japan.

Chapter 21
Ethical Evaluations of Clinical Trials in France: Towards European Standardization

Bettina Couderc

21.1   Introduction

21.1.1   The French Jardé Law and Clinical Research

21.1.1.1   Presentation of the Classification of Clinical Research

Clinical research encompassing all scientific studies carried out on the human person (grouped under the French acronym RIPH: Research Involving the Human Person) is framed in France by the Jardé law (2012) which was modified by ordinance in 2016.[1] RIPH encompasses research which involves the study of patients or healthy volunteers.

This research involves the evaluations of drugs, biotherapies, medical devices, and comparisons of different treatment methods. Furthermore, RIPH also includes, amongst other things, studies aimed at identifying metabolic pathways involved in pathologies, physiological responses to different treatments, and research into diagnostic or prognostic biomarkers.

Only these types of RIPH trials are evaluated at the national level. They are classified into the following three categories according to the nature of the intervention—involving answering the question: is the usual management of the patient modified?—and the level of risk and inconveniences.

---

[1] Jardé law, N° 2012–300 of 5/3/2012 which was amended by ordinance in 2016 N° 2016–800 of 16/9/2016. French public Health Code: L.1121.1.

B. Couderc (✉)
Université de Toulouse, Toulouse, France
e-mail: bettina.couderc@inserm.fr

Category 1 trials. These consist in interventional research involving interventions on human beings that are 'beyond' the usual scope of treatments (such as drugs, medical devices, etc.).

Category 2 trials. These consist in interventional research with minimal risks and inconvenience. For example, patients are treated with their usual treatment for their condition, but have the added intervention of medication, or minimally invasive procedures. These trials also include interviews and questionnaires whose results may lead to changes in the patients' usual treatment.

Category 3 trials. These trials consist of non-interventional research: research in which all procedures are performed and products are used routinely, without any additional or unusual diagnostic or monitoring procedures. It involves data contained in patients' records as well as data that is systematically collected during diagnostic and/or therapeutic management. If a file with personal data is created, it must be declared to the CNIL (Commission Nationale de l'Informatique et des Libertés)[2] (Levy et al. 2017).

21.1.1.2 Authorities Involved in the Evaluation of Clinical Research Applications

In order to carry out a Category 1 human research trial in France, the sponsor of the trial needs permission from the ANSM (National Agency for the Safety of Medicines and Health Products) and one of the 39 French CPPs (Committee for the Protection of Individuals, more commonly known as 'ethics committees')—assigned by drawing lots. In contrast, Category 2 and 3 trials are only evaluated by one of the CPPs, which will inform the ANSM of their evaluation opinion.

All trials, regardless of their designated categories, must request authorization from the National Commission of Informatique and Liberty (CNIL) concerning the processing of personal data in accordance with the general data protection regulation (*règlement général sur la protection des données* = RGPD)[3]. Furthermore, regardless of the category the trial falls under, sponsors must provide the CPPs with adequate proof that the trial adheres to the practices of the corresponding approved methodology (MR001 for Categories 1 and 2, MR003 for Category 3 obtained from the CNIL).

For Category 1—which is of particular interest within the scope of this chapter's exploration going forwards—the file provided to the two authorities (ANSM and CPP) includes the following administrative forms for regulatory registration of the

---

[2] Data Protection Act of 6 January 1978: The CNIL is an independent administrative authority, i.e. a public body that acts on behalf of the State, without being placed under the authority of the government or of a minister, and is responsible for ensuring the protection of personal data contained in computer files and processing, both public and private.

[3] The General Data Protection Regulation (RGPD) is a European regulatory text that frames data processing equally across the European Union. It came into force on May 25, 2018. The RGPD is a continuation of the French Data Protection Act of 1978 establishing rules on the collection and use of data on French territory.

trial: the detailed protocol, the certifications of each product involved in the trial, the investigator's brochure(s), the summary, the participant information sheets and consent forms intended for the patient, their legal guardian, or the healthy volunteer—depending on the case, the CVs of each investigator, the constitution of the monitoring committee, the insurance certificate, the certification of the adequacy of the human and material resources, the modalities of reimbursement of the expenses incurred by the patient (travel)—amongst others.

The timeframe for the evaluation for authorisation by the ANSM and CPP is strictly regulated. An opinion must be given within 45 days of the admissibility of the file. Since the request for authorisation of the research trial is evaluated across different European countries simultaneously, the final decision must be conferred by all bodies at the same time.

21.1.2 The Ethical Evaluation of a Clinical Research Project

The ethical evaluation of clinical research is performed by the CPP. Whilst members of the ANSM are exclusively paid experts (scientific or medical), the members of CPPs are volunteers who are supposed to represent the society at large. The CPPs are composed of 28 members equally divided into two groups. The first group is made up of persons qualified in scientific matters (doctors, researchers, nurses, pharmacists, epidemiologists). The second is made up of persons qualified in legal or ethical matters (such as lawyers, jurists), as well as psychologists, social workers, and members of patients' associations.

21.1.2.1 Project Evaluation

In the context of Category 1 human research, trials are evaluated under different criteria depending on the evaluating authority. The ANSM focuses more on the scientific character of the trial, addressing questions such as, for example: Which intervention is being studied? What effects does it have on all cell types? How was it produced? Where? How was it certified as complying with precise specifications? Is this certification verified? What excipients? What interactions occur between the active product and the excipients? Were the experiments carried out in vitro (cell culture) or in vivo (animal experimentation) before the first administration to humans?—and so on. On the other hand, the CPP is more interested in the protection of the participants in the trial. The CPP address issues such as, for example: How were these participants recruited? Are they fully informed? What is it they commit to if they consent to this trial? What is the benefit/risk balance for these participants?

Of course, both authorities will assess the entirety of the trial's application file, as mentioned previously (part 1 and part 2), and jointly evaluate the objectives of the project, its justification, the methodology, the qualifications of the investigators, the location of the research and so on. The evaluations are made in a cross-referenced and global way. Each file is evaluated at the CPP by 3 rapporteurs (one from group 1, one from group 2 and a methodologist). For more specific files (pediatrics, the first administration of a molecule in humans, ionizing rays, etc.), a fourth report, called an expert report, is requested from a specialist of the discipline, chosen either from the members of the CPP, if there is one, or from outside the CPP.

The Jardé law stipulates that the ANSM must concentrate on the scientific questions, whereas the CPPs concentrade on the ethical questions raised by the research. However, in practice, the CPPs are keen to evaluate the scientific part as well (hence why the CPP is divided into the two aforementioned groups which evaluate scientific and ethical matters respectively). Indeed, this dual system is crucial as it is not possible to evaluate an information letter intended for the patient without having read the scientific protocol. It is necessary to ensure that the interventions and sampling to which the patient will consent corresponds to what is indicated in the protocol, and that there are no unnecessary (or overly invasive) techniques implemented that are in contrast to the objectives of the study. Thus, it is necessary for the members of the CPP to evaluate the protocol from a scientific perspective. Moreover, their role is also to evaluate if what is foreseen in the protocol conforms with the French law, hence the presence of the 4 lawyers in group 2. Finally, and similarly to the ANSM, the methodologists of the CPP also evaluate the methodology—addressing issues such as: Are the statistical analyses and the evaluation criteria adapted to the objectives? Does the planned number of participating patients correspond to what is necessary?

In the new European regulation, it is foreseeable that the ANSM experts and the 3 CPP rapporteurs will communicate in order to write a common scientific evaluation of the protocol. However, the "evaluation of the information and consent documents" and "compliance with the French legislation" will only be carried out by the CPP. The proceedings are as follows. The rapporteurs present their work in plenary committees. Their conclusions are debated collegially amongst all the committee members. The votes of all members carry the same weight if there is a vote. At the end of the discussion, a common report is written with either a favourable or unfavourable decision. Or a request for additional information to the promoter in order to give an educated opinion.

In the following section, we will detail the criteria that are evaluated in France in order to authorize a research project and discuss some points that differ from other countries. We will conclude with some claims from the CPP for potential improvements to be made to the evaluations, and a plea for ethical evaluations to be carried out in a homogeneous way throughout the European Union.

21.2 Evaluation Criteria that May Entail Ethical Issues

21.2.1 Ethical Problem of Scientific Fraud: Cases Wherein Research on the Human Person is Authorized if Scientific Knowledge and Pre-clinical Experience are Sufficient to Attest to the Merits of the Study

When sponsors submit a clinical trial application, they provide the regulatory authorities with a scientific file including, among other things, a description of the state of the case (description of the pathology, description of the molecular mechanisms involved in the pathology, current treatments and possible therapeutic impasses), the working hypothesis, the proposed solution, the experiments carried out in vitro, ex vivo, and in vivo in animals, and the results of previous clinical phases if applicable. The rapporteurs of the dossier, as well as the experts, rely on publications and the results of the experiments presented. The rewievers assume that the sponsor and the investigators approached to participate in the research provide us with reliable and honest data—this is a fiduciary relationship. If publications from peer-reviewed journals are included, we assume that the results presented have been verified.

However, it turns out that 20% of the world's publications are fraudulent. Some of them contain invented results; others contain arranged or plagiarized data. The fraud is carried out either by academic research laboratories or by the Research and Development departments of pharmaceutical industries (Fanelli 2009; Gökçay and Arda 2017; Haug 2015; Khajuria and Agha 2014). This is a real problem for the authorities that are supposed to grant authorization for research trials since they have to decide on the merits of a trial and the safety of the patients based on the scientific results presented.

We can cite the scandals that have involved such fraudulent publications, for example, acalabutinib (Acerta—Astrazeneca) (lex Keown, n.d.) or the research of Dr. Piero Anversa (Michael O'Riordan, 2019). The latter published results on cardiac stem cells that were considered revolutionary by his peers. He presented results proving that it is possible to grow new heart cells that can be used to prevent heart attacks and treat heart failure. Whilst Phase 2 clinical trials were underway, a scandal broke out that revealed that the published work was all fraudulent. More than 30 studies published over nearly 10 years contained falsified or fabricated data. All those publications were withdrawn (Davis 2019) of course, but in the meantime, a Phase 1 and a Phase 2 had been completed, and patients involved received no benefit, nor was there any recieved by the company.

More recently, in the context of the current COVID-19 pandemic, we can cite a major fraud that has led to halting the acceptance of all new clinical protocols submitted evaluating hydroxychloroquine in France. This fraud was revealed following an article published in *The Lancet* by Mehra et al. on 22 May 2020 (Mehra et al. 2020a, 2020b). In that article, the authors evaluated the clinical evolution of 96,032 patients, treated vs untreated, worldwide for SARS-CoV-2 virus infection with hydroxychloroquine administered either alone or in association with a macrolide. The analysis of all their data led them to conclude that the administration of hydroxychloroquine (or chloroquine) did not lead to an improvement in the patients' health status (given that mortality and hospitalization time remained identical with or without hydroxychloroquine), even though that particular treatment was seemingly associated with a decrease in hospital survival and an increased frequency of ventricular arrhythmias when used for the treatment of COVID-19. Based on those results, all French CPPs systematically issued an unfavorable opinion to new proposals for chloroquine clinical trials for the treatment of COVID-19. However, three days after the press release, it turned out that the data used in the study presented in that article were fraudulent. The authors were provided with the article by Surgisphère who claimed to have collected them from various hospitals around the world. In the end, the origin of the data could not be confirmed (Catherine Offord 2020).

Even more recently, a suspected fraud has raised doubts about the development of the COVID-19 Gam-COVID-Vac vaccine (trade-named Sputnik V), developed by the Gamaleya Research Institute of Epidemiology and Microbiology in Russia. An article written by Logunov et al. (Logunov et al. 2020) and published in the Lancet concerning the Phase 1 and 2 clinical development showed that the vaccine is non-toxic and able to induce immunity. One week after the article was published, *The Lancet* published a letter written by Bucci et al. challenging the reliability of the Phase 1 and 2 results when using Sputnik (Bucci et al. 2020).

Thus, there is a real problem of scientific fraud throughout the world. Clinical trial evaluation committees rely on published results for clinical trials. Given the rate of fraudulence in the scientific community, does this entail that 20% of trials are therefore launched on fraudulent bases (Abbott 2019)? Mentioning scientific fraud and its consequences for the implementation of clinical research is not original to this chapter. Dozens of publications have described it, and numerous articles have been published by groups of doctors or researchers. The causes of such fraud are mentioned, and proposals to remedy it are made, but few concrete solutions have been found to truly remedy the situation. However, there are existing tools that alert reviewers as to whether publications have been certified as accurate, as well as alerting reviewers as to the reputability of the researchers. We can cite companies that propose to verify the integrity of published data, such as the firm Resis (Bucci),[4] which works with certain publishers to track down scientific fraud, and open access sites such as https://pubpeer.com/ and https://forbetterscience.com/, which can provide some information about the authors of publications. In the future, it may be possible that the publications used as the basis for the realization of clinical research at the European level will be subjected to a verification of authenticity at the European level; and that this validation will need to be attached to the file submitted for approval of the research.

---

[4] https://www.resis-srl.com.

21.2.2 Research Involving Humans is Permissible if the Foreseeable Risk to Those Who Participate in the Research is Not Disproportionate to the Expected Benefit or the Breadth of Knowledge to Be Gained From the Trial

All foreseeable risks inherent in participating in any clinical trial must be listed and presented in the information letter that will be given to the patient. Patients will need to certify that they understand and accept all risks in order to participate in the trial.

The role of CPP members is to judge whether the risks have been properly assessed and whether the list is complete. Again, of course, they will rely on the data provided by the sponsor and researchers, and on published data. It is necessary that nothing be hidden from these parties (i.e. the information must be clear, honest and sufficient for the participants to make an informed choice). For such CPP judgements, it is not only a question of listing all foreseeable risks and presenting them to the patients so that they can give informed consent; it is also a question of protecting patients and not letting them take risks that are deemed to be disproportionate.

Using the current pandemic as an example, people often wonder how the effectiveness of the vaccine is evaluated. Some think that the SARS-CoV-2 virus will be injected into vaccinated volunteers to verify that the vaccine is effective. This is called a "human challenge" study. The question is not so absurd when we remember that the great Louis Pasteur injected rabies virus to his first vaccinated patient to prove the effectiveness of his preventive treatment. It would seem that there are two different attitudes in Europe towards that possibility. Indeed, the more Latin countries (France, Spain, Italy) would consider that such practices are not ethically acceptable (Jamrozik and Selgelid 2020) while others would agree to accept them in the case of a well conducted research (Kirby 2020; McPartlin et al. 2020; Shah et al. 2020). For instance, even given that this practice was authorized in the United Kingdom, no French CPP committee has accepted such a high risk for healthy volunteers.

There is another type of research where French laws are much stricter than in some European countries, and this concerns medically assisted reproduction. In France, it is forbidden to modify in any way whatsoever within the context of such medically assisted reproduction, or to modify the human genetic heritage in a way that can be transmitted to the descendants—through modification of the zygote. Thus, it is forbidden, for example, to propose a treatment for mitochondrial diseases that would consist in performing an in vitro fertilization followed by implanting the fertilized nucleus in an enucleated oocyte from a woman without mitochondrial pathology (Sendra et al. 2021). Any clinical research on this type of treatment would be impossible in France. The question of treatment of embryos via genome editing was raised at the international level following Dr. Hé's communiqué on the birth of two babies whose genetic heritage was modified thanks to the CRISPR Cas9 technique (Wang and Yang 2019). An international moratorium has since been placed on the practice of offering such care for genetic disorders. However, the moratorium ends by saying that, under certain conditions and depending on the different governments, it would be possible to propose the genetic modification of the embryo for pathologies of particular gravity (Lander et al. 2019). Thus, there is an urgent need to define a common policy; not only for academic research on the embryo, but also for medical applications that could one day be the object of medical research.

21.2.3 Authorizable Human Research Designed to Minimize Any Harm Associated with the Disease or Research

Since the ultimate goal of research is to attain generalizable knowledge, many biological (activation of particular cell populations by proposed treatments), sociological (quality of life), physical (physical capacity, imaging) and even genetic parameters must be evaluated. Each sample (biopsy) and each collection of data (e.g. ethnic origin) must be justified by the sponsor. The CPP evaluates the merits of the requests made. The CPP tries to keep the number of hospital visits to a minimum; it ensures that there are no imaging or excessive samples taken (of blood, bone marrow, etc.). It pays particular attention to the genetic analyses requested when genetic analysis is part of the main aim of the study, which may be the case with some forms of personalized medicine; through studying the consent of the individual patient, where consent to genetic analyses often appears as an appendix to the general consent. In fact, it is not rare that, under the cover of a clinical trial aimed at particular categories of patients (e.g. sexual offenders), the sponsor plans to identify "the" gene for a human characteristic such as paraphilia (Alanko et al. 2016) but also aims to identify genetic resistance to effort (GEFOS Any-Type of Fracture Consortium et al. 2017), sportive, or academic success (Smith-Woolley et al. 2018). Such research is prohibited in France.

Given the number of international publications on the identification of genes involved in human characteristics (empathy, taste for travel, anger, etc.) (Pickering et al. 2019), some of which are conducted by European researchers, it seems clear that this type of genetic analysis is authorized outside French borders. Harmonising of practices and the acceptability of such research aiming at identifying the genetic part of human behaviors would be desirable for Europe.

When whole-genome sequencing or whole-exome sequencing is included in a protocol, CPPs require a strict procedure for incidental findings. We demand that patients have the choice to know or to refuse being told the result. We demand medical follow-up for the patient (physical and psychological) in case of incidental findings. We demand a written procedure regarding informing the family in the case of discoveries of potentially hereditary genetic traits (constitutional analysis). All those regulations are binding for the sponsors. These requirements for genetic analysis are not as stringent in different European countries. The attitude of the French CPP is based on a desire to protect patients and their families against potential discrimination by insurance companies, employment agencies, or banks in the case of loans following the identification of a genetic polymorphism. The standardization of the requirements between all European countries would be desirable in order to ensure the same protection to all European patients.

Finally, French legislation, unlike in other European countries, does not allow the collection of ethnic data unless explicitly justified. This is done in order to not induce discrimination in the development of drugs that might only benefit economically well-off populations (Sankar and Kahn 2005). It would be appropriate for European ethical bodies to agree to adopt a common attitude towards the analyses of genetic characteristics, their availability and the collection of ethnic data. As suggested in the conclusion, it might be appropriate for a European body to issue recommendations concerning the collection of ethnic or genetic data of persons participating in clinical trials that would be appropriate for all European cultures (Latin, Anglo-Saxon, Northern or Eastern countries). A common European reflection has already been carried out on data-sharing between European countries and outside our borders. The same question could be envisaged concerning the mode of collection: which data and why? Could this future collection generate a certain eugenics towards patients (or even populations) less fortunate from the genetic point of view, both for predisposition to certain pathologies and for aptitudes (sport, memorization etc.)?

21.2.4 Research on the Human Person Permissible if the Means of Assessing the Outcome of the Research Are Appropriate

As previously mentioned, research underlying a clinical trial is intended to demonstrate the efficacy of drugs or biotherapies or medical devices, or to compare different treatment methods. That research also includes, among other things, studies aimed at identifying metabolic pathways involved in pathologies, physiological responses to different treatments, and research into diagnostic or prognostic biomarkers, etc. Efficacy is defined as a better management, or at least as good management of the disease, as compared to the control treatment.

The clinical trial must have only one primary objective. It must include enough patients (neither too few nor too many, with statistical tests to be planned in advance) to be able to reach a conclusion, and the criteria for assessing the expected clinical response must be precisely and adequately defined. The analysis of the methodology of the proposed research is the responsibility of both the ANSM and the ethics committee in France. That is not the case in all European countries (e.g., Portugal). 50% of the unfavourable opinions issued by a CPP come from a judgement of weakness in the methodology. It would be advisable to have a common attitude in Europe for this evaluation of the methodology.

The gold standard of clinical trials is to conduct randomized double-blind trials against placebos (Douglas de Oliveira et al. 2016; Gomberg-Maitland et al. 2003)). To elaborate, this means that we evaluate the efficacy of a new treatment compared to a placebo drug. Neither the patient nor the physician knows who is in the new therapy group and who is in the placebo group. This methodology is more reliable for obtaining conclusive results. It is evidence-based medicine that was first developed by Paul Ricoeur (Moher et al. 2010; Schulz and Grimes 2002).

During the COVID-19 pandemic, this gold standard was questioned on various grounds, including the following:

(a) In trials conducted involving different medicines that cannot be taken according to the same schedule (e.g., one treatment is taken by mouth in the morning and evening, the other is injected weekly, the third is taken by mouth twice a week…), it is not possible to conduct a double-blind trial. In this case, an open trial is carried out. The doctor and the patient know in which group the patient is in. This was the case of the trial promoted by INSERM ("Trial of Treatments for COVID-19 in Hospitalized Adults", also known as DisCoVeRy), where different anti-COVID-19 treatments were tested. In that trial, there were major recruitment problems due to the controversy caused by the media interventions of Prof. Raoult advocating the use of hydroxychloroquine. Patients who agreed to participate in DisCoVeRy withdrew from the trial if they were not included in the hydroxychloroquine group.

(b) When a single medication was being tested with regard to its ability to combat SARS-CoV-2 infection, some researchers asked the evaluating authorities for permission to evaluate the treatments of interest by conducting a double-blind study which does not use a placebo as a control, but instead to uses the hydrox-ychloroquine treatment for the same reasons as previously mentioned (patient interest in Pr. Raoult's proposals). More than 500 clinical research protocols were proposed between February 2020 and June 2020 to French evaluation authorities. Most of them concerned the redistribution of known medications (Tocilizumab, interferon, remdesivir, hydroxychloroquine…) from the treatment of other ailments to the treatment of COVID-19. Because of the strong media coverage of a clinician from Marseille, the French wanted to be treated with hydroxychloroquine. The CPPs refused that proposal last spring, justifying their decision by the fact that hydroxychloroquine treatment was not the standard treatment validated by the international community. Those decisions have been widely contested in France by investigators and the general public, who have used so-called "ethical" arguments, saying that it was not "ethical" to offer no treatment at all to patients when there exists a potentially effective treatment. Here, two counterarguments were raised against these contradictors. First, for research to be rigorous, it is necessary to compare a new treatment with the reference treatment, even if the latter is non-existent. Secondly, the goal of a treatment is to do no harm, and the data in the literature available at the time making decisions regarding these clinical trials had not shown these proposed medications to be either efficiently or even safe in the context of treating patients with COVID-19.

The debate on the use of hydroxychloroquine has revived the debate on the placebo group. Is it still ethical to demand it? During the testing of certain targeted cancer therapies, some patients' associations request that certain drugs be evaluated in Phase 2 without a placebo group, if a medication presents spectacular therapeutic effects in Phase 0 (administrating a single dose of drugs to a limited number of patients of in order to test the biological properties of a treatment and possibly its therapeutic effect) or Phase 1. They argue that it is unethical to not offer a visibly effective treatment to the control patient group for Phase 2 (creating chances for treatment to be lost for those patients which are in the control group).

Consideration of the need for a control group in every clinical trial cannot be done on a case-by-case basis (Kube and Rief 2017). It requires a broader reflection on clinical research and on the identification of very specific cases where it would not be "ethical" to propose to certain patients to participate in a control group when we are sure—but what is certainty?—that the experimental treatment is really beneficial (Kube and Rief 2017; Smania et al. 2016; Watts 2004).

In order to propose not including a control group in certain trials, we can cite clinical research on innovative therapy drugs (gene and/or cell therapy) where the treatment is administered to a group of patients without a control group (as it is not conceivable to inject an empty gene transfer vector into a patient suffering from a serious gene pathology). In the same vein, the development of Phase 0 clinical trials, where a small number of patients are offered a dose of medication that is supposed to be therapeutic, in order to determine whether or not the medication is effective against a disease before starting Phase 1, 2, etc., is also an interesting avenue for early detection of whether or not a therapy should be developed, and for adapting the subsequent clinical phases. The debate remains open.

21.2.5 Research on the Human Person that is Permissible if the Patient's Consent is Truly Free and Informed

The CPPs are particularly involved on this front, and provide multidisciplinary aid in the reviewing of a trial's information leaflet (NICE) and consent form. In France, as in all countries, every participant must be able to consent to participating in a trial in a free and informed manner. That applies not only to adults and persons deprived of liberty, but also to children, persons with a disability, and persons under guardianship and curatorship. For the latter three, the authorization of both parents (in the case of minors), of the trusted person or of the tutor (or curator) will be required for the inclusion of the patient in a trial. Nevertheless, patients must be informed with due consideration of their degree of understanding, and their informed consent must be sought (Palmeirim et al. 2020).

(a) Voluntary consent: Researchers must ensure that patients are not pressured by their physician, health care team, or others to make a decision on whether or not to participate in a clinical trial. The CPP ensures that there is sufficient time between the clinician's proposal of the trial and the time the patient signs the consent letter. That time period allows patients to seek the advice of those close to them, to reflect, and not feel obliged to sign. The NICE also states that patients will not be subject to reprisals from the health care team if they refuse to participate. After the commitment to participate, a patient may withdraw from the study at any time without explanation and without reprisals from the investigators.

There are also points that the researcher really needs to consider when proposing that a patient join a clinical trial. These are the notions of intimacy, confidentiality (not sending medication or reports to the patients' homes without their authorization), and the integrity and inviolability of the person. The researcher must ensure that the patient will not suffer psychologically from participating in the trial (i.e. no humiliating or degrading medical acts). The researcher must be interested not only in the physical aspects, but also in the mental or spiritual aspects (i.e. no proposal of a trial where patients would have to consume drugs whose excipients would come from pork for example, and protecting the possibility of continuing religious worship during their inclusion in the trial). Patients must be respected in their choices, lifestyle, and aspirations even if they are included in a trial.

(b) Informed consent. The trial package insert should be clear, complete and written in a language that is understandable to the patient being recruited. The leaflet will therefore be different depending on whether it is addressed to an 8-year-old child, a teenager, a competent adult, or a cognitively impaired person. Discussions on the content of the leaflet are ongoing. To date, French regulations required that all potential side effects of the study drugs or ancillary drugs administered in the clinical trial be mentioned. As a result, patients often received a presentation leaflet that could be up to 40 pages long. The patient does not bother (or does not have the concentration to read it) and signs a consent that is not truly informed. Thus, changes in the regulations could be suggested so that the potential side effects of the drugs be listed in an appendix to the package insert, so that the patient can focus on the trial. We recall here that consent is not truly informed if the sponsor has concealed anything about potential serious side events due to the taking of the drug.

As mentioned above, the sponsor and investigators undertake the duty of notifying patients (via the addendum to the trial package insert and seeking confirmation of consent) of new adverse events that may be identified during the course of the trial (beyond what is first included in the protocol).

(c) Special cases of vulnerable persons: children, persons under guardianship or curatorship, persons incapable of giving consent (e.g. life-threatening emergency, coma). French law indicates that consent may be given by the person with legal responsibility for the patient (parents, guardians, curators), or their medical proxy, in the event of a life-threatening emergency or coma. In the latter case, however, the patient's consent must be sought as soon as the patient has either come out of the coma or is no longer in a period of vital emergency. Given a transition to European regulations, we would like the same criteria concerning patient information and inclusion to be required throughout Europe. The French CPPs have been asked to accept that the consent of only one of the two parents is sufficient in case of inclusion of a child in a protocol. The CPPs, although reluctant, have accepted. However, we would like the other criteria to be respected.

21.2.6 *Human Research Permissible if Patient Recruitment is Fair*

Developers have a duty to treat people fairly. The same respect and concern for each person must be observed. This applies to each individual in the research population (singular scale). Researchers must take an interest in the economic, material, and social conditions of their patients. They must take care, for example, that the patients' participation in the trial will not lead to significant familal, social, or material problems for them. It would not be fair if a patient could not participate in a trial for economic reasons. Compensation for the costs of the research should be systematically offered. The distribution of the benefits and drawbacks of the research must be fair. It would be unethical to offer participation in a trial only to either socially advantaged or disadvantaged populations. There should be no "concentration" on one segment of the population; there ought to be fair selection of trial subjects.

In France, apart from clinical research on rare diseases, trials are not allowed to include adults and minors in the same trial. All initial administrations of therapeutic medications to humans are performed on adults, except in very special cases. There is no need for research into maximum tolerated doses of a drug in minors if it is possible to conduct this research on adults. In the same way, a clinical trial is only authorized in children (minors) if it is not possible to determine the efficacy based on trials conducted on adults (for example if the drug is metabolized in a particular way). It would be appropriate for all European ethical structures to have the same strategy for the evaluation of clinical research in the very specific case of pediatrics.

21.3 And Afterwards…

When creating a bibliography on clinical trials carried out in France, one detects some anomalies. For example, although there is a legal limitation on the inclusion of underage patients in clinical trials, Gautret et al. (2020) reports having included children under the age of 12 for the evaluation of the benefits of taking hydroxychloroquine.

One ethical problem in the evaluation of clinical trials is the fact that the authorities that give their approval in France have no control over the proper conduct of the trial. When the trial is over, the CPP never receives the result of the study, and is even less likely to receive the resulting scientific publication. Ethics committees only receive information telling us that the trial is closed. Therefore, we have no way of verifying that patient recruitment has been carried out in accordance with the protocol (number of patients, compliance with the inclusion criteria), that the duration of the trial has been respected, that data protection has complied with the legislation (no transmission of personal or identifying data outside our borders, no genetic data), that the proposed methodology is indeed the one that has been carried out (number of groups, doses of drugs used), and that analyses not foreseen in the protocol have not been carried out.

A second ethical problem that was revealed during the COVID-19 pandemic is that the multiplicity of ethical authorities involved in the evaluation of trials (39 in France) leads to a lack of knowledge of the state of research as a whole. During the pandemic, 500 clinical trials were proposed in France between February and June 2020, including 200 on hydroxychloroquine alone. Taken individually and if carefully written, there was no reason why some were not accepted. As a result, French researchers opened too many trials for to few patients. The majority of them did not include enough patients to reach a conclusion or even did not start. As a result, the small number of patients that were included suffered the inconvenience of participating in a trial without contributing to the development of knowledge. Faced with this problem, France has reacted by creating a body that will now evaluate trials prior to their submission to the CPPs (Reactin and Capnet) in order to limit competition between trials. Could an authority of this type be set up at the European level so that we have a broader view of all the tests carried out Europe wide?

Finally, France has stricter legislation concerning clinical trials than some European countries (e.g., limitation of analysis of genetic characteristics, limits in the collection of ethnic data or sexual practices, unavailability of the human body). Our government fears that our country may not be able to lead clinical trials for the benefit of other European countries because of its limitations. It would be a good idea to standardize ethical requirements at the European level as much as possible in order to protect the patients and volunteers who take part in clinical research and to whom we are so indebted.

In order to harmonize the ethical evaluation, it is necessary to know how clinical research is evaluated in all European countries at the ethical level. Indeed, France has the particularity that the ethical evaluation of a clinical research project is decisional. If the CPP gives an unfavorable opinion, the project cannot be conducted. In some countries, the opinion of the ethics committee that evaluates the project is consultative. This means that the opinion may or may not be followed for the acceptance of clinical research. In all countries, the opinion of the ethics committee should be decisional.

In France, the ethical evaluation includes the evaluation of the methodology. That is not the case in all countries, yet it should be noted that when a French CPP gives an unfavourable opinion on a research project, it is more than 60% due to inadequate methodology (too few patients, unsuitable evaluation criteria, poor statistical analysis, etc.). Of course, the law differs from one country to another and it will be difficult to standardize certain practices that are legal in some countries (analysis of genetic characteristics, research on the embryo, research on the end of life, etc.), but the ethical evaluation is there to protect the patient in all circumstances and throughout Europe.

A European ethics committee was created in 2002 under the impetus of Dr. François Chapuy, the EUREC network (European Network of Research Ethics Committees). It aims at harmonizing ethical evaluations throughout Europe and works on the development of evaluation standards in order to propose quality clinical trials and to ensure the safety of human beings in these trials. This network is currently headed by Dr. D. Lanzerath. Perhaps the national authorities could rely on this network, which has a good knowledge of the network of European ethics committees and the particularities of the functioning of each one. An evaluation of the criteria examined by the different ethics committees should be carried out in order to determine the relevant differences and similarities in order to propose the most common methodology possible. We propose these changes so that patients can benefit from the same protection within the European Union, and in particular so that the choice of a country through which to promote a clinical trial is not made according to the permissiveness of the evaluations of that country, but for higher and better reasons.

References

Abbott, A. 2019. The science institutions hiring integrity inspectors to vet their papers. *Nature* 575 (7783): 430–433. https://doi.org/10.1038/d41586-019-03529-w.

Alanko, Katarina, Annika Gunst, Andreas Mokros, and Pekka Santtila. 2016. Genetic variants associated with male pedophilic sexual interest. *The Journal of Sexual Medicine* 13 (5): 835–842. https://doi.org/10.1016/j.jsxm.2016.02.170.

Bucci, Enrico, Konstantin Andreev, Anders Björkman, Raffaele Adolfo Calogero, Ernesto Carafoli, Piero Carninci, Paola Castagnoli, et al. 2020. Safety and efficacy of the Russian COVID-19 vaccine: more information needed. *The Lancet* 396 (10256): e53. https://doi.org/10.1016/S0140-6736(20)31960-7.

Davis, Darryl R. 2019. Cardiac stem cells in the post-anversa era. *European Heart Journal* 40 (13): 1039–1041. https://doi.org/10.1093/eurheartj/ehz098.

de Oliveira, Douglas, Dhelfeson Willya, Olga Dumont Flecha, Leandro Silva Marques, and Patricia Furtado Gonçalves. 2016. A Commentary on randomized clinical trials: how to produce them with a good level of evidence. *Perspectives in Clinical Research* 7 (2): 75. https://doi.org/10.4103/2229-3485.179432.

Fanelli, D. 2009. How many scientists fabricate and falsify research? A systematic review and meta-analysis of survey data. *PLoS ONE* 4 (5): e5738. https://doi.org/10.1371/journal.pone.0005738.

GEFOS Any-Type of Fracture Consortium, Sara M. Willems, Daniel J. Wright, Felix R. Day, Katerina Trajanoska, Peter K. Joshi, and John A. Morris, et al. 2017. Large-Scale GWAS identifies multiple Loci for hand grip strength providing biological insights into muscular Fitness. *Nature Communications* 8 (1): 16015. https://doi.org/10.1038/ncomms16015.

Gökçay, B., and B. Arda. 2017. A review of the scientific misconduct inquiry process, Ankara Chamber of medicine Turkey. *Science and Engineering Ethics* 23 (4): 1097–1112. https://doi.org/10.1007/s11948-016-9824-8.

Gomberg-Maitland, M., L. Frison, and J.L. Halperin. 2003. Active-control clinical trials to establish equivalence or noninferiority: Methodological and statistical concepts linked to quality. *American Heart Journal* 146 (3): 398–403. https://doi.org/10.1016/S0002-8703(03)00324-7.

Haug, C.J. 2015. Peer-review fraud-hacking the scientific publication process. *The New England Journal of Medicine* 373 (25): 2393–2395. https://doi.org/10.1056/NEJMp1512330.

Jamrozik, Euzebiusz, and Michael J. Selgelid. 2020. COVID-19 human challenge studies: ethical issues. *The Lancet Infectious Diseases* 20 (8): e198-203. https://doi.org/10.1016/S1473-309 9(20)30438-2.

Khajuria, A., and R. Agha. 2014. Fraud in scientific research—birth of the Concordat to uphold research integrity in the United Kingdom. *Journal of the Royal Society of Medicine* 107 (2): 61–65. https://doi.org/10.1177/0141076813511452.

Kirby, T. 2020. COVID-19 human challenge studies in the UK. *The Lancet Respiratory Medicine* 8 (12): e96. https://doi.org/10.1016/S2213-2600(20)30518-X.

Kube, T., and W. Rief. 2017. Are placebo and drug-specific effects additive? Questioning basic assumptions of double-blinded randomized clinical trials and presenting novel study designs. *Drug Discovery Today* 22 (4): 729–735. https://doi.org/10.1016/j.drudis.2016.11.022.

Lander, E.S., F. Baylis, F. Zhang, E. Charpentier, P. Berg, C. Bourgain, B. Friedrich, J.K. Joung, J. Li, D. Liu, L. Naldini, J.-B. Nie, R. Qiu, B. Schoene-Seifert, F. Shao, S. Terry, W. Wei, and E.-L. Winnacker. 2019. Adopt a moratorium on heritable genome editing. *Nature* 567 (7747): 165–168. https://doi.org/10.1038/d41586-019-00726-5.

Levy, C., A. Rybak, R. Cohen, and C. Jung. 2017. La loi Jardé, un nouvel encadrement législatif pour une simplification de la recherche clinique? *Archives De Pédiatrie* 24 (6): 571–577. https://doi.org/10.1016/j.arcped.2017.03.012.

Logunov, Denis Y., Inna V. Dolzhikova, Olga V. Zubkova, Amir I. Tukhvatullin, Dmitry V. Shcheblyakov, Alina S. Dzharullaeva, Daria M. Grousova, et al. 2020. Safety and immunogenicity of an RAd26 and RAd5 vector-based heterologous prime-boost COVID-19 vaccine in two formulations: two open, non-randomised phase 1/2 studies from Russia. *The Lancet* 396 (10255): 887–897. https://doi.org/10.1016/S0140-6736(20)31866-3.

McPartlin, S. O., Morrison, J., Rohrig, A., and Weijer, C. 2020. Covid-19 vaccines: Should we allow human challenge studies to infect healthy volunteers with SARS-CoV-2? *BMJ*, m4258. https://doi.org/10.1136/bmj.m4258.

Mehra, Mandeep R., Sapan S. Desai, SreyRam Kuy, Timothy D. Henry, et Amit N. Patel. 2020a. Retraction: Cardiovascular disease, drug therapy, and mortality in covid-19. N Engl J Med. https://doi.org/10.1056/NEJMoa2007621. *New England Journal of Medicine* 382(26): 2582–2582. https://doi.org/10.1056/NEJMc2021225.

Mehra, Mandeep R, Sapan S Desai, Frank Ruschitzka, et Amit N Patel. 2020b. RETRACTED: Hydroxychloroquine or chloroquine with or without a macrolide for treatment of COVID-19: A multinational registry analysis. *The Lancet*, S0140673620311806. https://doi.org/10.1016/S0140-6736(20)31180-6.

Moher, D., S. Hopewell, K.F. Schulz, V. Montori, P.C. Gotzsche, P.J. Devereaux, D. Elbourne, M. Egger, and D.G. Altman. 2010. CONSORT 2010 Explanation and elaboration: Updated guidelines for reporting parallel group randomised trials. *BMJ* 340 (1): c869–c869. https://doi.org/10.1136/bmj.c869.

Michael O'Riordan. 2019. Stem cell research—shattered after fabrication scandal—needs to rebuild, says EHJ editor. https://www.tctmd.com/news/stem-cell-research-shattered-after-fabric ation-scandal-needs-rebuild-says-ehj-editor.

Offord, Catherine. 2020. Lancet, NEJM retract surgisphere studies on COVID-19 patients. *The Scientist*. https://www.ncbi.nlm.nih.gov/search/research-news/10099/.

Palmeirim, M.S., A. Ross, B. Obrist, U.A. Mohammed, S.M. Ame, S.M. Ali, and J. Keiser. 2020. Informed consent procedure in a double blind randomized anthelmintic trial on Pemba Island, Tanzania: Do pamphlet and information session increase caregivers knowledge? *BMC Medical Ethics* 21 (1): 1. https://doi.org/10.1186/s12910-019-0441-3.

Pickering, C., Kiely, J., Grgic, J., Lucia, A., & Del Coso, J. (2019). Can genetic testing identify talent for sport? *Genes*, *10*(12). https://doi.org/10.3390/genes10120972.

Sankar, Pamela, and Jonathan Kahn. 2005. BiDil: Race medicine or race marketing?: Using race to gain a commercial advantage does not advance the goal of eliminating racial/ethnic disparities in health care. *Health Affairs* 24 (Suppl1): W5–455-W5–464. https://doi.org/10.1377/hlthaff. W5.455.

Schulz, K.F., and D.A. Grimes. 2002. Blinding in randomised trials: Hiding who got what. *Lancet (London, England)* 359 (9307): 696–700. https://doi.org/10.1016/S0140-6736(02)07816-9.

Sendra, L., A. García-Mares, M.J. Herrero, and S.F. Aliño. 2021. Mitochondrial DNA replacement techniques to prevent human mitochondrial diseases. *International Journal of Molecular Sciences* 22 (2): 551. https://doi.org/10.3390/ijms22020551.

Shah, S.K., F.G. Miller, T.C. Darton, D. Duenas, C. Emerson, H.F. Lynch, E. Jamrozik, N.S. Jecker, D. Kamuya, M. Kapulu, J. Kimmelman, D. MacKay, M.J. Memoli, S.C. Murphy, R. Palacios, T.L. Richie, M. Roestenberg, A. Saxena, K. Saylor, A. Rid, et al. 2020. Ethics of controlled human infection to address COVID-19. *Science* 368 (6493): 832–834. https://doi.org/10.1126/science.abc1076.

Smania, G., Baiardi, P., Ceci, A., Magni, P., and Cella, M. 2016. model-based assessment of alternative study designs in pediatric trials. Part I: Frequentist approaches. *CPT: Pharmacometrics and Systems Pharmacology*, 5(6), 305–312. https://doi.org/10.1002/psp4.12083.

Smith-Woolley, Emily, Ziada Ayorech, Philip S. Dale, Sophie von Stumm, and Robert Plomin. 2018. The genetics of university success. *Scientific Reports* 8 (1): 14579. https://doi.org/10.1038/s41598-018-32621-w.

Wang, H., and H. Yang. 2019. Gene-edited babies: What went wrong and what could go wrong. *PLOS Biology* 17 (4): e3000224. https://doi.org/10.1371/journal.pbio.3000224.

Watts, N.B. 2004. Is it ethical to use placebos in osteoporosis clinical trials? *Current Rheumatology Reports* 6 (1): 79–84. https://doi.org/10.1007/s11926-004-0087-z.

Bettina Couderc is Professor of Biotechnology at the University of Toulouse 3 Paul Sabatier. After 30 years of research work in oncology, focused on cell and gene therapy, she has been a member of the Inserm Bioethics research team (UMR1027) for two years and works on ethical issues concerning the end of medicalized life. She is president of the French Committee for the Protection of Persons CPPSOOM 2 in charge of the national ethical evaluation of clinical projects including French patients. She is also President of the Ethics Committee of the University Cancer Institute of Toulouse (IUCT-O) and has just been appointed referent deontologist of the University Toulouse 3.

Chapter 22
Research Ethics and Research Ethics Committees in Europe

Dirk Lanzerath

22.1 Introduction

In Europe, topics like "ethics" and "ethics committees" are becoming increasingly present in publications, university seminars, public debate, but also in connection with codes of conduct of organizations, companies or non-governmental organizations. The reasons for this are to be found, in particular, in the fact that *technological developments are* dramatically changing society *and* nature. This is because natural boundaries are being shifted or overcome, which raises the question of new moral boundaries in order to control or minimize the negative consequences or risks of the application of technologies. Morality thus becomes the "price of modernity" (Höffe 1993) that societies must pay through increasing dissolution of boundaries in the form of self-limitations. Ethical reflections structure this field of possible normative reorganizations and can help to open up attitudes of the actors in order to articulate ethical commitments independent of legal regulations. For the field of science and research, an ethos of researchers would ideally emerge as a common scientific ethos (Merton 1973) before state-imposed legal regulations intervene in an orderly manner.

Currently, in addition to new methods, for example in genetic engineering (e.g., genome editing by Cripr-Cas9), the challenge posed by the corona COVID-19 pandemic (e.g., development of drugs and vaccines against COVID-19), the focus is particularly on information technology research and innovations. Data collection, automation, and robotic and autonomous systems are significantly transforming not only biomedical research but also medical practice and health care systems. Moreover, these developments are not selective or regionally limited; rather, they encounter internationally interconnected societies. However, a global view of the ethical debates also reveals different views of values and norms, which may vary in detail even within

D. Lanzerath (✉)
University of Bonn, Bonn, Germany
e-mail: lanzerath@drze.de

© The Author(s), under exclusive license to Springer Nature Switzerland AG 2023 423
T. Zima and D. N. Weisstub (eds.), *Medical Research Ethics: Challenges in the 21st Century*, Philosophy and Medicine 132,
https://doi.org/10.1007/978-3-031-12692-5_22

Europe despite many moral similarities. This is already the case, for example, with ideas about the appropriate handling of personal data (Wachter et al. 2017).

In addition to the ethical debates on new fields of research and their applications, a system of ethical *consultation by ethics committees has been* established worldwide within the scientific community for medical research. Against the background of abusive medical research, not only under the Nazi regime in Germany, the involvement of ethics committees is intended to ensure compliance with ethical standards. The importance of the work of ethics committees is not exhausted in the review of the protection of subjects who make themselves available for research purposes, but also represents a genuine contribution to the stabilization of the moral credibility of research. Ethics committees are therefore often also addressed as "intermediary institutions" between science and society that contribute to societal trust in research and contribute to the ethical infrastructure of a research institution (Mejlgaard and Bloch 2012; Owen et al. 2012; Tallacchini 2012; Tsipouri 2012; European Commision 2007).

22.2   Science as Social Practice

While ethics committees have long been established in medical research, scientists are increasingly requesting advice from an ethics committee outside of clinical trials. Not only both funding agencies, but also and scientific publication bodies often expect a prior ethical evaluation by an independent ethics committee before studies take place and before their results are published. Some universities and research institutions have provisions for this in their statutes; nevertheless, this has not yet become a general standard outside medical research (Heinrichs and Lanzerath 2017). As a result of this development, medical ethics committees at universities are increasingly confronted with requests to provide a vote on a research protocols outside of medicine. Many universities in Europe have now established additional nonmedical ethics committees, for example, for the fields of psychology, social sciences, or technology. These are requested in particular when the intended research projects directly involve human subjects—for example, in the case of new techniques that are used directly on individuals,—or when personal data are collected—for example, in the form of interviews. Outside of medical research, however, only a few countries have legally regulated tasks of ethics committees for these research areas through *research laws*. The Finnish National Commission on Scientific Integrity (TENK), for example, also acts as the central research ethics committee for the humanities and social sciences (Finnish National Board on Research Integrity TENK 2019). The consequence of this inconsistent situation in Europe is that approaches to ethical evaluation vary considerably and are difficult to compare. Standards differ not only from country to country but also between institutions within a country. Multicenter studies are therefore often subject to widely varying requirements in Europe with regard to ethics consultation.

The fact that ethics committees are being expanded, and new ethics committees are being established outside of medical research, is guided by the conviction that *ethical reflection* can *improve science* and that science is not simply understood as a method, but as a *social practice.* This is because *science and research should be seen as integral parts of societies,* rather than as a space opposed to the living world. An ethics committee consultation—which is all too easily perceived as nothing more than an administrative hurdle—can instead contribute very productively to scientists and scholars thinking about normative aspects of their own research and its social implications from a perspective other than just the methodological issues of the discipline. This forms an ethos of research in the first place (Lanzerath D. Ethos (n.d.)). The less an ethics committee is involved in the implementation of regulatory requirements, such as in connection with the German Medicines or Medical Devices Act, the more it has an essentially advisory function.

Ethics committees can also help to further develop the form of ethics consultation. In the meantime, beyond medical research, representatives of ethics committees are working, for example, on the development of criteria for guidelines for the ethical evaluation of research into digital applications, robotics or artificial intelligence. Platforms for this are provided, for example, by the SHERPA, PANELFIT, and SIENNA projects funded by the European Commission (Shaping the ethical dimensions of smart information systems (SIS) 2019; Cordis 2019a; Stakeholder-Informed Ethics for New technologies with high socio-ecoNomic and human rights impAct (SIENNA) 2019). Ethics committees are also particularly intensively involved in the development of teaching and learning materials, as is the case, for example, in the Vir2tue and Path2Integrity networks (Cordis 2019b; Path2Integrity 2019). The goal of these efforts is to increase awareness of research ethics issues among students, researchers, and ethics committee members through training opportunities and educational materials.

22.3 Logics of Healing, Research, and Economy

Research Ethics Committees in medicine encounter an interweaving of actions with very different objectives within the health care system. These objectives must be kept apart, especially from an ethical perspective, and may well lead to conflicts of interest between the parties involved. For modern medicine, advances in diagnosis and therapy are inconceivable without systematic experimental research achievements in this field. This includes the development and advancement of new drugs by public research institutions and industry, but also research into new methods in the various individual disciplines, such as surgery or neurology. Since modern times, the goal of medicine, to effectively help sick people, can only be achieved through an experimental understanding of medicine. This development leads into the current debates about the possibilities and limits of evidence-based medicine (Vos 2005; Wessling 2011). However, the coincidence of a *logic of research* and a *logic of healing in* medical research requires a very fundamental reflection on research ethics, especially in the context of human experimentation, if human experimentation is to become an ethically legitimized practice in which the partial instrumentalization of human beings must be ethically justified (Heinrichs 2000).

By combining the *logic of research* with the *logic of healing,* a tension arises between the *individual benefit of* medical diagnosis and therapy on the one hand and the principle *external benefit* inherent in research on the other. While, in curative treatment, the focus is on the individual, the individual case has almost no significance for research. Scientific knowledge is rather determined by generalization and general validity. Wolfgang Wieland emphasizes in his medical-ethical and scientific-theoretical contributions to medical diagnosis that an individual experiment is always of interest only "insofar as it is representative of a general. In contrast, the patient is of interest for his own sake, if complaints and findings are to be explained and, if possible, eliminated by a diagnosis" (Wieland 1975; Lanzerath and Tambornino 2020). Medical research ethics and guidelines governing medical research must therefore address, in this tension between the logic of healing and the logic of research, the conditions under which the researcher is legitimized to make a human being a participant in an experiment and thus an *object of* the logic of research. Striking a balance between adequate protection of subjects and research that enables medical progress for the benefit of patients is the central theme of ethics and the law of medical research.

In the meantime, however, another aspect is becoming increasingly dominant in medicine and research, namely the economic aspect. Beyond public funding, scientific innovations to improve the health care of the population are very much tied to economic conditions that open up scope for investment. In the end, investments by industry must also lead to economic gains and not just to medical successes. If developments take too long to produce commercially viable results, or if there are too few buyers for products such as drugs, then investors lack the incentives to get involved, even though this would certainly be in the interests of patient welfare. In the meantime, the healthcare industry has developed into one of the major growth areas in the industrialized nations. As a result, the *logic of business is* gaining influence not only on medical care, but also on the forms of medical research. It joins the logic of *research* and the *logic of healing.* The consequence of this is, for example, that under aspects of economic efficiency, without government support, there is often no development for drugs that affect only a small group of people. From the point of view of research ethics, this economic rationality violates the social requirements arising from the research ethics principle of justice.

22.4 Ethical Reviews and Legal Frameworks

As medical research is increasingly organized across states in multicenter studies, the European legislative bodyor has decided to formulate a common European legal rule beyond the internationally agreed ethical obligations. In 2001, the "EU Directive 2001/20 EC" therefore came into force for the field of pharmaceutical research. This directive very explicitly requires the vote of an ethics committee. The Directive defines in Article 2 (k) an "Ethics Committee" as an "an independent body in a Member State, consisting of healthcare professionals and non-medical members, whose responsibility it is to protect the rights, safety and wellbeing of human subjects involved in a trial and to provide public assurance of that protection, by, among other things, expressing an opinion on the trial protocol, the suitability of the investigators and the adequacy of facilities, and on the methods and documents to be used to inform trial subjects and obtain their informed consent" (Europäisches Parlament, Europäischer Rat. Richtlinie 2001).

In April 2014, the European Parliament and Council finally adopted the "Regulation (EU) No. 536/2014" on clinical trials on medicinal products for human use and repealing Directive 2001/20/EC. "This set of regulations also in no way denies the need to comply with research ethics standards. But a new regulation is necessary—according to the European Commission in its introduction to the 2012 draft regulation—because the currently existing directive has had a negative impact on "the *costs* and *feasibility of* clinical trials" so that, "activity in the field of clinical trials in the EU has declined. "This assessment marks the motive behind the rewrite of the regulatory regime for this area, namely to accommodate the *economic concerns of* the pharmaceutical industry. The procedures for approving research protocols are to be accelerated. Thus, behind the political motivation, the increasing importance of the described *logic of economic activity* becomes apparent. The focus is on *embedding the European research landscape in the European economic area*, which must assert itself globally. It is less a matter of improving subject protection. It is undeniable that industry-supported European research also faces fierce international competition in the field of drug and medical device development. Nor should it be forgotten how much society benefits from new medicines and therapies. Patient organizations in particular are calling for more medical research. That is why speedy approval procedures are a legitimate concern not only for industry and research.

However, a no less justified desideratum is to guarantee the protection of those people who voluntarily make themselves available for such experiments. In this context, the function and operation of independent ethics committees plays an important role. In its own funding instrument, the research framework program HORIZON 2020, similar to the expired Framework Program 7, the European Commission even makes great efforts to ensure that research proposals undergo a process of ethical evaluation by an ethics committee. The EU Commission describes in its research funding programs that for all activities it funds, "ethics is an integral part of research from beginning to end, and ethical compliance is seen as pivotal to achieve real research excellence "(European Commission 2018). The "Ethics Appraisal Procedure" mandated by the European Commission is often much more extensive and rigorous than those in individual EU member states, as is the case when compared to the procedures mandated in Germany, Italy or France. While many member states do not have national regulations for the ethical evaluation of research projects in the social, engineering or environmental sciences, for example, and only a few ethics committees are set up for these areas at research institutions, these research fields are the subject of an ethics review by the European Commission as standard. This is especially true when research involves personal data, environmentally hazardous substances, or endangered species (European Commission 2019). Although there are European and international framework legislations such as the General Data Protection Regulation or the Convention on Biological Diversity, none prescribes the involvement of ethics committees outside of medical research. The regulations governing the establishment of ethics committees are national for this purpose. In Germany, for example, there is the phenomenon that, if there are no special regulations at the respective university, a qualitative study that, for example, asks patients in a format of a survey, must be submitted to an ethics committee due to the professional regulations for physicians if this study is carried out by physicians. If, on the other hand, the same study is conducted by non-physician researchers, whether or not an ethics vote is required depends solely on the circumstances of the research institution in question. In the meantime, however, more and more research funding organizations or scientific publication bodies require a positive opinion by an ethics committee. In practice, this creates the problem that often no ethics committee feels responsible for this type of evaluation.

The experiences of the extensive ethics review process at the European Commission within HORIZON2020 have probably had little influence on the first draft of the new EU regulation published by the European Commission in July 2012. For this draft not only made the ethics committees sit up and take notice, because it leaves the execution of an ethical evaluation to the member states. Ethics committees, as defined in the 2001 Directive, are no longer mentioned in the draft. Many experts concerned with research ethics have come to the conclusion that this marginalized position of ethics committees, which contribute significantly to the protection of the subjects and patients involved, would weaken this important protective element as a procedural principle in the new legislative practice. The great achievement of civilization, namely that medical research involving human subjects is scientifically, ethically and legally evaluated by an independent body before a competent authority grants approval for it, is up for grabs with the draft. Although the ethics committees have their place again in the final version of the regulation of April 2014 compared to the draft, there remains a tendency of marginalization, because a detailed description of their responsibility, as in the directive, is no longer found in the regulation. This legislative decision does not fit in with the development in research ethics to demand an even more conscientious review by ethics committees as a central procedural element, because even longer informed consent forms—which essentially transfer the burden of decision-making and risk assessment to the subject—do not contribute significantly to improving the design of informed consent. The more difficult an informed consent practice becomes, the more important *procedural elements* become as confidence-building measures in a complex procedure (Manson and O'Neill 2007). These elements include an active role of ethics committees.

It can be seen as a genuine political-social matter and not only as a *research-ethical* but also as a *social-ethical* concern that a society does not forget what has been learned from past abuses, but rather takes care to implement what has been learned in law, research funding and testing practice so that such occurrences do not repeat. Well-functioning and independent ethics committees as well as transnational ethical standards are an important building block in this regard, which in no way impedes research, but rather makes research safer for test subjects and thus promotes successful research. Only medical research that adheres to "good practice" rules and is balanced in the logic of research, healing and business is good research.

22.5 Exchange of Information and Harmonization

The justified concern to rapidly develop new therapies and drugs for patients in Europe and to remain internationally competitive in the process must not be at the expense of safety for trial subjects, who must be able to rely on an independent ethics opinion. It should therefore certainly be a concern of the European legislator to strengthen the position of ethics committees and to support them in harmonizing decision-making processes and streamlining procedural issues. In the meantime, the ethics committees have joined together in a European network—especially thanks to the support of the European Commission—in order to better exchange information on a European level. This *European Network of Research Ethics Committees* (EUREC) promotes in particular cooperation between the national representations of ethics committees (European Network of Research Ethics Committees 2018). Among other things, the European association has resulted in the formation of national networks that were previously lacking, such as in Spain, Italy, and Poland. The primary purpose of this is to harmonize the multi-disciplinary ethical evaluation of medical research projects involving human subjects or bio-materials taken from them or identifiable data. This is intended to promote the protection of participants in research projects and the quality of research at the European level. Especially for the exchange on the impact and implementation of the new EU Regulation as well as on the related establishment of a central EU portal by the European Medical Agency (EMA) in the course of the Regulation, EUREC has provided a forum for a qualified debate. The European ethics committees thus have a forum for the joint formation of opinions and coordination among themselves Lanzerath (2016). This is why *national network building* as part of a harmonization process is so important to the European debate. For without such a national network, no adequate representation at the European level can be derived, since there are usually too many ethics committees in one country. In Italy, for example, the number of ethics committees has been reduced from about 250 to about 60. Especially for ethics committees of smaller EU countries, which cannot be adequately represented in every European board, the information channel via EUREC is the essential platform, not only to keep up to date with the development process in the implementation of the EU regulation, but also to articulate their own positions on this topic. Therefore, in addition to the representation by the national representatives of the Ethics Committees in the Expert Committee at EMA, which is reserved for the member countries, EUREC also participates in the stakeholder consultations on the EU portal at EMA.

The implementation of the EU Regulation is primarily determined by the responsible ministries and authorities. A large part of medical research—in many countries even the majority—is not covered by the directive or the new regulation. However, even this type of medical research is in most cases supervised by ethics committees and is in part prescribed by professional law. Here, the ethics committees have a greater scope of action, but the need for harmonization is different because a European framework legislation is missing. EUREC also supports research institutions from non-European countries here, which are looking for the appropriate contacts among ethics committees in Europe for their research projects. Votes of ethics committees are currently only available nationally from the respective responsible committee.

EUREC has set a special focus through its cooperation with the European Network of Paediatric Research at the European Medicines Agency (Enpr-EMA). Pediatric researchers and representatives of ethics committees exchange views on the sensitive topic of research involving children and adolescents. In joint workshops, also with representatives of the Young Persons' Advisory Groups (YPAG) from different European countries, to improve research with children in a responsible way and to specifically address the requirements of informed consent in an age-appropriate way. The goal is to strike a successful balance between protecting young subjects and achieving health-related research objectives. Cases are discussed together and, in the case of differing assessments, it is examined why there are variances in the assessment here and whether these can be resolved (shared decision making exercise). Especially the off-label use of drugs in children is still widespread due to a lack of alternatives, but the risks, for example due to additives unsuitable for children or due to over- or underdosing, are high and are shifted to the treating pediatricians. Therefore, considerable research is still needed in this area (Lanzerath and Rietschel 2018; Lepola et al. 2016; Ruperto et al. 2012). EUREC participates therefore in the pediatric research project connect4children (c4c: https://conect4children.org).

Despite many harmonization efforts, which have undoubtedly been successful in recent years, such as the exchange of different concepts to meet the requirements of EU regulation, the models of ethics committee in Europe vary. The practical organization of ethics committees is still a national task.

22.6 Variants of Models of European Ethics Committees

Different models can be identified for how countries have implemented the requirement for ethics review into their legal and organizational systems. Ethics committees vary according to the scope of the review (type of research being evaluated) and their affiliation (national, regional, or institutional). Legal status, composition, functions, areas of responsibility, operating principles, and duration of issuance of a "positive opinion" also vary from country to country. For example, in some countries, ethics committees conduct follow-up studies to review long-term effects of clinically active substances within approved trials, whereas in some countries they do not (e.g., Finland) (Hemminki 2016). Whether, in the case of a negative evaluation within the ethics committee system, possible *appeal procedures* take effect, leading to a further instance or a second opinion, also varies widely. In some countries, a national ethics committee acts as an appeal body; in other countries, for example, nationally appointed ombudsmen mediate.

Some EU Member States have established a *reporting practice on* the part of ethics committees to an authority that is absent in other countries (Veerus et al. 2014). In countries such as the United Kingdom, Lithuania, or Norway, the ethics committees are subject to a supervisory body (authority or ministry). In Finland, for example, the National Ethics Committee is responsible for accompanying the local committees, offering training, but also providing a second opinion for a protocol. However, it has no formal supervisory function (Hemminki 2016). The number of ethics committees proportional to the number of inhabitants varies considerably among Member States (Druml et al. 2009). A disproportionate number of ethics committees results in fewer research protocols per ethics committee and correspondingly less developed experience and infrastructure. In addition, the frequency of meetings is lower. This leads to the difficulty that the deadlines set in the new EU regulation may not be met. Some countries have therefore set a minimum number of protocols for the accreditation of an ethics committee (Halila 2014).

An important question for the trustworthiness of ethics committees is that of *independence*. It concerns first of all the members, but also the institutional anchoring. Here, too, different models have developed in Europe. In the case of committees set up at the research institutions themselves, the personal proximity to the researchers being evaluated is criticized in particular. However, this is not necessarily a disadvantage, because the more the circumstances of a research are known outside the paper form, the easier it is to evaluate. In the case of commissions, which in more centralized systems are closely tied to a ministry or regulatory agency, this institutional proximity to the politically controlled infrastructure is again criticized (Tenti et al. 2018; Gefenas et al. 2018; Stahl 2017; Hasford 2017).

22.7 Current Challenges

22.7.1 EU Regulation

For all European ethics committees, EU Regulation (EU) No 536/2014 represents a major challenge. The responsibilities they still have under the new legislation are interpreted quite differently and implemented nationally. In particular, the scope of the ethical review carried out by the ethics committees varies considerably in the views of the member countries. There is agreement that procedural processes need to be significantly streamlined and professionalized to meet the requirements of the new deadlines set by the regulation. Not all currently existing ethics committees will be able to meet these standards, which increases the pressure for structural adjustments in the member countries.

Irrespective of the various national interpretations regarding the *scope of the future competence of* ethics committees, one particularly worrying consequence of the EU regulation can already be identified with regard to the role of ethics committees as an intermediary link between society and science. The practical importance of ethics committees is marginalized by the strong dependence on national regulatory authorities, a less clear definition of tasks compared to the still valid directive, and the ever-present threat of *silent approval in* case of missed deadlines. This tendency was even more evident in the European Commission's 2013 draft regulation than in the version finally adopted by the European Parliament. Moreover, there is a structural difference between the scope of competence of the ethics committees as defined by the still current directive, which clearly includes method evaluation and risk–benefit analysis, and the provisions in the new regulation. This is because in the future it will be up to the Member States themselves to determine the *scope of the ethics evaluation.* Therefore, it is possible that some Member States will choose a narrow review model and evaluate only Part II (e.g., envisaged in France and Italy), such as informed consent documents, suitability of investigators and investigational sites, etc. This model would exclude an assessment of the methodology and risk of a study (Part I). If this becomes generally accepted, questions such as the choice of control drug or the use of placebos cannot be addressed, but also, for example, the protection of research participants at risk (decision on a minimum risk standard) would not necessarily fall within the remit of ethics committees. This breadth of interpretation does not contribute to European harmonization.

European organizations such as EUREC or the European Group of Ethics (EGE), which advises the European Commission, have clearly pointed out in their statements that the marginalization of the ethics committees also removes a protective factor for the subjects (The European Group on Ethics in Science and New Technologies (EGE) 2020). The freedoms that member states have at this point do not lead to an increase in harmonization processes. Since only one access to the EU portal is established per country for the competent national regulatory authority and no direct access for the ethics committees, the ability of the ethics committees to work with regard to the EU portal is closely tied to the cooperation with the respective regulatory authority (Tenti et al. 2018).

For subject protection, the evaluation of *both parts of* the application dossier (Part I, which is evaluated together with the Competent Authorities, and Part II, for which the Ethics Committee has sole responsibility) by the Ethics Committees would be desirable. However, since the time limits foreseen for the evaluation are relatively short, especially for multinational studies, and communication with the sponsor is only in writing, the committees will have to engage in procedures that also meet these changed requirements in practice.

The *deadlines for* processing the applications and the need for *exchange* with the competent authority via the central electronic EU portal require very disciplined workflows at the ethics committees. Those committees that meet only at longer intervals (such as only monthly) cannot meet these standards. However, as a compensatory measure, it is also suggested that the *real meetings of* the ethics committees can and must be replaced or supplemented by *virtual* meetings or by electronic exchange of documents and opinions. In addition, it has been common practice for many ethics committees to also discuss problematic issues orally with the sponsor. The new framework provides exclusively for written communication with the sponsor via the EU portal (Hasford 2017). In many countries, the secretariats of the ethics committees will have to work much more effectively in their processes and, in particular, professionalize the area of electronic data processing. Some member states expect that, in addition to the new *EU portal/database system*, a national IT system will be required for national collaboration and interaction (Stahl 2017). It is also anticipated that the institutional model of self-regulation based on the participation of honorary members may need to be replaced or at least expanded by a full-time commission (Druml et al. 2009). Some EU countries have moved to having studies covered by the regulation evaluated by only one or a few commissions in order to concentrate the new effort (Petrini and Garattini 2016). However, such an approach may conflict with the Regulation's requirement for *"lay"* representation as well as multidisciplinary consultation as a key element of what has previously been considered "appropriate" ethics review, because it also "professionalizes" lay people.

In the EU member states, but also in Europe as a whole, a diversification of ethical review for biomedical research is therefore to be expected, i.e., a tendency against harmonization processes. Different modalities of ethics review may be established for protocols covered by the Regulation versus review of other types of biomedical research. It is likely that centralized commissions or a few commissions per country will be the more common model, rather than professionalizing all currently existing commissions. In addition, it is likely that much of the work will be conducted or prepared *online* by *full-time* members. For biomedical research outside the jurisdiction of the regulation, including the addition of review of studies from other disciplines, a more "traditional" multidisciplinary meeting in the field may be retained. In this regard, there may be few changes in those countries that already have a comprehensive system in place for ethical evaluation of human subjects research projects.

22.7.2 The COVID-19 Pandemic

The COVID-19 pandemic is an enormous and extraordinary challenge for societies, economies, politics, health systems, and especially for medical research. The multiple measures to contain the pandemic in Europe and worldwide must include the development and testing of effective drugs and vaccines. This is a particularly urgent matter. However, medicines to be approved in the future to cure COVID-19 must be as effective and safe as possible. For this reason, the European Commission, the European Medicines Agency (EMA), and the national heads of medicines agencies (HMAs) have published a "Guidance document for sponsors on the management of clinical trials during the COVID-19 (coronavirus) pandemic" (European Medicines Agency (EMA) 2020) that describes how to manage the conduct of clinical trials in this particular context and how to address issues of safety, risk assessment, and informed consent.

This new mode in medical research also leads to an enormous challenge for the European ethics committees. The ethics committees are aware that they have to make a corresponding contribution. This contribution is based in particular on the fact that the administrative processes for reviewing research protocols must be accelerated and simplified when these protocols are related to the treatment, prevention or diagnosis of infections caused by SARS-CoV-2. However, all of this must be guided by the principle that ethics committees should not compromise the quality of review even in these special circumstances; an expedited process should not come at the expense of safety, particularly the safety of research participants. The accepted ethical principles of autonomy, beneficence, non-harm, and justice must always be respected.

In particular, EUREC has agreed on the following rules.

1. The submitted studies related to the prevention or treatment of diseases associated with COVID-19 and COVID-19 should be given a clear priority in the ethical evaluation. The evaluation of studies related to other serious diseases without satisfactory treatment options should not be neglected.
2. Informed consent must be consistent with European and national regulations, recognizing that national regulations and their application may vary across Europe. The suggestions in Sect. 8 of the above-mentioned European "Guide" on the appropriate and simplified handling of informed consent under the conditions of COVID-19 should be considered by the European Ethics Committees.
3. In the current pandemic situation, the traditional meetings of the ethics committees cannot necessarily be organized in the usual way, often face to face. Ethics committees should therefore adopt new working methods, such as secure video-conferencing, that are appropriate to the current situation and respect the new rules of conduct regarding the pandemic. If necessary, appropriate provision should be made for changes in the rules of procedure.
4. It should be possible to hold extraordinary meetings outside the regular meetings to evaluate research protocols on the treatment, prevention, or diagnosis of infections caused by SARS-CoV-2.
5. The responsible ethics committees must be composed of experts with the relevant expertise. With regard to the evaluation of studies on COVID-19, relevant experience and expertise must also be ensured within the REC.
6. The information and communication technologies used must, however, be designed in such a way as to ensure data transfer that complies with the European General Data Protection Regulation.
7. In the course of the clinical trial, the recording of damage events and their effects as well as their forwarding and evaluation by the investigator must also be ensured. The responsible REC must also be appropriately involved in upcoming decisions and changes, e.g. of the protocol in case of subsequent changes. If necessary, a new informed consent may be required.

The overarching task of all ethics committees is to protect the dignity, rights, safety and well-being of research participants, i.e. patients and healthy volunteers, in medical studies. This is also true in the context of the current pandemic situation. Therefore, the pressure currently being exerted on medical research must not lead to drugs being researched or tested on humans without complying with the ethical standards applicable to medical research (European Network of Research Ethics Committees(EUREC) 2020).

Vaccine development is particularly challenging here. This usually takes many years and requires different phases. Currently, researchers are therefore proposing to replace the last phase with a human challenge trial (HCT) to shorten the time (Eyal et al. 2020; Jamrozik and Selgelid 2020]. Volunteers should be intentionally infected with SARS-CoV-2 to see how a vaccine candidate works. As early as June 2020, about 30,000 people from 140 countries signed up as volunteers on an online platform called "1Day Sooner" (1Day Sooner 2020). However, intentionally infecting a healthy person with a potentially deadly virus goes against the highest ethical standards that researchers in medical science must follow, especially the no-harm principle. Therefore, the benefits must be enormous, and the hurdles to conducting an HCT must be set very high, if the exception of intentional infection is to be justified while protecting volunteers.

If enough people are willing to be vaccinated with a "safe" vaccine in the future, it could reduce the global burden of the coronavirus pandemic. However, the risks to participants could be high, and the long-term risks cannot be predicted. Most clinical trials involve potential benefits and risks, and it is never easy for RECs to decide whether the benefits can justify the risks or whether the risks are acceptable. Nevertheless, the review of a coronavirus HCT is very challenging because, on the one hand, the societal benefits could be very high under certain conditions and, on the other hand, in the worst case, participants could die after infection with COVID-19. From a research ethics perspective, infecting a healthy volunteer with a potentially lethal virus should be prohibited. However, RECs should not reject such studies per se. Because of the lack of scientific knowledge about COVID-19, there is no way to adequately reduce the risks to participants. This means that it is not currently possible to conduct a coronavirus HCT in an ethical manner. However, even if the risks could be minimized, the social benefits of the results could be small if the vaccine only works in members of the groups tested (in terms of age, sex, etc.).

New research may make risk minimization feasible for HCT. If so, RECs must demonstrate risk minimization strategies and, in addition, review very carefully the methods described for the informed consent process. In addition, researchers must ensure that participants receive the best possible care during an infection. In countries where the health care system cannot guarantee adequate care for infected individuals, coronavirus HCTs should never be performed. Although participation in an HCT is demanding for volunteers and certain conditions of isolation must be maintained for an extended period of time, monetary compensation should be weighed appropriately. Under no circumstances should a monetary incentive be the motivation for participation (Lanzerath and Tambornino 2020).

22.8   Conclusion: Interim Balance Sheet in a Phase of Upheaval

At the present time, only a very preliminary and cautious interim assessment can be made of the direction in which ethics committees will develop during this period of upheaval. It can be assumed that the need to introduce elements of "professional" management of ethics committees and the possibility of exchanging information during the evaluation process will also have a positive effect on the quality of ethics review in other research areas. However, the marginalization tendency that can be seen through the EU Regulation reads contrary to the other observation described for Europe, that more and more non-medical research projects are being ethically evaluated. However, there is a lack of frameworks that could be internationally or Europeanly binding for this. At the same time, it cannot be denied that the development and further development of the *Declaration of Helsinki* of the World Medical Association also has an impact on research *outside medicine*. Especially recently, the World Medical Association has repeatedly expressed the view that the Declaration of Helsinki should also be the *legitimizing standard for research on humans in* other disciplines. Yet the Declaration of Helsinki itself has a legitimizing problem. The World Medical Association is merely a representative body for physicians and also only for the national associations represented in it and in no way represents *the* researchers.

Therefore, irrespective of the significance of the content of the individual provisions of the Declaration of Helsinki, other approaches will have to be taken—for example, in cooperation with the professional societies—in order to standardize the content and procedures of ethical review in Europe and internationally and to determine the task of ethics committees accordingly. It is hardly desirable that in the future every form of research is evaluated by an ethics committee. This is neither administratively nor factually appropriate and would unnecessarily impede the freedom of research. Nevertheless, there is research on humans outside of medicine that is ethically questionable. For example, the UNICEF Office of Research in its report "Best Practice Requirements by the Ethical Research Involving Children (ERIC)" (Graham et al. 2013) rightly points out how children can also be at risk as subjects in qualitative studies. The nature and content of a scientific survey may imply a high level of exposure for participating children. To treat such studies differently from comparable studies in medicine with regard to ethical evaluations seems hardly understandable. Part of the common research ethos is then to weigh up exactly where more and where less ethical evaluation by ethics committees is necessary. Under the impression of the SARS-CoV2 pandemic, it is important for the ethics committees to pay attention to the respect of ethical standards in medical research and especially to the protection of the participants. This is especially true when there are forays into Human Challenging Trials to accelerate vaccine development, in which healthy volunteers allow themselves to be infected in order to test a vaccine. At the time of writing, not enough is known about the potentially deadly virus that one can responsibly conduct such clinical trials from the perspective of ethics committees. This is true even for low-risk groups until the risk can be minimized and a standardized treatment for COVID-19 is exited.

References

Commision, European. 2007. *Taking European knowledge society seriously: Report of the expert group on science and governance to the science, economy and society directorate, directorate-general for research, European commission[R]*. Brussels: European Commission DG Research Science, Economy and Society.

Cordis. 2019b. Virtue based ethics and Integrity of Research: Train-the-Trainer program for Upholding the principles and practices of the European Code of Conduct for Research Integrity (VIRT2UE)[EB/OL]. https://cordis.europa.eu/project/rcn/214892/factsheet/en.

Cordis. 2019a. Participatory Approaches to a New Ethical and Legal Framework for ICT (PANELFIT)[EB/OL]. https://cordis.europa.eu/project/rcn/218355/factsheet/en.

Diagnose, Wieland W. 1975. *Überlegungen zur medizintheorie [M]*, 70–71. Berlin: De Gruyter.

Druml, C., M. Wolzt, J. Pleiner, and E.A. Singer. 2009. Research ethics committees in Europe: Trials and tribulations[J]. *Intensive Care Medicine* 35 (9): 1636–1640.

Europäisches Parlament, Europäischer Rat. 2001. Richtlinie 2001/20/EG des Europäischen Parlaments und des Rates zur Angleichung der Rechts- und Verwaltungsvorschriften der Mitgliedstaaten über die Anwendung der guten klinischen Praxis bei der Durchführung von klinischen Prüfungen mit Humanarzneimitteln[EB/OL]. [17.12.2018]. http://ec.europa.eu/health/files/eudralex/vol-1/dir_2001_20/dir_2001_20_de.pdf.

European Commission. 2019. HORIZON2020 online manual: Ethics Appraisal Procedure[EB/OL]. http://ec.europa.eu/research/participants/docs/h2020-funding-guide/cross-cutting-issues/ethics_en.htm.

European Commission. Science With And For Society (Swafs), Research Ethics[EB/OL]. [2018–12–17]. http://ec.europa.eu/research/swafs/index.cfm?pg=policy&lib=ethics.

European Medicines Agency (EMA), Heads of Medicines Agency (HMA). GUIDANCE ON THE MANAGEMENT OF CLINICAL TRIALS DURING THE COVID-19 (CORONAVIRUS) PANDEMIC, Version 3 [EB/OL]. [2020–06–25]. https://ec.europa.eu/health/sites/health/files/files/eudralex/vol-10/guidanceclinicaltrials_covid19_en.pdf.

European Network of Research Ethics Committees—EUREC[EB/OL]. [2018–12–17]. http://www.eurecnet.org/index.html.

European Network of Research Ethics Committees(EUREC). Position of the European Network of Research Ethics Committees (EUREC) on the Responsibility of Research Ethics Committees during the COVID-19 Pandemic[EB/OL]. (2020–04–27) [2020–06–25]. http://www.eurecnet.org/documents/Position_EUREC_COVID_19.pdf.

Eyal, N., M. Lipsitch, and P.G. Smith. 2020. Human challenge studies to accelerate coronavirus vaccine licensure[J]. *The Journal of Infectious Diseases* 221 (11): 1752–1756.

Finnish National Board on Research Integrity TENK. Ethical review in human sciences [EB/OL]. [2019–03–27] https://tenk.fi/en/ethical-review/ethical-review-human-sciences.

Gefenas, E., A. Cekanauskaite, J. Lekstutiene, and V. Lukaseviciene. 2018. Application challenges of the new EU Clinical Trials Regulation[J]. *European Journal of Clinical Pharmacology* 73 (7): 795–798.

Graham, A., Powell, M., Taylor, N., Anderson, D., and Fitzgerald, R. 2013. Ethical research involving children[M], Vol. 13. Florence: UNICEF Office of Research—Innocenti.

Halila, R. 2014. Evaluation of the work of hospital districts' research ethics committees in Finland[J]. *Journal of Medical Ethics* 40(12).

Hasford, J. 2017. The impact of the EU Regulation 536/2014 on the tasks and functioning of ethics committees in Germany[J]. *Bundesgesundheitsblatt, Gesundheitsforschung, Gesundheitsschutz* 60 (8): 830–835.

Heinrichs, J.H., and D. Lanzerath. 2017. Nichtmedizinische Forschung am Menschen—Probandenschutz jenseits der Medizin[J]. *Forschung, Politik, Strategie, Management* 10 (3/4): 90–94.

Heinrichs, B. 2006. Forschung am Menschen. Elemente einer ethischen Theorie biomedizinischer Humanexperimente[M]. Berlin: De Gruyter: 15; Lanzerath, D. 2000. Krankheit und ärztliches Handeln. Zur Funktion des Krankheitsbegriffs in der medizinischen Ethik[M]. Karl Alber: Freiburg i. Br.:68.

HEmminki, E. 2016. Research ethics committees in the regulation of clinical research: comparison of Finland to England, Canada, and the United States[J]. *Health Res Policy Sys 14*: article no. 5.

Höffe, O. 1993. Moral als Preis der Moderne: Ein Versuch über Wissenschaft, Technik und Umwelt[M]. Frankfurt a.m: Suhrkamp.

Jamrozik, E., Selgelid, M.J. 2020. COVID-19 human challenge studies: ethical issues[J]. *The Lancet Infectious Diseases*. Published Online.

Lanzerath, D., Rietschel, M. eds. 2018. Ethics of research involving minors. A European Perspective [M]. LIT: Münster

Lanzerath, D., and Tambornino, L. 2020. COVID-19 human challenge trials—what research ethics committees need to consider. *16*(3–4) (in print).

Lanzerath, D. Ethos. *Handbuch der Bioethik[M]*, ed. Sturma, D., and Heinrichs, B., 35–43. Stuttgart: Metzler.

Lanzerath, D. 2000. Krankheit und ärztliches Handeln. Zur Funktion des Krankheitsbegriffs in der medizinischen Ethik[M], 60–62. Karl Alber: Freiburg i. Br.

Lanzerath, D. ed. 2016. Forschungsethik und klinische Forschung. Zur Debatte um die EU-Verordnung zu klinischen Studien [M]. LIT: Münster

Lepola, P., Tansey, S., Dicks, P., Preston, J., and Dehlinger-Kremer, M. 2020. Pharmaceutical industry and paediatric clinical trial networks in Europe—how they communicate [EB/OL]? [2020–06–30]. http://www.appliedclinicaltrialsonline.com/pharmaceutical-industry-and-pediatric-clinical-trial-networks-europe-how-do-they-communicate?.

Manson, N.C., O'Neill, O. 2007. Some conclusions and proposals, eds. Manson, N.C., and O'Neill, O., 199. Cambridge: Cambridge University Press

Mejlgaard, N., and C. Bloch. 2012. Science in society in Europe[J]. *Science and Public Policy* 39 (6): 695–700.

Merton, R.K. 1973. The normative structure of science. *The Sociology of Science: Theoretical and Empirical Investigations*, ed. Merton, R.K., 267–278 [M]. Chicago: University of Chicago Press.

Owen, R., P. Macnaghten, and J. Stilgoe. 2012. Responsible research and innovation: From science in society to science for society, with society[J]. *Science and Public Policy* 39 (6): 751–760.

Path2Integrity. 2019. Rotatory role-playing and role-models to enhance the research integrity culture[EB/OL]. https://www.path2integrity.eu.

Petrini, C., and S. Garattini. 2016. Trials, Regulation and tribulations[J]. *European Journal of Clinical Pharmacology* 72 (4): 503–505.

Ruperto, N., I. Eichler, R. Herold, et al. 2012. A European network of pediatric research at the European Medicines Agency (Enpr-EMA)[J]. *Archives of Disease in Childhood* 97: 185–188.

Shaping the ethical dimensions of smart information systems (SIS)—a European perspective (SHEREPA)[EB/OL]. [2019–03–27]. https://www.project-sherpa.eu.

1Day Sooner. [2020–05–25]. https://1daysooner.org/.

Stahl, E. 2017. Implementation status of Regulation EU 536/2014 in the member states [J]. *Bundesgesundheitsblatt, Gesundheitsforschung, Gesundheitsschutz* 60 (8): 836–840.

Stakeholder-Informed Ethics for New technologies with high socio-ecoNomic and human rights impAct (SIENNA)[EB/OL]. [2019–03–2019]. http://www.sienna-project.eu.

Tallacchini, M. 2012. Epistemology of the European identity [J]. *The Journal of Biolaw and Business*, Suppl Ser: 60–66.

Tenti, E., G. Simonetti, M.T. Bochicchio, and Martinelli. 2018. Main changes in European clinical trials regulation (No 536/2014) [J]. *Contemporary Clinical Trials Communication* 12: 99–101.

The European Group on Ethics in Science and New Technologies (EGE). Statement of the European Group on Ethics in Science and New Technologies (EGE) on the Proposal for a Regulation of the European Parliament and of the Council on Clinical Trials on Medicinal Products for Human Use, and repealing Directive 2001/20/EC (COM 2010) 369 final[EB/OL]. [2020–06–30]. https://ec.eur opa.eu/research/ege/pdf/statement_of_the_ege_on_the_clinical_trials_directive_revision.pdf.

Tsipouri, L. 2012. Comparing innovation performance and science in society in the European member states[J]. *Science and Public Policy* 39 (6): 732–740.

Veerus, P., J. Lexchin, and E. Hemminki. 2014. Legislative regulation and ethical governance of medical research in different European Union countries[J]. *Journal of Medical Ethics* 40 (6): 409–413.

Vos, R. 2005. Coordinating the norms and values of medical research, medical practice and patient Worlds. The ethics of evidence-based medicine in 'Boundary Fields of Medicine'. *Evidence-based practice in medicine and health care: a discussion of the ethical issues [M]*, eds. Meulen, R.T., Biller-Andorno, N., Lenk, C., and Lie, R., 87–95. Berlin: Springer.

Wachter, S., Mittelstadt, B., Floridi, L. 2017. Transparent, explainable, and accountable AI for robotics[J]. *Science Robotics*, 2(6).

Wessling, H.W.A. 2011. Theorie der klinischen Evidenz. Versuch einer Kritik der evidenzbasierten Medizin[M]. Münster: LIT.

Dirk Lanzerath , Professor of Ethics and Research Ethics, graduated in biology, philosophy and education; he holds a Ph.D. and the venia legendi (habilitation) of the faculty of philosophy of the University of Bonn. Dirk Lanzerath is managing director of the German Reference Centre for Ethics in the Life Sciences (DRZE) at the University of Bonn (Research Centre of the Northrhine Westfalian Academy of Sciences, Humanities and the Arts). He is secretary general of EUREC (European Network of Research Ethics Committees). Moreover, he is a member of the Central Ethics Committee at the German Physician Association; member of the Ethics Committee of the North-Rhine Medical Association; member of the Ethics Committee of the University of Maastricht, member of the Editorial Board of the Journal "Research Ethics Review". Dirk Lanzerath is a 'study abroad' professor for ethics/bioethics/environmental ethics/research integrity/ethics and the arts at the Study Abroad Program of the Loyola Marymount University, Los Angeles, Ca. (USA) and Honorary-Professor at the University of Applied Sciences Bonn Rhein-Sieg.

Chapter 23
Conflicts of Interest in Biomedical Research in French Law

Guillaume Rousset

There are several reasons to be interested in the issue of conflicts of interest in the field of health. First, headlines regularly demonstrate the relevance of the subject, such as the Mediator scandal[1] (Laude 2012a, b; Bloch 2021; Prévost 2021), or, to a lesser extent, the H1N1 influenze vaccination campaign (Godlee 2010; WHO 2011). Secondly, it is not overstatement to say that the question of conflicts of interest arises regularly in the health sector. On this point, various reports from public authorities show that a significant proportion of sectoral examples of conflicts of interest relate to the health field (Sauvé 2011). Finally, the stakes in this sector seem high. Of course, the stakes are high whatever the field (legal and judicial professions, business, administration), but they seem to be particularly high here because the health sector is characterized by a certain spirit, by several kind of rules (ethics, [Byk 1998]; deontology) which seem to be naturally opposed to conflicts of interest (for example, medicine *"must not be practiced as a business"*, article R. 4127–19 of the public health code). This type of conflict in a sector which is regulated in significant ways by a diversity of social norms is potentially more problematic, especially when the values impacted by these conflicts are fundamental (life, body and integrity).

---

[1] The Mediator affair is a French health and legal scandal which broke out in 2010. It involved users of benfluorex, marketed under the name "Mediator" by the Servier laboratories. Following a warning given by a pneumologist (Irène Frachon), the company was put on trial on the charge of aggravated deception and endangering the life of others (the amphetamine nature of the drug and its known risks were hidden). This lawsuit targeted both the Servier laboratories, for deliberately hiding the risks, and the National Agency for the Safety of Medicines and Health Products (ANSM) for not having suspended and withdrawn the marketing authorization for the drug quickly enough.

G. Rousset (✉)
Jean Moulin Lyon 3 University, Lyon, France
e-mail: guillaume.rousset@univ-lyon3.fr

© The Author(s), under exclusive license to Springer Nature Switzerland AG 2023   441
T. Zima and D. N. Weissub (eds.), *Medical Research Ethics: Challenges
in the 21st Century*, Philosophy and Medicine 132,
https://doi.org/10.1007/978-3-031-12692-5_23

Within the field of health care, many areas may be affected by conflicts of interests. This is the case in the relations between prescribing physicians and pharmaceutical laboratories, in the training of medical students or even in the independence of experts participating in the work of the various health authorities (National authority for health, *Haute autorité de santé, HAS*; National agency for the safety of medicines and health products, *Agence nationale de sécurité des médicaments et des produits de santé, ANSM*) (Moret-Bailly 2004; IGAS 2011). In this last case, while a number of disputes could be cited (Duguet 2012; Pissaloux 2011), one example may be enlightening: a conflict of interest was noticed within what is called the Transparency commission (*Commission de la transparence*), which is an internal body of the French National authority for health. It is responsible for evaluating the "medical benefit of drugs" (*Service médical rendu, SMR*) and this evaluation serves as the basis for the ministerial authorization of reimbursement of the drug by the health insurance. This commission was called into question in a decision of the State Council (the highest administrative jurisdiction in France, *Conseil d'Etat*) on February 12, 2007 (Mascret 2007) when it delivered an opinion that allowed the delisting of a drug, even though the opinion was based on the expertise of a evaluator whose links with a pharmaceutical company, which was an interested party in this evaluation, were considered sufficiently important to affect his impartiality. This case is interesting because it highlights one of the first positions taken by the judge on this subject and it sheds light on the characteristics of conflicts of interest: the links were declared; the person in question was not a member of the Commission but an external evaluator; the conflict did not involve the laboratory that sold the aforementioned drug, but a competing laboratory, likely to have an interest in this evaluation. This example relates to a French case, but obviously the international level is just as concerned, as demonstrated by the suspicion of a conflict between the expert panel of the World Health Organization (WHO) and the pharmaceutical laboratories in charge of preparing the vaccines during the H1N1 flu epidemic. It was indeed argued by some that the alert level and vaccine recommendations were disproportionate to the reality of the health threat. These claims were later refuted in an official report (Godlee 2010; WHO 2011).

Beyond these cases, the field of health is also strongly concerned by conflicts of interest through biomedical research, to which we now turn our attention. This case sets itself apart from the previous ones because of the specific issues it introduces. Biomedical research is indeed characterized by a situation of experimentation and therefore of uncertainty. This induces specific issues in patients' rights since the persons on whom this research is carried out are given a product, the quality and safety of which are, by design, yet to be confirmed. Of course, the protocols and rules make it possible to respect the safety of individuals as best as possible. Nonetheless, the stakes are even higher than in the context of legally sanctioned treatments known to all and therefore are governed by a different set of legal norms. The legislative and regulatory framework for biomedical research was initially set up by the law of December 20 1988, known as the Huriet-Sérusclat law. Since that date, it has been profoundly modified by law n° 2004–806 of August 9, 2004 on public health policy, a text that also transposed Directive 201/20CE of April 4, 2001 on clinical trials on medicinal products in humans. The law n° 2012–300 of March 5, 2012 relating to research involving the human person integrated a certain number of modifications (Dinisi-Peyrusse 2012; Leroyer 2012), in particular through the renaming of biomedical research as "research involving the human person".[2]

Before discussing it in the specific field of healthcare (Hermitte and Le Coz 2014), it is important to first define what is meant by "conflict of interest". This definition should be brief as reflections have already been carried out on this subject (Moret-Bailly 2011; Moret-Bailly and Rodwin 2012; Dondéro 2012). It appears that conflicts of interest *"develop in relationships in which a person, charged with defending or representing or protecting the interests of others, could betray them for the benefit of another interest, his own or that of a third party"* (Moret-Bailly 2011).

One of the key ideas to remember is that a distinction must be made between interest, link of interest and conflict of interests. It is not the interest or the relationship of interest that is the problem. We all have interests, which are most often perfectly laudable (the patient's, the researcher's…), and are frequently interwoven. This is the case, for example, of an investigating physician in a biomedical research, he may have a link of interest with the promoter, i.e. the laboratory. This link does not lead to conflict and does not in itself pose a problem. In fact, it is not the interest or the permitted relationship that is problematic, and it may not even be the conflict between several interests that is problematic. Conflicting interests do not necessarily constitute a conflict of interests. There are often conflicting interests to be managed, for example, between the interest of patients and the interest of the community that funds the care in the case of an expensive treatment with limited therapeutic efficacy but that relieves pain, as can be seen in cancer care. The problem mainly lies in the inappropriate prioritization of interests.

In view of all this, what can be said about conflicts of interests in biomedical research? Two things seem essential. First, it is important to know what conflicts of interests arise in this field. Second, it is fundamental to deal with the mechanisms available to fight against these conflicts and their unwelcome effects.

23.1 The Extent of the Problem of Conflicts of Interests in Biomedical Research

Conflicts of interest in health care are commonly associated with the situation of health care professionals, most often physicians, who have a conflict of interest involving pharmaceutical companies. This is certainly a relevant illustration of the subject, but it is essential to understand that these conflicts go far beyond this single situation. Indeed, the variety of potential conflicts of interest is as wide as the number

---

[2] For convenience and in view of the international and intercultural perspective of the work in which this chapter is included, we will refer to it as "biomedical research".

of actors involved and of the diversity of the relationships that are established between them. To understand the variety of possible conflicts, it is necessary to first explain the multiplicity of actors involved in biomedical research and how it relates to the diversity of their relationships.

23.1.1   The Multiplicity of Involved Actors

To raise awareness of the multiplicity of actors concerned with this issue, it is useful to list and group several actors who may be involved in conflicts of interest in four categories.

The first category is the promoter. This is the natural person or legal entity responsible for initiating the research, managing it and ensuring that it is funded. He is the interlocutor of the authorities that will be introduced further in this text. In practice, there are two main categories of promoters. The first is composed of companies (pharmaceutical industry, manufacturers of medical devices, cosmetics, etc.) who are private organizations that initiate research, in the sense that they themselves develop the research protocols to demonstrate the effectiveness of their product. The second category of promoters is composed of institutional actors, such as hospitals or clinics, and large public research organizations, such as the National Center for scientific research (*Centre national de la recherche scientifique, CNRS*) or the National institute for health and medical research (*Institut national de la santé et de la recherche médicale, INSERM*). In most cases, this type of actor manages and finances the research, but the initiative is taken by an investigator, a researcher, who writes the protocol.

The second category deals with investigators. In practice, these are the health professionals, most often physicians, who carry out the research. As such, they inform the volunteers about the research, collect their written consent before it begins and follow them throughout the research by applying the procedures set out in the research protocol.

The third category refers to the healthy or sick people participating in the biomedical research. Quite naturally, they are imagined as the potential victims of conflicts of interests. It is the interest of the patient that will be overshadowed by inappropriate links between the physician and the pharmaceutical industry for example. Nevertheless, they can also be key stakeholders in a conflict of interests, and even start it. This is especially true, not directly for individual patients, but for patients' representatives when they are a part of control bodies such as the research ethics committee (*Comité de protection des personnes, CPP*), as will be discussed further in this chapter.

The last essential category of actors in the field of biomedical research is composed of the various public authorities that exist at the local, regional and national levels.

At the local level, the key actor is the research ethics committee wich must be consulted (article. L. 1123–1 et seq. and R. 1123–1 et seq. of the Public health code). Its mission is to deliver a authorization on any research protocol involving human beings before it is implemented. It is a legal person of public law, allowing it to carry out its duties in complete independence. It is approved by the Ministry of Health, which has authority over its territorial jurisdiction while being funded by government. The coordination, harmonization and evaluation of the practices of the research ethics committees are ensured by a national commission (*Commission nationale des recherches impliquant la personne humaine*), placed under the responsibility of the Minister of Health. Appointed at random to examine research projects submitted by promoters, the research ethics committee is responsible for ensuring that the conditions for the research viability are fulfilled, especially regarding the protection of participants and the obtaining of their informed consent, the relevance of the research, the adequacy of the pursued objectives, the implemented means, as well as the terms of compensation for participants. It must also give its opinion on the scientific and ethical relevance of projects aiming at building up collections of biological samples during the course of the research, as well as when substantial changes arise in purpose of protocols involving the use of elements and products of the human body for scientific purposes for which participants already gave their consent. If the committee's opinion is unfavorable, the promoter of the research may ask the Ministry of Health to submit the project to another research ethics committee for a second examination. In any case, the opinion of the research ethics committee is necessary but not sufficient: the research must also be authorized by the competent authority of the French national agency for medicines and health products safety for any research involving the human person.

At the regional level, biomedical research also involves regional health agencies (*Agence régionale de santé, ARS*). These agencies play multiple parts. First, they deliver authorizations to medical facilities performing potentially risky procedures (article L. 1121–13 of the Public health code). Secondly, these agencies officially appoint the members of the research ethics committee (article L. 1123–1 of the Public health code).

Finally, at the national level, the French national agency for medicines and health products safety is the relevant authority. It intervenes alongside the research ethics committee to examine research projects before they are carried out (article. L. 1123–12 of the Public health code). It has the power to prohibit or suspend a research project at any time in the event of a risk to public health, if the promoter fails to respond to its observations, or in the event of non-compliance with legal provisions.

Now that these actors have been introduced, how is it that their relationships can lead to the emergence of a conflict of interests?

23.1.2 A Variety of Cases of Conflicts of Interests

As is the case for other health situations, biomedical research is subject to a wide variety of relationships that may lead to conflicts of interests.

This is the case, as some authors have specifically demonstrated, of the physician-investigator at work in public hospitals (Thouvenin 2015). For example, in the situation of an attending hospital doctor who works full-time in a public hospital and who is, at the same time, the lead investigator for a private research project carried out in partnership with a pharmaceutical company. In this example, there are two types of agreements governing the research: an agreement between the industry and the investigator on the one hand, and an agreement between the industry and the hospital on the other. Despite this framework, it is not always easy for the physician who plays the role of investigator to respect all the interests involved, as the conflict of loyalty can lead to a conflict of interests. Indeed, being part of an institution with a defined and official policy on biomedical research and doing research in this same field with a different outlook can create certain difficulties.

A conflict of interests can also arise from the confrontation of the two professional identities of a health professional, such a physician. Thus, for example, practicing both medicine and medical research can lead to divergent needs and practices, as the clinical practice of medicine is carried out at the same time as research involving patients.

Other conflicts of interest involve research ethics committee. Several problematic situations are liable to arise. One of them has been solved for several years, but it is interesting enough to be mentioned here. Before the allocation process of the research ethics committee was changed in order to make reviews more random (change initiated by law n° 2012–300 of 5 March 2012 on research involving the human person), the research protocols were submitted to the research ethics committee in the jurisdiction the investigator or, whenever possible, the lead investigator, carried out his or her activities (article L. 1123–6 of the Public health code). This was indeed simple in practical terms, but it could lead to conflicts of interests, for example when the member of the committee who was to evaluate the protocol knew or worked directly with the investigator of the research, often a fellow physician.

However, other situations that could lead to conflicts of interests may still present themselves within the research ethics committee. Such is the case when the research promoter (a health care institution) is the institution that physically hosts the committee, as research ethics committee often do not have independent premises. It is also not impossible to imagine a case in which a research ethics committee's president is the investigator of the research submitted to the body he chairs. Of course, as we shall see later, provision is made for limiting the risks of conflict, since the president is not the evaluator on this dossier and leaves the room during the deliberation, but one may wonder what impact this situation will have. Since the chair occupies a powerful position, will members treat this research in a completely neutral manner? The question is worth asking.

It is therefore clear that there are many examples of conflicts of interest in biomedical research. Beyond the diversity of cases of conflict, what is probably more important is the measures and mechanisms that the law can put in place to combat these elements.

23.2 The Difficulty of Dealing with Conflicts of Interests in Health Care

When one wants to combat conflicts of interests, different types of social norms can be usefully invoked. There are, of course, ethics and morality, but also deontology, which, through a certain number of major principles (for example, *the physician may not alienate his professional independence in any form whatsoever*", article R. 4127–5 Public health code); the physician is *"at the service of the individual and of public health"*, article R. 4127–2 Public health code); he must *"uphold the principles of morality, probity and devotion essential to the practice of medicine"*, article R. 4127–3 Public health code) or specific rules (in particular, article R. 4127–105 paragraph 2 of the Public health code: *"A physician must not accept an expert mission in which his own interests, those of one of his patients, one of his relatives, one of his friends or a group that usually calls on his services are at stake"*), can be applied.

These rules may also include the specific issue of informed consent. Indeed, beyond its fundamental role in biomedical research in general (Weisstub 1998a, b), informed consent is presented as a tool for preventing conflicts of interests, since the disclosure of interests is the most basic transparency requirement. Researchers are required to disclose their interests, for example financial interests, to research ethics committees and in their publications, but even more so to research participants. Respect for participants requires that they be informed of all aspects that might affect the research, including possible financial interests that might influence the researchers' behaviour. Even if it is clearly not always sufficient, information as a condition of consent is therefore a necessary element of the mechanisms for managing conflicts of interest, an analysis that is not specific to France since international experience and other countries are converging (Poisson 2002; Collège des médecins du Québec 2021). However, this is certainly not enough and accounts for the intervention of the law.

Among the various legal norms, the law n° 2011–2012 of December 29, 2011 on the reinforcement of the safety of medicines and health products (Laude 2012a, b; Peigné 2012), an attempt to respond to the Mediator affair, is often presented as a central text in the fight against conflicts of interests. This particular focus leads one to wonder what the content of the previous system was since, fortunately, France did not wait for the law of December 29, 2011 to address these conflicts.

23.2.1 The Classic Tools

Prior to the 2011 law, the enforcement of the regulation of conflicts of interests was mainly based on four legal instruments.

The first relates to the system of incompatibilities and incapacities set up for the conditions of exercise of the members of a body. This is the case, for example, for the research ethics committee. Thus, a person cannot be a member of this committee if he or she also holds executive functions within a research sponsoring institution (article R. 1123–5 of the Public health code). These incapacities or incompatibilities are a classic mechanism which is not specific to the healthcare sector and even less so to the field of biomedical research, although the latter makes extensive use of them. In any case, it makes it possible to prevent a certain number of conflicts of interests.

The second instrument is the result of what has been called the "anti-gift law". Since the law n° 93–121 of January 27, 1993 concerning various measures of a social nature, members of the medical professions can no longer receive direct or indirect advantages (except of negligible value) from companies providing services, producing or marketing reimbursable health products, unless this is done within the framework of agreements concluded in the field of research or hospitality offered on the occasion of scientific events, this exception being very well supervised (agreement with the real aim of research, control by the medical authorities…; article L. 4113–6 of the Public health code as it stood prior to 2011).

In addition to these elements, another essential instrument is the explicit prohibition of links of interest that would call into question the independence of the members of the various health agencies. This prohibition is quite old and concerns a fairly large number of bodies. In addition to agencies, certain health professionals are also targeted, especially among the medical professions since law n° 2002–303 of 4 March 2002 (article L. 4113–13 of the Public health code). In addition to this prohibition, there is also an obligation to declare interests.

The last mechanism is of a criminal nature. It is based on the qualification of forgery that can be applied to a declaration of interest (article 441–1 of the Penal code) and on the offence of illegal conflicts of interests (article 432–12 of the Penal code), an offence explicitly provided for in certain cases in the health field (article L. 1414–4 paragraph 2 of the Public health code concerning the National authority for health; article L. 5323–4, concerning the AFSSAPS, these two rules providing for this before the 2011 reform).

These four instruments were not deemed sufficient due to several identified dysfunctions. The first relates to the scope of application of the anti-gift law, which wasn't extensive enough since it only covered physicians, and not the other health professions. Other difficulties related to the explicit prohibition of conflicts of interest for the physicians. It took five years for the decree implementing the provisions of the law of 4 March 2002 to be issued (Decree no. 2007–454 of March 25, 2007 on agreements and links between members of certain health professions and companies and amending the Public health code), and then, although the legal machinery had finally been completed, a study showed the lack of compliance with the 4 March 2002 law and its instruments. In April 2008, the FORMINDEP association, which fights for independent training and information in health matters, studied the interventions of 150 health professionals (doctors, paramedics, pharmacists) in about thirty media and found that no declaration of interest had been made, even though about fifty of these professionals had relationships with the pharmaceutical industry that should have been declared. This investigation even led to a number of proceedings before the Departmental Councils of the Order of Physicians and the Ministry of Health. The third dysfunction was noted in a report by the National audit office (Cour des comptes 2007), which identified a number of significant shortcomings regarding compliance with the obligations to declare interests. For example, the Agency for the sanitary safety of health products (AFSSAPS, which was succeeded by the National agency for the safety of medicines and health products) noted that declarations of interest were made at a slow pace, and in some cases not at all (in 2006, eight years after the law on the obligation to declare competing interests came into force, 7% of experts still did not declare them). As to the National authority for health, the declarations of competing interest by external rapporteurs before the Transparency Commission were neither systematic nor updated. The last element showing the limits of traditional tools is the Mediator affair, in which significant suspicions of conflicts of interests were raised regarding certain experts from AFSSAPS, the structure that authorized and evaluated this drug, who had allegedly worked in the past for the Servicer laboratory, which marketed this product.

In view of these difficulties, is the law of December 29, 2011 the start of a genuine evolution?

23.2.2   Recent Tools: Break or Continuity?

The law of December 29, 2011, like many, is a mix of reinforcing already existing approaches and implementing more innovative elements. It is worth considering each of these points.

As regards to the sources of reinforcement, the experts' obligation to declare competing interest has been improved (Decree No. 2012–745 of May 9, 2012 on the public declaration of interest and transparency in public health and health security; Order of July 5, 2012 setting the standard document for the public declaration of interest mentioned in article L. 1451–1 of the Public health code; Instruction No. DAJ/2012/307 of July 30, 2012 on the implementation of the provisions relating to the public declaration of interests in the regional health agencies). It now applies to all those involved in health agencies, members of commissions, councils, ministerial cabinets and working groups under the authority of the Ministry of Health, who must make a declaration of interest when they take up their duties (article L. 1451–1 of the Public health code. V. Order of August 2, 2012 setting the list of bodies whose members are subject to the obligation of public declaration of interest). Criminal sanctions are to be applied in case of violation (offence punishable by a fine of 30,000 €, if deemed necessary, with additional penalties, articles L. 1454–2 and L. 1454–4 of the Public health code). It should be noted that, while this system already existed, it has been expanded. Whereas previously, only specific situations were covered, leading to the risk that it would not consider all involved parties, there is now a global principle. Furthermore, the information contained in these declarations is now subject to the control of an ethics committee set up within each body. Moreover, failure to declare competing interests and changes to the protocols exposes the infringer to criminal penalties. The innovation lies in the shift in the punishment's nature.

The reinforcement of the previous system achieved by the 2011 law is also expressed through the expansion of the scope of application of the anti-gift law system. The prohibition now also applies to students intending to enter the medical professions as well as to associations representing these students and these professions. It should be noted that the prohibition of illegal taking of interest is generalized and applies to a wide variety of actors: stakeholders of health agencies, members of commissions, councils, ministerial cabinets and working groups dependent on the Ministry of Health (article L. 1451–1 of the Public health code).

The new elements are diverse. One of them is quite original and consists in the compulsory hearing of the heads of the health agencies in front of the Parliament before they are appointed. In addition, there are some new features of note. This is the case of the publicity of the meetings and the recording, notably audiovisual, of the debates of the commissions, councils and collegiate bodies of health expertise which are consulted within the framework of administrative decision procedures (article L. 1451-1-1 of the Public health code for the organs mentioned in I of Article L. 1451–1 of the Public health code). It also includes the setting up of a charter for health expertise (article L. 1452–2 of the Public health code; See also DG Decision No. 2012–170 of 24 May 2012 *portant création d'un comité de déontologie à l'Agence nationale de sécurité du médicament et des produits de santé*, JORF No. 0152 of 1 July 2012, p. 10 839). In a more notable way, an obligation of transparency also weighs on pharmaceutical companies (Mascret 2012; Chimonas et al. 2017). They must make public the existence of agreements concluded with health actors, but also declare all the benefits in kind or in cash that they provide, under penalty of criminal prosecution (article L. 1453–1 I of the Public health code). The scope of application is quite broad, since it covers the agreements that these companies enter into with a variety of stakeholders: "*health professionals* [...]; *associations of health professionals; students studying for the professions covered by part four of this code, as well as the associations and groups representing them; associations of health system users; health establishments* [...]; *academies, foundations, learned societies and companies or consultancy bodies involved in the products or services sector* [...]; *legal persons publishing press, radio or television services and online public communication services; persons who, in the media or on social networks, present one or more health products in such a way as to influence the public; publishers of software to assist in prescribing and dispensing; legal persons providing or participating in the initial or continuing training or continuing professional development of health professionals* [...]". This is an important principle, even if it depends on the nature of the information that must be made public, the deadlines and procedures for publishing and updating this information and, even more so, the threshold at which the obligation to declare benefits applies. As is often the case, the devil is in the detail and it is on this that the effectiveness of tools to combat conflicts of interests in biomedical research, as in other areas of health, may depend.

Faced with the critical analysis of these different tools, the reaction may be pessimistic, suggesting that there are no effective ways to combat conflicts of interest.

This is not the case and the solution may lie in innovation. Law No. 2017–1836 of 30 December 2017 has indeed set up what is officially called "experiments for innovation in the health system" (Lemaire et al. 2020). They correspond to the possibility of experimenting with new health organisations based on new financing methods with the aim of improving patient pathways, the efficiency of the health system, access to care or even the relevance of the prescription of health products. The idea is to test new approaches by derogating from the various existing rules on regulation. The issue of conflicts of interest is not at the heart of this experimentation, but it is possible to hope that this "door opened" by the legislator will allow for innovation in order to fight more effectively against these conflicts, which are detrimental to all stakeholders in biomedical research.

References

Bloch, L. 2021. Condamnation pénale dans l'affaire du Médiator: des chiffres et des êtres (T. corr. Paris, 29 mars 2021). *Responsabilité Civile et Assurances (RCA)*, 6: 3

Byk, Ch. 1998. French law and biomedical research: a practical experiment. In *Research on human subjects: ethics, law and social policy*, ed. Weisstub, D.N, 158–174, Pergamon.

Chimonas, S., N.J. DeVito, and D.J. Rothman. 2017. "Bringing transparency to medicine: Exploring physicians' views and experiences of the sunshine" act. *Am J Bioeth* 17 (6): 4–18.

Collège des médecins du Québec. 2021. *Le médecin et la recherche clinique – Guide d'exercice*, autoédition, 2021, 86 p.

Cour des comptes. 2007. *La Sécurité sociale*, 508.

Dinisi-Peyrusse, A. 2012. La loi du 5 mars 2012 relative aux recherches impliquant la personne humaine: à la recherche d'une conciliation entre protection de la personne et développement de la recherche en vue de progrès médicaux. *Revue Juridique Personnes et Famille (RJPF)* 7: 6.

Directive 2001/20/EC of the European Parliament and of the Council of April 4 2001 on the approximation of the laws, regulations and administrative provisions of the Member States relating to the implementation of good clinical practice in the conduct of clinical trials on medicinal products for human use, OJ L 121 of 1 May 2001, p. 34.

Dondéro, B. 2012. Le traitement juridique des conflits d'intérêts: Entre droit commun et dispositifs spéciaux. *Dalloz*, 1686.

Godlee, F. 2010. Conflicts of interest and pandemic flu. *BMJ* 340: c2947.

Hermitte, M.-A., and L. Le Coz. 2014. La notion de conflit d'intérêts dans les champs de la santé et de l'environnement : Regards philosophique et juridique. *Journal International de Bioéthique* 25: 15.

IGAS/Bas-Theron, F., C. Daniel, and N. Durand. 2011. Place de l'expertise dans le dispositif de sécurité sanitaire; "Les saisines et le lancement de l'expertise sanitaire"; "Les experts et la valorisation de l'expertise sanitaire, and "L'association des parties prenantes à l'expertise sanitaire".

Laude, A. 2012. La nouvelle régulation des produits de santé—À propos de la loi du 29 Décembre 2011. *La semaine juridique édition G (JCP G)* 6: 123.

Laude, A. 2012b. La loi du 29 Décembre 2011 sur le médicament: Quels sont les enjeux ?. *Dalloz* 272.

Law n° 2004–806 of 9 August 2004 on public health policy, JORF n° 185 of August 11 2004, text n° 4.

Law n° 2012–300 of 5 march 2012 on research involving the human person, JORF n°0056 of 6 March 2012, p. 4 138.

Law n° 88–1138 of 20 December 1988 relating to the protection of persons who undergo biomedical research, JORF of 22 December 1988.

Lemaire, N., D. Polton, and A. Tajahmady. 2020. Article 51: Expérimenter pour innover mais aussi innover dans la façon d'expérimenter. *Les Tribunes de la santé* 63: 35.

Leroyer, A.-M. 2012. Note sous Loi n° 2012–300 du 5 mars 2012 relative aux recherches impliquant la personne humaine. *Revue trimestrielle de droit civil (RTD Civ.)* 2: 384.

Mascret, C. 2007. State council, 12 February 2007, Laboratoires Jolly-Jatel. *Revue De Droit Sanitaire et Social (RDSS)* 290164: 338.

Mascret, C. 2012. Analyse critique de l'obligation de publication des conventions et avantages unissant les professionnels ayant un lien avec la santé et l'industrie pharmaceutique à l'épreuve du droit communautaire. *Les Petites Affiches (LPA)* 64: 3.

Moret-Bailly, J., and M.-A. Rodwin. 2012. La qualification de conflits d'intérêts des médecins en France et aux Etats-Unis. *Revue De Droit Sanitaire et Social (RDSS)*, 501.

Moret-Bailly, J. 2004. Les conflits d'intérêts des experts consultés par l'administration dans le domaine sanitaire. *Revue de droit sanitaire et social (RDSS)* 855.

Moret-Bailly, J. 2011. Définir les conflits d'intérêts. Dalloz, 1100.

Peigné, J. 2012. Du Médiator aux prothèses PIP en passant par la loi du 29 décembre 2011 relative à la sécurité sanitaire des produits de santé. *Revue de droit sanitaire et social (RDSS)*, 315.

Poisson, D. 2002. Déclaration d'Helsinki, quelles nouveautés ? *Laennec* (Tome 50), 44.

Prévost, J.B. 2021. Du pharmakon au médicament: Pour une critique de la déraison pharmaceutique. *La Gazette Du Palais* 2021 (3): 79.

Sauvé, J.M. 2011. *Commission de réflexion pour la prévention des conflits d'intérêts dans la vie publique, Pour une nouvelle déontologie de la vie publique*, 121.

Thouvenin, D. 2015. Les conflits d'intérêts du médecin hospitalier public investigateur en France. In *Les conflits d'intérêts à l'hôpital public: Le débat*, eds. J. Moret-Bailly and D. Thouvenin, 204. Presses de l'EHESP.

Weisstub, D.N. 1998a. Roles un clinical and research ethics. In *Research on human subjects: Ethics, law and social policy*, ed. D.N. Weisstub, 56–72. Pergamon.

Weisstub, D.N., ed. 1998b. The ethical parameters of experimentation. In *Research on human subjects: Ethics, law and social policy*, 1–34, Pergamon.

World Health Organization. 2011. *Report of the review committee on the operation of the international health regulations (2005) with regard to pandemic influenza A (H1N1) 2009*, 200.

Guillaume Rousset is Senior lecturer at the University Jean Moulin Lyon 3 (CRDMS/IFROSS) and Secretary General of the French Association of Health Law (AFDS). Dr. Rousset was Vice-President of International Relations at the University Jean Moulin from 2017 to 2020. Specializing in the field of health law and medical consumerism, he is an expert on patients' rights and the regulation of the pharmaceutical industry. Active as a member of ethics committees in the areas of insurance and biomedical research, he is also a member of editorial boards of several French health law journals. Guillaume Rousset is a Chevalier in the order of Palmes Académiques.

Part X
Finances, Technology, and Public Policy

Chapter 24
Structural Problems in the Practice of Psychiatric Research

Heather Stuart

24.1 Introduction

Medical research has come under fire for its heavy reliance on funding from the pharmaceutical industry. Critics charge that the medical profession's culture and its public health mission have been undermined by commercial interests. Psychiatry is particularly vulnerable to commercial influences owing to a confluence of several historical factors. First, the nature of the research enterprise in all disciplines has changed dramatically over the past few decades, placing increasing emphasis on the acquisition of funds, at almost any cost. Secondly, categories of mental disorders have accumulated over time not necessarily because they reflect clear operational criteria pertaining to underlying disease states, but because of market pressures. Critics of the DSM system fear that decisions about what disorders to include and what revisions to make have been made largely on the basis of political expediency, supported by conflicted expert opinions that are debated behind closed doors under the veil of confidentiality agreements. As the majority of clinical experts on DSM panels are known to have ties to large pharmaceutical companies, there is concern that the system has been open to bias and manipulation as a result of these potential conflicts of interest. Finally, unlike other branches of medicine, psychiatry has fallen victim to system-level inequities that have left large gaps in funding for programs, professional development, and research, which funding from drug companies has largely stop-gapped. This chapter will examine these system-level conditions in more detail and consider the implications for the field.

H. Stuart (✉)
Queen's University, Kingston, Canada
e-mail: heather.stuart@queensu.ca

© The Author(s), under exclusive license to Springer Nature Switzerland AG 2023   455
T. Zima and D. N. Weisstub (eds.), *Medical Research Ethics: Challenges in the 21st Century*, Philosophy and Medicine 132,
https://doi.org/10.1007/978-3-031-12692-5_24

24.2   Historical Antecedents

While no part of medicine has been spared the influence of large pharmaceutical companies, psychiatry has shown itself to be particularly open to corruption because of its greater dependence on pharmaceutical funding to meet its growing research mandate. One important historical driver has been the changing ecology of the research environment, which places greater importance on the acquisition of funding, both at the individual and institutional levels. A second important driver that is unique to psychiatry has been the quasi-scientific and, some would argue entirely corrupt process used to codify psychiatric disorders. These processes have resulted in an improper dependence on the pharmaceutical industry with policies, financial incentives, and practice patterns that have compromised scientific integrity and incentivised the profession to act as an extension of pharmaceutical marketing at the expense of its public health mandate (Cosgrove and Wheeler 2013).

24.3   The Changing Ecology of Research Funding

The nature of the research environment has changed substantially in recent decades and this has important implications for scientific integrity across the board. Polster (2007) has analyzed the growing importance of research grants to Canadian universities and academics and explores how this has transformed the research landscape. The pressure to acquire research funding has largely outstripped the pressure to publish. Federal government policy with respect to research funding has also moved away from curiosity-based research to targeted initiatives, and has reduced the amount of funding available to support the research enterprise. As university operating budgets have dwindled, research grants have become an important means of offsetting these shortfalls. University stature is now based largely on funding from the national research councils and increasingly viewed as a primary measure of excellence. Funding levels are used by the government to assign additional specialized funding and grants are important for academics to move through the ranks. The amount of money obtained through grants increasingly determines an academic's standing within the university and the university's standing amongst its peers.

At the same time, it has become increasingly difficult to obtain research funds from national granting agencies. The growing pressure to get grants has meant that universities and academics have become responsive to the needs of the private sector and other groups that can co-sponsor research or otherwise financially support the university. In this competitive environment, industry-sponsored clinical trials are an easy and generous source of research funding—one that does not require academics to write a competitive grant proposal, conduct analyses, or write up manuscripts and reports. In their survey of academic departments in Canada and the US, Balon and Singh (2001) have noted that most academic departments of psychiatry do not have dedicated researchers and most psychiatric trainees are not provided with the fundamental research skills that would allow them to conceptualize a problem, design a study, analyze data, or write up a report. Indeed, the National Institute of Mental Health noted a decline in the number of psychiatrist-researchers disproportionate to any other branch of medicine (Hamoda et al. 2011).

Mental illness related stigma has also worked against the mental health research enterprise because it has made it difficult to attract psychiatrists and psychiatric researchers (Yager et al. 2004). In the UK, for example, only about 4% of medical students choose psychiatry as a career, leading to headlines such as "Psychiatry in crisis". Stigma remains a major issue as the professional status of psychiatrists is poor relative to other medical specialities (Stuart et al. 2015). Further, a poor public image, low status, and low morale have contributed to doctors leaving psychiatric practice after choosing it as a career (Brown et al. 2009). Within this small and diminishing pool of psychiatric professionals, clinical researchers represent a particularly "endangered species" (Kupfer et al. 2002). Structural or system-level stigma that has created large inequities in funding for psychiatric research is a major problem. In Australia, for example, the proportion of public funding for mental health related research is lower than for medical conditions such as cardiovascular disease or cancer. When adjustments are made for the burden of disease (measured as disability adjusted life years), the areas of diabetes, asthma, cancer, arthritis, and cardiovascular disease all receive proportionally more research dollars per attributable disability-adjusted life-year. Between 2009 and 2010 there was a 2.8% decrease in funding for mental health research with the result that there has been no "narrowing of the gap" in the proportion of funding provided to this field (Christensen et al. 2011).

Pharmaceutical companies sponsor significant amounts of medical research and psychiatry has become increasingly dependent on this source of research funding. In addition to clinical trials, pharmaceutical companies also support meta-analyses, reviews, epidemiology, laboratory science, and health economics. They have long recognized the importance of reporting their research in scientific journals, but in the last three decades have increased their efforts to use research as a resource that must be carefully developed and deployed in order to influence the opinions of researchers and practitioners. Now, industry sponsored research, publications, presentations, and continuing medical education is governed by a carefully thought out publication plan designed to maximize commercial value. As Sismondo (2009) describes, the ideal publication plan starts even before the research so that it can contribute to the research design, including creating a new market or understanding of unmet need, mapping out of key messages, identifying different papers for different audiences and journals, and targeting potential authors for these papers.

Because the basis for deciding which knowledge is worth pursuing is increasingly defined by criteria related to corporate demand rather than academic curiosity, critics suggest that academic researchers have largely given up their research independence and academic integrity (Brownlee 2020). Drug companies tend to focus on those drugs for which there is a lucrative market, or where a market can be created, rather than on the drugs for which there is the greatest need. Illnesses that persist in target groups that have limited resources (such as less-developed countries or people with orphan conditions) are neglected (Duval 2004; Lexchin 2012). Brownlee reports that less than 3% of research funding in the top 54 Canadian and US research universities is directed to diseases that affect the world's poorest people. More than a billion people currently suffer from 'neglected diseases' or diseases that are not researched by private sector companies because most of those affected are too poor to provide a market for new drugs (Brownlee 2020).

24.4   Classification Drift and Expansion

Many of the thorniest ethical dilemmas for the practice of psychiatry and psychiatric research have arisen as a result of the way in which disorders are classified. Unlike other branches of medicine that classify *diseases*, psychiatric classification systems do not. Rather than diseases, they reflect groupings of clinical manifestations (or syndromes). Mostly, they do not reflect a single, mutually exclusive underlying biological reality that is known to be externally valid. While they provide a useful framework for clinicians in organizing and explaining the clinical experience and give guidance for treatment decisions, once identified, there is a tendency to view disorder categories as quasi-disease entities. Over time, fragmentation of psychopathology into an ever-increasing set of symptom clusters (termed "disorders" in DSM), has led to the proliferation of spurious comorbidities, further challenging the validity of the system (Jablensky and Kendell 2002).

In the DSM system, mental disorders have accumulated greatly over time, now numbering more than 300. Decisions about what to include in the DSM system are made on the basis of expert consensus, behind closed doors, without transparency, objective tests, or validity testing of new categories. Widespread conflicts of interest between subject matter experts and pharmaceutical companies remains a problem for DSM development, the creation of treatment guidelines, and clinical research in general. Overdiagnosis, and the recent expansion of childhood disorders, for example, may be as much a result of drug companies pushing into new markets as new developments in scientific understanding. Of greater concern is the fact that the American Psychiatric Association, the organization that produces the DSM, receives substantial funding from the drug industry so is potentially open to bias and manipulation (Cosgrove and Vaswani 2019; Cosgrove 2010).

The DMS process has been sharply criticized for enlarging diagnostic categories in a way that has facilitated pharmaceutical interests. According to Caplan (Caplan 2014) the largest increase in the number of mental disorders listed in the DSM system occurred with the release of DSM-IV, where the number of disorder categories rose from 297 to 374, resulting in many more millions of people meeting the criteria for a diagnosis (see Fig. 24.1). Frances and Widiger (2012), co-chairs of the DSM-IV task force, have reflected on the impact of DSM revisions on public health and identified a number of fundamental problems and issues centred on the "elusive definition" of mental disorders. They note that conditions have become mental disorders largely through historical and political processes (including clinical expediency), not because disorder categories have met some objective, scientific, or biologically based set of criteria; growing from six, in the initial census of mental patients in the mid-nineteenth century to over 300 today. Against a backdrop of overdiagnosis, they describe normality as an endangered species and identify a number of diagnostic epidemics that have arisen as a result of revisions. Among the potential drivers of psychiatric epidemics, they include the success of the pharmaceutical industry in expanding the market for existing drugs into new populations, notably children. They describe having made what they thought were "small, cosmetic" changes to Attention Deficit Disorder (ADD) diagnostic criteria which were expected to increase the rate of diagnosis by 15%. However, the rates doubled owing to Federal Drug Administration approval for new ADD drugs and direct-to-consumer advertising campaigns. Aggressive drug marketing of the ADD diagnosis targeted not only clinicians, but parents, teachers, and therapists.

Chappell and colleagues (2016) note that the reach of the pharmaceutical industry is so strong that it is able to define diseases, shape guidelines for drug therapy, influence physicians regarding what constitutes best prescribing habits, and affect patient care. According to Gopal and colleagues, 68% of the DSM-V task force members had ties to the pharmaceutical industry, representing a 20% increase over that of DSM-IV (Gopal et al. 2010). It is interesting to note that the year after DSM-IV was released, Johnson and Johnson were alleged to have paid close to half a million dollars (US) to the chair of the DSM task force and two colleagues to create treatment guidelines to promote one of their anti-psychotic drugs as a first line treatment. A year later an additional half million dollars were paid to continue and expand the marketing campaign (Chappell et al. 2016).

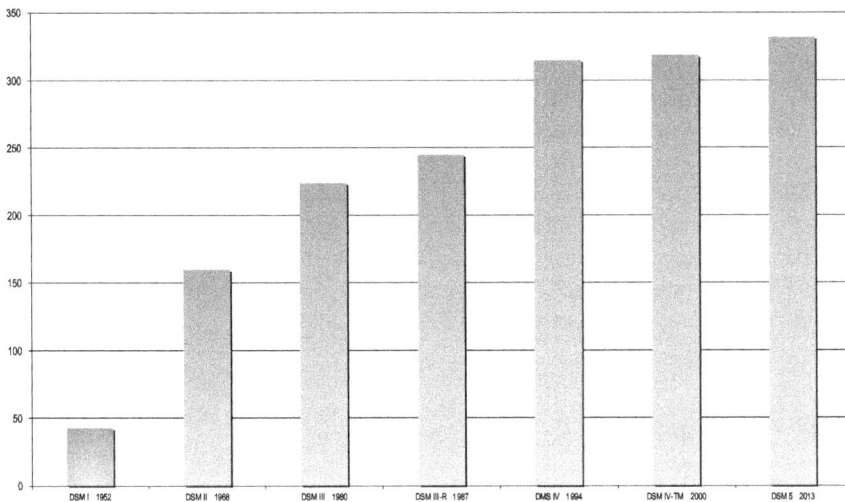

Fig. 24.1 Estimated growth in the number of DMS disorder categories 1952 to 2013; used with permission from Shu-Ping Chen and Heather Stuart, 2020

Gopal and colleagues (2010) have further criticized DSM-IV and V for creating new disorder categories that can be addressed with new drugs and a lack of transparency regarding the serious side effects of these drugs. They give the example of the proposed inclusion of Attenuated Psychosis Syndrome (which was subsequently included in DSM-V as a disorder needing more research). It describes symptoms that are thought to appear in individuals at risk for schizophrenia. The rationale for treatment of these symptoms was that it would prevent at-risk individuals from developing full blown schizophrenia. However, they point out that various studies demonstrated that the majority of people with these 'soft' symptoms (70–85%), never went on to develop schizophrenia. Neither has treatment been demonstrated to prevent its onset. They also point to the various adverse side effects including movement disorders, weight gain, and diabetes. For these reasons, researchers have concluded that the risk/benefit ratio does not support treating those at risk with anti-psychotic medications. Despite the lack of evidence, sponsored research publications indicate that Attenuated Psychosis Syndrome is a valid diagnostic category and advocate the use of antipsychotic medications as the main treatment intervention. This would increase the likelihood that more children and adolescents would be prescribed anti-psychotic medications with potentially damaging short- and long-term health effects. DSM's silence on the issue of metabolic side effects, and the push to diagnose and medicate an ever-larger number of patients, reflects a serious lack of rigour and balance. Benefits were highlighted and negative side effects were minimized. Not only are adolescents especially vulnerable to cumulative metabolic risks, but also to the humiliation and stigmatization that may accompany premature diagnostic labelling, which was never considered.

In discussing the influence of drug companies on scientific integrity in psychiatry, Cosgrove and Vaswani (2019) draw a distinction between "bad apples" and the "bad barrel". A bad apple is an individual who is consciously and explicitly corrupt for reasons such as greed or weakness. The bad barrel refers to system-level corruption that occurs when institutions fail, even if the individuals working within them believe they are acting in ways that are congruent with the organization's mission. Institutional corruption is the result of implicit bias, problematic incentive structures, and a lack of procedural or organizational integrity. The result has been that the evidence base in psychiatry has been distorted and has become a driver for non-rational prescription practices, poor quality of care, the medicalization of normal human experiences, overdiagnosis, and overtreatment.

24.5   Controlling the Scientific Narrative

Historically, the role of Big Pharma in scientific research in the field of psychiatry has been important, spending much more than traditional granting agencies (Rafols et al. 2014). While the value of sponsored research is difficult to know, it has been estimated that between 1980 and 2003, research and development expenditures for US pharmaceutical companies increased from $2 billion to $33 billion. Further it has been estimated that more than 70% of the funding for clinical trials comes from industry rather than government and private foundations (Duval 2004).

Because they are the largest funders, pharmaceutical companies have the power to control the scientific narrative. They use scientific publications as a means of marketing. Clinical trials sponsored by the pharmaceutical industry are overwhelmingly favourable to the sponsor's product, partly because unfavourable results are not published, and partly because the best 'spin' is put onto the data by company-sponsored ghostwriters (Baker et al. 2003; Smith 2005). A large favourable clinical trial published in a leading journal is worth millions of dollars of advertising. Editors also know that companies will often purchase thousands of dollars' worth of reprints, where the profit margin may be as high as 70% (Smith 2005).

Fanelli (2012) has noted the significant drop in negative findings over time (between 1990 and 2007) with an accompanying increase in positive results of approximately 6% per year. In 2007, the proportion of papers that reported positive results was 86%, up from 70% in 1990–91. Of interest is that papers published in psychology and psychiatry journals had the second highest odds of having positive results (Odds ratio = 2.992, 95% Confidence interval = 1.9–4.8), the highest odds ratio across all papers being 3.2 (95% CI = 1.6–6.2).

Hughes, Cohen, and Jaggi (2014) compared serious adverse events reported in industry-funded drug trial summaries for antipsychotic and antidepressants that were posted on an industry-sponsored online trial registry with the same trials reported in the scientific literature. Of the 244 trial summaries identified from the registry, just over half (58%) had a corresponding journal publication. Almost half (43%) of the serious adverse events described in the registry summaries were not reported in the journal articles, especially in the case of death (62% not reported) and suicide (53% not reported). The trial registry that was used in this research was subsequently dismantled and removed from the internet. Complete information, including serious adverse events, for the bulk of these studies (92%) never appeared in other trial registries despite a clear legislative mandate from the regulatory agency.

Finally, large pharmaceutical companies control the narrative by controlling the independence of researchers. Whistleblowers are discouraged by confidentiality clauses, and if they stand up, they may experience systematic harassment and character assignation. In 1993, for example, Dr. Nancy Olivieri signed a confidentiality contract with Apotex giving them the right to control communications of trial data for one year after termination of the trial. The trial was testing an experimental iron-chelation drug used to treat thalassaemia. Olivieri identified several medical risks and, on the advice of the Chair of the University of Toronto Ethics committee, she notified her trial patients and regulatory authorities. What happened next has been described as a "horror story" of corruption, including false allegations of scientific misconduct and covering up of significant conflicts of interest. At one point, the President of the University of Toronto appealed to the Prime Minister of Canada asking the government not to change drug-patent regulations that would be damaging to Apotex because it could jeopardize the University's planned new biomedical research centre. Apotex would have made the largest single donation that the university had ever received. In the end, after much tribulation, Olivieri was entirely exonerated (Somerville 2002).

In a second prominent case, the University of Toronto and the Centre for Addiction and Mental Health (CAMH) successfully recruited an eminent psychiatrist, Dr. David Healy, from the United Kingdom, only to abruptly terminate him after he mentioned in a lecture that Prozac, manufactured by Eli Lilly, might be responsible for increasing risk of suicide in certain types of patients. Within a week, he received an email rescinding the offer of employment. Prior to Healy's presentation, Eli Lilly had donated 1.5 million dollars to CAMH and a new wing of the hospital that was built with funding from Lilly was supposed to open soon after. Like the Olivieri case, this incident raised legitimate concerns about conflicts of interest and system-wide corruption. Schafer (2004) notes that the Olivieri and Healy scandals shared many characteristics including wealthy and powerful drug companies hovering in the background providing significant donations to universities and research centres. Both Olivieri and Healy experienced negative consequences, including dismissal from their positions, because they spoke publicly about potential dangers of drugs. Damaging rumours (amounting to a smear campaign) were circulated about each. Olivieri was alleged to be scientifically incompetent, guilty of stealing money from research grants, unethical in her patient care, and sleeping with scientists who looked favourably on her research findings. Healy was alleged to be a bad clinician, a racist, and a member of Scientology (an infamous anti-psychiatry cult). Each appealed to senior administrators, the University of Toronto, and the Faculty of Medicine for assistance but to no avail. In some cases, senior administrators and hospital staff were among those circulating the negative rumours.

A similar example from the US raises concern that such cases are not isolated events, but indicative of broader, system-wide problems. In the mid 1990's a prac-ticing psychologist in Virginia conducted a series of epidemiological studies that documented high rates of ADHD diagnoses and drug treatment (Watson et al. 2014). She found that 8–10% of children in southeastern Virginia, including 17% of white boys, were being medicated in school for ADHD, and in one district, 63% of children who were young for their grade were medicated. Students who reported ADHD were 3 to 7 times more likely to experience adverse educational outcomes, and outcomes were significantly worse for students who had been medicated compared to those that had not. The team mounted a series of community-oriented interventions which corresponded with a 32% decrease in the rate of ADHD diagnosis. This touched off a series of largely anonymous and aggressive attacks against her research charging scientific misconduct. One of her strongest critics was closely involved with the pharmaceutical industry as a sponsored opinion leader with a sizeable portion of his income (perhaps as high as a third) coming from them. Under growing pressure from outside sources and the media, the Dean of her medical school terminated her research, placed her on administrative leave, and promised that the study results would never be used again. Subsequently, in 2010, it was estimated that nearly 20% of American boys aged 15–17 years were diagnosed with ADHD making it the most prevalent diagnosis among children 3–17 years of ages. Between 2005 and 2007, the state of Florida witnessed a 250% increase in the prescription of anti-psychotic drugs for children, many for children with the diagnosis of ADHD.

Owing to DSM-V revisions, a growing number of children are now being prescribed an ever-widening formulary of powerful psychotropic drugs; many of whom are now sharing, swapping, stealing, and abusing these medications. According to Watson et al. (2014) these were not isolated incidents. There are many examples of scientists who have been harassed, bullied, or had their jobs threatened because their research on psychosocial interventions raised questions about the effectiveness of drug treatments. In this case, interventions that raise questions about the effectiveness of ADHD treatments were terminated and study findings were suppressed.

24.6 Ghosts in the Machine

Ghostwriting occurs when someone who has made a substantial contribution to writing a manuscript is not identified as an author. It often occurs alongside guest authorship, when someone (usually an identified opinion leader) who has not contributed to the manuscript in a meaningful way, is identified as an author (Gøtzsche et al. 2009). Although it is difficult to quantify precisely, because it is largely covert, it has been estimated that 11% to 25% of journal publications have honorary authors and background ghostwriters, though this is likely a serious underestimate (Egilman and Druar 2011).

There have been increasing concerns about pharmaceutical ghostwriting because of their selective reporting of clinical trial outcomes to fit company marketing schemes. Critics allege that the purpose of ghostwriting is to conceal conflicts of interest and promote misinterpretation of the research literature (Moffatt and Elliott 2007; McHenry and Jureidini 2008). It has been estimated that discrepancies between the primary outcome listed in the clinical trial documentation and those listed in the published paper occurred in 62% of 112 trials, in 50% of efficacy outcomes and 35% of harm outcomes reported. Selective reporting has been particularly problematic in antidepressant research (Jureidini and McHenry 2011).

A case in point is Study 329 commissioned by SmithKline Beecham examining the safety of using antidepressants for adolescents. The majority of the 22 authors listed did not make any substantial contributions to the article and were listed as a courtesy. Misleading findings were subsequently published (and widely cited) showing positive effects of antidepressant therapy, when in reality, there were a number of significant adverse consequences including suicidal behaviours. Disguised authorship and manipulation of results to achieve marketing ends undermines scientific integrity and carries serious consequences for people receiving these medications (McHenry and Jureidini 2008). An independent reanalysis of the Study 329 data (Noury 2015) identified a number of problems with the methods and analysis that could have been used to deliberately bias results. For example, there were four outcome variables in the published paper that were not specified in the original protocol and no official amendments to the protocol were made to accommodate these additional measures.

These were the only measures that were reported and they were all statistically significant. Adverse events that were recorded in the participant's case reports were not transcribed into the study data that were used for the analysis, thereby seriously underestimating the number of adverse events occurring in the trial. In addition, adverse events were coded idiosyncratically; one time in one category, another time in a different category, thereby masking their frequency. The original study reported only on events that occurred in 5% or more of the sample. The idiosyncratic coding allowed the adverse events to be sprinkled across categories so that the 5% threshold was not met. In the reanalysis, all adverse events were reported. The conclusion from the reanalysis was that there was no advantage of paroxetine or imipramine over placebo in adolescents with symptoms of depression on any of the prespecified outcome variables. In addition, safety concerns were raised by the increases in adverse events, including suicide related events, when the data were systematically coded and all adverse events were identified from patient-level data.

24.7   Off-Label Prescribing

Finally, there is the problem of off-label prescribing (Rodwin 2013a). Off-label prescribing occurs when medications are prescribed for conditions that have not been approved by regulating bodies, for which there are no data to support their safety or efficacy, and it is common in the case of psychiatric medications. Off-label prescribing can place patients at risk of harm without adequate knowledge. Clinical trial protocols typically exclude children, pregnant women, elderly, and patients who are more vulnerable to drug reactions. Therefore, clinical trials cannot provide information about the safety and efficacy of a drug across the board; only in the highly circumscribed study populations. Drug manufacturers promote off-label prescribing because it increases revenues. Off-label prescribing is promoted by constructing clinical trials to generate publications, paying physicians to be consultants, paying providers to prescribe drugs, distorting the presentation of risks and benefits, influencing continuing medication education programs, ghostwriting articles, recruiting clinical investigators, and counselling clinical investigators how to obtain reimbursements for off-label prescriptions.

24.8   Discussion

This chapter has examined select structural and institutional loop-holes and market incentives that allow pharmaceutical companies to maximize profits at the expense of public health, create institutional corruption, and undermine the scientific integrity of psychiatric (and other medical) research. Practices include illegal off-label promotion of prescription drugs, financial incentives and kick-backs to physicians to alter their prescribing behaviours, financial incentives to academic centres, the creation and dissemination of biased research information to health professionals through ghostwriting, suppression of negative results, and widespread conflicts of interest.

The wide-spread systemic nature of these problems suggests that solutions will have to create unprecedented system-wide changes across multiple sectors. Toward this end, a number of structural solutions have been offered. For example, Gagnon (2013) has identified a three pronged approach that could reduce, but not eliminate, the problems associated with "market-based medicine". First, he argues that fines and criminal penalties for illegal conduct are insufficient to act as a deterrent. Though fines against drug companies have been mounting in recent years, they are still minor in comparison to the billions of dollars of profits that are made by breaking the rules. Increased fines coupled with financial penalties and criminal prosecution of corporate officers, directors, and managers may reduce financial incentives for harmful practices. Secondly, he argues that taxes on promotion activities could be used to create a more rational use of medicines. In Italy, for example, companies must contribute 5% of their yearly expenditures for promotional activities to a central fund, which is then used to sponsor independent research and support the development of orphan drugs needed to treat rare diseases. Finally, he argues that pricing should be based on added therapeutic value, which would then make evidence-based medicine central to market incentives. A difficulty with this is determining the added therapeutic value of drugs when much of the published literature has been biased by drug companies' ghost writers. Even if implemented, such reforms likely would have limited impact without greater transparency in drug-company-sponsored clinical trials and reduced publication bias.

Turner (2013) has noted that the development of clinical trial registries where trial plans are made public has been a significant step in the attempt to address publication bias. However, only about half of the clinical trials registered with one registry (ClinicalTrials.gov) were published so this still leaves considerable room for bias. In areas where trial results are also required to be posted, such as in the US, the majority are not posted within the required time frame and many still remain unpublished in medical journals where they cannot be viewed by prescribing physicians. He argues that journals can do more by making a concerted effort to publish the results of negative trials and where articles are reviewed without knowledge of their results (as positive results are typically favoured), based only on the rationale and methods. It is interesting to note that one study that revisited meta-analyses of new molecular compounds including unpublished data from the Food and Drug Regulation found that 46% of the recalculated meta-analyses showed decreased efficacy, but 46% showed an increase in efficacy (Hart et al. 2012). The psychiatric drugs included in this review showed the most consistent changes. Four of the five outcomes changed to show decreased efficacy when unpublished FDA trial data were included.

Rothman and colleagues (2009) have produced an extensive series of recommendations for professional medical associations to control conflict of interest situations. They note that professional medical associations have an important role to provide evidence-based information and treatment recommendations, and to uphold scientific integrity. Anything that undermines or appears to undermine the promise of scientific integrity or unqualified commitment to patient well-being, must be avoided. Thus, their recommendations are stringent. For example, they recommend that professional medical associations work toward the goal of zero contributions from industry so that they do not collaborate or profit from industry marketing. All leaders of professional medical associations and executive staff should be completely free of conflicts of interest and, over time, this should apply to board members and members of practice guideline committees. They suggest that industry funding used to support professional medical association activities (such as student support, research, conference participation, conferences, and continuing medical education) should be collected in a separate fund and dispersed by a committee that is free from any conflict of interest. Individual researchers or students who receive funding should never know the name of the company providing the funding to eliminate any perceived pressure. Finally, they noted that such changes would require great sacrifice, but were considered to be the only way to meet mounting criticism and calls for fundamental reforms from critics and government leaders who are certain that past practices have undermined scientific integrity and patients' best interests.

Smith (2005) is not optimistic that any solutions will be entirely effective so long as journals publish the results of industry sponsored clinical trials without the knowledge of what other unpublished studies exist. Many journals favour randomized trials as offering the most compelling evidence, but they also provide an important source of income for many journals as companies will purchase large quantities for distribution. To minimize editorial conflicts of interest, Smith calls on editors to insist that all trials are registered, that the role of pharmaceutical companies be made explicit, and refuse publication if trials are not under the complete control of the researchers. A more radical step would be for journals to stop publishing trials altogether (where results are made available on regulated web sites). Finally, he argues for greater public funding for trial research.

Also focussing on publication bias, Van Lent and colleagues (2013) have argued for standardized reporting of clinical trials, particularly with respect to primary outcomes and sponsorship, so that it is possible to improve transparency and better assess publication bias. In the 472 manuscripts of randomized trial results that they reviewed, they noted problems related to the classification of outcomes as a result of poorly described methods and vaguely described endpoints. They also determined that it was difficult to tell whether trials were investigator initiated or initiated by pharmaceutical companies. They offer an extensive list of recommendations pertaining to classification of positive and negative outcomes in trials, as well as definitions for types of sponsorship ranging from pharmaceutical sponsored trials to trials in which the industry is not in any way involved.

Similarly, Gøtzsche and colleagues (2009) argue for the use of a checklist that could be imposed by journal editors in the 'instructions to authors' and used by all authors who have used medical writers. This would include prompts to acknowledge professional medical writers, their funding sources, and to confirm that they (the authors) controlled the study outcomes and the data reported in the manuscripts. This approach favours full disclosure, rather than an outright ban. They considered that an outright ban could have unanticipated consequences by increasing non-publication and reducing the quality of manuscripts (as ghost written manuscripts tend to be of high quality).

Finally, Rodwin (2013b) has considered how to manage the problem of off-use drug prescriptions, but recognizes that the measures suggested likely would be insufficient to entirely curb this problem. Nevertheless, he suggests that off-label prescriptions should be tracked, by physician, by use, and by frequency of use. Not only would such data help set priorities for the evaluation of risks and benefits of off-label use, it would help target education to physicians and, when appropriate, warn the public. Secondly, he suggests that we find ways of reducing the profits gained by pharmaceutical companies with off-label marketing by establishing different reimbursement strategies for on and off-label use. Finally, he suggests that when use of an off-label drug passes a certain threshold, as measured by the registry, it should be systematically evaluated for efficacy and safety. Funding for these evaluations could come from a cut from off-label profits which would be mandated to be contributed to a central government-supervised fund.

Against this backdrop a number of medical schools have eliminated industry support for continuing medical education and an increasing number of medical journals are banning pharmaceutical and medical device advertising. Many also have strict policies against ghostwriting. Professional medical associations are increasing aware of the corrosive influence of pharmaceutical funding and are taking steps to protect the public's trust. While not a complete solution—significant market forces are still at play—they point to the growing recognition that academic medicine and pharmaceutical funding should not mix, and the increasing divestiture of pharmaceutical sponsorship (Brownlee 2020). While most critics would argue that this has not been enough, it is a promising start.

24.9 So, Where Do We Go from Here?

In light of what has been described as wholesale corruption, critics have either called for massive restructuring of academic, educational, and health care systems, which may be impractical or even impossible to implement, or band-aid solutions such as stiffer penalties and higher fines, recognizing that these will have limited overall impact. A common response has been to increase transparency, such as requiring journal authors to disclose potential conflicts of interest.

A more pragmatic approach may be to work toward limiting the amount of influence pharmaceutical companies have on medical opinions—specifically, their perceptions of illnesses and their perceptions of the efficacy of drug treatments to manage these illnesses (Sismondo 2013). DSM disorders have had historically poor reliability and little or no attention to validity, leaving them open to manipulation. The monumental expansion of DMS disorder categories and corresponding reduction in 'normalcy' has led to one at least one tongue-in-cheek proposal to classify happiness as a mental disorder. Bentall (1992) highlights the absurdity of the DSM criteria for expanding disorder categories by showing that happiness meets all of these criteria: it is statistically abnormal; it consists of a discrete cluster of symptoms; there is some evidence that it reflects the abnormal functioning of the central nervous system; and it is associated with various cognitive abnormalities, including at times, a lack of contact with reality. On this basis he proposes happiness be included in the DSM under the formal description of *major affective disorder, pleasant type.* The constant expansion of the DSM categories to include what some would consider normal emotional states, and the massive conflicts of interest that have been uncovered, have discredited the entire process. At the same time, they have shaped (and expanded) clinicians' perceptions of what types of affective, cognitive, and behavioural states may respond to pharmacological treatments.

Careful scientific validation of psychiatric diagnoses would do much to establish psychiatric disorders as 'real', rather than politically contrived and market driven entities. A number of authors have recommended that more attention be given to validation studies and a number of processes have been identified to conduct validation research, including symptomatologic, biochemical, genetic, prognostic, and neurobiological factors. Current (and past) statistical techniques are more than able to identify discrete clinical groupings while allowing for fuzzy (i.e. overlapping) boundaries. Therefore, the problem is not that there are no methods to conduct validation research. The problem is that these studies have not been conducted (Jablensky and Kendell 2002). Building a stronger scientific foundation for DSM categories unfettered from pharmaceutical funding would do much to alleviate current criticisms and pathways to corruption in the DSM apparatus. Validated disorder categories could alter (and hopefully narrow) the range of illness states that would be viewed by clinicians as amenable to drug interventions. Independent drug research may also highlight disorders that do not respond well to pharmacological agents.

One solution to managing the misperceptions propagated by drug research has been the emergence of online trial registries which are designed to make trial results publicly available. In some countries, such as the US, legislation requires companies to post trial results. While registries have been found to contain more information than scientific papers, particularly with respect to adverse events (Hughes et al. 2014), their use has been undermined by poor compliance. The majority of trials (80%) are not posted within the required time frame (Turner 2013), presumably prior to publication of misleading results in the scientific literature. While rules governing the registration of clinical trials have become more stringent (and should continue in this direction), negative results may still be hidden from public and professional view.

As more academic journals move to online formats, it is now possible to attach considerable supplemental information to published articles. In the case of drug research, publication could be made contingent on the availability of detailed supplemental information. For example, some journals, such as *Science*, now require that raw data be made available to readers. More detailed supplemental information could also be a requirement of publication. This might include the original proposal clearly outlining the study's primary and secondary endpoints (often not clearly specified), the raw data, detailed data dictionaries describing variable coding formation, a listing of all adverse outcomes (with a clear definition of what constitutes an adverse outcome), and the summary of the results filed with the relevant regulatory agency (such as the Food and Drug Administration). Also, though not all journal editors agree (citing concerns with methodological rigor of unpublished work), journals could refuse publication of systematic reviews of clinical trials unless unpublished data (available through regulatory agencies) were included (Hart et al. 2012). Alternatively, journals could refuse publication of all industry sponsored research thereby breaking the crosswalk between marketing and science. Research could be made available on pharmaceutical company websites and mandated trial registries, where it would be clear that results were part of a marketing enterprise. Finally, university-based ethics committees, which typically require investigators to report adverse events that occur during a research project, could form networks to monitor and report adverse events occurring in the context of drug sponsored clinical trials.

With respect to off-label prescribing, Rodwin has argued that more systematic monitoring of prescriptions would allow for the tracking of unsafe off-label drug use (Rodwin 2013a). As many physician offices and pharmacies move to online platforms and electronic records, linking data across databases is increasingly manageable and could be relatively straightforward if unique anonymous personal identifying information were part of standard reporting. However, implementation of population-based monitoring on this scale, though technically possible, may be prohibitively expensive. Targeted surveillance systems set up in large population centres could be a lower cost, pragmatic alternative.

Sentinel surveillance systems located in hospitals, family physician practices and laboratories, have been widely used in public health in both high- and low-income countries to monitor disease outbreaks. (see Jernigan et al. (2001) for an example). Thus far, sentinel surveillance systems have not been set up to monitor off-label prescription use of psychotropic medications, or post-marketing surveillance of adverse outcomes, but they could be developed with these goals in mind. Indeed, there is considerable expertise in the public health field concerning the development and management of these systems and a strong evidence base to support their use.

One example of a primary care sentinel system that could lend itself to monitoring off-label prescriptions of psychiatric medications is the Canadian Primary Care Sentinel Surveillance Network (see cpcssn.ca). Established in 2008, it is Canada's first multi-disease electronic medical record surveillance system. Data comes from physicians participating in 10 practice-based research networks across Canada. Data are extracted from electronic records on a quarterly basis, then cleaned and coded using diagnostic and other algorithms. There are currently almost 5 million records from across the country including information on such variables as providers, health conditions, encounter diagnoses, medications, physical signs, and lab results, to name a few. Such networks could be regionally specific or located in large cities, and expanded to include other key health provider sites such as hospitals and pharmacies. Large health systems, such as those operated by the US Department of Defence or large employee and family health care providers may also lend themselves to electronic monitoring of off label prescription use.

Most would agree that it is not possible for one solution to reduce publication bias, prevent pharmaceutical companies from using scientific findings as a marketing tool, or curb off-label prescriptions. And, it is equally unlikely that drug companies will take the lead in developing monitoring systems that will certainly cut into their profits. The main impetus for change must come from the academic and medical communities to make a concerted effort to disentangle drug marketing from science. Only in this way will psychiatry ultimately reclaim its scientific integrity and public trust.

References

Baker, C.B., M.T. Johnsrud, M.L. Crismon, R.A. Rosenheck, and S.W. Woods. 2003. Quantitative analysis of sponsorship bias in economic studies of antidepressants. *The British Journal of Psychiatry* 183 (DEC): 498–506.

Balon, R., and S. Singh. 2001. Status of research training in psychiatry. *Academic Psychiatry* 25 (1): 34–41.

Bentall, R.P. 1992. A proposal to classify happiness as a psychiatric disorder. *Journal of Medical Ethics* 18: 94–98.

Brown, N., C. A. Vassilas, and C. Oakley. 2009. Recruiting psychiatrists—A Sisyphean task?, *Psychiatric Bulletin*.

Brownlee, J. 2020. The corporate corruption of academic research. *Alternate Routes: A Journal of Critical Social Research* 26, no. Retrieved from http://alternateroutes.ca/index.php/ar/article/view/22311.

Caplan, P.J. 2014. Diagnostisgate: Conflict of interest at the top of the psychiatric aparatus. *Aporia Journal* 7 (1): 30–41.

Chappell, N., A. Cassels, L. Outcalt, and C. Dujela. 2016. Conflict of interest in pharmaceutical policy research: An example from Canada. *International Journal of Health Governance* 21 (2): 66–75.

Christensen, H., P.J. Batterham, I.B. Hickie, P.D. McGorry, P.B. Mitchell, and J. Kulkarni. 2011. Funding for mental health research: The gap remains. *Medical Journal of Australia* 195 (11): 681–684.

Cosgrove, L., and A. Vaswani. 2019. *Critical psychiatry. Controversies and Clinical Implications.* Springer Nature Switzerland.

Cosgrove, L., and E.E. Wheeler. 2013. Drug firms, the codification of diagnostic categories, and bias in clinical guidelines. *Journal of Law, Medicine and Ethics* 41 (3): 644–653.

Cosgrove, L. 2010. Psychiatric taxonomy, psychopharmacology and big pharma. *Couns. Sch. Psychol. Fac. Publ. Ser. Pap. 7.*

Duval, G. 2004. Interest: Protecting accountability. *Int Comp Heal Law Ethics A 25-Year Retrosp* Winter: 613–626

Egilman, D.S., and N.M. Druar. 2011. Corporate versus public interests: Community responsibility to defend scientific integrity. *International Journal of Occupational and Environmental Health* 17 (2): 181–185.

Fanelli, D. 2012. Negative results are disappearing from most disciplines and countries. *Scientometrics* 90 (3): 891–904.

Frances, A.J., and T. Widiger. 2012. Psychiatric diagnosis: Lessons from the DSM-IV past and cautions for the DSM-5 future. *Annual Review of Clinical Psychology* 8 (1): 109–130.

Gagnon, M.A. 2013. Corruption of pharmaceutical markets: Addressing the misalignment of financial incentives and public health. *Journal Law, Medicine and Ethics* 41 (3): 571–580.

Gopal, A.A., L. Cosgrove, and H.J. Bursztajn. 2010. Commentary: The public health consequences of an industry-influenced psychiatric taxonomy: 'Attenuated Psychotic Symptoms Syndrome' as a case example. *Accountability in Research* 17 (5): 264–269.

Gøtzsche, P.C., et al. 2009. What should be done to tackle ghostwriting in the medical literature? *PLoS Medicine* 6 (2): 0122–0125.

Hamoda, H.M., M.S. Bauer, D.R. Demaso, K.M. Sanders, and E. Mezzacappa. 2011. A competency-based model for research training during psychiatry residency. *Harvard Review of Psychiatry* 19 (2): 78–85.

Hart, B., A. Lundh, and L. Bero. 2012. Effect of reporting bias on meta-analyses of drug trials: Reanalysis of meta-analyses. *BMJ* 344 (7838): 1–11.

Hughes, S., D. Cohen, and R. Jaggi. 2014. Differences in reporting serious adverse events in industry sponsored clinical trial registries and journal articles on antidepressant and antipsychotic drugs: A cross-sectional study. *BMJ Open*, 4(7).

Jablensky, A., R. E. Kendell. 2002. Criteria for assessing a classification in psychiatry. In *Psychiatric diagnosis and classification*, eds. M. Maj, W. Gaebel, J. J. Lopez-Ibor, and N. Sartorius, 1–24. Wiley.

Jernigan, D.B., L. Kargacin, A. Poole, and J. Kobayashi. 2001. Sentinel surveillance as an alternative approach for monitoring antibiotic-resistant invasive pneumococcal disease in Washington State. *American Journal of Public Health* 91 (1): 142–145.

Jureidini, J.N., and L.B. McHenry. 2011. Conflicted medical journals and the failure of trust. *Accountability in Research* 18 (1): 45–54.

Kupfer, D.J., S.E. Hyman, A.F. Schatzberg, H.A. Pincus, and C.F. Reynolds. 2002. Recruiting and retaining future generations of physician scientists in mental health. *Archives of General Psychiatry* 59 (7): 657–660.

Lexchin, J. 2012. Sponsorship bias in clinical research. *The International Journal of Risk and Safety in Medicine* 24 (4): 233–242.

Le Noury, J., et al. 2015. Restoring study 329: Eficacy and harms of paroxetine and imipramine in treatment of major depression in adolescence. *BMJ* 351.

McHenry, L.B., and J.N. Jureidini. 2008. Industry-sponsored ghostwriting in clinical trial reporting: A case study. *Accountability in Research* 15 (3): 152–167.

Moffatt, B., and C. Elliott. 2007. Ghost marketing: Pharmaceutical companies and ghostwritten journal articles. *Perspectives in Biology and Medicine* 50 (1): 18–31.

Polster, C. 2007. The nature and implications of the growing importance of research grants to Canadian universities and academics. *Higher Education* 53 (5): 599–622.

Rafols, I., et al. 2014. Big Pharma, little science?. A bibliometric perspective on Big Pharma's R&D decline. *Technological Forecasting and Social Change* 81 (1): 22–38.

Rodwin, M.A. 2013. Drug Use. *Journal of Law, Medicine and Ethics* Fall: 654–664.

Rodwin, M.A. 2013. Introduction: Institutional corruption and the pharmaceutical policy. *Journal Law, Medicine and Ethics* 41 (3): 544–552.

Rothman, D.J., et al. 2009. A proposal for controlling conflict of interest. *JAMA, the Journal of the American Medical Association* 301 (13): 1367–1372.

Schafer, A. 2004. Biomedical conflicts of interest: A defence of the sequestration thesis—Learning from the cases of Nancy Olivieri and David Healy. *Journal of Medical Ethics* 30 (1): 8–24.

Schott, G., H. Pachl, U. Limbach, U. Gundert-Remy, W.D. Ludwig, and K. Lieb. 2010. The financing of drug trials by pharmaceutical companies and its consequences. *Deutsches Ärzteblatt International* 107 (16): 279–285.

Sismondo, S. 2009. Ghosts in the machine: Publication planning in the medical sciences. *Social Studies of Science* 39 (2): 171–198.

Sismondo, S. 2013. Key opinion leaders and the corruption of medical knowledge: What the sunshine act will and won't cast light on. *Journal Law, Medicine and Ethics* 41 (3): 635–643.

Smith, R. 2005. Medical journals are an extension of the marketing arm of pharmaceutical companies. *PLoS Medicine* 2 (5): 0364–0366.

Somerville, M.A. 2002. A postmodern moral tale: The ethics of research relationships. *Nature Reviews. Drug Discovery* 1 (4): 316–320.

Stuart, H., et al. 2015. Images of psychiatry and psychiatrists. *Acta Psychiatrica Scandinavica* 131(1).

Turner, E.H. 2013. Publication bias, with a focus on psychiatry: Causes and solutions. *CNS Drugs* 27 (6): 457–468.

van Lent, M., J. Overbeke, and H.J. Out. 2013. Recommendations for a uniform assessment of publication bias related to funding source. *BMC Medical Research Methodology* 13: 120.

Watson, G.L., A.P. Arcona, D.O. Antonuccio, and D. Healy. 2014. Shooting the messenger: The case of ADHD. *Journal of Contemporary Psychotherapy* 44 (1): 43–52.

Yager, J., J. Greden, M. Abrams, and M. Riba. 2004. The institute of medicine's report on research training in psychiatry residency: Strategies for reform—background, results, and follow up. *Academic Psychiatry* 28 (4): 267–274.

Heather Stuart is Professor in the Departments of Public Health Sciences, Psychiatry, and the School of Rehabilitation Therapy at Queen's University. She also holds the Bell Canada Mental Health and Anti-stigma Research Chair at Queen's. Dr. Stuart is also the Senior Consultant to the Mental Health Commission of Canada's Opening Minds Anti-stigma initiative and the past Chair of the World Psychiatric Association's Stigma and Mental Health Scientific Section. She is a Fellow of the Royal Society of Canada and a recipient of the Order of Canada.

Chapter 25
The Place of Digital and Artificial Intelligence in Medical Research

Anne-Marie Duguet

A   The contribution of digital technologies in clinical trials

In preclinical studies: development of "in silico" electronic methods for the quantitative activity structure relationship (QSAR) and alternative methods for the use of laboratory animals. Simulation of diseases: From databases combining imaging, clinical and environmental data, modelling makes it possible to electronically simulate diseases or the functioning of organs (simulation of a brain developing Alzheimer's disease, simulation of pulmonary pathologies, lymphoma data lake and data hub).

B   Digital technologies for other health research

For public health research, the intelligent analysis medical data of cohort studies the collective behavior of patients and automatically aggregates health trajectories.

Research in medical imagery is certainly the area in which digital is developing most rapidly through knowledge modelling, algorithms and recognition from 3D digitized images using neural networks and deep learning.

   Biology has become a data science. Machine learning methods attempt to provide predictive analysis from large collections of collected data (Precision medicine).

C   Ethical considerations

There are questions about the use of digital artificial intelligence tools and methods in health research. The first question is the scientific validation of the instruments and the relevance of their use. The quality and choice of annotated data is a permanent concern. The human conception of software supposes choices to avoid the biases likely to create discriminations. The explicability of the methods is a major questioning because the automatisms must not be without control or justifications. They

A.-M. Duguet (✉)
University of Toulouse-Paul Sabatier, Toulouse, France
e-mail: aduguet@club-internet.fr

© The Author(s), under exclusive license to Springer Nature Switzerland AG 2023   473
T. Zima and D. N. Weisstub (eds.), *Medical Research Ethics: Challenges
in the 21st Century*, Philosophy and Medicine 132,
https://doi.org/10.1007/978-3-031-12692-5_25

should allow the human guarantee, that is to say the intervention of the professional to validate or update the system.

The possibility of replacement, in the future, of humans by a model of disease and patients or the creation of virtual patients must be analyzed according to the potential risks.

Finally in data sciences, data collection presupposes public-private cooperation, with data sharing which raises questions of intellectual property of research data.

25.1 Introduction

Digital and artificial intelligence are widely used in multiple fields, including economics, marketing, banking, insurance. Digital and artificial intelligence have developed through software and databases and have since proliferated into all areas of health. In Europe, the future of health is digital. However, the use of digital and artificial intelligence in medical research is a key challenge of the twenty-first century. In the brochure *Connected for a Healthy Europe* (2019), the European Commission considers the proposition that the digital transformation of health and healthcare can improve the well-being of millions of citizens, and radically change the way health and healthcare services are delivered to patients. Digital health opportunities are real: 93% of EU citizens want to manage their health data, and 80% agree to share their data. Digital intelligence is everywhere in health, manifesting in the use of software and databases.

Recent advances—including the increasingly skilled performance capabilities of computers and new algorithms—now allow faster management of large amounts of medical data, as well as the analysis of denser datasets, such as genomic data. The field of data science has seen developments through increased server storage capacity, improved computational speeds, and knowledge sharing. Massive public and private databases are now available ('big data'). With the advent of Artificial Intelligence, digital technology is taking the next step by creating machines that can be human operated in their use but can also operate autonomously entirely or in part. Machine learning allows computers to improve through learning. The use of a set of algorithms improves the performance of the machine as it recovers data. The machine is informed of the quality of the result, when the recognition is correct, and when the result is wrong. Gradually, with an increasing number of datasets, the machine 'learns' and can output correct conclusions with increased frequency.

Deep learning is a technology based on artificial neural networks capable of managing a large amount of data. The term "neural network" is used to draw a comparison to neurons in the human brain that combine multiple types of information, produce multiple answers, and improve as new knowledge is acquired. Deep learning can be used generally for language processing, text analysis, and image recognition. Within healthcare, deep learning can be used for genomics and predictive medicine, and targeted chemotherapy and molecule screening for pharmaceutical research and development.

When these technologies are applied to the domain of human health, this raises questions surrounding the ethical consequences of their use and design. Algorithms learn on their own by integrating user data. The initial coding of this data can be influenced by unconscious representations and prejudices (O'Neil Cathy 2016). Decisions and actions arising from these types of algorithms should not affect human beings or our humanistic principles to a complete or even partial degree. Taking into account ethical questions, "ethics by design" protects citizens from voluntary or involuntary artificial intelligence biases. The use of these technologies facilitates the work of professionals, and their technical performance is considered a benefit. They have the potential to add important value. However, without careful vetting, they may exacerbate health disparities amongst the most vulnerable (Nebeker 2019). As algorithms make decisions, these decisions—as all decisions are—are prone to bias and potential mistakes (Martin 2019).

A consensus has been formed for the use of AI in research with minimal transparency and dependability, which cites the need for the promotion of the following principles: promoting well-being, minimizing harm, respecting dignity, privacy, human rights, and freedoms. This presentation falls within the framework of research ethics. The first part is devoted to exploring the role of digital and artificial intelligence in medical research: describing research that already uses digital and artificial intelligence in their development of new medicines. The second part deals with medical algorithms as research tools; through exploring the legal and ethical aspects of the construction of health databases and the design of algorithms used by health professionals. Though this work is a general contribution based on some elements of the international literature currently published on the subject, it also refers to some specific elements of the European and French context.

25.2 Part I—Numerics and Artificial Intelligence in Research: The Example of Drugs

Research on drugs and new technologies requires human participation in clinical trials, so, the question is whether the experimental product is efficient and relevant in its use on humans. Traditionally, the research process for a new drug begins with the development of a molecule, with its selection being made according to its therapeutic properties. Next, toxicological tests on animals are conducted, and finally the drug is available for human trials. The research of the future is already here; with digital and artificial intelligence influencing all areas of medical research. Currently, digital research is conducted on hospital medical records datasets, and on the data collected by laboratories during their clinical trials. This data is aggregated into health databases.

In future, the discovery of new drugs will increasingly rely on digital technologies and artificial intelligence tools. The treatment of patients will be in conjunction with the administration of drugs and the use of tests and medical devices. This 'personalized medicine' is targeted—using diagnostic tests to adapt the drug for the patient and implanted devices to measure pharmacological effectiveness and dosage adaptation, which serves to increase the effectiveness of the product and limits adverse effects. Digital technology is currently being used in drug development research in preclinical trials: the first step in the search for new drugs, in which humans are not involved. The next steps in research—clinical trials on animals or humans—are long and expensive, and any possible substitution with or assistance from digital tools has advantages in terms of saving both time and money.

"In silico" clinical trials are a development of "patient-specific models to form virtual cohorts for testing the safety and/or the efficacity of new drugs and new medical services" (Francico and Giulia 2019).

(1) Preclinical methods in silico

The term "in silico" refers to digital methods based on the laws of physics and chemistry that, using mathematical approaches, can simulate or model a biological phenomenon using computational tools. These methods are complementary to in vitro and in vivo studies but cannot replace human trials.

(a) Analysis of the QSAR (Quantitative Structure Activity Relationship) allows a chemical structure to be correlated with an effect, such as chemical reactivity and toxicity, via mathematical models integrated with specialized software. Similar chemical molecules have similar biological activities, which is why structural analogues are sought. The choice of the experimental data to create the database is decisive; the data must be reliable. The software becomes more efficient and able to inform many pharmacokinetic criteria: absorption, distribution, metabolism, excretion. They are used for screening new ingredients. The European Chemical Agency ECHA (2016) published a Practical Guide on how to use and report QSARs.

(b) In silico methods in toxicology: In the search for new drugs, preclinical toxicology studies play a major role. No clinical trials can be authorised without toxicology tests. Traditionally, in vivo studies are long and expensive; these trials require many animals and are considered unethical, which is why in silico toxicity prediction studies save time and prove efficient in the development of a substance—or in its evaluation—before the marketing authorization. Benedette Cuffari (2020) states, "computational tools that can be utilized for in silico toxicology studies include databases that contain data on the toxicity and chemical properties of chemicals etc. … and a wide variety of models … have been encoded".

(2) Substitution for optimizing the use of animals in preclinical studies

To reduce the number of laboratory animals used, Isabelle Fabre (2009) describes alternative methods to the experiments on laboratory animals including the use of databases made up of experimental results and in vitro methods (physicochemical methods or biological elements prepared from living organisms such as bacteria, cells, and reconstructed tissue models). These methods were developed following the publication of European Directive 86–609 on the protection of animals used for scientific purposes. More recently, prospective models of molecular biology and stem cell use have been developed. The development of these methods has greatly expanded to protect animals and the new REACH Regulation 1907/2006 for Registration, Evaluation, Authorisation and Restriction of Chemicals encourages the use of animal trials only as a last resort. Cell cultures and reconstructed tissues, despite demonstrating a good level of differentiation, cannot simulate the complexity of reactions occurring in an entire organism; thus, these methods must be validated, and animal tests are still necessary.

(3)   Data mining of medical records reduce the time of clinical trials

In research involving testing new uses for older medications, pharmaceutical companies have begun to replace parts of clinical trials with mining medical records to speed up the drug approval process. Three U.S. laboratories, Pfizer, Johnson and Johnson, and Amgen and Roche, have obtained marketing approvals from the Food and Drug Administration based on data mining of medical records to evaluate a new use for existing cancer drugs in the treatment of breast cancer, bladder cancer, and leukemia (Loftus 2019). They found correlations in the data on adverse events, tumor improvement, etc. These correlations are used instead of clinical trials. These tests can be done in a few months instead of the several years required for clinical trials.

Johnson and Johnson looked for correlations between genetic characteristics and the use of these cancer drugs to explain the ineffectiveness of bladder cancer treatment. A new treatment, Belvera, was then tested on patients with these genetic traits in a clinical trial. Pfizer used Iqvia to see if a treatment for breast cancer for women, Ibrance, could be tested on men with the same cancer. Roche received FDA approval for its tumor drug Rozyltek based on a combination of clinical trials involving 53 patients and data of 69 patients provided by Flatiron Health Analytics Database. (Fuang Scarlet 2019). For the next stages of such research, the use of digital and artificial intelligence is still experimental and new advances are widely reported in the mainstream press. Large amounts of public and private funds are invested in this research, and the announcement of possible innovations have an impact on the stock market listing of major pharmaceutical companies.

(4)   Promising research for the future
(a)   Screening molecules with databases and Artificial Intelligence

The analysis (screening) of compounds listed in databases saves considerable time. Zinc 15 data base is a free data base of 230 million purchasable chemical compounds prepared especially for virtual screening. (http://ZINC15.dockin.org).

Halicine is an antibiotic molecule discovered through AI analysis of the ZINC15 database which identified 23 promising candidates in 3 days. Researchers (Stokes and Coll, 2020) trained a deep neural network to predictively identify molecules with antibacterial activity from the Drug Repurposing Hub—a collection of FDA approved drugs. This new molecule has a different structure from conventional antibiotics and, in mouse models, demonstrates bactericidal activity against a wide spectrum of germs including: Mycobacterium tuberculosis, Clostridioides difficile and Acinetobacter baumanii. This work highlights the utility of deep learning approaches to expand the antibiotic arsenal through the discovery of structurally distinct antibacterial molecules.

(b) Artificial intelligence techniques

Through artificial intelligence techniques in the drug creation process and based on the analysis of patents and previous research, the determination of potentially suitable molecules for new research favours those that can be synthesized in the laboratory. Gentrl is a deep generative model, which performs generative tensorial reinforcement learning, for de novo small-molecule design. It was used in the discovery of the potent inhibitor of discoidin domain receptor1, a kinase target implicated in fibrosis and other diseases. 30,000 models of molecules were identified within 21 days. 6 of these molecules were synthesized in the laboratory and tested on stem cells, with the most promising being tested on mice. The resulting drug was prepared within 46 days (Zhavoronkov 2019). These examples are important indicators that, despite the very rapid acceleration of the first steps, the rest of the development process will be the same, and it will take several years before the drug can be put on the market.

25.3 Part II—Research on Digital Health Tools: Regulatory Framework and Ethical Reflection

Algorithms can only work with digitized data, and cannot be dissociated from the data they are working on. It is therefore necessary to gather more and more data to feed Artificial Intelligence systems. Health-related data is personal data that is protected, depending on the relevant country the research is conducted in.

(1) Building databases for research
(a) Data protection, the European Union example

In Europe, in accordance with the GDPR (2016), health professionals and researchers must ensure the protection of digitized data collected during healthcare proceedings, which includes: diagnosis, treatment, and patient medical records. This regulation applies to all EU member states, and to all operators outside EU borders who collect data on European citizens living in the EU countries. The GDPR defines the rights of data holders as including the following: to get information, to have access and rectification to personal data, to limit treatment, to object to data processing, and the right not to be subject to individual automated decision. The data controller must at the time of processing provide appropriate measures which are designed to uphold data protection principles in order to meet the GDPR requirements. Each member state must designate at least one independent public authority to be responsible for monitoring the application of the GDPR. If member states do not comply, they are liable to financial penalties of up to 4% of their global annual turnover with a maximum of EUR 20 million.

(b) Ethical reflection on data bases for research

– Confidentiality

Health professionals and researchers must ensure the protection of digitized data collected in the course of care (diagnosis, treatment, medical records), but may use this data for research purposes, provided that the subject has been informed of this secondary use and has not objected to it. However, such annotated data is valuable and sharing it with other researchers, whether in Europe or outside of Europe, requires the consent of the patient. In multi-centre research projects involving several countries, individuals are informed of where their data is processed and how well they are protected. The GDPR applies equally to all EU countries and allows free exchange of data, but outside the EU, exchanges are only allowed when the level of data protection in the recipient country is equivalent to the level imposed in Europe. This is not the case in the United States, since the Privacy Shield does not provide sufficient protection (Schrems II decision, 17 July 2020, EU Court of Justice).

The analyzed data is made anonymous. This raises the question: is data anonymization sufficient protection? De-identification may be reversible (pseudonymisation) or irreversible (anonymisation); and in the latter case the data is no longer considered personal, and therefore can escape the GDPR (no consent and free access), provided that there is no risk of re-identification through cross-referencing of the data. However, if reducing this risk would require the removal of a large amount of potentially re-identifying data, the value of such "anonymized" data would be greatly diminished. Standard anonymization procedures are necessary to respect the content of the data. The persistence of a risk of re-identification makes it necessary to control and regulate the possibilities of access to such sensitive data.

– The quality of the sources

Not all health-related data are of the same quality. Data collected by connected objects is not validated by health professionals and cannot be considered health data. By using connected objects, the approach of these patients is a desire to manage individual well-being from a prevention and development perspective. In its publication, *The body new connected object*, the National Commission Digital and Freedoms (in French CNIL 2013) argues that the data collected by the connected objects is not validated by health professionals, and some developers maximize the collection of information, which results in the collection of data that is irrelevant to the purpose of the service and the absence of explicit consumer consent. This raw data produced without validation cannot reasonably be used for medical or predictive purposes. Conversely, data collected by physicians is valuable because it is validated and reliable—especially the data collected in public hospitals that can be easily shared. In France, the Law 2016–1321 on a Digital Republic says that all public data is accessible because French law considers that "all data produced by the State are public goods". This data is made available under certain conditions and cannot be the subject of an onerous transfer. Consequently, the Health data hub—http://healthdatahub.sol idarites-sante.gouv.fr—has been created. This platform for accessing data is a trusted third-party for data sharing (a secure, non-exclusive, one-stop shop which guarantees the quality of the shared data).

– Prospective research in population uses public and private data

Health population research uses AI methodology to extrapolate predictions from clinical and non-clinical data. This data comes from social media which falls under the public domain, and from anonymous secondary health data. This data is exempt from ethics committee scrutiny. Nowadays, little attention has been given to the ethics governance, raising the question: how can we assure the public that researchers are acting ethically? To set standards for the use of algorithms for predictive purposes, Gariele Samuele (2020) suggests creating national open-science repositories for open-source algorithms, and to organize a sector-specific validation of research processes and algorithms by a committee of academics and stakeholders.

– Data sharing

The sharing of research data according to the principles of FAIR (FAIR 2016)—Findable, Accessible, Interoperable, Reusable—is a crucial element for the international development of scientific research in health and helps to ensure the reproducibility and validity of results. Access to data from health care, health systems, or other personal databases represents a major progress by avoiding duplicated data collection for clinical trials or observational data (large cross-sectional surveys, cohorts). For this reason, consent for the secondary sharing of data in the interest of research should be sought at the time of data collection. In these circumstances, precautions should be taken to avoid the risk of re-identification from databases from which direct identifiers have been removed (www.go-fair.org).

(2) Design of algorithms for health professionals: ethical principles

Algorithms and trained models must be transparent, explainable, able to build patient and provider trust. Algorithms should also ensure protection from malicious attacks and privacy breaches and be monitored to perform as predicted and do not degrade over time. A major issue is the scientific validation of algorithms and the relevance of their use. The quality and choice of the annotated data is an ongoing concern. The human design of software requires decision-making choices to avoid bias. To prevent the algorithm from changing during learning, safeguards must be integrated to regulate future mutations (Debray Christine 2018), and avoid automatic decisions that could impact a person or group of people. This is why we must maintain some representation for humanity in medical decision-making.

(a) Explainability and human guarantee of software and algorithms

Correlations and automatisms must not be uncontrolled and unjustified. In his report, Cedric Villani (2018) considers that "as a society, we cannot accept that certain important decisions can be made without explanation". The results of algorithmic decisions must be explainable (explainability) and allow the intervention of the professional to validate or update the system (human guarantee). One should know how the decision is made in case any uncertainty or unforeseen situation arises. To more broadly understand how one arrives at the result, it must be possible to specify the criteria used to prioritize and classify the information. The self-learning character does not prevent an a posteriori discovery of the factors considered in the proposed solution. Therapeutic decision-making software uses statistical prediction methods to assess the time it takes to develop a complication, cancer recurrence, or death, and provide targeted treatment or prevention. (Seroussi 2014). Clinical decision support systems generate probable diagnoses for a patient based on their medical data (diagnostic decision making). They use expert systems that replicate the opinions of experts with statistical and probabilistic numerical approaches. The algorithm must be taught using a reliable and representative database to avoid the production of design biases.

(b) Governance of algorithmic decisions

Given that all algorithmic decisions make mistakes, an ethical algorithm will offer a mechanism through which to identify, judge and correct mistakes. (Martin 2019). Category mistakes occur when an incorrect assignment is made (false positive) or when someone is incorrectly excluded (false negative). The medical community tends to avoid false negatives: cases wherein a patient is not labeled as sick when he is in fact sick. In such medical judgement mistakes, the preference may to be mistakenly identify someone as having a cancer rather than letting cancer go undetected. Process mistakes appear when an inappropriate factor is used in the decision making (machine learning, or neural network). This process mistake may be by design or learned by the algorithm from biased training data. Algorithms simply reproduce the discrimination in the data they are provided. The decision-making algorithm reproduces social inequalities (minorities… rare cases…). Some categories are not represented because of ignorance of weak signals (the elements that are in the data but which the statistical processing does not show). To achieve the collective interest of the program, some studies do not take into account particular situations. A bias could result from the systematization of model-based reasoning that does not have the ability to take into account all the characteristics of each patient. Adding reflection helps to correct mistakes. Algorithmic decision-making can incorporate the ability to revisit the answers to ensure that the classification is working as desired and not creating mistakes. The degree to which an algorithm is inscrutable contributes to our ability to identify, judge, and correct mistakes in algorithmic decisions.

(c) Design of imaging algorithms: multidisciplinary ethical proposals

For several years, image recognition has been used to develop computer-assisted diagnostic methods from a very large amount of data collected from digitized imaging techniques (radiology, dermatology, oncology). Machines are trained to improve their performance. Gradually the machine learns and outputs more and more correct conclusions. With this system, smaller volume abnormalities are detected when they were not detectable, for example, with the eye of an experienced radiologist. The performance of the system depends on the number and quality of the annotations carried out by the radiology specialists to train the machine. A multidisciplinary statement has been proposed by North American and European radiologists, from which some recommendations are highlighted here (Geis et al. 2019). These recommendations are as follows. AI has a great potential to increase efficiency and accuracy throughout radiology, but also carries inherent pitfalls and biases. These are stand-alone software systems that can generate systemic errors. The automation bias is a tendency for humans to favor machine-generated decisions. However, to what degree can physicians delegate this task to the machine? The liability ultimately falls to the human operator. Radiologists are learning about ethical AI at the same time we invent and implement it.

The statement highlights the consensus that ethical use of AI in radiology should promote well-being, minimize harm, and ensure that the benefits and harms are distributed among stakeholders in a just manner. AI should respect human rights and freedoms, including dignity and privacy. It should be designed for maximum transparency and dependability. The radiology community should begin, as soon as possible, to develop codes of ethics and block uses of radiological data and algorithms for financial gain which lacks these ethical attributes. Ultimate responsibility and accountability for AI remains with its human designers and operators for the foreseeable future. Radiologists will remain ultimately responsible for patient care.

25.4   Concluding Remarks

What benefits does Artificial Intelligence bring to health research, to the society, the pharmaceutical companies and to patients? Artificial Intelligence tools are created to help practitioners, not to replace them. The role of digital platforms is to provide computing power and tools to improve treatments, and to have doctors implement these technologies.

The benefit of AI systems to society is the possibility for a better knowledge of diseases with quicker "medicine by evidence" checks. Are the results made available to the public for the common good? Pharmaceutical laboratories and research teams that collect data from their studies have total control over the storage and exchange of data. While some private organizations provide free access to certain datasets, it is understandable that data management in this case is subject to the rules of industrial secrecy and intellectual property, which private companies wish to control.

The benefits of AI to manufacturers: Artificial Intelligence saves time in the choice of molecules to be tested. There is a definite benefit for animal welfare—fewer laboratory animals are used in experiments. Pharmaceutical companies compete fiercely to discover new products and bring them to market before their competitors.

The benefits of AI to patients: The need to establish a fiduciary relationship with the algorithms that will be used to assist in the research can be emphasized. Will subjects agree to be included in a research protocol after a decision has been made exclusively by an algorithm? Some have excessive confidence in the machine and believe it is infallible. Others, on the other hand, doubt the AI's real capacity, knowing that the choices made have been programmed by man from the sum of individual experiences.

The virtualization of the person through his or her data with the prospect of replacing the human with a virtual patient includes a risk of reification. Only a truly informed consent of the subject, which gives the patient the capacity to control the future of his/her data by informing him/her of all these consequences, remains the guarantee of respect for the dignity of the person. For the time being, it is not possible to accept that artificial intelligence alone can create innovation. This hypothesis is not accepted in Europe, which has refused to grant patents for inventions created by artificial intelligence such as DABUS, on the grounds that the protection of intellectual property can only apply to humans (Crimino Valentin, 2020). In France, the Comité Consultatif National d'Ethique (CCNE, 2018) in its opinion 129 wishes to maintain final control of the healthcare professional in their interactions with the patient to ensure appropriate decision-making. Guarantees are provided by regular verification procedures, and the seeking of a second medical opinion in case of doubt. The CCNE has come out in favour of prohibiting a decision-making act performed solely by artificial intelligence without human intervention.

References

Brochure «Connected for a healthy Europe» (2019) https://ec.europa.eu/digital-single-market/en/news/connected-healthy-future-brochure.

Nebeker Camille, John Torous, and Rebecca J Bartlett Ellis. 2019. Building the case for actionable ethics in digital health research supported by artificial intelligence. *BMC Medicine* 17: 137 10./s12916-019-1377-7.

O'Neil Cathy. 2016. Weapons of maths destruction: How Big Data Increases Inequality and threatens Democracy Crown Publishing Group, 2016.

CCNE. 2018. Opinion 129 contribution of the CCNE to to revision of the bioethics law https://www.ccne-ethique.fr/en/publications/contribution-comite-consultatif-national-dethique-revision-bioethics-law

Debray, Christine, Pour une intelligence artificielle éthique dès sa conception Les échos, 4 avril 2018 https://www.lesechos.fr/2018/04/pour-une-intelligence-artificielle-ethique-des-sa-conception-988121.

CNIL. 2013. Cahier IP innovation and prespectives n°2 « Le corps, nouvel objet connecté » Rapport de la Commission Informatique et libertés 13 Octobre 2013 https://www.cnil.fr/sites/default/files/typo/document/CNIL_CAHIERS_IP2_WEB.pdf.

Cuffari, B. 2020. What is in silico toxicology. 10 March 2020 News Medical life Sciences https://www.news-medical.net/life-sciences/What-is-In-Silico-Toxicology.aspx. Consulté le juin 13 2020.

ECHA (European CHemical Agency) Practical Guide on How to use and report QSARs. 2016. ECHA-16-B-09-EN. https://doi.org/10.2823/81818 http://echa.eu/contact.

Directive Européenne 86–609 sur la protection de l'animal utilisé à des fins scientifiques.eur-lex.europa.eu https://eur-lex.europa.eu/legal-content/FR/TXT/?uri=LEGISSUM%3Al28104

Fabre, I. 2009. Les méthodes substitutives à l'expérimentation animale. *Bulletin De L'académie Nationale De Médecine* 193 (8): 1783–1791.

Pappalardo Francico, Giulia Russo, Flora Tshimanou Musuamba, and Marco Viceconti. 2019. In silico clinical trials: Concepts and early adoptions. *Briefing in Bioinformatics* 20(5): 1699–1708. https://doi.org/10.1093/bib/bby043. https://pubmed.ncbi.nlm.nih.gov/29868882/

GDPR. 2016. Regulation (EU) 2016/679 on the protection of natural persons with regard to the processing of personal data and on the free movement of such data and repealing Directive 95/46/EC. https://gdpr-info.eu/.

Geis, J., Brady, R., Adrian, P., Wu, C., Spencer, J., and Coll. 2019. Ethics of artificial intelligence in radiology: Summary of the joint European and North American Multisociety Statement. *Canadian Association of Radiologist Journal* 329 Elsevier B.V.

https://www.nature.com/articles/sdata201618.

Martin, Kristen. 2019. Desining ethical algorithms. *MIS Quaterly Executive* 18 (2). https://doi.org/10.17705/2msqe.00012.

Loftus, P. 2019. Drugmakers turn to data mining to avoid expensive, lengthy drug trials. *Wall street Journal* 23 https://www.wsj.com/articles/drugmakers-turn-to-data-mining-to-avoid-expensive-lengthy-drug-trials-11577097000.

Loi 2016–1321 du 7 octobre 2016 pour une république numérique https://www.gouvernement.fr/action/pour-une-republique-numerique.

REACH Regulation 1907/2006 https://echa.europa.eu/regulations/reach/understanding.

Fuang Scarlet, Karma December, 23, 2019h https://karmaimpact.com/medical-data-mining-expediting-new-drug-approval-from-fda-arouses-private-investors/

Séroussi, B., Bouaud, J. 2014. Computerized clinical decision support systems: Overview of data- and knowledge-based approaches. In Pratique Neurologique—FMC. 5(4): 303–316. Language: French. https://doi.org/10.1016/j.praneu.2014.09.006, Base de données: Science Direct.

Stokes Jonathan M., and Coll. 2020. A deep learning approach to antibiotic discovery. Cell 180(4): P688–702.E13.

Valentin, Crimino. 2020. L'office Européen des brevets rejette deux demandes dont l'inventeur était une intelligence artificielle le Siècle Digital 4 janvier 2020 https://siecledigital.fr/2020/01/04/lunion-europeenne-rejette-deux-brevets-dont-linventeur-etait-une-ia/.

Villani, C. 2018. For a meaningful artificial intelligence. http://www2.assemblee-nationale.fr/15/les-delegations-comite-et-office-parlementaire/office-parlementaire-d-evaluation-des-choix-scientifiques-et-technologiques/secretariat/a-la-une/intelligence-artificielle-presentation-du-rapport-de-cedric-villani

Wilkinson, M., M. Dumontier, I. Aalbersberg, et al. 2016. The FAIR guiding principles for scientific data management and stewardship. *Sci Data* 3: 160018. https://doi.org/10.1038/sdata.2016.18.

Zhavoronkov, A., A. Ivanenkov, Y.A. Aliper, et al. 2019. Deep learning enables rapid identification of potent DDR1 kinase inhibitors. *Nature Biotechnology* 37: 1038–1040.

Anne-Marie Duguet has been a Senior Lecturer for a long career in the fields of Medical Law and Bioethics at Paul Sabatier University while being attached to INSERM, specifically in the field of Epidemiology and Public Health. As well, she has been a core researcher in the International Institute of Research in Bioethics (IIREB www.iireb.org), a network established in Canada where she has coordinated the area of "Research Ethics". She has served as President of the ARFDM (arfdm.asso.fr), a European Summer School in Bioethics and Medical Law. Dr. Duguet has been a central person in evolving the field of bioethics in China, having organized seven liaison Symposia in France and China. In her extensive career, she has published almost 200 scientific articles and edited or contributed to 25 books.

Chapter 26
A Framework to Govern the Use of Health Data for Research in Africa: A South African Perspective

Ciara Staunton, Rachel Adams, Lyn Horn, and Melodie Labuschaigne

26.1 Introduction

The global push towards "open science" and the open sharing of samples, data, research results and other research processes is based on the premise it promotes the optimal use of resources, results in more reproducible research, increases statistical power and encourages new research on existing data bases (Walport and Brest 2011). In parallel, there has been a strengthening of the protection of personal information of data subjects—which includes research participants—through the introduction of legal frameworks such as the General Data Protection Regulation (GDPR) and the Health Insurance Portability and Accountability Act of 1996 (HIPAA). These frameworks impact the use and sharing of data for research, and the open science movement must be balanced against and adhere to the rules and procedures inherent in them.

Many jurisdictions in Africa have similarly introduced general data protection regulations that will impact on the use of data for research. They were largely introduced in response to the GDPR and many of the regulations are thus broadly similar in style and structure to the GDPR. This strengthening of data protection overlaps with a gradual increase in research investment on the continent, with genomic research specifically seeing an exponential growth through large consortia such as HapMap,

C. Staunton (✉)
Middlesex University, London, England
e-mail: Ciara.Staunton@eurac.edu

R. Adams
Research ICT Africa, Cape Town, South Africa

L. Horn
Stellenbosch University, Stellenbosch, South Africa

M. Labuschaigne
University of South Africa, Gauteng, South Africa

© The Author(s), under exclusive license to Springer Nature Switzerland AG 2023   485
T. Zima and D. N. Weisstub (eds.), *Medical Research Ethics: Challenges in the 21st Century*, Philosophy and Medicine 132,
https://doi.org/10.1007/978-3-031-12692-5_26

MalariaGen, H3Africa, B3Africa and PHA4GE (Gibbs et al. 2003; Ciara Staunton and Moodley 2013; The H3Africa Consortium et al. 2014). Institutions such as the Africa Centre for Disease Control established in 2016 and the African Academy of Sciences are also strengthening and promoting research across the continent. These collaborations are complemented by other local and national initiatives aimed at increasing research and building research capacity (Ciara Staunton et al. 2020).

Much of the research currently undertaken on the Continent is funded by international organisations such as the US National Institutes of Health, the European Commission and the Wellcome Trust, that often include the open sharing of data as a condition of their funding. This requirement poses several challenges for research conducted in Africa. Africa has a history of exploitative research and 'safari research', whereby researchers in Africa were simply viewed as sample collectors and the power asymmetries meant that any collaboration was generally for the benefit of researchers in high income countries (HICs) (Chu et al. 2014; Hardy et al. 2008; Wonkam et al. 2011). Many of the regulations governing research in jurisdictions in Africa have been developed in response to a need to guard against such exploitative research practices and may not support the open sharing of samples and data (Ciara Staunton and de Vries 2020). It is against this background that the need to consider the impact of the new and emerging data protection regulations in Africa on the use and sharing of data for research arises.

South Africa is one country that recognises the need to balance the drive towards more open science with a legal and ethical framework that protects its research participants. The development of science, technology and innovation (STI) has been identified by the South African government as an enabler to overcome the legacies of inequality of apartheid and driving economic growth (White Paper on Science, Technology and Innovation 2019). Health, together with education, is recognised as a key area for STI investment (White Paper 2019: 2). In particular, genomic research has been identified as important in developing a strong bio-economy (Bio-economy Strategy 2013) that has the potential to both improve population health and drive job creation (ASSAf 2018). As part of this development, the Department of Science and Innovation (DSI), formerly the Department of Science and Technology, has identified data sharing as important in its 2019 White Paper on Science, Technology and Innovation, through making available publicly funded research and research data, providing access to this data, and open access journal publishing (White Paper 2019). The draft National Data and Cloud Policy of 2021 recognises the importance of data and the digital economy across all sectors and seeks to develop "citizen centric frameworks" that will harness the potential of data and cloud computing. However, this faces some changes particularly as currently the regulation of use, access to, and sharing of health data for research in South Africa is governed by a patchwork of laws, regulations, and policies. A coherent framework to regulate the use of health data for research is thus lacking and became more complex with the entering into force of most of the provisions of the Protection of Personal Information Act (POPIA) No 4 of 2013 on 1 July 2021. Unsurprisingly a Code of Conduct for research was called for (Staunton et al. 2019) and the Academy of Sciences of South Africa (AASAf) heeded this call and has embarked on a process to develop such a Code. It is the framework through which this Code should be developed that is the focus of this chapter.

Broadly, this chapter will explore the existing South African regulatory framework for data protection, with specific focus on the use of health data for research. "Health data" for the purpose of this chapter refers to personal, identifiable data relating to health and not anonymised health data that falls outside the remit of POPIA. To contextualise this broad aim, general legal and ethical issues relating to the use of health data for research in Africa, followed by a discussion of the emergence of data protection regulations and their impact on the use of health data for research generally will be considered, before turning to the South African framework. This chapter will argue that the application of some of POPIA's principles and provisions to the use of health data for research are unclear and require further guidance and interpretation. One pertinent example on which clarity is necessary and which this chapter will explore, is the legal status of broad consent for the use and sharing of special personal information in genomic research under POPIA. The chapter argues that the human research ethics framework as it is currently understood and applied in South Africa may not necessarily be the best single framework to interpret POPIA's provisions for genomic data sharing and data governance. Rather, this chapter proposes a framework that could strike a justifiable balance between human research ethics, participants' rights, and the need for scientific advances against the background of a communitarian ethic that also includes considerations drawn from both a public health ethics and research integrity perspective. We will submit that this can be achieved through a regulator-approved Code of Conduct for Research that includes inputs from both academic and commercial sectors.

26.2   Ethical and Legal Issues in the Use of Health Data for Research in Africa

The emergence of genomic research has resulted in a re-think of some key principles in research ethics. It is trite that the principle of informed consent is seen as essential in protecting the autonomy of research participants and a critical requirement is that participants are informed of the risks and benefits prior to participation in research. If they consent, they are consenting to that study only. However, the ease with which samples and data can be shared for research has challenged this "one-study, one-informed consent" paradigm. As such, other consenting models, such as broad consent, tiered consent, dynamic consent (Kaye et al. 2015) and meta-consent (Ploug and Holm 2017) have been considered as alternatives to maximize the use of the resources. As the debate in South Africa has very much focused on broad and tiered consent and both expressly mentioned in the South African Department of Health Ethics Guidelines, this chapter will focus on broad consent and tiered consent only.

The exact definition of broad consent may vary, but it generally refers to a process whereby participants consent to the use of their samples and data for future unspecified research. The future use of these samples for research is generally subject to approval by a research ethics committee (REC). As such, it can be described as "consent to governance" (Koenig 2014; Tindana and de Vries 2016). This makes it distinct from blanket consent whereby participants would donate their samples without any restrictions (Wendler 2013).

The use of broad consent raises particular concerns in the African context owing in part to the history of so-called parachute research whereby samples were collected and sent for research in HICs (Hardy et al. 2008). As a result, researchers in Africa not only lost a valuable resource, but the research was often not for the benefit of the local population, nor was the use of the samples subject to local oversight or used in line with the local legislative framework (Ciara Staunton and de Vries 2020). It is thus unsurprising that certain countries, such as Malawi and Zambia, have provisions that specifically prohibit the use of broad consent. The majority of countries in Africa that regulate the use of biological samples generally permit broad consent (de Vries et al. 2017), but there appears to be a disconnect with the legislative frameworks that permit the use of broad consent and the acceptance of broad consent in practice. Engagement with RECs on the continent demonstrates that some RECs have been hesitant to approve genomic studies that seek to use broad consent, even when it is legally permitted (de Vries et al. 2015). Despite this, some empirical research does suggest acceptance of broad consent (van Schalkwyk et al. 2012). The H3Africa Framework for Best Practice for Genomics Research and Biobanking in Africa also views broad consent as a "best compromise" and accepts it as a model provided that it is supported by community engagement and accompanied by a mechanism that supports accountability and equity in the use of the resources (Yakubu et al. 2018).

Despite legislative frameworks across Africa broadly permitting broad consent, the ethical debate continues and tiered consent has been proposed as an alternative to broad consent (Nembaware et al. 2019). Under this model, participants are presented with a range of consent preferences, including use of their sample in a specific study, use for a particular illness (such as diabetes) or use in future unknown research. This approach provides more choice to participants. It has been proposed as an ethically preferable consent model to broad consent in Africa on the basis that providing participants with more options enables them to make a more autonomous—and more informed—choice. However, two points are worth noting here. First, although tiered consent has been advocated as a more ethically acceptable consent model in Africa, one of the options as part of a tiered consent model is consent to the use of their samples in future research. Tiered consent thus incorporates broad consent and the acceptability of tiered consent as a consent model is in part contingent on the acceptability of broad consent. Second, tiered consent has also been criticised as although the range of options enables participants to more readily make an autonomous decision, it provides participants with too much choice (Ram 2008). Concerns have already been raised that due to low levels of education in many African countries, specific consent is already a challenge (de Vries et al. 2015) and the provision of too many options may overly complicate the consenting process. More empirical work is needed to fully understand and unpack attitudes towards broad consent and tiered consent, as well as the operation of both in practice.

26.3   Data Protection and the Use of Health Data for Research

26.3.1   General Data Protection Regulation (GDPR)

There has been a global strengthening of personal data globally, notably with the introduction of the GDPR. The GDPR is a general legal framework that provides high level principles on the use of personal information. In addition to many of the strict processing requirements, there is a general ban on the processing of special personal data, and this includes genetic data, unless it comes within one of the exempted grounds under Article 9(2). There are exemptions to many of the processing requirements and the general ban if the processing of the personal data is for scientific research (Chassang 2017). It is for this reason that scientific research in the GDPR is described as having "a lighter data protection regulatory touch" than other industries (Dove and Chen 2020), but this is aimed at enabling the protection of personal information to be balanced against the need for scientific research, in particular health research aimed at advancing human well-being.

There is, however, some uncertainty in the application of some of these high-level principles on personal data processing in the context of health research, and for the purposes of this chapter, we will focus on consent. First, there is often the misconception that consent must be the lawful basis of processing for research. Although informed consent is required as part of the ethical conduct of research and legally required in the case of clinical trials, consent is but only one lawful basis for the use of personal data for research under the GDPR (Dove and Chen 2020).

Turning to the legal status of broad consent, Article 1(b) requires that personal data be collected for "specified, explicit and legitimate purposes" and that it cannot be used again for a matter that is incompatible with the purpose for which it was originally collected. On the face of it, this would preclude broad consent, but Article 1(b) further provides an exception for scientific research. Recital 33 of the GDPR recognises that it is "often not possible to fully identify the purpose of personal data processing for scientific research purposes at the time of data collection" and states that data subjects can give consent to certain areas of scientific research, in line with ethical standards for research. Broad consent is thus permissible under the GDPR, and this has been endorsed by the European Data Protection Board Supervisor (EDPS) in his *Preliminary Opinion on data protection and scientific research*. The EDPS does, however, stresses that when the research purposes cannot be fully specified, it is essential that more is done to ensure that the essence of the research participants' rights are honoured.

Article 1(b) provides an indication of how this may be done and states that the secondary use of personal data for research must be in line with Article 89(1). This requires that safeguards, that includes technical and organisational measures, be put in place to protect the rights and freedoms of the data subject. Article 89(1) specifically states that one of those measures may include pseudonymisation, which involves the removal of certain information so that the data subject is not identifiable. The EDPS cites access limitations as another example, but other than these there is limited guidance as to possible safeguards or measures. It must be stressed that while broad consent is legally permitted, this does not mean it can be used in every case for scientific research. Its use must be considered on a case-by-case basis, subject to appropriate oversight by an REC and it must be demonstrated that there are safeguards and measures in place to respect the data subjects' rights.

26.3.2 Protection of Personal Information Act no 3 of 2013

Turning to South Africa, POPIA is the first comprehensive data protection regulation in South Africa. Similar to the GDPR, it is an omnibus and general legal framework that seeks to regulate the processing of personal information, and to ensure greater transparency and oversight in protecting the personal information of the data subject through the eight conditions that must be met for the lawful processing of data. The development of POPIA was informed by the GDPR and it is quite likely that the interpretation of POPIA will be informed by the interpretation of the GDPR. However, the application of the special provisions for research in POPIA are less clear when compared those in the GDPR.

Similar to the GDPR, under POPIA consent is not a preferred basis for the processing of personal information but one of eight lawful grounds under which to process personal information. Equally, there is a general ban under s.27 on the processing of "special personal information", which includes genomic data. This general ban does not apply if the use of this data is for research that is either in the public interest *or* where it would be "impossible or would involve a disproportionate effort to ask for consent". If genomic data is to be used for research under either of these two provisions, there must be *sufficient* guarantees in place that the privacy of the participant will not be disproportionately affected. Thus, similar to the GDPR, the processing of genomic information can be processed under POPIA if it is for research or in the public interest under certain conditions. Unlike the GDPR, POPIA is silent on each of these crucial points and the rules and procedures to be followed to comply with s.27 are unclear. It is furthermore uncertain whether this exemption applies to scientific research only or all types of research. Guidance or a framework as to *what* is considered to be in the public interest is not provided and there is similarly lack of a framework to guide a Responsible Party as to what would be considered to be a "disproportionate effort", "sufficient guarantees" and how a balance may be struck with a participant's privacy. POPIA not only lacks the limited guidance that is found within the GDPR, but no guidance has also been forthcoming from the Office of the

Information Regulator on the application of the provisions to research, nor has there been any indication whether such guidance would be forthcoming.

A second point of uncertainty relates to broad consent. Under s.13, personal information must be collected for a "specific, explicitly defined and lawful purpose". In isolation, this provision would appear to require specific consent. Section 15(3)(e) permits the *further* use of the personal information beyond the purpose for which it was collected if such further use is *solely* for research. Thus it has been argued that a purposive interpretation of POPIA permits broad consent, but there has been academic disagreement on this point and clarity is essential (Staunton et al. 2019). A further point to note is that even if broad consent is permitted under POPIA by virtue of s.15(3)(e), this can only occur if the personal information is not published in an identifiable form. The question thus facing genomic researchers in South Africa is what is meant by "published in an identifiable form" in the context of genomic research.

POPIA defines personal information as information relating to an "identifiable, living, natural person". No definition of identifiable is provided in the Act, but de-identify is defined in s.1 as to delete any information that identifies the data subject, can be manipulated by a reasonably foreseeable method to identify the data subject, or can be linked by a reasonably foreseeable method to other information that identifies the data subject. There are also two types of de-identified information: de-identified information that can be re-identified and de-identified information that can no longer be re-identified.

De-identified information that can no longer be re-identified is most likely to be what has traditionally been known as anonymous information. De-identified information that can be re-identified is most likely similar to what the GDPR considers as pseudonymised data, which is defined as the removal of information that will identify the data subject. While it is clear under the GDPR that pseudonymised data is considered to be identifiable data, it is less clear whether de-identified data that can be re-identified is considered to be "identifiable". These definitional issues become even more complex when applied to genomic research, as a growing number of studies demonstrating that individuals may be re-identified from genetic data (Shapiro 2015; Segert 2018). Shabani and Marelli argue that the type of data, the sample size and the rareness of the genetic variant are all key factors to be considered in determining whether genomic data can re-identified (Shabani and Marelli 2019). This raises the question of whether, in the context of POPIA, we must look at the personal information only to determine whether it is identifiable, or if we should also consider the context in which it is held.

A final point of uncertainty is what is meant by "published". One would assume that it would preclude the publication of identifiable personal information in any publications that would include academic journals, but would this include the depositing of individual level data in a publicly accessible depository? Considering this may be a funding requirement of by the National Institutes of Health (*NIH Data Sharing Policies* n.d.), which funds much of genomic research in South Africa, it is essential that there is clarity on this point.

The application of data protection principles to genomic research is complex and a problem that is not unique to South Africa (Juengst and Meslin 2019; Mascalzoni et al. 2019; Shabani and Marelli 2019). However, the application of POPIA to genomic research in South Africa is mired with uncertainty due to the lack of clarity and guidance on many key definitional issues and in particular how they should be defined in the context of genomic research.

26.4   A Code of Conduct for Research in South Africa

It is clear from the preceding discussion that clarity is necessary for researchers and RECs on POPIA's impact on genomic research, but these issues will equally impact other forms of scientific research. Of concern to many researchers is the potential impact it will have on international funding conditions. In Europe, a Code of Conduct to guide genomic data sharing has been suggested (Molnár-Gábor and Korbel 2020) and in South Africa, work on a Code of Conduct is ongoing and led by ASSAf. Chapter 7 of POPIA provides for the development of codes of conduct that can apply to specified information or a class of information, such as health information, a specific activity such as research, or a specific industry, profession or vocation. Once approved by the Information Regulator after a period of public consultation, this Code becomes legally binding. Chapter 7 enables the development of a sector-specific response to guide the implementation of POPIA to research.

Research in South Africa is regulated through a complex web of laws, regulations and policies that includes the National Health Act No. 61 of 2003; the National Health Act's 2012 Regulations; the Promotion of Access to Information Act (PAIA) No. 2 of 2000, the 2018 Material Transfer Agreement; the legally enforceable 2015 Department of Health Ethics in Health Research Guidelines, and now POPIA. As there is no specific law on research in South Africa, the development of a Code of Conduct for research under Chap. 7 of POPIA cannot occur in isolation and must consider all these instruments that currently apply to research. To the extent possible, it should complement existing regulations, guidelines, and policies so that there is a coherent and consistent approach to research in South Africa.

The development of this Code does, however, require a framework under which it is developed. The current framework for ethics review of research in South Africa is grounded within the research ethics framework that is underpinned by respect for persons, beneficence and justice, as laid out in the Belmont Report (Belmont Report 2010). The culmination of these principles into the relevant norms and standards contained in the Department of Health Ethics Guidelines is very much weighted in favour of concerns for the protection of research participants and individual interests and the guidelines on the use and re-use of health data in research focus on issues of consent and protection of the privacy of research participants. Such an approach is not uncommon but has been criticised by Ballantyne who has noted that there appears to be a disproportionate focus on issues of consent in the research ethics literature, as well as during REC discussions (Ballantyne 2019).

A framework primarily grounded in research ethics may not be wholly suitable for the regulation on the use of health data in research in South Africa, on three grounds. First, it is does not adequately reflect the concept of privacy against constitutional jurisprudence that emphasises Ubuntu and the role of the individual in the community as opposed to its role as individual over the community (Himonga et al. 2013). Ubuntu is an African philosophy particularly dominant in southern Africa and it is through our connection with others that our humanity exists. As stated by Langa J in *S v Makwanyane*, the communal spirit of Ubuntu "places emphasis on communality and on the interdependence of the members of a community" (Makwanyane 1995, paragraph 224). Reviglio and Alunge have detailed how both understandings and protections of privacy in the contemporary digital and data-led paradigm can be advanced through notions of Ubuntuism (Reviglio and Alunge 2020).[1] Specifically in relation to consent, the authors note:

> If the self is dependent on the community, then it can be argued that protecting the community amounts to protecting the individual, or that protecting the individual should start from protecting the community. In the big data context, Ubuntu would therefore support a top-down regulatory paradigm of privacy, as a right which protects an individual through principles and rules regulating the collection and sharing of data relating to entire communities or groups of individuals. Individual informed consent, for example, may not be sufficient at all. Ideally, "collective informed consent" may be much more appropriate.

This leads us to our second point that while POPIA's primary purpose is to give effect to the constitutional right to privacy by safeguarding the processing of personal information, s. 2 recognises that this right can be limited if the purpose is to balance the right to privacy with other rights and protect "important interests". Such interests are expressly stated in POPIA as 'including the free flow of information within the Republic and across international borders". Data sharing is thus seen as important under POPIA and as we have outlined, data sharing for research and genomic research has been identified as important in addressing the burden of disease in South Africa and elsewhere on the continent, developing the South African bio-economy and driving economic growth. Arguably, genomic research and the sharing of genomic data for research as an important interest that can justifiably put some limits on the right to privacy in this context. However, a research ethics framework that is primarily focused on individual rights and interests will unlikely consider these other interests.

Third, insufficient attention is given in the research ethics framework to the importance of research integrity and its role in this domain. We need not only look at the research itself, but also ensure the integrity of the researcher, as it is the researcher who will carry responsibility for data management and data sharing. The Singapore Statement on Research Integrity is the first international effort to encourage the development of unified policies, guidelines, and codes of conduct, with the ultimate goal of fostering greater integrity in research across the globe. It has been widely accepted. And in some cases, formally endorsed by universities in South Africa, for example Stellenbosch University and the University of the Witwatersrand have incorporated

---

[1] Reviglio, U., Alunge, R. "I Am Datafied Because We Are Datafied": an Ubuntu Perspective on (Relational) Privacy. Philos. Technol. (2020). https://doi.org/10.1007/s13347-020-00407-6.

the statement into their research integrity policy frameworks. The Singapore Statement requires researchers to share data and findings openly and promptly, where possible, but a key focus is on the responsible conduct of researchers. The use of genomic data for research requires us to consider the ongoing use of genomic data and not just at the outset of the research.

Finally, while the protection of research participants will be a key factor, there is a need to consider other values that can and should underpin the genomic research and associated data sharing. Calls for a shift in approach for genomic research is not new in the context of biobank research and genomic research. Prainsack and Buyx have argued that biobank governance should be guided by solidarity (Prainsack and Buyx 2013). Chen and Pang identify a number of key principles that should underpin biobank research, in addition to respecting and protecting research participants and wider issues of trust, transparency, quality and equity (WHO | A Call for Global Governance of Biobanks n.d.). This shift in approach to genomic research must also be considered in the South African context where there has been historically exploitative research against the background of the recognition of the need to promote non-exploitative research in LMICs. The TRUST *Global Code of Conduct for Research in Resource-Poor Settings* (TRUST Global Code n.d) that now applies to all EU funded research, emphasises the importance of fairness, respect, care, and honesty in any collaborative research. The *San Code of Research Ethics* (San Code 2017), developed after the publication of exploitative and discriminatory research in *Nature* (Chennells and Steenkamp 2018), identify respect, honesty, justice, and fairness, care and due process as important values in conducting research on the San community, who are the oldest inhabitants of southern Africa. Looking specifically at genomic research, the H3Africa *Ethics and Governance Framework for Best Practice in Genomic Research and Biobanking in Africa* is guided by the principles of solidarity or communal-based worldviews, fairness, equity, and reciprocity (H3Africa Ethics and Governance Framework 2017).

26.5   A Framework to Govern the Use of Genomic Data in South Africa: Charting a Way Forward

It is for these reasons that we submit that a research ethics framework is not the appropriate framework to guide the development of a Code of Conduct for research under POPIA. This critique of the research ethics framework for genomic research is not new. Staunton and de Vries have argued the regulation and oversight of genomic research in Africa currently adopts a precautionary approach that at times restricts the research, that is part due to previous exploitative research on the continent (Ciara Staunton and de Vries 2020). They argue that genomic research should be welcomed as important in improving the health and research capacity in Africa and that the 'regulatory tilt' or the focus of the regulations should be tilted in favour of guarding against exploitative research and exploitative collaborations. They assert that this shift in focus can be achieved if the governance of genomic research in Africa is guided by the principles of solidarity, reciprocity, justice, and trust (Ciara Staunton and de Vries 2020). The question that arises is under which framework a POPIA Code of Conduct that addresses the specific challenges for genomic research should be developed.

On the issue of the use of health data for research, Ballantyne has argued that public health ethics that places public benefit, proportionality, equity, trust, and account-ability as its core principles, is a more suitable framework in which to consider the use of health data for research. She argues that this framework will refocus the debate from consent to broader issues that must be considered when deciding on the use of data in the public interest (Ballantyne 2019). Such an approach is much more grounded in in a communitarian, or Ubuntu ethic, as has been adopted by the Consti-tutional Court as a jurisprudential lens. It also complements POPIA which seeks to embed accountability and proportionately in the governance of personal information and also envisages that some of the strict processing requirements can be exempted on grounds of public interest.

Developing a Code of Conduct for research under a public health ethics framework can allow us to reorient the focus from the protection of the research participant to the management of personal information in a manner that sees the individual as part of the collective. This is not to erode the voice of the individual, but rather enable a balance to be struck between public interest and individual privacy. Drafters of this Code can be guided by the 2016 OECD Recommendation of the Council on Health Data Governance, 2009 *OECD Guidelines on Human Biobanks and Genetic Research Databases,* the World Medical Association (WMA) Declaration of Taipei on Ethical Considerations Regarding Health Databases and Biobanks and any Codes of Conduct for Research that may be developed under the GDPR. The South African Code must, however, be appropriate in light of the jurisprudential, social and cultural context in which genomic research is occurring.

It is for the drafters of the Code to specify how genomic research may be conducted under a public health ethics framework. Such a framework, in our view, would permit broad consent for genomic research. The focus should thus be on how best to enable the use of genomic data in a manner that best serves the interests of South Africa.

26.6   Conclusion

We contend that a narrow approach to the protection and sharing of human research data (specifically genetic and genomic research) that places all emphasis on individual protection, is inappropriate for the South African context and could hamper important research. Balancing human research ethics with other ethical frameworks, including the promotion of research integrity and a more communitarian ethic, as embodied in public health ethical frameworks, should be used as a framework to interpret and apply POPIA. This can be achieved via a regulator-approved Code of Conduct for Research that would include inputs from both academic and commercial sectors.

The implications for this approach for the use of broad consent strategies would mean that broad consent per se is not irreconcilable with POPIA's provisions, if its use were approved by a REC that recognises the need to balance the protection of individual rights to privacy and autonomy with due diligence by research teams in the context of data stewardship, whilst being mindful of the need to also consider the common good and the values of reciprocity and solidarity that are inherent in this approach.

References

Academy of Science of South Africa (ASSAf). Human genetics and genomics in South Africa: ethical, legal and social implications. 2018. Retrieved on 20 September 2020 from http://res earch.assaf.org.za/handle/20.500.11911/106.

Ballantyne, A. 2019. Adjusting the focus: A public health ethics approach to data research. *Bioethics* 33 (3): 357–366. https://doi.org/10.1111/bioe.12551.

Chassang, G. 2017. *The impact of the EU general data protection regulation on scientific research.* https://doi.org/10.3332/ecancer.2017.709.

Chennells, R., and Steenkamp, A. 2018. International genomics research involving the San people. In *Ethics Dumping*, eds. D. Schroeder, J. Cook, F. Hirsch, S. Fenet, and V. Muthuswamy, 15–22. Springer International Publishing. https://doi.org/10.1007/978-3-319-64731-9_3.

Chu, K.M., S. Jayaraman, P. Kyamanywa, and G. Ntakiyiruta. 2014. Building research capacity in Africa: Equity and global health collaborations. *PLoS Medicine* 11 (3): e1001612. https://doi.org/10.1371/journal.pmed.1001612.

de Vries, J., A. Abayomi, K. Littler, E. Madden, S. McCurdy, O.O.M. Oukem-Boyer, J. Seeley, C. Staunton, G. Tangwa, and P. Tindana. 2015. Addressing ethical issues in H3Africa research–the views of research ethics committee members. *The HUGO Journal* 9 (1): 1–4.

de Vries, J., S.N. Munung, A. Matimba, S. McCurdy, O.O.M. Oukem-Boyer, C. Staunton, A. Yakubu, and P. Tindana. 2017. Regulation of genomic and biobanking research in Africa: A content analysis of ethics guidelines, policies and procedures from 22 African countries. *BMC Medical Ethics* 18 (1): 8.

Dove, E.S., and J. Chen. 2020. Should consent for data processing be privileged in health research? A comparative legal analysis. *International Data Privacy Law* 10 (2): 117–131. https://doi.org/10.1093/idpl/ipz023.

Gibbs, R. A., J. W. Belmont, P. Hardenbol, T. D. Willis, F. Yu, H. Yang, L.-Y. Ch'ang, W. Huang, B. Liu, Y. Shen, P. K.-H. Tam, L.-C. Tsui, M. M. Y. Waye, J. T.-F. Wong, C. Zeng, Q. Zhang, M. S. Chee, L. M. Galver, S. Kruglyak, and Methods Group. 2003. The international hapmap project. *Nature* 426(6968): 789–796. https://doi.org/10.1038/nature02168

H3Africa. Ethics and Governance Framework for Best Practice in Genomic Research and Biobanking in Africa. 2017. Retrieved on 20 September 2020 from H3africa.org/wp-content/uploads/2018/05/Final-Framework-for-African-genomics-and-biobanking_SC-.pdf.

Hardy, B.-J., B. Séguin, R. Ramesar, P.A. Singer, and A.S. Daar. 2008. South Africa: From species cradle to genomic applications. *Nature Reviews Genetics* 9 (1): S19–S23. https://doi.org/10.1038/nrg2441.

Himonga, C., M. Taylor, and A. Pope. 2013. Reflections on judicial views of *Ubuntu. Potchefstroom Electronic Law Journal/potchefstroomse Elektroniese Regsblad* 16 (5): 369–427. https://doi.org/10.4314/pelj.v16i5.8.

Juengst, E.T., and E.M. Meslin. 2019. Sharing with strangers: Governance models for borderless genomic research in a territorial world. *Kennedy Institute of Ethics Journal* 29 (1): 67–95. https://doi.org/10.1353/ken.2019.0000.

Kaye, J., E.A. Whitley, D. Lund, M. Morrison, H. Teare, and K. Melham. 2015. Dynamic consent: A patient interface for twenty-first century research networks. *European Journal of Human Genetics* 23 (2): 141–146. https://doi.org/10.1038/ejhg.2014.71.

Koenig, B.A. 2014. Have we asked too much of consent? *The Hastings Center Report* 44 (4): 33–34. https://doi.org/10.1002/hast.329.

Makwanyane, S v. 1995. 3 SA 391 (CC).

Mascalzoni, D., H. Beate Bentzen, I. Budin-Ljøsne, Andrew L. Bygrave, J. Bell, S. Dove, E., Fuchsberger, C., Hveem, K., M. Th. Mayrhofer, V. Meraviglia, D. R. O'Brien, C. Pattaro, P. Pramstaller, V. Rakić, A. Rossini, M. B. Shabani, D. J. Svantesson, M. Tomasi, L. Ursin, and J. Kaye. 2019. Are requirements to deposit data in research repositories compatible with the European union's general data protection regulation? *Annals of Internal Medicine*. https://doi.org/10.7326/M18-2854.

Molnár-Gábor, F., and J. O. Korbel. 2020. Genomic data sharing in Europe is stumbling—Could a code of conduct prevent its fall? *EMBO Molecular Medicine, 12*(3). https://doi.org/10.15252/emmm.201911421.

Nembaware, V., K. Johnston, A.A. Diallo, M.J. Kotze, A. Matimba, K. Moodley, G.B. Tangwa, R. Torrorey-Sawe, and N. Tiffin. 2019. A framework for tiered informed consent for health genomic research in Africa. *Nature Genetics* 51 (11): 1566–1571. https://doi.org/10.1038/s41588-019-0520-x.

NIH Data Sharing Policies. (n.d.). [Product, Program, and Project Descriptions]. U.S. National Library of Medicine. Retrieved 22 September 2020, from https://www.nlm.nih.gov/NIHbmic/nih_data_sharing_policies.html.

Nuntius, S. (n.d.). *Singpore Statement on Research Integrity*. Retrieved 22 September 2020, from https://www.wcrif.org/guidance/singapore-statement.

Ploug, T., and S. Holm. 2017. Eliciting meta consent for future secondary research use of health data using a smartphone application—A proof of concept study in the Danish population. *BMC Medical Ethics* 18 (1): 51. https://doi.org/10.1186/s12910-017-0209-6.

Prainsack, B., and A. Buyx. 2013. A solidarity-based approach to the governance of research biobanks. *Medical Law Review* 21 (1): 71–91. https://doi.org/10.1093/medlaw/fws040.

Ram, N. (2008). Tiered consent and the tyranny of choice. *Jurimetrics, 48*(3), 253–284. JSTOR.

Reviglio, U., R. Alunge. 2020. I am datafied because we are datafied: An Ubuntu perspective on (relational) privacy. *Philosophy and Technology*. Retrieved on 20 September 2020 from https://doi.org/10.1007/s13347-020-00407-6

San Code of Research Ethics. South African San Institute. 2017. Retrieved on 20 September from http://trust-project.eu/wp-content/uploads/2017/03/San-Code-of-RESEARCH-Ethics-Booklet-final.pdf.

Segert, J. Understanding ownership and privacy of genetic data. Retrieved on 20 September 2020 from ohttp://sitn.hms.harvard.edu/flash/2018/understanding-ownership-privacy-genetic-data/.

Shabani, M., and L. Marelli. 2019. Re-identifiability of genomic data and the GDPR: Assessing the re-identifiability of genomic data in light of the EU General Data Protection Regulation. *EMBO Reports, 20*(6). https://doi.org/10.15252/embr.201948316.

Shapiro, Z. 2015. Big data, genetics, and re-identification. Retrieved on 20 September 2020 from https://blog.petrieflom.law.harvard.edu/2015/09/24/big-data-genetics-and-re-identification/.

Staunton, Ciara, and K. Moodley. 2013. Challenges in biobank governance in Sub-Saharan Africa. *BMC Medical Ethics* 14 (1): 35.

Staunton, Ciara, R. Adams, D. Anderson, T. Croxton, D. Kamuya, M. Munene, and C. Swanepoel. 2020. Protection of Personal Information Act 2013 and data protection for health research in South Africa. *International Data Privacy Law* 10 (2): 160–179. https://doi.org/10.1093/idpl/ipz024.

Staunton, Ciara, and J. de Vries. 2020. The governance of genomic biobank research in Africa: Reframing the regulatory tilt. *Journal of Law and Biosciences*, 1–20.

Staunton, C., R. Adams, M. Botes, E. S. Dove, L. Horn, M. Labuschaigne, G. Loots, S. Mahomed, J. Makuba, A. Olckers, M. S. Pepper, A. Pope, M. Ramsay, N. N. Loideain, and J. De Vries. 2019. Safeguarding the future of genomic research in South Africa: Broad consent and the protection of personal information act no. 4 of 2013. *South African Medical Journal 109*(7): 468. https://doi.org/10.7196/SAMJ.2019.v109i7.14148.

The Belmont Report. 2010. [Text]. HHS.Gov. https://www.hhs.gov/ohrp/regulations-and-policy/belmont-report/index.html.

The Bio-economy Strategy. Pretoria: Department of Science and Technology. 2013. Retrieved on 20 September from nnovus.co.za/media/Bioeconomy_Strategy.pdf.

The H3Africa Consortium, E. Matovu, B. Bucheton, J. Chisi, J. Enyaru, C. Hertz-Fowler, M. Koffi, A. Macleod, D. Mumba, I. Sidibe, G. Simo, M. Simuunza, B. Mayosi, R. Ramesar, N. Mulder, S. Ogendo, A. O. Mocumbi, C. Hugo-Hamman, O. Ogah, and C. Rotimi. 2014. Enabling the genomic revolution in Africa. *Science, 344*(6190): 1346–1348. https://doi.org/10.1126/science.1251546

Tindana, P., and J. de Vries. 2016. Broad consent for genomic research and Biobanking: Perspectives from low- and middle-income countries. *Annual Review of Genomics and Human Genetics* 17 (1): 375–393. https://doi.org/10.1146/annurev-genom-083115-022456.

TRUST. Global Code of Conduct for Research in Resource-poor Settings. Retrieved on 20 September 2020 from globalcodeofconduct.org/wp-content/uploads/2018/05/Global-Code-of-Conduct-Brochure.pdf.

van Schalkwyk, G., J. de Vries, and K. Moodley. 2012. 'It's for a good cause, isn't it?'—Exploring views of South African TB research participants on sample storage and re-use. *BMC Medical Ethics* 13 (1): 19. https://doi.org/10.1186/1472-6939-13-19.

Walport, M., and P. Brest. 2011. Sharing research data to improve public health. *The Lancet* 377 (9765): 537–539. https://doi.org/10.1016/S0140-6736(10)62234-9.

Wendler, D. 2013. Broad versus blanket consent for research with human biological samples. *The Hastings Center Report* 43 (5): 3–4. https://doi.org/10.1002/hast.200.

White Paper on Science, Technology and Innovation. March 2019. Department of Science and Innovation. South Africa. Retrieved 2 July 2020 from https://www.dst.gov.za/images/2019/White_paper_web_copyv1.pdf.

WHO | A call for global governance of biobanks. n.d. WHO. Retrieved 19 February 2020, from https://www.who.int/bulletin/volumes/93/2/14-138420/en/.

Wonkam, A., M.A. Kenfack, W.F.T. Muna, and O. Ouwe-Missi-Oukem-Boyer. 2011. Ethics of human genetic studies in Sub-Saharan Africa: The case of cameroon through a bibliometric analysis. *Developing World Bioethics* 11 (3): 120–127. https://doi.org/10.1111/j.1471-8847.2011.00305.x.

Yakubu, A., P. Tindana, A. Matimba, K. Littler, N. S. Munung, E. Madden, C. Staunton, and J. De Vries. 2018. Model framework for governance of genomic research and biobanking in Africa—A content description. *AAS Open Research, 1*: 13. https://doi.org/10.12688/aasopenres.12844.1.

Ciara Staunton Prior to starting her academic career, Dr. Staunton was a Legal Researcher at the Law Reform Commission of Ireland. From 2010-2013 she co-ordinated the Advancing Research Ethics in Southern Africa (ARESA) Program and was a member of the H3Africa Ethics and Regulatory Issues Working Group. She is a Senior Lecturer in Law at Middlesex University and a Senior Researcher at the Institute for Biomedicine, Eurac Research (Italy). Her research focuses on the governance of new and emerging technologies, in particular stem cell research, genomic research and biobanking. Her current research focuses on the sharing of health data for research, with a particular focus on Africa and is a consultant to the South African National Institute for Communicable Diseases on strengthening its data protection frameworks. She has received grants from the Wellcome Trust, the National Institutes of Health (NIH) and the Irish Research Council and has been involved in the development of policy in Ireland, Bahrain and a number of countries in Africa.

Dr. Rachel Adams in addition to her role as Principal Researcher at Research ICT Africa where she directs the AI4D Africa Just AI Centre, Dr. Adams is also the Project Lead of the African Observatory on Responsible AI(AI4D) and is the Principal Investigator of the Global Index on Responsible AI. She is a member of the UNESCO Expert Committee for the implementation of the UNESCO Recommendation on the Ethics of Artificial Intelligence, an Associate Fellow at the Leverhulme Centre for the Future of Intelligence at the University of Cambridge, a Research Associate with the Information Law and Policy Centre at the University of London, and a Research Associate of the Tayarisha African Centre of Excellence on Digital Governance at the University of Witwatersrand, Johannesburg. Rachel was previously a Chief Research Specialist at the Human Sciences Research Council, South Africa. Rachel has published widely in areas such as AI and society, gender and AI, transparency, open data, and data protection. Dr. Adams is the author of Transparency: New Trajectories in Law (Routledge, 2020), and the lead author of Human Rights and the Fourth Industrial Revolution in South Africa (HSRC Press, 2021).

Dr. Lyn Horn is currently the Director of the Office of Research Integrity at UCT and an Honorary Associate Professor in the Centre for Applied Ethics at Stellenbosch University. She is a medical doctor with a PhD in Bioethics. She has been actively involved in either chairing or running research ethics committees (in both the biomedical and humanities domains, teaching research ethics and integrity, and policy development for many years. Dr. Horn currently serves as the bioethics consultant for four international EDCTP funded clinical trials relating to Tuberculosis and HIV that are active in several African countries. She is a member of the South African Protection of Information Act implementation team at her institution and was a Co-Chair of the 2022 7th World Conference on Research Integrity.

Prof. Melodie Labuschaigne is Professor in Medical Law and Ethics in the Department of Jurisprudence in the School of Law, University of South Africa. She is a former Director of the School of Law and Deputy Executive Dean of the College of Law at UNISA. Her field of specialisation is medico-legal issues and the legal regulation of biotechnology applications. She has been involved with the revision and drafting of health legislation for many years. She serves on several boards and councils involved with health research, most recently the SA National Health Research Ethics Council and the Bioethics Advisory Panel of the SA Medical Research Council. She also chairs the Genomics working group of the Bioethics Advisory Panel of the SA Medical Research Council.

CPSIA information can be obtained
at www.ICGtesting.com
Printed in the USA
LVHW082252260123
738055LV00005B/60

9 783031 126918